ANNOTATED
INSTRUCTOR'S
EDITION
LABORATORY
EXPERIMENTS

CHEMISTRY

THE CENTRAL SCIENCE

NINTH EDITION

John H. Nelson Kenneth C. Kemp

ANNOTATED INSTRUCTOR'S EDITION
LABORATORY EXPERIMENTS

THE CENTRAL SCIENCE

Ninth Edition

Brown LeMay Bursten

Upper Saddle River, NJ 07458

Project Manager: Kristen Kaiser
Senior Editor: Nicole Folchetti
Editor in Chief: John Challice
Executive Managing Editor: Kathleen Schiaparelli
Assistant Managing Editor: Dinah Thong
Production Editor: Natasha Wolfe
Supplement Cover Management/Design: Paul Gourhan
Manufacturing Buyer: Ilene Kahn
Cover Image: Ken Eward/Biografx

Pearson Education, Inc.
Upper Saddle River, NJ 07458

Printed in the United States of America

10 9 8 7 6 5 4

ISBN 0-13-009803-5

Pearson Education Ltd., *London*
Pearson Education Australia Pty. Ltd., *Sydney*
Pearson Education Singapore, Pte. Ltd.
Pearson Education North Asia Ltd., *Hong Kong*
Pearson Education Canada, Inc., *Toronto*
Pearson Educacíon de Mexico, S.A. de C.V.
Pearson Education—Japan, *Tokyo*
Pearson Education Malaysia, Pte. Ltd.
Pearson Education, *Upper Saddle River, New Jersey*

Contents

*Approximate time required to complete experiment.

**The numbers in brackets after experiment titles refer to chapter(s) in the ninth edition of *Chemistry: The Central Science* by Brown, LeMay, and Bursten that are relevant to the experiment.

To the Instructor

Whenever a new laboratory program is introduced at any institution, it inevitably involves considerable effort on the part of the instructor, the stockroom, and the laboratory personnel. The purpose of this guide is to minimize that effort upon adoption of *Laboratory Experiments for Chemistry: The Central Science, Ninth Edition.*

This Instructor's Edition is divided into five parts. The first section offers general guidelines for disposal of laboratory wastes. The next two sections are concerned with laboratory desk items and necessary equipment and materials for the experiments. For your convenience, the equipment is generally given for a laboratory section of 25 students, and the chemicals are listed on a per-student basis. This is followed by a complete version of the student edition of the laboratory manual. This AIE version includes exemplary data and answers to the questions at the ends of the experiments. The last portion of this guide (Appendix P) gives answers to the review questions that were not given in Appendix J at the back of the student's edition of the manual. There are many review questions at the end of the experiments. Instructors using this lab manual should not necessarily feel obliged to have their students answer all of these questions.

Although each of these experiments has been repeatedly tested and found to work well, problems that we have not anticipated or encountered may arise. We would sincerely appreciate receiving any comments and suggestions for improvement from users of the manual. Please send these to jhnelson@equinox.unr.edu. We are very grateful to the many individuals who have made comments on the first eight editions. Where possible, these suggestions have been incorporated into the ninth edition.

We are particularly appreciative of the many helpful suggestions made by the following reviewers and accuracy checkers:

Robert Cloney	Fordham University
Sesi McCullough	University of Nevada
Robert C. Pfaff	Saint Joseph College
David Shinn	University of Hawaii, Hilo
Richard S. Treptow	Chicago State University
John B. Vincent	University of Alabama, Tuscaloosa
Mary Walczak	St. Olaf College
Barry L. Westcott	Central Connecticut State College
David White	University of North Carolina, Wilmington

We gratefully acknowledge the help of James Buckley at Black Diamond Graphics and the help of the Prentice-Hall editorial and production staff, especially John Challice, Kristen Kaiser, Dinah Thong, and Natasha A. S. Wolfe. The quality of this manual is the result of their effective skills and expertise.

JOHN H. NELSON
jhnelson@equinox.unr.edu

KENNETH C. KEMP
kemp@chem.unr.edu

University of Nevada, Reno

General Guidelines for the Disposal of Laboratory Waste

Environmental pollution is, or should be, of concern to everyone. Prudent practices should be followed for the disposal and handling of chemical wastes, and environmental legislation mandates the same. No chemicals should be washed down the drain without consideration of their effect upon the environment. Instead, they should be collected in suitable containers and disposed of either by incineration or underground burial at a designated disposal site in concert with local practice and law.

Chemical wastes that are potentially harmful and require proper disposal may be broadly classified into essentially four categories: (1) flammable materials, (2) corrosive materials, (3) strong oxidizing or reducing agents, and (4) toxic substances. Before each laboratory session, instruct your students how and where they are to dispose of the chemicals that are used or produced during the lab session. At the end of each laboratory period, unused reagents should be returned to the storeroom or laboratory instructor, who will either store them for future use or dispose of them. Discard only dry paper into ordinary waste baskets. Never discard into the same container chemical wastes that will react with each other. If in doubt concerning which chemicals may react with others, have a separate container for each kind of chemical. Mark each container clearly as to its contents. Only nonflammable, neutral, nontoxic, water-soluble liquids should be flushed down the drain. Acidic or basic solutions may be flushed down the drain only after they have been neutralized. It is desirable to have separate drainage lines for laboratory sinks and sanitary drainage.

In several of the experiments in this laboratory manual, we suggest that you use waste containers for the collection of individual products. In most of these experiments, the quantities of materials used are quite small, so waste disposal is not too great a task. Experiments 2, 6, 8, 9, 10, 35, and 39 produce combustible organic wastes: alcohols, hydrocarbons, ethers, halocarbons, and acetone. A metal container should be provided for the collection of these materials, and it should be labeled as flammable organics. Experiments 4, 14, 21, 24, 25, 28, 29, 33, and 34 have toxic substances as wastes (compounds of copper, chromium, cadmium, mercury, or oxalates). These should be collected in a labeled metal container and disposed of properly.

The above information is presented as a guideline based upon sources considered to be reliable and accurate. Although every reasonable effort has been made to provide dependable information, the authors or publisher cannot assume responsibility for the completeness or correctness of the information. We strongly urge you to consult your chemical suppliers for detailed instructions for the safe disposal of individual chemicals. In addition, the following books are helpful:

Prudent Practices for the Disposal of Chemicals from Laboratories, 2nd Ed., National Academy Press, Washington, D.C., 1983.

Armour, M. A., Browne, L. M., and Weir, G. L., *Hazardous Chemicals Information and Disposal Guide,* University of Alberta, 1982.

Armour, M. A., Browne, L. M., and Weir, G. L., *Hazardous Chemicals Information and Disposal Guide,* University of Alberta, 1984.

Solid Waste Information, U.S. Environmental Protection Agency, *Disposing of Small Batches of Hazardous Wastes,* Cincinnati, OH 45268, 1976.

Flinn Scientific, Inc., *Chemical Catalog/Reference Manual,* P.O. Box 231, 917 W. Wilson Street, Batavia, IL 60510, 1985.

Department of Educational Activities, American Chemical Society, *Health and Safety Guidelines for Chemistry Teachers,* 1155 Sixteenth Street, N.W., Washington, D.C. 20036, 1979.

Department of Educational Activities, American Chemical Society, *Safety in Academic Chemistry Laboratories,* 1155 Sixteenth Street, N.W., Washington, D.C. 20036, 1979.

U.S. Product Safety Commission, *School Science Laboratories, A Guide to Hazardous Substances,* Washington, D.C. 20207, 1984.

Kareful, H.B., *Working Safely with Chemicals in the Laboratory,* 2nd Ed., Genium Publishing Corporation, Dept. GLSG94B, Genium Plaza, Schenectady, NY, 12304-4690.

Lunn, G. and Sansone, E.B., Safe Disposal of Highly Reactive Chemicals, *J. Chem. Educ.*, 71, 972, 1994.

Farr, A. Keith, *CRC Handbook of Laboratory Safety,* 4th Ed., CRC Publishing Co., Boca Raton, FL, 1995.

Urbem, P.G., Editor, *Brethericks Handbook of Reactive Chemical Hazards,* Butterworth, Stoneham, MA, 1995.

There are many Web sites on the Internet regarding Manufacturers Safety Data Sheets (MSDS). This information can be obtained by searching the Internet for "chemical MSDS."

Student Laboratory Desk Items

GENERAL COMMENTS

We suggest that each student have an apron or lab coat and protective gloves, and it is imperative that each student have approved eye protection (including splash guards). Each student should also have soap or detergent (stored in a small bottle) in his or her lab desk.

COMMON ITEMS

The following items should be located in common lockers available to all students in the laboratory. One common locker for every four students is usually sufficient. The students should be cautioned not to place these items in their individual lockers.

- 4 Burners* w/24 in. tubes
- 4 Clamps, single, utility
- 2 Clamps, double, buret
- 8 Clay triangles
- 4 Crucible tongs
- 4 Funnel holders
- 4 Gas bottles, 8 oz.
- 1 Gas lighter with flint
- 2 Mortar and pestles
- 4 Ring stands
- 4 Iron rings
- 2 Test tube brushes; 1 large, 1 small
- 4 Wingtop flame spreaders
- 4 Wire gauzes, w/heat-resistant center

CONSUMABLE ITEMS

To be replaced at the end of the term, if used. Glass tubing, 6 mm × 16 inches; 1 towel; 1 sponge

ACCOUNTABLE ITEMS

- 9 Beakers: 1-100; 1-150; 3-250 2-400; 2-600 mL
- 2 Bottles, one-pint, narrow mouth
- 2 Evaporating dishes
- 5 Flasks, Erlenmeyer: 1-50; 3-250; 1-500 mL
- 3 Funnels, long or short stem
- 3 Gas bottles, 4 oz.
- 2 Graduated cylinders: 10 and 50 mL
- 2 Medicine droppers, complete with top
- 4 Medicine dropper bulbs
- 1 Pinch clamp
- 1 Pipet, volumetric, 10 mL
  Rubber tubing-3 ft.
- 1 Nichrome wire
- 1 Scoopula, stainless steel
- 3 Stirring rods, 5 mm
- 12 Stoppers:
  - Two hole: 1-No. 2: 1-No. 5; 1-No. 6
  - One hole: 1-No. 1; 1-No. 2; 1-No. 5; 1-No. 6
  - Solid: 2-No. 3; 1-No. 5; 1 No. 6
  - 1-Two hole No. 3 split one side
- 15 Test tubes, Pyrex: 8-10 × 75 mm; 6-20 × 150 mm; 1-22 × 175 mm
- 1 Test tube block
- 1 Test tube holder, wire
- 1 Thermometer, 100° Celsius
- 1 Thistle tube
- 4 Shell vials, 2 dram
- 3 Watch glasses, 100 mm
- 2 Vials, screw cap, 2-1/4"
- 1 Plastic wash bottle

*Bunsen burners are used as heat sources in many of the experiments. If hot plates are available, they should be used instead of the burners, where feasible.

Necessary Equipment and Materials and Helpful Hints

The apparatus lists are given at the beginning of each experiment and are not repeated here.

EXPERIMENT 1 BASIC LABORATORY TECHNIQUES [1]*

The following items should be provided for a class of 25 students:

- 6 glycerine dropper bottles (30-mL Barnes Bottles with droppers)
- 12 meter sticks
- 6 rubber bulbs for use with pipets

Care should be stressed in this first experiment in the proper use of balances. Do not drop the weighing beam suddenly into full weigh position. Do not add items to or remove items from the balance while it is in the weigh position; arrest the beam (partially) whenever adding or subtracting weight. Do not force the partial- or full-release knobs.

Other cautions for the students include: *Be sure to place the thermometer in the slit side of the stopper and use glycerine as a lubricant!* Demonstrate this to avoid unnecessary injuries. Have some students do Part D before Part C to avoid long lines at the balances at the end of the period. In Procedure C, have students heat water while doing other measurements (A and B) to save time.

The following gives a rationale for the calibration procedures performed in this experiment. The students generally don't understand why they are doing it. Explain the rationale to the students.

The thermometer calibration (Procedure C, p. 6) is performed to check the accuracy of each individual thermometer. These measurements will show how measured temperatures (read from thermometer) compare with true temperatures (the boiling and freezing points of water). The freezing point of water is 0°C, the boiling point depends upon the atmospheric pressure and is calculated as shown in Example 1.2.

The procedure for the calibration of a pipet (Procedure D, p. 7) will permit the student to compare the measured volume (10 mL) with that calculated from the measured mass of the water and its known density

*The numbers in brackets after the experiment titles refer to the chapter(s) in the ninth edition of *Chemistry: The Central Science* by Brown, LeMay, and Bursten that correlate with the experiment.

(Table 1.6, p. 18). Such a calculation is shown in Example 1.3. Record the *calculated* volume on the report sheet (p. 12). Because mass can be measured very accurately on an analytical balance and the density of water is very accurately known, the actual volume delivered by the pipet can be determined very accurately in this manner. In actual practice, however, this procedure may give low results because of the evaporation of water between the time the water is dispensed and when its mass is measured. Nevertheless, this procedure does give practice at measuring masses, in doing calculations involving density, and in seeing how the volume can be determined from mass and density. The students will later use these calibrated values in subsequent experiments. Instruct and caution the students in the use of a pipet bulb so they do not force the pipet into the bulb, causing it to break and thus injuring themselves.

EXPERIMENT 2 IDENTIFICATION OF SUBSTANCES BY PHYSICAL PROPERTIES [1,2]

Each student will require: melting-point capillaries open at one or both ends, rubber bands (cut 1/4″ rubber tubing into small segments), 20 mL cyclohexane, 15 mL ethanol, 5 mL toluene, 1 g naphthalene, and boiling chips (marble chips are adequate). Place a refuse container in the laboratory for disposal of the solvents and unknown; otherwise, the students will throw them in the sink.

SUITABLE UNKNOWNS

Liquids: isopropyl alcohol, methanol, chloroform, cyclohexane

Solids: α-naphthol, biphenyl, *p*-dichlorobenzene, *m*-dinitrobenzene, *p*-dibromobenzene

Some explanation of the use of the nomograph (Figure 2.5) may also be advisable. To avoid long lines at the balances, have half of the students start on Part A and the other half on Part B. Remind students about the proper methods of handling chemicals.

a. Do not take reagents from the reagent shelf to the student desk.
b. Do not pour excess chemicals back into reagent bottles.
c. Do not place ground glass stoppers on the table where they can be contaminated.
d. Do not dispose of chemicals in the sink—use refuse containers.

EXPERIMENT 3 SEPARATION OF THE COMPONENTS OF A MIXTURE [1,2]

Each student will require 20 to 25 g of an unknown mixture (a 6-inch test tube, 1/2 to 3/4 full is adequate). The following mixtures are suggested:

	No. 1	No. 2	No. 3	No. 5
NH_4Cl	255 g (25.5%)	253 g (25.3%)	358 g (35.8%)	207 g (20.7%)
NaCl	205 g (20.5%)	207 g (20.7%)	358 g (35.8%)	693 g (69.3%)
SiO_2	540 g (54.0%)	540 g (54.0%)	284 g (28.4%)	100 g (10.0%)
	1000 g	1000 g	1000 g	1000 g

Unknowns can obviously be varied to suit local conditions. Use fine sand such as Monterey—fine builders' sand, Otawa sand, etc., which is white and blends well with other ingredients. It should be well mixed so as to eliminate lumps and large particles and ensure homogeneity. Remind students to carry out the NH_4Cl sublimation in the hood. Also, the larger the size of the unknown, the longer the evaporation and sublimation processes take. A 2- to 3-g sample is adequate, but the sample calculations are based on a 6-g sample size. Answers will probably provide good results for percent NaCl, low results for percent NH_4Cl, and high results for percent SiO_2. Results improve if watch glasses are washed off into an evaporating dish near the end of the evaporation procedure and then the evaporation carefully continued. Use of a hot-water bath for the initial evaporation lessens the splattering so common with the direct heating procedure described in this experiment, but it is slow.

EXPERIMENT 4 CHEMICAL REACTIONS [3,4]

Each student will require: $KMnO_4$ (0.5 g), Na_2CO_3 (0.5 g), Na_2SO_3 (0.5 g), ZnS (0.5 g), 3 g sulfur, 8 g mossy zinc, 2 in. copper wire (14, 16, or 18 gauge).

The following reagents should be made up in liter or pint bottles and placed in the laboratory (requirements per 25 students):

(50 mL) *Concentrated nitric acid:* 15.8 *M* commercial reagent.

(200 mL) *0.1* M *sodium oxalate:* 13.4 g $Na_2C_2O_4$ per liter of solution

(100 mL) *0.1* M *potassium permanganate:* 15.8 g $KMnO_4$ per liter of solution. Dissolve the $KMnO_4$ in water by gently heating for 1 hour. Cool, filter through an M-grade sintered glass filter, and dilute to the necessary volume. Clean the filter by soaking in 0.05 *M* oxalic acid solution (4.5 g $H_2C_2O_4 \cdot 2H_2O$ per liter).

(100 mL) *10* M *sodium hydroxide:* 400 g NaOH per liter of solution. Add H_2O all at once to 400 g NaOH in a 2-liter beaker, and after the mixture has cooled, transfer to a volumetric flask and dilute to the mark with CO_2-free distilled water.

(100 mL) *0.1* M *lead nitrate:* 33.1 g $Pb(NO_3)_2$ per liter.

(100 mL) *0.1* M *barium chloride:* 24.4 g $BaCl_2 \cdot 2H_2O$ per liter.

(100 mL) *1.0* M *potassium chromate:* 194.0 g K_2CrO_4 per liter.

(100 mL) *0.1* M *sodium bisulfite:* 9.5 g $Na_2S_2O_5$ (sodium metabisulfite) per liter. $Na_2S_2O_5 + H_2O \longrightarrow 2NaHSO_3$ *must be freshly made.*

(200 mL) *6.0* M *ammonium hydroxide:* 400 mL concentrated NH_4OH diluted to 1 liter.

(5 L) *6.0* M *hydrochloric acid:* 500 mL concentrated HCl diluted to 1 liter.

(50 mL) *0.01* M *copper sulfate:* Catalyst not normally required; 2.5 g $CuSO_4 \cdot 5H_2O$ per liter.

(100 mL) *3.0* M *ammonium carbonate:* 288 g $(NH_4)_2CO_3$ per liter.*

(1 L) 6 M *sulfuric acid:* 333.3 mL concentrated acid diluted to 1 liter.

Reclaim the unused mossy zinc by washing with distilled water and air drying. Place a receptacle in the laboratory for this purpose.

EXPERIMENT 5 CHEMICAL FORMULAS [3,4]

The following are required for each student:

6 *M* HCl: dilute 500 mL concentrated acid to 1 liter

Granular zinc: 1 g but student should not use more than 0.5 g.

Copper wire (2 g) scrap 12-, 14-, 16-gauge or whatever is available, cut into 2-inch lengths

Powered or precipitated sulfur (3 g)

Caution the students to not heat the $ZnCl_2$ too much or it will sublime if overly heated, and not to weigh hot objects. Also, have students place their hot objects on heat-resistant pads, not the desktop. If possible, perform $ZnCl_2$ preparations in the hood, as HCl fumes build up during the evaporation step. Copper sulfide should be prepared in the hood because of the formation of SO_2. As alternatives, ZnS, NiS, or PbS may be prepared from Zn, Ni, or Pb powder and elemental sulfur, and $NiCl_2$, $PbCl_2$, $CaCl_2$, or $CrCl_3$ may be prepared from Ni, Pb, Ca, or Cr powder and concentrated HCl. The quantities of materials are similar to those used for the Cu_2S and $ZnCl_2$ preparations.

EXPERIMENT 6 CHEMICAL REACTIONS OF COPPER AND PERCENT YIELD [3,4]

Each student will require 0.5 g copper wire (16- or 18-gauge), 5 mL concentrated HNO_3 (15.8 *M*), 30 mL (3.0 *M* NaOH), 15 mL (6.0 *M* H_2SO_4), 2.0 g (granular zinc), about five 1-in aluminum squares, 10 mL methanol, 10 mL acetone. Use glass beads or carborundum chips as boiling chips.

Reagent preparation: *3.0* M *sodium hydroxide*: 120 g NaOH per liter
6.0 M *sulfuric acid*: 333.3 mL concentrated H_2SO_4 diluted to 1 liter

Throughout the introductory discussion of the experiment, brackets are used to refer to equation numbers. In the experimental procedures, parentheses are employed to relate observations to the proper blanks to be filled in on the answer sheet. Remind students to perform the reaction of nitric acid with copper metal in the fume hood and to use approved eye protection.

---

*Should be fresh because it decomposes (see Merck Index).

EXPERIMENT 7 CHEMICALS IN EVERYDAY LIFE [4,13]

Each of the test reagents should be supplied in dropper bottles. The household reagents are most effectively dispensed from the commercial (grocery store) packages. The following comments should be helpful:

Do not use sudsy ammonia.

Any commerical fertilizer is acceptable, as is any chlorine bleach (Purex, Clorox, etc).

Do not use iodized salt.

The reagent preparations and their consumption (per 25 students) follow:

(50 mL) *mineral oil*: in dropper bottles

(50 mL) *1.0* M *ammonium chloride*: 53.5 g NH_4Cl per liter

(50 mL) *8.0* M *sodium hydroxide*: 320 g NaOH per liter

(50 mL) *18* M *sulfuric acid*: concentrated H_2SO_4 in dropper bottles

(50 mL) *saturated barium hydroxide*: 60 g $Ba(OH)_2 \cdot 8H_2O$ per liter

(50 mL) *3.0* M *nitric acid*: 189.9 mL concentrated HNO_3 diluted to 1 liter

(50 mL) *0.1* M *silver nitrate*: 17.0 g $AgNO_3$ per liter in a dark dropping bottle

(50 mL) *0.2* M *barium chloride*: 49.0 g $BaCl_2 \cdot 2H_2O$ per liter

SUITABLE UNKNOWNS (0.5 g)

$Na_2CO_3 \cdot H_2O$, NaCl, $Na_2SO_4 \cdot 10H_2O$, KI

A section of 25 students will consume: 250 mL household ammonia, 25 g NH_4Cl, 50 g fertilizer, 50 g baking soda, 50 mL vinegar, 50 g chalk, 50 g NaCl, 50 g Epsom salts, 50 g NaI, 50 mL bleach. Some chalks do not contain $CaCO_3$ but rather $CaSO_4$. An-du-septic low-dust white chalk and P/C polychromatic alphasite chalk work well. Spot plates can be used in this experiment to reduce reagent consumption and waste generation.

EXPERIMENT 8 GRAVIMETRIC ANALYSIS OF A CHLORIDE SALT [3,4,13]

The following will be required per student:

(1.5 g) *chloride unknown*: may be acquired from Thorn Smith, Inc., 7755 Narrow Gauge Road, Beulah, MI 49617.

(75 mL) *0.5* M *silver nitrate*: 85 g $AgNO_3$ per liter

(6 mL) *6.0* M *nitric acid*: 381 mL concentrated HNO_3 diluted to 1 liter

(60 mL) *acetone*

The silver nitrate solution should be stored in and dispensed from a dark bottle. Caution the students not to get it on their hands, as it will turn the skin black, which will eventually wear away. The skin color change, however, is harmless.

Sharkskin filter paper is recommended.

The AgCl should be collected so that expensive Ag may be recovered. Provide a receptacle for this purpose. Recovery of $AgNO_3$ is described in N. C. Thomas, *J. Chem. Ed.*, 67, 794 (1990).

EXPERIMENT 9 GRAVIMETRIC DETERMINATION OF PHOSPHORUS IN PLANT FOOD

Each student will require:

10 to 15 g of plant food

(250 mL) *10% magnesium sulfate:* 100 g $MgSO_4 \cdot 7H_2O$ per liter

(300 mL) 2 M *ammonium hydroxide:* dilute 135 mL 14.8 *M* NH_3 to 1 liter

(300 mL) *75% isopropyl alcohol:* 750 mL isopropyl alcohol diluted with water to 1 liter

Epsom salts, household ammonia (non-sudsy), and rubbing alcohol, respectively, may be substituted for these reagents.

EXPERIMENT 10 PAPER CHROMATOGRAPHY: SEPARATION OF CATIONS AND DYES [1,4,5]

The following reagents are required for 25 students:

(1 L) 0.5 M *copper nitrate:* 120.8 g $Cu(NO_3)_2 \cdot 3H_2O$ per liter; add 10 mL concentrated HNO_3 per liter

(1 L) 0.5 M *ferric nitrate:* 202.0 g $Fe(NO_3)_3 \cdot 9H_2O$ dissolved in 100 mL concentrated HNO_3 diluted to 1 liter

(1 L) 0.5 M *nickel nitrate:* 145.4 g $Ni(NO_3)_2 \cdot 6H_2O$ per liter

(1250 mL) *Chromosolvent:* 90% acetone, 10% 6 *M* HCl—prepare as needed, it is very perishable

(250 mL) 15 M *ammonium hydroxide*

(125 mL) *1% dimethylglyoxime:* 1 g dimethylglyoxime in 100 mL absolute ethanol; put in a sprayer such as "Sprayon" Jet Pak sprayer

(125 mL) *isopropyl alcohol*

Unknown solutions are made from the above standard (0.5 *M*) metal solutions by mixing equal volumes as shown in the following table:

	Fe^{3+}	Ni^{2+}	Cu^{2+}
No. 1	X	X	X
No. 2	X	X	—
No. 3	—	X	X
No. 4	X	—	X

*X means the ion is present, and (—) means the ion is absent.

Add equal amounts of distilled water to solutions 2, 3, and 4 to make uniform dilutions.

EXPERIMENT 11 MOLECULAR GEOMETRIES OF COVALENT MOLECULES

Suggest using Prentice-Hall Molecular Model set for General and Organic Chemistry, or Styrofoam balls available from Snow Foam Products, Inc., 9917 W. Gidley St., El Monte, CA 91731, and pipe cleaners.

EXPERIMENT 12 ATOMIC SPECTRA AND ATOMIC STRUCTURE [6]

Each student will require:

(50 mL) 6 M *hydrochloric acid:* 500 mL concentrated HCl diluted to 1 liter

(5 mL) *0.1* M *NaCl:* 5.84 g NaCl per liter

(5 mL) *0.1* M $CaCl_2$: 14.7 g $CaCl_2 \cdot 2H_2O$ per liter

(5 mL) *0.1* M $SrCl_2$: 19.5 g $SrCl_2 \cdot 2H_2O$ per liter

(5 mL) *0.1* M *LiCl:* 4.24 g LiCl per liter

(5 mL) *0.1* M *KCl:* 7.46 g KCl per liter

(5 mL) *0.1* M $BaCl_2$: 20.8 g $BaCl_2$ per liter

Five milliliters each of two unknown solutions, one containing only one of the above cations except K^+ and the other containing one of the following combinations: Li and Ba; Sr and Ba; Ca and Ba; or Sr and Na. All unknown solutions should be 0.1 *M* in each cation. Spectroscopes, diffraction gratings, gaseous discharge tubes, and power supplies are available from Central Scientific Co. (CENCO), 2600 S. Kostner Avenue, Chicago, IL 60623 and other laboratory supply companies. CAUTION THE STUDENT ABOUT THEIR USE.

Students may work in groups of 4 or 6, or this may be done as a demonstration to reduce the amount of equipment required.

EXPERIMENT 13 BEHAVIOR OF GASES: MOLAR MASS OF A VAPOR [10]

A demonstration apparatus needs to be constructed for this experiment. We have found that the following apparatus is easy to construct and use and gives very good results:

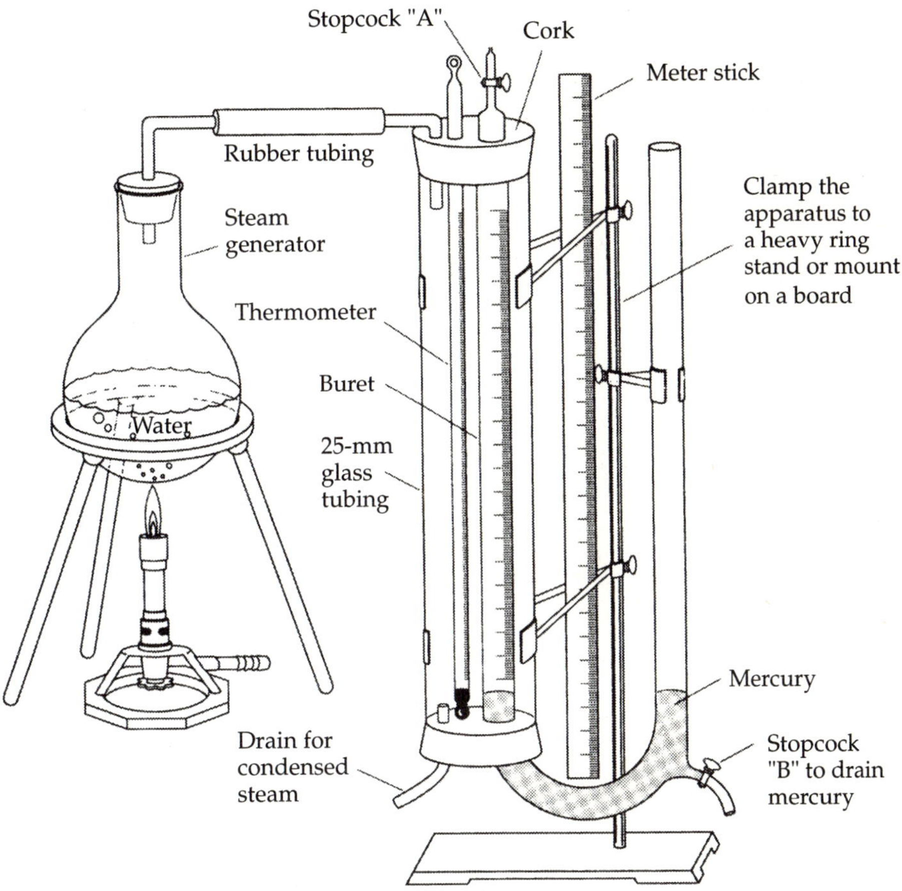

With the apparatus assembled, close stopcock "B." Pour mercury into the open end of the U-tube until it covers the marked region of the buret. Record the atmospheric pressure and the volume of the air in the buret. Now close stopcock "A" and add additional mercury. Record the new air volume and the difference in heights of the mercury columns, using the meter stick to measure the heights. This demonstrates Boyle's law. Now record the temperature and begin generating steam. When the temperature within the steam jacket has equilibrated, record it and open stopcock "B" to equalize the mercury levels and the pressure. Record the new volume. This demonstrates Charles's law. If stopcock "B" were not opened and the mercury levels not equalized, the temperature change would have caused both a volume and pressure change, and the ideal-gas law could have also been demonstrated.

For the molar mass of a vapor, the following are required: aluminum foil cut into 2-in. squares (one per student), common pins (five for 25 students), rubber bands (one per student), thin-walled 125-mL Erlenmeyer flasks (one per student), and 2 mL of a volatile liquid unknown. The following have proved satisfactory:

	Boiling point	*Molar mass*
acetone	56°	58
1,1,1-trichloroethane	74°	133
1,2-dichloroethane	60.0°	99
chloroform	58° to 61°	119.4
2-butanone	80°	72
methanol	65°	32
ethanol	78.5°	46
1-chlorobutane	78°	92.6
tetrahydrofuran	65°	72.1
methylene chloride	40.1°	84.9
cyclohexane	81.4°	84
ethyl acetate	77°	88

Caution the students not to heat the volatile liquid for too long or their molar mass will be too low. Before making final weighings of flasks, cap, rubber band, and condensed vapor, make sure the outside of the flask and foil are completely dry. Drops of water may be trapped under that part of the foil below the rubber band. Immediately after removing the flask from the boiling water, carefully bend this part of the foil away from the flask and wipe it dry.

EXPERIMENT 14 DETERMINATION OF *R*: THE GAS-LAW CONSTANT [10]

Each student requires about 0.3 g $KClO_3$ and 0.02 g MnO_2. Warn students not to return excess MnO_2 into the $KClO_3$ bottle.

EXPERIMENT 15 ACTIVITY SERIES [4,20]

Each student will require the following:

(30 mL) 6 M *hydrochloric acid:* 500 mL concentrated HCl diluted to 1 liter

(25 mL) 0.2 M *calcium nitrate:* 47.2 g $Ca(NO_3)_2 \cdot 4H_2O$ per liter

(25 mL) 0.2 M *magnesium nitrate:* 51.2 g $Mg(NO_3)_2 \cdot 6H_2O$ per liter

(25 mL) 0.2 M *zinc nitrate:* 59.2 g $Zn(NO_3)_2 \cdot 6H_2O$ per liter (add 5 mL concentrated HNO_3 to clear)

(25 mL) 0.2 M *ferric nitrate:* 80.7 g $Fe(NO_3)_3 \cdot 9H_2O$ and 10 mL concentrated HNO_3 diluted to 1 liter

(25 mL) 0.2 M *ferrous sulfate:* 55.6 g $FeSO_4 \cdot 7H_2O$, 5 mL 3 *M* H_2SO_4 and 2 to 3 g iron wire per liter; make fresh as needed

(25 mL) *0.2 M stannic chloride:* 70.0 g $SnCl_4 \cdot 5H_2O$ per liter

(25 mL) *0.5 M cupric nitrate:* 120.8 g $Cu(NO_3)_2 \cdot 3H_2O$ and 10 mL concentrated HNO_3 per liter

The following should be placed in bottles in the laboratory:

calcium shavings	4 g per student
magnesium ribbon	seven 2-in. strips per student
granular zinc	4 g per student
steel wool	1 g per student
granular tin	4 g per student
copper wire	seven 2-in. pieces per student

Spot plates can be used in this experiment to reduce reagent consumption and waste generation.

EXPERIMENT 16 ELECTROLYSIS, THE FARADAY, AND AVOGADRO'S NUMBER [20]

A DC source of electricity with a known voltage and an ammeter need to be provided for this experiment. The setup can be constructed as follows:

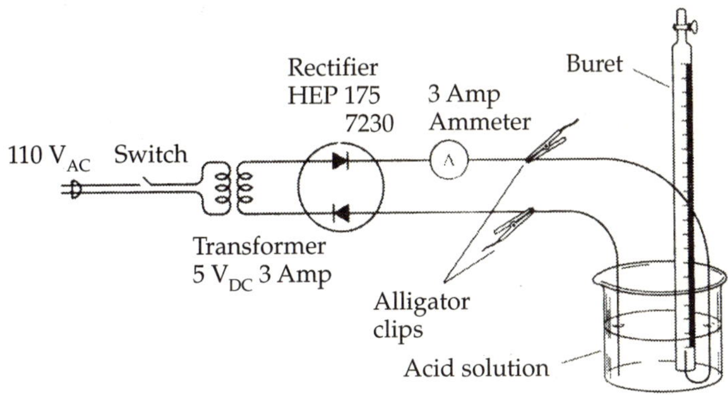

Electrodes can be two pieces of 12-gauge insulated copper wire—one bent to fit up inside the buret—with about 2 in. of insulation removed. The parts for the current source can be procured from any radio shop such as Lafayette or Radioshack. The cathode, not the anode, needs to be in the buret, for you want to collect H_2 not O_2. Students can work in teams to reduce the number of current sources needed as required.

EXPERIMENT 17 ELECTROCHEMICAL CELLS AND THERMODYNAMICS [19,20]

Each student will need the following items: a 6-in. strip of lead, of tin, and of copper, emery cloth, two alligator clips and lead wires, and 3 glass U tubes.

(150 mL) *1 M $Pb(NO_3)_2$*: 331.2 g $Pb(NO_3)_2$ + 5 mL HNO_3 per liter

(150 mL) *1 M $Cu(NO_3)_2$*: 241.60 g $Cu(NO_3)_2 \cdot 3H_2O$ + 5 mL HNO_3 per liter

(150 mL) *1* M *$SnCl_2$*: 225.63 g $SnCl_2 \cdot 2H_2O$ + 500 mL HCl + a small piece of Sn per liter

(150 mL) *0.1* M *KNO_3*: 10.11 g KNO_3 per liter

(3 g) agar

A DC voltmeter or potentiometer capable of measuring mV is required for every four students. Do not move any part of the apparatus when determining the temperature dependence of the EMF, as movement changes the potentials. When making the U-tube agar bridges, there should be no air bubbles in the tube and the cotton plugs should protrude out of the end of the U-tube. Each U-tube may only be used in one cell, as contamination will change the potential. This experiment was adapted from L. P. Beaulieu, *J. Chem. Ed.* 55, 53 (1978).

EXPERIMENT 18 THE CHEMISTRY OF OXYGEN: BASIC AND ACIDIC OXIDES AND THE PERIODIC TABLE [16,22]

Each student will need the following items: 1 deflagrating spoon, 1 pneumatic trough, 15 pieces of pH paper, 1 g calcium shavings, 0.5 g steel wool, 1 2-in. piece of magnesium ribbon, 0.5 g red phosphorus, 1 g powdered or precipitated sulfur, 0.25 g charcoal, 10 g $KClO_3$, 0.2 g MnO_2, 0.5 g H_3BO_3, 0.5 g Na_2O_2, 1 mL 0.1 *M* $HClO_4$ (9 mL of 70% $HClO_4$ diluted to 1 L with H_2O), 5 mL 6 *M* HCl, 0.75 g ZnO and 5 mL 19 *M* NaOH (add 500 mL H_2O to 500 g NaOH all at once in a 2-liter beaker and allow to cool; use reagent-grade low-carbonate NaOH). Do not let burning magnesium drop into or touch the sides of the bottle, as the heat from it will crack the glass.

EXPERIMENT 19 COLLIGATIVE PROPERTIES: FREEZING-POINT DEPRESSION AND MOLAR MASS [13]

Each student will need 1.5 g roll or precipitated (not flowers of) sulfur and 25 g naphthalene.

Provide a container in the laboratory for the disposal of the sulfur-naphthalene. DO NOT THROW IT IN THE SINK! To clean the test tube, warm it until the mixture melts, and pour into the container. A slight residue of naphthalene will sublime, leaving a residue of sulfur that can be brushed away. Do *not* heat the naphthalene to boiling, as it will catch fire, producing volumes of carbon soot. Organic compounds with a molecular weight of about 200 can also be used as unknowns. Suitable unknowns are: *p*-dichlorobenzene, camphor, *p*-dibromobenzene, benzil, benzophenone, biphenyl, 1-bromo-4-chlorobenzene and diphenylmethane. Cyclohexane may be substituted for naphthalene, but it must be cooled with an ice bath. See: M. J. Steffel, *J. Chem. Educ.*, 58, A133, (1981) for details.

EXPERIMENT 20 TITRATION OF ACIDS AND BASES [4,13,17]

Caution students that 19 *M* NaOH is hazardous. Each student will require:

3 mL of 19 *M* NaOH. To prepare 19 *M* NaOH, add 500 mL distilled water at room temperature (not above) all at once to 500 g low-carbonate NaOH in

a 2-liter beaker. Much time can be saved if students are provided with carbonate-free 0.1 *M* NaOH (300 mL for each student).

2 grams potassium hydrogen phthalate (KHP) primary standard, 1 mL of 1% phenolphthalein solution—dissolve 1 g phenolphthalein in 90 mL absolute ethanol and add 10 mL H_2O.

2 grams of unknown—these may be obtained from Thorn-Smith, Inc., 7755 Narrow Gauge Rd., Beulah, MI 49617; phone: 616-882-4672. Twenty-five varieties are available.

Caution: Ensure that the burets are completely cleaned of NaOH solution after use, or the stopcocks will freeze and ruin the buret. Time may be saved if the students weigh their unknown directly from their sample vials and if they begin boiling water early in the laboratory period. Reagent consumption and waste production can be reduced if 10- or 25-mL burets are used.

EXPERIMENT 21 REACTIONS IN AQUEOUS SOLUTION: METATHESIS REACTIONS AND NET IONIC EQUATIONS [4,13,15,17]

Each class of 25 students will require 500 mL of the following reagents:

1.0 M *HCl*: 83.4 mL concentrated HCl diluted to 1 liter

1.0 M *H_2SO_4*: 55.6 mL concentrated H_2SO_4 diluted to 1 liter

1.0 M *ammonium chloride*: 53.5 g NH_4Cl per liter

0.1 M *barium chloride*: 24.4 g $BaCl_2 \cdot 2H_2O$ per liter

0.1 M *cadmium chloride*: 18.33 g $CdCl_2$ per liter

0.1 M *lead nitrate*: 33.1 g $Pb(NO_3)_2$ per liter

0.1 M *nickel chloride*: 23.77 g $NiCl_2 \cdot 6H_2O$ per liter

0.1 M *potassium chloride*: 7.46 g KCl per liter

0.1 M *silver nitrate*: 17.0 g $AgNO_3$ per liter in a dark bottle

1.0 M *sodium acetate*: 82 g $NaC_2H_3O_2$ per liter

1.0 M *sodium carbonate*: 124 g $Na_2CO_3 \cdot H_2O$ per liter

1.0 M *sodium hydroxide*: 40 g NaOH per liter

0.1 M *sodium nitrate*: 8.5 g $NaNO_3$ per liter

0.1 M *trisodium phosphate*: 38.01 g $Na_3PO_4 \cdot 12H_2O$ per liter

0.1 M *sodium sulfide*: 24.02 g $Na_2S \cdot 9H_2O$ per liter

These reagents should be prepared for dispensing from dropper bottles. A class of 25 students will also require 250 g sodium nitrate and 200 g potassium chloride. Part A can be performed with spot plates to reduce reagent consumption and waste production.

When no apparent reaction occurs, the student should consider what the possible products might be and if there is a way to determine if a reaction occurred. Have students *cautiously* smell the NaOH + NH_4Cl and $NaC_2H_3O_2$ + HCl reactions.

EXPERIMENT 22 DETERMINATION OF DISSOCIATION CONSTANT OF A WEAK ACID [16,17]

Each student will require a pH meter with electrodes, 3 mL of 19 *M* NaOH (see Experiment 20), 2 grams primary standard potassium hydrogen phthalate, and 100 mL of an unknown acid. For the unknown acid, acetic acid solutions having concentrations of 0.05 to 0.2 molar are satisfactory. These may be made by dilution of glacial acetic acid, which is 17.4 molar.
For example:

mL glacial acetic acid diluted to 1 liter; *M* HOAc

2.87 mL - - - - - - - - - - - - 0.05 *M*
4.31 mL - - - - - - - - - - - - 0.075 *M*
5.75 mL - - - - - - - - - - - - 0.10 *M*
7.18 mL - - - - - - - - - - - - 0.125 *M*
8.62 mL - - - - - - - - - - - - 0.15 *M*

A class of 25 sudents will also require 100 mL of pH 7.00 and pH 4.70 buffers, which can be purchased commercially or made up in the stockroom. Reagent consumption and waste production can be reduced if 10- or 25-mL burets are used.

EXPERIMENT 23 TITRATION CURVES OF POLYPROTIC ACIDS [17]

Each class of 25 students will require 100 mL of pH 7.00 and pH 4.70 standard buffers, 50 g of primary standard potassium hydrogen phthalate, 75 mL of 19 *M* NaOH (see Experiment 20) or 5 liters of 0.1 *M* NaOH and 2.5 liters of unknown polyprotic acid (0.1 *M* H_3PO_4 is suggested as the unknown acid, but any concentration between 0.025 and 0.1 *M* is satisfactory). These unknowns can be prepared by diluting 85% phosphoric acid that is 14.8 *M*. Maleic and malic acids are also satisfactory unknowns. Reagent consumption and waste production can be reduced if 10- or 25-mL burets are used.

EXPERIMENT 24 HYDROLYSIS OF SALTS AND pH OF BUFFER SOLUTIONS [16,17]

The following reagents and their quantities (per student) will be required (the indicators should be placed in dropper bottles*):

(2 mL) *methyl orange*: 1 g methyl orange powder per liter of distilled water

(2 mL) *methyl red*: 1 g methyl red dissolved in 600 mL ethanol diluted to 1 liter with H_2O

(2 mL) *bromothymol blue*: 1 g bromothymol blue ground in 16 mL 0.1 *M* NaOH diluted to 1 liter with H_2O

(2 mL) *phenol red*: 0.4 g phenosulfonthalein in 45 mL 0.025 *M* NaOH diluted to 1 liter

(2 mL) *phenolphthalein (1%)*: 1 g phenolphthalein in 90 mL 95% ethanol + 10 mL H_2O

*pH paper may be used in place of indicators.

(2 mL) *alizarin yellow R*: 1 g *p*-nitrobenzeneazosalicylate-sodium salt in 1 liter H_2O

(40 mL) *0.1* M *sodium chloride*: 5.8 g NaCl per liter

(40 mL) *0.1* M *sodium acetate*: 8.2 g $NaC_2H_3O_2$ per liter

(40 mL) *0.1* M *copper nitrate*: 24.2 g $Cu(NO_3)_2 \cdot 3H_2O$ per liter

(40 mL) *0.1* M *ammonium chloride*: 5.4 g NH_4Cl per liter

(40 mL) *0.1* M *zinc chloride*: 13.6 g $ZnCl_2$ per liter

(40 mL) *0.1* M *potassium aluminum sulfate*: 47.5 g $KAl(SO_4)_2 \cdot 12H_2O$ per liter

(40 mL) *0.1* M *sodium carbonate*: 12.4 g $Na_2CO_3 \cdot H_2O$ per liter

(8.8 mL) *3.0* M *acetic acid* (172.4 mL glacial acetic diluted to 1 liter)

(1 mL) *6* M *hydrachloric acid* (496 mL of concentrated HCl diluted to 1 liter)

(1 mL) *6* M *sodium hydroxide* (240 g NaOH per liter)

(3.5 g) *sodium acetate trihydrate*

(50 mL) *buffer standard*, pH 7

EXPERIMENT 25 DETERMINATION OF THE SOLUBILITY-PRODUCT CONSTANT FOR A SPARINGLY SOLUBLE SALT [15,17]

Each student will require the following reagents:

(50 mL) *0.0024* M *K_2CrO_4*: 0.4656 g K_2CrO_4 per liter

(15 mL) *0.0040* M *$AgNO_3$*: 0.6795 g $AgNO_3$ per liter

(25 mL) *0.25* M *$NaNO_3$*: 21.2475 g $NaNO_3$ per liter

Waste chromate solutions may be disposed of by adding sufficient solid $BaCl_2$ to completely precipitate $BaCrO_4$. Remove the $BaCrO_4$ by filtration or decantation.

EXPERIMENT 26 HEAT OF NEUTRALIZATION [5,19]

Each student will require two 7- or 9-oz. styrofoam cups, one 4-in. cardboard square with hole for thermometer (use cork borers to make the hole), rubber band (cut small sections 3/16″ rubber tubing to make these), and the following quantities of the following reagents:

(50 mL) *1* M *hydrochloric acid*: dilute 82 mL concentrated HCl to 1 liter.

(100 mL) *1* M *sodium hydroxide*: 40 g NaOH per liter.

(50 mL) *1* M *acetic acid*: dilute 57.5 mL glacial acetic acid to 1 liter

EXPERIMENT 27 RATES OF CHEMICAL REACTIONS: A CLOCK REACTION [14]

Each student will require the following quantities of the following reagents:

(200 mL) *0.2* M *potassium iodide*: 33.2 g KI per liter

(50 mL) *0.4* M *sodium thiosulfate*: 99.3 g $Na_2S_2O_3 \cdot 5H_2O$ per liter (prepare fresh)

(5 mL) *1% boiled starch solution*: mix 10 g soluble starch in a little cold water to make a slurry, then pour into 1 liter boiling water; stir; cool and refrigerate (in dropper bottles)

(300 mL) *0.2* M *potassium nitrate*: 20.2 g KNO_3 per liter

(3 mL) *0.1* M *EDTA*: 37.2 g (ethylene dinitrilo) tetraacetic acid disodium salt per liter (in dropper bottles)

(200 mL) *0.2* M *ammonium persulfate (ammonium peroxydisulfate)*: 45.6 g $(NH_4)_2S_2O_8$ per liter

The ammonium persulfate must be fresh because it decomposes with time to produce ammonium sulfate and liberate oxygen.

See pages 316 and 317 of lab manual for helpful comments.

EXPERIMENT 28 RATES OF CHEMICAL REACTIONS II: RATE AND ORDER OF H_2O_2 DECOMPOSITION [14]

Each pair of students will require the following reagents: 20 mL of commercial 3% hydrogen peroxide or 30% solution diluted to approximately 3% (the H_2O_2 should be fresh) and 30 mL of 0.1 *M* KI (16.60 g KI per liter). A separatory funnel or large thistle tube may be substituted for the leveling bulb. Remind students that the reason for keeping the levels of the liquids at the same height during the evolution of the oxygen is to keep the pressure on the gas constant. HgI_2 will form a soluble complex, HgI_4^{2-}, in the presence of excess iodide. If no precipitate forms, add more $Hg(NO_3)_2$. Remind students that they should keep track of cumulative not incremental times for this experiment.

EXPERIMENT 29 INTRODUCTION TO QUALITATIVE ANALYSIS [15-17]

The following reagents should be provided in dropper bottles. Each student will require about 4 mL.

Al^{3+} 0.1 M *aluminum nitrate*: 37.5 g Al $(NO_3)_3 \cdot 9H_2O$ per liter

NH_4^+ 0.1 M *ammonium nitrate*: 8.0 g NH_4NO_3 per liter

Ca^{2+} 0.1 M *calcium nitrate*: 23.6 g $Ca(NO_3)_2 \cdot 4H_2O$ per liter

Cr^{3+} 0.1 M *chromium nitrate*: 40.0 g $Cr(NO_3)_3 \cdot 9H_2O$ per liter

Fe^{3+} 0.1 M *ferric nitrate*: 40.4 g $Fe(NO_3)_3 \cdot 9H_2O$ per liter

Mg^{2+} 0.1 M *magnesium nitrate*: 25.6 g $Mg(NO_3)_2 \cdot 6H_2O$ per liter

Ni^{2+} 0.1 M *nickel nitrate*: 29.0 g $Ni(NO_3)_2 \cdot 6H_2O$ per liter

Ag^+ 0.1 M *silver nitrate*: 17.0 g $AgNO_3$ per liter

Na^+ 0.1 M *sodium nitrate*: 8.5 g $NaNO_3$ per liter

Zn^{2+} 0.1 M *zinc nitrate*: 29.8 g $Zn(NO_3)_2 \cdot 6H_2O$ per liter

18 M *sulfuric acid*: commercial concentrated reagent

6 M *sulfuric acid*: dilute 334 mL concentrated H_2SO_4 to 1 liter

6 M *hydrochloric acid*: 500 mL concentrated reagent diluted to 1 liter

16 M *nitric acid*: commercial concentrated HNO_3 is 15.8 *M*

6 M *nitric acid*: dilute 380 mL concentrated HNO_3 to 1 liter

3 M *sodium hydroxide*: 120 g NaOH per liter

15 M *ammonium hydroxide*: commercial concentrated reagent is 14.8 *M*

0.1% aluminon: 1 g aurintricarboxylic acid in 1 liter H_2O

0.2 M *barium chloride*: 48.9 g $BaCl_2 \cdot 2H_2O$ per liter

0.1 M *barium hydroxide*: 31.5 g $Ba(OH)_2 \cdot 8H_2O$ per liter

0.2 M *ammonium oxalate*: 24.8 g $(NH_4)_2C_2O_4$ per liter

5.0 M *ammonium chloride*: 267.4 g NH_4Cl per liter

1% dimethylglyoxime: 10 g dimethylglyoxime in 1 liter 95% ethanol

0.2 M *ferrous sulfate* (make it fresh as required): 55.6 g $FeSO_4 \cdot 7H_2O$, 5 mL 3 *M* H_2SO_4 and 5 g soft iron wire per liter

0.01% magnesium reagent: 0.1 g *p*-nitrobenzeneazoresorcinal + 1.0 g NaOH per liter

0.2 M *potassium ferrocyanide*: 84.5 g $K_4Fe(CN)_6 \cdot 3H_2O$ per liter

0.1 M *silver nitrate*: 17 g $AgNO_3$ per liter

chlorine water: use commercial Clorox or Purex or slowly bubble chlorine gas through 0.1 *M* NaOH

mineral oil

3% hydrogen peroxide (make it relatively fresh as it has a poor shelf life): 10 mL of 30% H_2O_2 diluted to 100 mL with H_2O

The following reagents should be placed in the laboratory as solids. Each student will require about 0.5 g: NaCl, NaBr, NaI, $NaNO_3$, Na_2CO_3, $Na_2SO_4 \cdot 10H_2O$.

Unknowns: Both a solid (1 g) and a solution (6 mL) unknown are required. The solution is for the cations, and the solid for the anions analyses. The following are suitable solid unknowns: $Al_2(SO_4)_3$, NH_4Cl, $CaCO_3$, $Cr(NO_3)_3 \cdot 9H_2O$, $Fe(NO_3)_3 \cdot 9H_2O$, $MgCO_3$, $NiSO_4 \cdot 6H_2O$, $Na_2SO_4 \cdot 10H_2O$, NaBr, KBr, $ZnSO_4 \cdot 7H_2O$, NaCl or NaI.

The solution unknowns may be prepared by mixing 10-mL portions of 0.1 *M* solutions of each of the unknowns (as nitrates) listed below with sufficient water to make 60 mL solutions. In this way, solutions that are 0.016 *M* in each metal ion will be obtained. This concentration is adequate for detection.

SOLUTION UNKNOWNS FOR QUALITATIVE ANALYSIS

1. NH_4^+; Fe^{3+}; Al^{3+}; Ni^{2+}; Mg^{2+}
2. Ag^+; Cr^{3+}; Ca^{2+}; Zn^{2+}; Na^+
3. NH_4^+; Fe^{3+}; Al^{3+}; Ca^{2+}; Ni^{2+}; Mg^{2+}
4. Ag^+; Fe^{3+}; Cr^{3+}; Ca^{2+}; Zn^{2+}; Na^+
5. NH_4^+; Cr^{3+}; Al^{3+}; Ni^{2+}; Mg^{2+}
6. Ag^+; Al^{3+}; Ca^{2+}; Zn^{2+}; Na^+
7. Fe^{3+}; Ca^{2+}; Ni^{2+}; Zn^{2+}; Mg^{2+}
8. Fe^{3+}; Cr^{3+}; Al^{3+}; Ni^{2+}; Zn^{2+}
9. NH_4^+; Ag^+; Fe^{3+}; Al^{3+}; Mg^{2+}
10. Ag^+; Fe^{3+}; Al^{3+}; Ca^{2+}; Mg^{2+}; Na^+
11. Ag^+; Cr^{3+}; Ca^{2+}; Zn^{2+}; Mg^{2+}
12. NH_4^+; Cr^{3+}; Ca^{2+}; Ni^{2+}; Mg^{2+}
13. Cr^{3+}; Al^{3+}; Zn^{2+}; Mg^{2+}; Na^+
14. Fe^{3+}; Cr^{3+}; Al^{3+}; Ni^{2+}; Zn^{2+}; Mg^{2+}
15. Fe^{3+}; Cr^{3+}; Al^{3+}; Ca^{2+}; Ni^{2+}; Zn^{2+}
16. NH_4^+; Ag^+; Fe^{3+}; Al^{3+}; Mg^{2+}
17. NH_4^+; Ag^+; Ni^{2+}; Zn^{2+}; Na^+
18. NH_4^+; Ag^+; Cr^{3+}; Al^{3+}; Zn^{2+}
19. NH_4^+; Ag^+; Fe^{3+}; Ca^{2+}; Ni^{2+}
20. Fe^{3+}; Cr^{3+}; Al^{3+}; Ni^{2+}

EXPERIMENT 30 ABBREVIATED QUALITATIVE ANALYSIS SCHEME [15-17]

The following reagents should be provided in dropper bottles. Each student will require about 4 mL.

6 M *acetic acid* ($HC_2H_3O_2$): dilute 350 mL of glacial acetic acid with water to 1 liter

12 M *hydrochloric acid* (HCl): use C.P. 38% acid (sp. gr. 1.19) without dilution

6 M *hydrochloric acid* (HCl): dilute 500 mL of 12 *M* HCl with water to 1 liter

2 M *hydrochloric acid* (HCl): dilute 167 mL of 12 *M* HCl with water to 1 liter

16 M *nitric acid* (HNO_3): use C.P. 69% acid (sp. gr. 1.42) without dilution

6 M *nitric acid* (HNO_3): dilute 375 mL of 16 *M* HNO_3 with water to 1 liter

3 M *nitric acid (HNO_3)*: dilute 188 mL of 16 *M* HNO_3 with water to 1 liter

18 M *sulfuric acid (H_2SO_4)*: use C.P. 94% acid (sp. gr. 1.83) without dilution

6 M *sulfuric acid (H_2SO_4)*: dilute 333 mL of 18 M H_2SO_4 with water to 1 liter

15 M *ammonia (NH_3)*: use C.P. 28% ammonia (sp. gr. 0.90) without dilution

6 M *ammonia (NH_3)*: dilute 400 mL of 15 *M* NH_3 with water to 1 liter

barium hydroxide [$Ba(OH)_2$], saturated solution: add 60 g of $Ba(OH)_2 \cdot 8H_2O$ to 1 liter of water; warm and stir, cool to room temperature, filter or decant clear solution after settling

6 M *sodium hydroxide (NaOH)*: dissolve 240 g of NaOH in 200 mL of water, then dilute with water to a total volume of 1 liter

0.2 M **tin (II) chloride ($SnCl_2$)*: dissolve 45 g of $SnCl_2 \cdot 2H_2O$ in 500 mL of 6 *M* HCl; dilute to 1 liter with water; keep in a well-stoppered bottle containing a few pieces of granular tin

0.1% *aluminon*: dissolve 1 g of the ammonium salt of aurintricarboxylic acid in water and dilute to 1 liter

aluminum wire: 26-gauge, cut to 25–mm lengths

1 M *ammonium acetate ($NH_4C_2H_3O_2$)*: dissolve 77 g of $NH_4C_2H_3O_2$ in water and dilute to 1 liter

2 M *ammonium chloride (NH_4Cl)*: dissolve 107 g of NH_4Cl in water and dilute to 1 liter

**ammonium molybdate [$(NH_4)_2MoO_4$]*: dissolve 20 g of MoO_3 in a mixture of 60 mL of water and 30 mL of 15 *M* NH_3; add this solution slowly and with constant stirring to a mixture of 230 mL of water and 100 mL of 16 *M* HNO_3

ammonium sulfide [$(NH_4)_2S$]: add 1 volume of reagent grade $(NH_4)_2S$ to 2 volumes of water, or saturate 6 *M* NH_3 with H_2S gas

0.2 M *barium chloride ($BaCl_2$)*: dissolve 49 g of $BaCl_2 \cdot 2H_2O$ in water and dilute to 1 liter

mineral oil

**chlorine water, saturated*: in the hood, bubble chlorine gas into 500 mL cool water until saturated

diethyl ether [$(CH_3CH_2)_2O$]: commercial reagent; fire hazard

1% *dimethylglyoxime [$(CH_3)_2C_2(NOH)_2$]*: dissolve 10 g of dimethylglyoxime in 1 liter of 95% ethanol

ethanol, ethyl alcohol (C_2H_5OH), 95% ethyl alcohol

0.2 M *ferric nitrate, iron(III) nitrate [$Fe(NO_3)_3$]*: dissolve 80 g of $Fe(NO_3)_3 \cdot 9H_2O$ in water and dilute to 1 liter

0.2 M **ferrous sulfate, iron(II) sulfate ($FeSO_4$)*: dissolve 55.6 g of $FeSO_4 \cdot 7H_2O$ in 1 liter of water; place a few clean scraps of iron metal in the solution and acidify with 1 mL of 6 *M* H_2SO_4

---

*These solutions do not keep well; prepare a small amount at a time, and replace periodically.

3% **hydrogen peroxide* *(H_2O_2)*: use U.S.P. 3% hydrogen peroxide solution, or dilute 30% U.S.P. hydrogen peroxide by a factor of 10

0.2 M *lead acetate [$Pb(C_2H_3O_2)_2$]*: dissolve 76 g $Pb(C_2H_3O_2)_2 \cdot 3H_2O$ in water, add 10 mL of 6 *M* acetic acid and dilute to 1 liter

0.1 M *manganese(II) chloride ($MnCl_2$)*: dissolve 20 g of $MnCl_2 \cdot 4H_2O$ in 100 mL of 12 *M* HCl

0.1 M *mercuric chloride, mercury(II) chloride ($HgCl_2$)*: dissolve 27 g of $HgCl_2$ in water and dilute to 1 liter

1 M *potassium chromate (K_2CrO_4)*: dissolve 194 g of K_2CrO_4 in water and dilute to 1 liter

1 M *potassium oxalate ($K_2C_2O_4$)*: dissolve 184 g of $K_2C_2O_4 \cdot H_2O$ in water and dilute to 1 liter

0.2 M *potassium nitrite (KNO_2)*: dissolve 17 g KNO_2 in water and dilute to 1 liter

0.2 M *potassium thiocyanate (KSCN)*: dissolve 19 g KSCN in water and dilute to 1 liter

0.1 M *silver nitrate ($AgNO_3$)*: dissolve 17 g $AgNO_3$ in water and dilute to 1 liter

sodium bismuthate ($NaBiO_3$), solid reagent

sodium carbonate (Na_2CO_3), saturated solution: warm excess Na_2CO_3 with 1 liter of water, cool, and filter

sodium peroxide (Na_2O_2), solid reagent: provide small quantity

starch solution: triturate 2 to 4 g soluble starch with cold water to make a thick paste; slowly pour this into 200 mL boiling water in which 2 g boric acid crystals have been dissolved; continue boiling for 1 min, then cool and store in stoppered bottle

0.2 M *tin (II) chloride ($SnCl_2$)*: dissolve 45 g of $SnCl_2 \cdot 2H_2O$ in 500 mL of 6 *M* HCl, dilute with water to 1 liter; keep in a well-stoppered bottle containing a few pieces of granulated tin

1 M **thioacetamide* *(CH_3CSNH_2)*: dissolve 75 g of thioacetamide in water and dilute to 1 liter (Solutions normally last about a year, but it is best to prepare small quantities as needed. Decomposition is accompanied by deposition of sulfur and gradual formation of sulfate. If solution gives precipitate with $BaCl_2$, a fresh solution is needed.)

zinc, granular, 20 or 30 mesh

KNOWN CATION SOLUTIONS

In general, solutions should be about 0.2 *M*, with respect to each ion being tested. The group 1 cations, Ag^+, Hg_2^{2+}, and Pb^{2+} may be 0.1 *M* because they are easily detected. The concentration of NH_4^+ should be higher, about 0.5 *M*, to increase the success of the test for that ion.

Group 1

Place 28 g $HgNO_3 \cdot H_2O$ in a 2 L beaker; add 50 mL of 16 *M* HNO_3; then add water in nine 100 mL portions with frequent stirring to give a volume of about 950 mL. Stir until the $HgNO_3 \cdot H_2O$ dissolves. Then add 17 g of $AgNO_3$ and 33 g of $Pb(NO_3)_2$. Stir until the solids are completely dissolved, and then dilute to 1 L (each cation is 0.1 *M*).

Group 2

Place 1.0 g of $PbCl_2$, 34 g of $CuCl_2 \cdot 2H_2O$, 63 g of $BiCl_3$, and 70 g of $SnCl_4 \cdot 5H_2O$ in a 2 L beaker. Add 250 mL of 12 *M* HCl and 500 mL of water. Stir until the solids dissolve, warming as necessary. Dilute to 1 L. (Because of its limited solubility, the concentration of $PbCl_2$ is only 0.0036 *M*; other cations are 0.2 *M*.)

Group 3

Place 40 g of $MnCl_2 \cdot 4H_2O$, 48 g of $NiCl_2 \cdot 6H_2O$, 54 g of $FeCl_3 \cdot 6H_2O$, and 48 g of $AlCl_3 \cdot 6H_2O$ in a 2 L beaker. Add 50 mL of 12 M HCl and then 700 mL of water. Stir until the solids dissolve, then dilute to 1 L (each cation is 0.2 *M*).

Group 4

Place 49 g of $BaCl_2 \cdot 2H_2O$, 29 g of $CaCl_2 \cdot 2H_2O$, 12 g of NaCl, and 27 g of NH_4Cl in a 2 L beaker. Add 850 mL of water and stir until the solids dissolve. Dilute to 1 L. (NH_4^+ is 0.5 M, and other cations are 0.2 *M*.)

UNKNOWN CATION SOLUTIONS

The unknown solutions should be prepared in the same concentrations as the known solutions, and the same salts should be used.

ANION KNOWNS AND UNKNOWNS

Anion knowns and unknowns are best given out as solid salts. Usually sodium salts are most convenient. If desired, 0.2 *M* solutions of anions may be employed.

EXPERIMENT 31 COLORIMETRIC DETERMINATION OF IRON [25]

Each student will require the following quantities of the following reagents:

(20 mL) *standard iron solution*: 1 mL = 0.050 mg Fe, 361.6 mg $Fe(NO_3)_3 \cdot 9H_2O$ + 10 mL 6 M HNO_3 diluted to 1 liter

(10 mL) 1 M *ammonium acetate*: 77 g $NH_4C_2H_3O_2$ per liter

(10 mL) *10% hydroxylaminehydrochloride*: 10 g $NH_2OH \cdot HCl$ per liter

(100 mL) *0.3% o-phenanthroline*: 3 g 1,10-phenanthroline dissolved in 80 mL absolute ethanol diluted to 1 liter with H_2O

(10 mL) 6 M *sulfuric acid*: 334 mL concentrated H_2SO_4 diluted to 1 liter

UNKNOWNS

Each student will require about 0.1 g iron unknown. Thorn Smith analyzed ferrous ammonium sulfate containing 9% to 16% iron is suitable. Available from Thorn Smith, Inc., 7755 Narrow Gauge Road, Beulah, MI 49617.

EXPERIMENT 32 DETERMINATION OF ORTHOPHOSPHATE IN WATER [18]

Each student will require the following quantities of the following reagents:

(0.2 g) *KH_2PO_4*: dry commercial reagent at 110° C in an oven for several hours

(25 mL) 6 M *hydrochloric acid*: 500 mL concentrated HCl diluted to 1 liter

(2 mL) 18 M *sulfuric acid*: commercial reagent in dropper bottle

(5 mL) *chloroform*: commercial reagent in dropper bottle

(35 mL) *ammonium vanadomolybdate solutions*: see page 417 of the manual for its preparation

UNKNOWNS Unknown phosphates should be procured from local water sources such as irrigation ditches, sewage plant effluents, rivers, ponds, etc. Unknowns can also be made up by using diluted detergent mixtures; however, an antifoaming surfactant such as Dow Corning Antiform A spray must then be used as well.

EXPERIMENT 33 ANALYSIS OF WATER FOR DISSOLVED OXYGEN [18]

Each student will require the following quantities of the following reagents:

(1 mL) 18 M *sulfuric acid*: commercial concentrated H_2SO_4 in dropper bottle

(15 mL) 1 M *sulfuric acid*: 55.6 mL 18 *M* H_2SO_4 diluted to 1 liter

(1 mL) *2% sodium azide*: 2 g NaN_3 in 100 mL H_2O

(2 mL) 2.15 M *manganous sulfate*: 480 g $MnSO_4 \cdot 4H_2O$ or 402 g $MnSO_4 \cdot 2H_2O$ or 364 g $MnSO_4 \cdot H_2O$ per liter (filter this solution to clarify)

(4 mL) *alkaline iodide azide reagent*: 500 g NaOH + 135 g NaI per liter, then add to 10 g NaN_3 dissolved in 40 mL H_2O

(1 mL) *0.025* M *sodium thiosulfate solution* (This can be prepared and given to the student to save time if desired): carefully weigh 6.205 g $Na_2S_2O_3 \cdot 5H_2O$ and dissolve in 1 liter of H_2O; boil, cool, dilute to 1 liter, and add 0.4 g NaOH and 5 mL chloroform; the students may then standardize it

(0.50 g) *solid potassium iodate (KIO_3)*

(2 g) *solid potassium iodide (KI)*

(10 mL) *1% boiled starch solution* (no preservatives; refrigerate after preparation): boil 1 liter of water; while boiling, add 10 g soluble starch mixed in cold water to form a thin slurry; boil for 2 min, cool, decant, bottle, and refrigerate

UNKNOWNS The students should collect their own water sample from any source available.

EXPERIMENT 34 PREPARATION AND REACTIONS OF COORDINATION COMPOUNDS: OXALATE COMPLEXES

Each student is to prepare one coordination compound; the directions are given in the experiment. In addition, the reactions of *cis* and *trans* K $[Cr(C_2O_4)_2(H_2O)_2]$ are to be demonstrated, and the preparations for these two compounds are given on pages 440 and 441 of the lab text. These materials should be prepared in advance, as they require some time to prepare. The consumption of reagents depends upon which coordination compound is prepared. Consult the experiment for further details.

EXPERIMENT 35 OXIDATION-REDUCTION TITRATIONS I: DETERMINATION OF OXALATE [20,25]

Each student will require the following quantities of the following reagents:

(1.5 L) *1.0* M *sulfuric acid*: dilute 111.2 mL concentrated H_2SO_4 to 2 liters

(500 mL) *0.02* M *potassium permanganate*: dissolve 3.16 g $KMnO_4$ in 1 liter H_2O and simmer for 1 hour. Cool, filter through medium-porosity sintered glass filter, and dilute to 1 liter. Store in dark bottle. Clean filter by soaking in a 0.1 *M* oxalic acid solution (12.6 g oxalic acid dihydrate/liter) to dissolve the MnO_2 and flush with distilled H_2O.

(0.75 g) *sodium oxalate* ($Na_2C_2O_4$) primary standard.

UNKNOWNS

1 g standard oxalate samples are obtainable from Thorn Smith, Inc., or use the oxalate complex that the student prepared in Experiment 34. See Experiment 31 for the address of Thorn Smith.

Note Oxalate-permanganate titrations may be done at room temperature if ascorbic acid is used as a catalyst. See *J. Chem. Ed.*, 56, 659 (1979).

Reagent consumption and waste generation may be reduced if 10- or 25-mL burets are used with an appropriate reduction in the amount of the oxalate compounds.

EXPERIMENT 36 PREPARATION OF SODIUM BICARBONATE AND SODIUM CARBONATE

Each student will require the following quantities of the following reagents:

(20 g) *marble chips* (calcium carbonate): technical grade

(50 mL) *saturated* $(NH_4)_2CO_3$ ÷ *NaCl solution*: preparation given on page 453 of the manual

(150 mL) 6 M *HCl*: dilute 500 mL concentrated HCl to 1 liter

(6 mL) 6 M *ammonium hydroxide*: 426 mL concentrated ammonium hydroxide diluted to 1 liter

EXPERIMENT 37 OXIDATION-REDUCTION TITRATION II: ANALYSIS OF BLEACH

Each student will require the following amount of reagents:

(35 mL) *3 M H_2SO_4*: 167 mL 18 *M* H_2SO_4 diluted to 1 liter

(50 mL) *0.0100* M *KIO_3*: 2.140 g KIO_3 dissolved in water to make exactly 1 liter

(25 mL) *3* M *KI*: 498 g KI dissolved in 1 liter water. Protect from air and sun. Prepare shortly before use.

(15 mL) *1* M *$Na_2S_2O_3$*: 248 g $Na_2S_2O_3 \cdot 5H_2O$ disolved in 1 liter water

(2 mL) *3% ammonium heptamolybdate*: 32 g $(NH_4)_6Mo_7O_{24} \cdot 4H_2O$ dissolved in 1 liter water

(4 mL) *starch indicator*: mix 2 to 4 g of soluble starch with cold water to make a thick paste. Slowly pour this into 200 mL of boiling water in which 2 g of boric acid crystals have been dissolved. Continue boiling, then cool. Store in refrigerator

(2 mL) *commercial bleach* (about 5.25% OCl^-): dilute to make different concentrations. These solutions do not keep well. Prepare a small amount at a time and replace periodically.

EXPERIMENT 38 MOLECULAR GEOMETRY [26]

Suggest using Prentice-Hall Molecular Model Set for General and Organic Chemistry.

EXPERIMENT 39 PREPARATION OF ASPIRIN AND OIL OF WINTERGREEN [26]

Each student will require 4 g salicylic acid (*o*-hydroxybenzoic acid), 6 mL acetic anhydride, 50 mL absolute ethanol, 5 mL methanol, 2 mL concentrated H_2SO_4, and 1 mL of 1% ferric chloride (16.5 g $FeCl_3 \cdot 6H_2O$ per liter).

It is a good practice to collect the aspirin from the students!

EXPERIMENT 40 ANALYSIS OF ASPIRIN [17,26]

Each student will require 1.5 g of an unknown aspirin sample, which may either be a student preparation or an unbuffered commercial sample.

(250 mL) *0.1* M *NaOH*: 4 g NaOH per liter (standardized)

(250 mL) *0.1* M *HCl*: 8.33 mL concentrated HCl diluted to 1 liter (standardized)

The student may standardize both the NaOH and HCl solutions, but considerable time will be saved if standardized reagents are provided.

EXPERIMENT 41 ION-EXCHANGE RESINS: ANALYSIS OF A CALCIUM, MAGNESIUM, OR ZINC SALT [16,25]

Each student will require about 1 g of an unknown calcium, magnesium, or zinc sulfate. They also will require 400 mL of 0.1 *M* NaOH (4 g NaOH/liter), 0.6 g primary standard potassium hydrogen phthalate, phenolphthalein solution (consult Experiment 19), and 10 g of Dowex 50W-X-8 ion-exchange resin. A large container should be provided for the return of the ion-exchange resin, because it may be regenerated and reused as described in the experiment.

Name ______________________________ Desk ______________

Date ______________ Laboratory Instructor ______________________

LOCATION OF SAFETY EQUIPMENT NEAREST TO YOUR BENCH

In the rectangle below, indicate the location of the following items using the designated symbols for each.

1. All doors | (heavy line)
2. Your bench [bench]
3. Your desk x (on bench)
4. Fume hood [HOOD]
5. Safety shower (SS)
6. Eye wash fountain (EW)
7. Fire extinguisher (FE)

Your instructor may request additional items to be indicated, such as fire blankets, telephones, fire alarms, etc. Locate them as appropriate. Any additional safety information can be listed below.

____

____

____

____

____

____

____

Laboratory Safety and Work Instructions

LABORATORY SAFETY

The laboratory can be—but is not necessarily—a dangerous place. When intelligent precautions and a proper understanding of techniques are employed, the laboratory is no more dangerous than any other classroom. Most of the precautions are just common-sense practices. These include the following:

1. Wear *approved* eye protection (including splash guards) at all times while in the laboratory. (*No one will be admitted without it.*) Your safety eye protection may be slightly different from that shown, but it must include shatterproof lenses and side shields to provide protection from splashes.

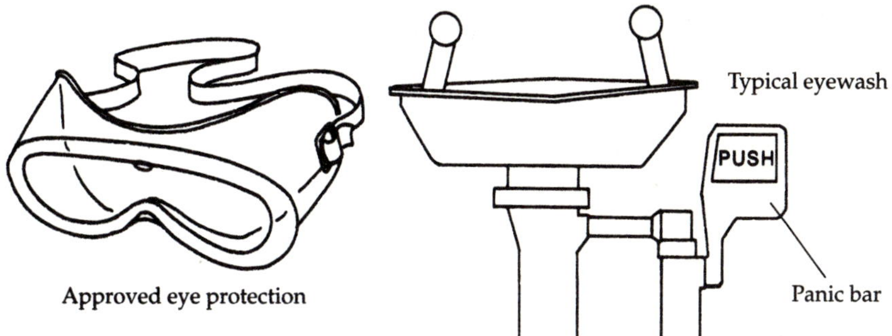

   The laboratory has an eyewash fountain available for your use. In the event that a chemical splashes near your eyes, you should use the fountain ***before the material runs behind your eyeglasses and into your eyes.*** The eyewash has a "panic bar," which enables its easy activation in an emergency.

2. Wear shoes at all times. (*No one will be admitted without them.*)
3. Eating, drinking, and smoking are strictly prohibited in the laboratory at all times.
4. Know where to find and how to use all safety and first-aid equipment (see the first page of this book).
5. Consider all chemicals to be hazardous unless you are instructed otherwise. ***Dispose of chemicals as directed by your instructor.*** Follow the explicit instructions given in the experiments.

6. If chemicals come into contact with your skin or eyes, wash immediately with copious amounts of water and then consult your laboratory instructor.
7. Never taste anything. Never directly smell the source of any vapor or gas. Instead, by means of your cupped hand, bring a small sample to your nose. Chemicals are not to be used to obtain a "high" or clear your sinuses.

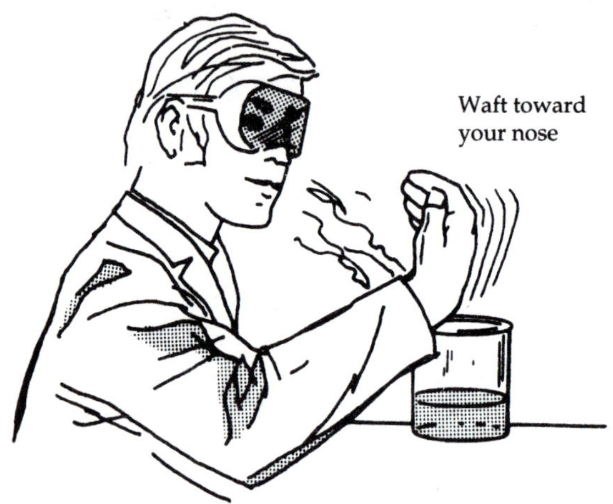

8. Perform in the fume exhaust hood any reactions involving skin-irritating or dangerous chemicals, or unpleasant odors. This is a typical fume exhaust hood. Exhaust hoods have fans to exhaust fumes out of the

hood and away from the user. The hood should be used when you are studying noxious, hazardous, and flammable materials. It also has a shatterproof glass window, which may be used as a shield to protect you from minor explosions. Reagents that evolve toxic fumes are stored in the hood. Return these reagents to the hood after their use.

9. Never point a test tube that you are heating at yourself or your neighbor—it may erupt like a geyser.

10. Do not perform *any* unauthorized experiments.
11. Clean up all broken glassware *immediately*.
12. Always pour acids into water, not water into acid, because the heat of solution will cause the water to boil and the acid to spatter. "Do as you oughter, pour acid into water."
13. Avoid rubbing your eyes unless you *know* that your hands are clean.
14. When inserting glass tubing or thermometers into stoppers, *lubricate the tubing and the hole in the stopper with glycerol or water.* Wrap the rod in a towel and grasp it as close to the end being inserted as possible. Slide the glass into the rubber stopper with a twisting motion. Do not push. Finally, remove the excess lubricant by wiping with a towel. Keep your hands as close together as possible in order to reduce leverage.

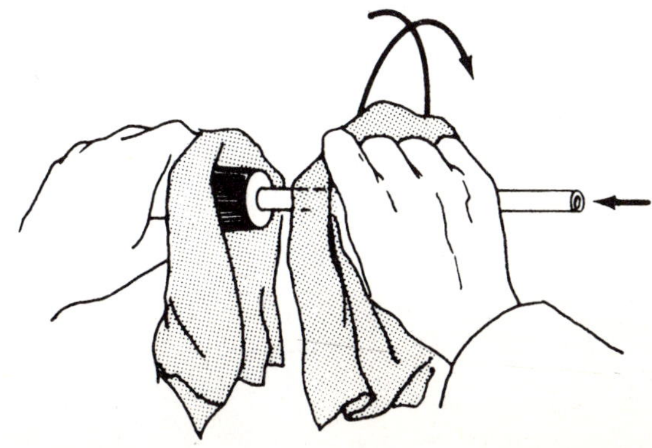

15. NOTIFY THE INSTRUCTOR IMMEDIATELY IN CASE OF AN ACCIDENT.
16. Many common reagents—for example, alcohols, acetone, and especially ether—are highly flammable. *Do not use them anywhere near open flames.*
17. Observe all special precautions mentioned in experiments.
18. Learn the location and operation of fire-protection devices.

    In the unlikely event that a large chemical fire occurs, carbon dioxide fire extinguishers are available in the lab (usually mounted near one of the exits in the room). A typical carbon dioxide fire extinguisher is shown below.

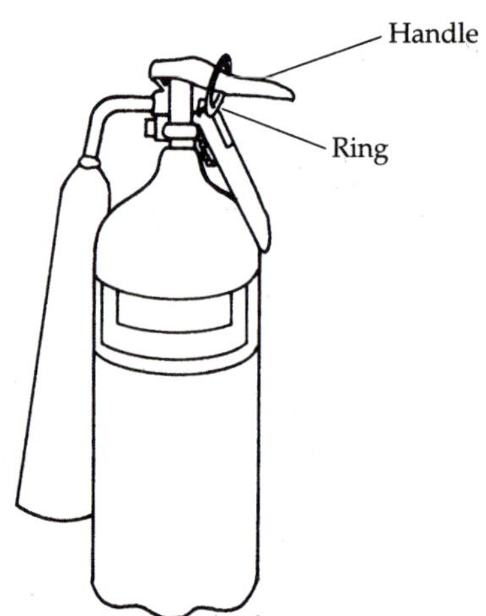

In order to activate the extinguisher, you must pull the metal safety ring from the handle and then depress the handle. Direct the output from the extinguisher at the base of the flames. The carbon dioxide smothers the flames and cools the flammable material quickly. If you use the fire extinguisher, be sure to turn the extinguisher in at the stockroom so that it can be refilled immediately. If the carbon dioxide extinguisher does not extinguish the fire, evacuate the laboratory immediately and call the fire department.

One of the most frightening and potentially most serious accidents is the ignition of one's clothing. Certain types of clothing are hazardous in the laboratory and must *not* be worn. Since *sleeves* are most likely to come closest to flames, ANY CLOTHING THAT HAS BULKY OR LOOSE SLEEVES SHOULD NOT BE WORN IN THE LABORATORY. Ideally, students should wear laboratory coats with tightly fitting sleeves. Long hair also presents a hazard and must be tied back.

If a student's clothing or hair catches fire, his or her neighbors should take prompt action to prevent severe burns. Most laboratories have a water shower for such emergencies. A typical laboratory emergency water shower has the following appearance:

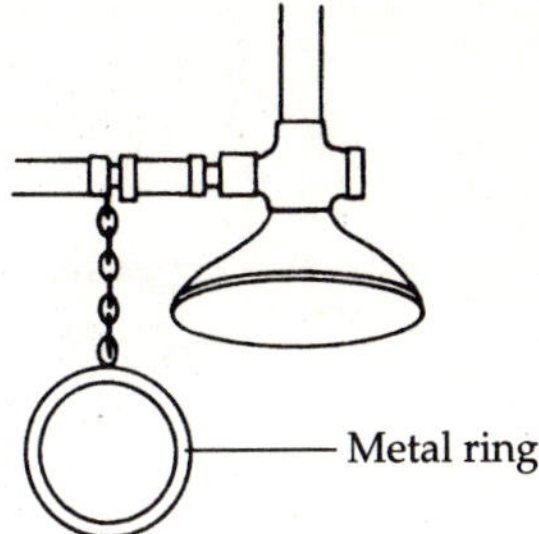

In case someone's clothing or hair is on fire, immediately lead the person to the shower and pull the metal ring. Safety showers generally dump 40 to 50 gallons of water, which should extinguish the flames. These showers generally cannot be shut off once the metal ring has been pulled. Therefore, the shower cannot be demonstrated. (Showers are checked for proper operation on a regular basis, however.)

19. Whenever possible, use hot plates in place of Bunsen burners.

BASIC INSTRUCTIONS FOR LABORATORY WORK

1. Read the assignment *before* coming to the laboratory.
2. Work independently unless instructed to do otherwise.
3. Record your results directly onto your report sheet or notebook. DO NOT RECOPY FROM ANOTHER PIECE OF PAPER.
4. Work conscientiously to avoid accidents.
5. Dispose of excess reagents as instructed by your instructor. NEVER RETURN REAGENTS TO THE REAGENT BOTTLE.
6. Do not place reagent-bottle stoppers on the desk; hold them in your hand. Your laboratory instructor will show you how to do this. Replace the stopper on the same bottle, never on a different one.
7. Leave reagent bottles on the shelf where you found them.
8. Use only the amount of reagent called for; avoid excesses.
9. Whenever instructed to use water in these experiments, use distilled water unless instructed to do otherwise.
10. Keep your area clean.
11. Do not borrow apparatus from other desks. If you need extra equipment, obtain it from the stockroom.
12. When weighing, do not place chemicals directly on the balance.
13. Do not weigh hot or warm objects. Objects should be at room temperature.
14. Do not put hot objects on the desktop. Place them on a wire gauze or heat-resistant pad.

COMMON LABORATORY APPARATUS

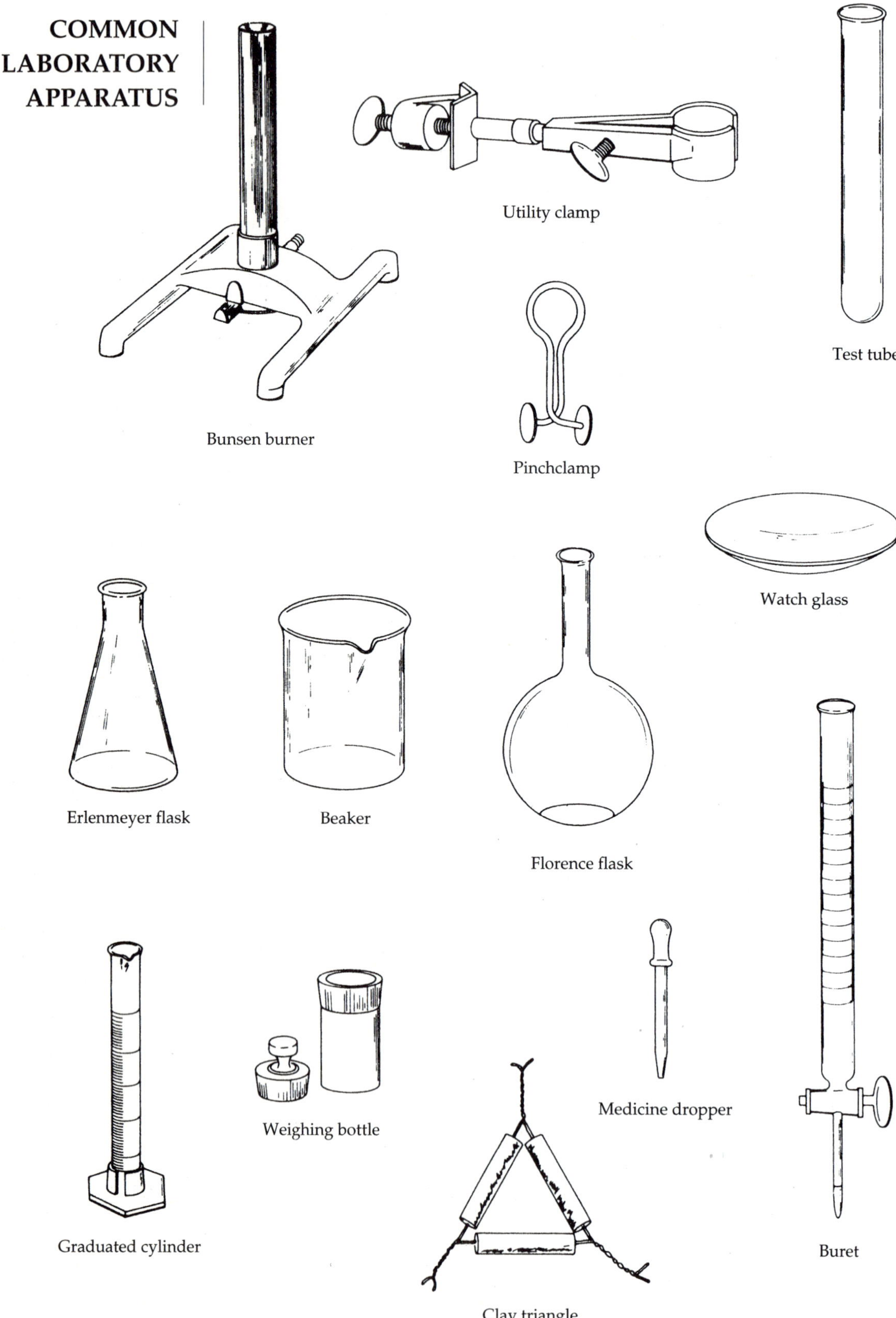

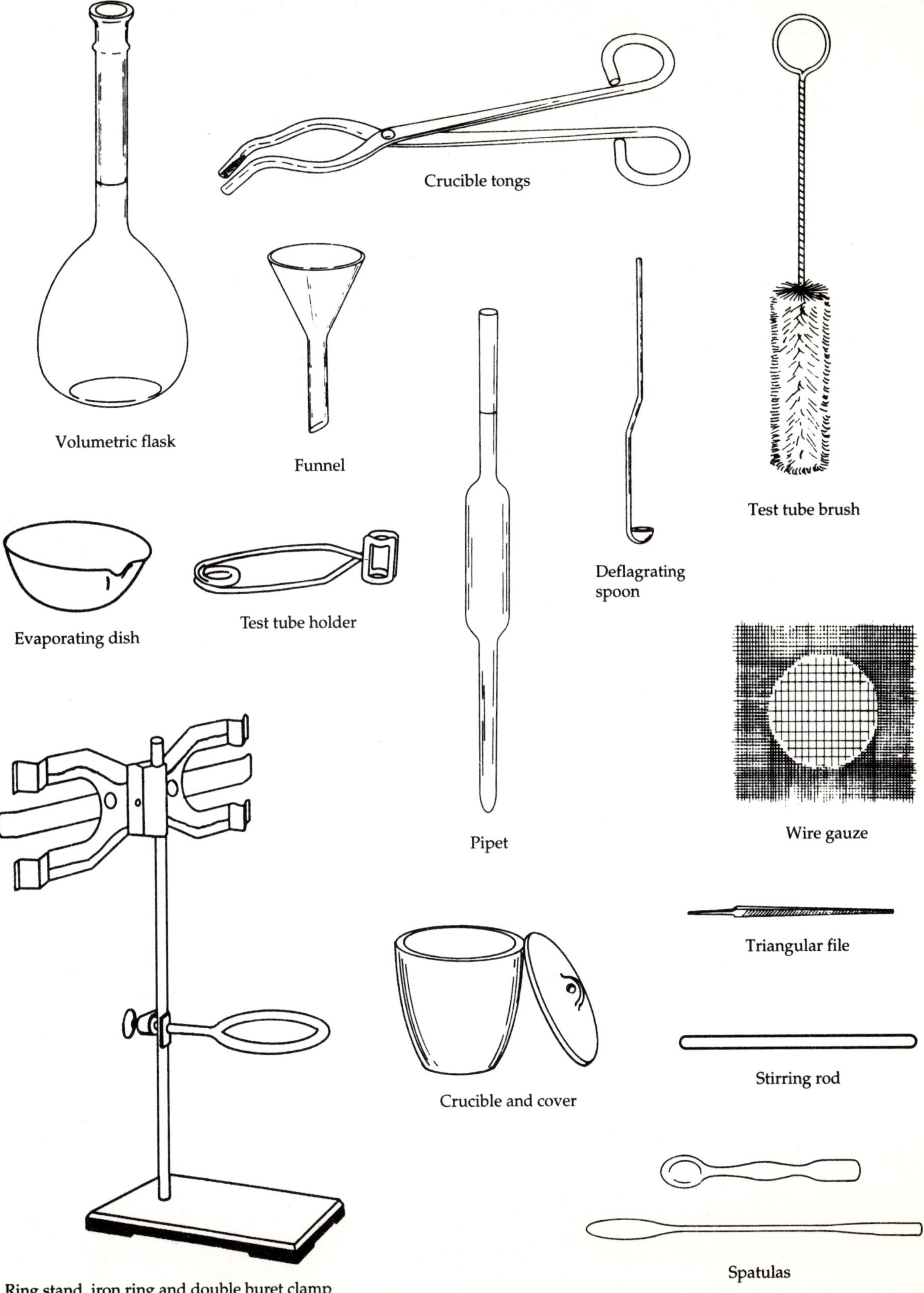
Volumetric flask
Crucible tongs
Test tube brush
Funnel
Pipet
Deflagrating spoon
Evaporating dish
Test tube holder
Wire gauze
Ring stand, iron ring and double buret clamp
Triangular file
Crucible and cover
Stirring rod
Spatulas

Experiment

Basic Laboratory Techniques

OBJECTIVE

To learn the use of common, simple laboratory equipment.

APPARATUS AND CHEMICALS

Apparatus

balance
150- and 250-mL beakers
50- and 125-mL Erlenmeyer flasks
50- or 100-mL graduated cylinder
barometer
clamp
large test tube
Bunsen burner and hose
meter stick
10-mL pipet
rubber bulb for pipet
ring stand and iron ring
thermometer

Chemicals

ice
antifreeze (ethylene glycol)

DISCUSSION

Chemistry is an experimental science. It depends upon careful observation and the use of good laboratory techniques. In this experiment, you will become familiar with some basic operations that will help you throughout this course. Your success as well as your safety in future experiments will depend upon your mastering these fundamental operations.

Because every measurement made in the laboratory is really an approximation, it is important that the numbers you record reflect the accuracy and precision of the device you use to make the measurement. Appendix A of this manual contains a section on significant figures and measurements that you may find helpful in performing this experiment. Our system of weights and measures, the metric system, was originally based mainly upon fundamental properties of one of the world's most abundant substances: water. The system is summarized in Table 1.1. Conversions within the metric system are quite simple once you have committed to memory the meaning of the prefixes given in Table 1.2 and you use dimensional analysis.

In 1960 international agreement was reached, specifying a particular choice of metric units in which the basic units for length, mass, and time are the meter, the kilogram, and the second. This system of units, known as the International System of Units, is commonly referred to as the SI system and is preferred in scientific work. A comparison of some common SI, metric, and English units is presented in Table 1.3.

In Table 1.1, the prefix *means* the power of 10. For example, 5.4 *centi*meters means 5.4×10^{-2} meter; *centi-* has the same meaning as $\times 10^{-2}$.

TABLE 1.1 **Units of Measurement in the Metric System**

Measurement	Unit and definition
Mass or weight	Gram (g) = weight of 1 cubic centimeter (cm^3) of water at 4°C and 760 mm Hg Mass = quantity of material Weight = mass × gravitational force
Length	Meter (m) = 100 cm = 1000 millimeters (mm) = 39.37 in.
Volume	Liter (L) = volume of 1 kilogram (kg) of water at 4°C
Temperature	°C, measures heat intensity: $°C = \frac{5}{9}(°F - 32)$ or $°F = \frac{9}{5}°C + 32$
Heat	1 calorie (cal), amount of heat required to raise 1 g of water 1°C: 1 cal = 4.184 joules (J)
Density	*d*, usually g/mL, for liquids and g/L for gases: $d = \frac{\text{mass}}{\text{unit volume}}$
Specific gravity	sp gr, dimensionless: $\text{sp gr} = \frac{\text{density of a substance}}{\text{density of a reference substance}}$

TABLE 1.2 **The Meaning of Prefixes in the Metric System**

Prefix	Meaning (power of 10)	Abbreviation
femto-	10^{-15}	f
pico-	10^{-12}	p
nano-	10^{-9}	n
micro-	10^{-6}	μ
milli-	10^{-3}	m
centi-	10^{-2}	c
deci-	10^{-1}	d
kilo-	10^{3}	k
mega-	10^{6}	M
giga-	10^{9}	G

TABLE 1.3 **Comparison of SI, Metric, and English Units**

Physical quantity	SI unit	Some common metric units	Conversion factors between metric and English units
Length	Meter (m)	Meter (m) Centimeter (cm)	1 m = 10^2 cm 1 m = 39.37 in. 1 in. = 2.54 cm
Volume	Cubic meter (m^3)	Liter (L) Milliliter (mL)*	1 L = 10^3 cm^3 1 L = 10^{-3} m^3 1 L = 1.06 qt
Mass	Kilogram (kg)	Gram (g) Milligram (mg)	1 kg = 10^3 g 1 kg = 2.205 lb 1 lb = 453.6 g
Energy	Joule (J)	Calorie (cal)	1 cal = 4.184 J
Temperature	Kelvin (K)	Degree celsius (°C)	0K = −273.15°C $°C = \frac{5}{9}(°F - 32)$ $°F = \frac{9}{5}°C + 32$

*A mL is the same volume as a cubic centimeter: 1 mL = 1 cm^3.

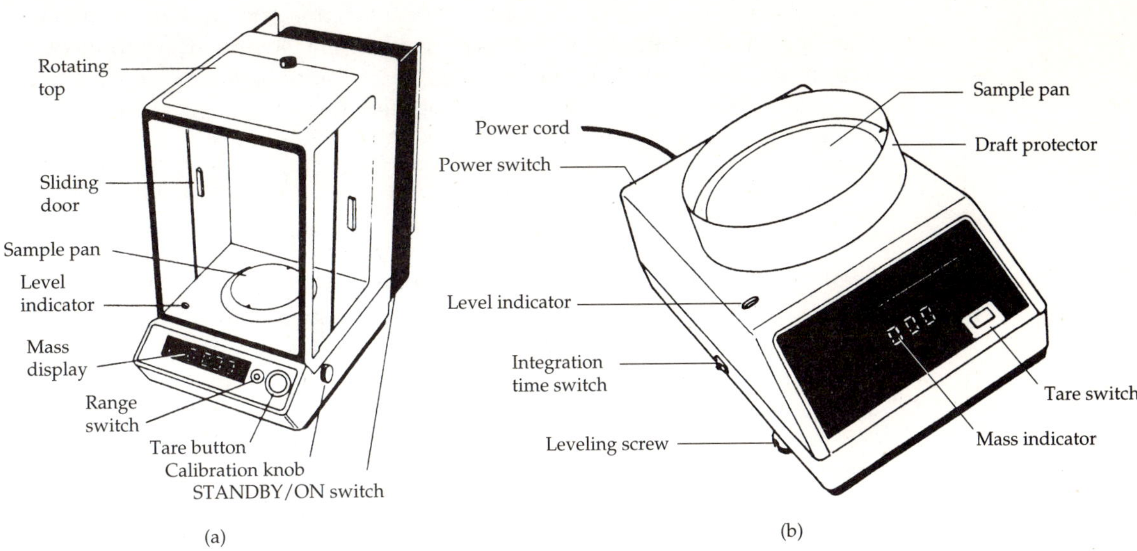

▲ **FIGURE 1.1** Digital electronic balances. The balance gives the mass directly when an object to be weighed is placed on the pan. (a) Analytical balance. (b) Top loader.

Example 1.1

Convert 6.7 nanograms to milligrams.

SOLUTION:

$$(6.7\ \cancel{ng})\left(\frac{10^{-9}\ \cancel{g}}{1\ \cancel{ng}}\right)\left(\frac{1\ mg}{10^{-3}\ \cancel{g}}\right) = 6.7 \times 10^{-6}\ mg$$

Notice that the conversion factors have no effect on the magnitude (only the power of 10) of the mass measurement.

The quantities presented in Table 1.1 are measured with the aid of various pieces of apparatus. A brief description of some measuring devices follows.

Laboratory Balance

A laboratory balance is used to obtain the mass of various objects. There are several varieties of balances, with various limits on their accuracy. Two common kinds of balances are depicted in Figure 1.1. These single-pan balances are found in most modern laboratories. Generally they are simple to use, but they are very *delicate* and *expensive.* The amount of material to be weighed and the accuracy required determine which balance you should use.

Meter Rule

The standard unit of length is the meter (m), which is 39.37 in. in length. A metric rule, or meterstick, is divided into centimeters (1 cm = 0.01 m; 1 m = 100 cm) and millimeters (1 mm = 0.001 m; 1 m = 1000 mm). It follows that 1 in. is 2.54 cm. (Convince yourself of this, since it is a good exercise in dimensional analysis.)

Graduated Cylinders

Graduated cylinders are tall, cylindrical vessels with graduations scribed along the side of the cylinder. Since volumes are measured in these cylinders by measuring the height of a column of liquid, it is critical that the cylinder have a uniform diameter along its entire height. Obviously, a tall cylinder with a small diameter will be more accurate than a short one with a large diameter. A liter (L) is divided into milliliters (mL), such that 1 mL = 0.001 L, and 1 L = 1000 mL.

Thermometers

Most thermometers are based upon the principle that liquids expand when heated. Most common thermometers use mercury or colored alcohol as the liquid. These thermometers are constructed so that a uniform-diameter capillary tube surmounts a liquid reservoir. To calibrate a thermometer, one defines two reference points, normally the freezing point of water (0°C, 32°F) and the boiling point of water (100°C, 212°F) at 1 atm of pressure (1 atm = 760 mm Hg).* Once these points are marked on the capillary, its length is then subdivided into uniform divisions called *degrees*. There are 100° between these two points on the Celsius (°C, or centigrade) scale and 180° between those two points on the Fahrenheit (°F) scale.

Pipets

Pipets are glass vessels that are constructed and calibrated so as to deliver a precisely known volume of liquid at a given temperature. The markings on the pipet illustrated in Figure 1.2 signify that this pipet was calibrated to deliver (TD) 10.00 mL of liquid at 25°C. *Always* use a rubber bulb to fill a pipet. NEVER USE YOUR MOUTH! A TD pipet should not be blown empty.

It is important that you be aware that every measuring device, regardless of what it may be, has limitations in its accuracy. Moreover, to take full advantage of a given measuring instrument, you should be familiar with or evaluate its accuracy. Careful examination of the subdivisions on the device will indicate the maximum accuracy you can expect of that particular tool. In this experiment you will determine the accuracy of your 10-mL pipet. The approximate accuracy of some of the equipment you will use in this course is given in Table 1.4.

Not only should you obtain a measurement to the highest degree of accuracy that the device or instrument permits, but you should also record the reading or measurement in a manner that reflects the accuracy of the instrument (see the section on significant figures in Appendix A). For example, a mass obtained from an analytical balance should be observed and recorded to the nearest 0.0001 g. This is illustrated in Table 1.5.

PROCEDURE

A. The Meterstick

Examine the meterstick and observe that one side is ruled in inches, whereas the other is ruled in centimeters. Measure and record the length and width of your lab book in both units. Mathematically convert the two measurements to show that they are equivalent.

B. The Graduated Cylinder

Examine the 100-mL graduated cylinder and notice that it is scribed in milliliters. Fill the cylinder approximately half full with water. Notice that the *meniscus* (curved surface of the water) is concave (Figure 1.3).

---

*1 mm Hg is also called 1 torr.

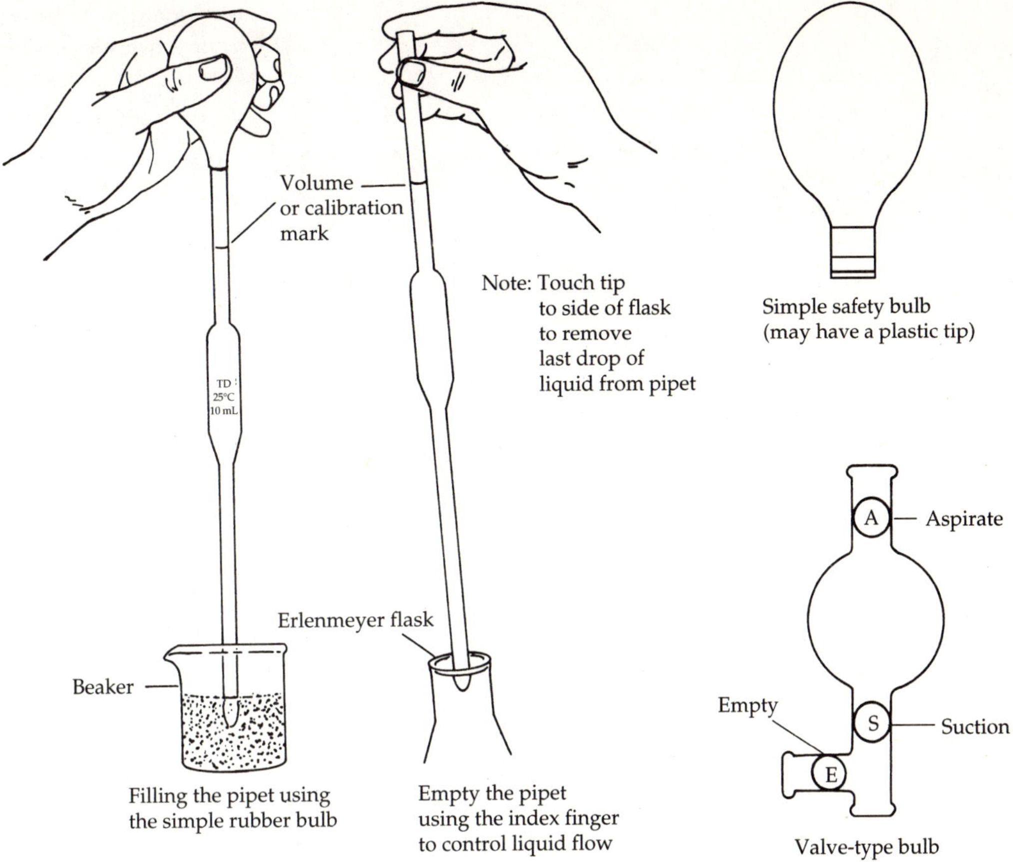

▲ **FIGURE 1.2** A typical volumetric pipet, rubber bulbs, and the pipet-filling technique.

TABLE 1.4 Equipment Accuracy

Equipment	Accuracy
Analytical balance	±0.0001 g (±0.1 mg)
Top-loading balance	±0.001 g (±1 mg)
Meterstick	±0.1 cm (±1 mm)
Graduated cylinder	±0.1 mL
Pipet	±0.02 mL
Buret	±0.02 mL
Thermometer	±0.2°C

TABLE 1.5 Obtaining Significant Figures

Analytical balance	Top loader
85.9 g (incorrect)	85.9 g (incorrect)
85.93 g (incorrect)	85.93 g (incorrect)
85.932 g (incorrect)	85.932 g (correct)
85.9322 g (correct)	

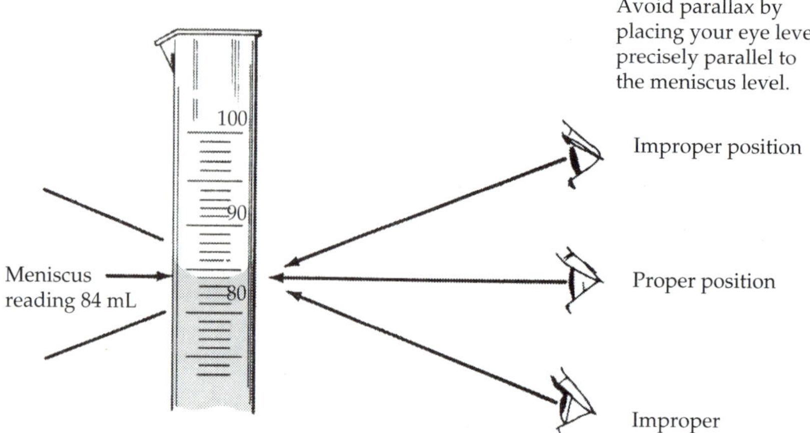

▲ **FIGURE 1.3** Proper eye position for taking volume readings.

The *lowest* point on the curve is always read as the volume, never the upper level. Avoid errors due to parallax; different and erroneous readings are obtained if the eye is not perpendicular to the scale. Read the volume of water to the nearest 0.1 mL. Record this volume. Measure the maximum amount of water that your largest test tube will hold. Record this volume.

C. The Thermometer and Its Calibration

Instructor: To save time, have the students begin the cooling and boiling procedures for Part C while they are doing Part B.

This part of the experiment is performed to check the accuracy of your thermometer. These measurements will show how measured temperatures (read from thermometer) compare with true temperatures (the boiling and freezing points of water). The freezing point of water is 0°C; the boiling point depends upon atmospheric pressure and is calculated as shown in Example 1.2. Place approximately 50 mL of ice in a 250-mL beaker and cover the ice with distilled water. Allow about 15 min for the mixture to come to equilibrium and then measure and record the temperature of the mixture. *Theoretically, this temperature is 0°C.* Now, set up a 250-mL beaker on a wire gauze and iron ring as shown in Figure 1.4. Fill the beaker about half full with distilled water. Adjust your burner to give maximum heating and begin heating the water. *(Time can be saved if the water is heated while other parts of the experiment are being conducted.)* Periodically determine the temperature of the water with the thermometer, but be careful not to touch the walls of the beaker with the thermometer bulb. Record the boiling point (b.p.) of the water. Using the data given in Example 1.2, determine the *true boiling point at the observed atmospheric pressure*. Obtain the atmospheric pressure from your laboratory instructor.

EXAMPLE 1.2

Determine the boiling point of water at 659.3 mm Hg.

SOLUTION: Temperature corrections to the boiling point of water are calculated using the following formula:

$$\text{b.p. correction} = (760 \text{ mm Hg} - \text{atmospheric pressure}) \times (0.037°\text{C/mm})$$

The correction at 659.3 mm Hg is therefore

$$\text{b.p. correction} = (760 \text{ mm Hg} - 659.3 \text{ mm Hg}) \times (0.037°\text{C/mm}) = 3.7°\text{C}$$

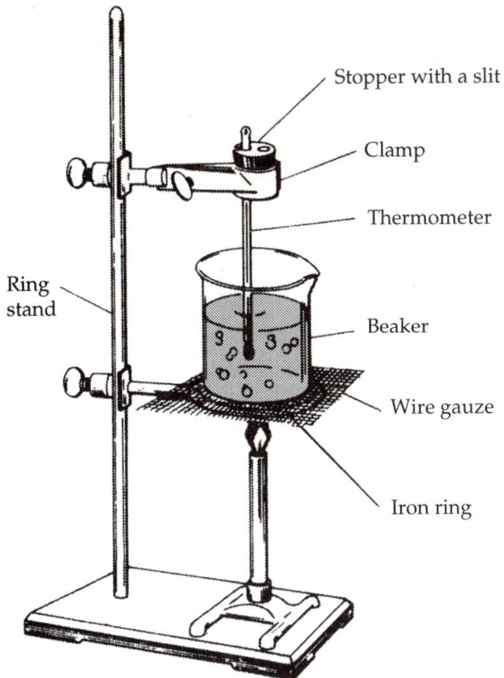

▲ **FIGURE 1.4** Apparatus setup for thermometer calibration.

The true boiling point is thus

$$100.0°C - 3.7°C = 96.3°C$$

Using the graph paper provided, construct a thermometer-calibration curve like the one shown in Figure 1.5 by plotting observed temperatures versus true temperatures for the boiling and freezing points of water.

D. Using the Balance to Calibrate Your 10-mL Pipet

Weighing an object on a single-pan balance is a simple matter. Because of the sensitivity and the expense of the balance (some cost more than $2500), you must be careful in its use. Directions for operation of single-pan balances vary with make and model. Your laboratory instructor will explain how to use the balance. Regardless of the balance you use, proper care of the balance requires that you observe the following:

Instructor: Demonstrate how to use the balance and how to fill and empty the pipet.

1. Do not drop an object on the pan.
2. Center the object on the pan.
3. Do not place chemicals directly on the pan; use a beaker, watch glass, weighing bottle, or weighing paper.
4. Do not weigh hot or warm objects; objects must be at room temperature.
5. Return all weights to the zero position after weighing.
6. Clean up any chemical spills in the balance area.
7. Inform your instructor if the balance is not operating correctly; do not attempt to repair it yourself.

The following method is used to calibrate a pipet or other volumetric glassware. Obtain about 40 mL of distilled water in a 150-mL beaker. Allow the

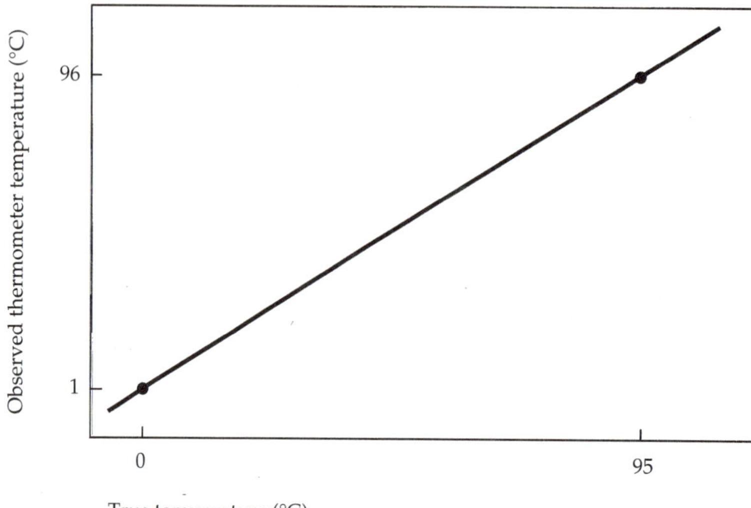

▲ **FIGURE 1.5** Typical thermometer-calibration curve.

TABLE 1.6 Density of Pure Water at Various Temperatures

T (°C)	*d* (g/mL)	T (°C)	*d* (g/mL)
15	0.999099	22	0.997770
16	0.998943	23	0.997538
17	0.998774	24	0.997296
18	0.998595	25	0.997044
19	0.998405	26	0.996783
20	0.998203	27	0.996512
21	0.997992	28	0.996232

water to sit on the desk while you weigh and record the mass of an empty, dry 50-mL Erlenmeyer flask (tare) to the nearest 0.1 mg. Measure and record the temperature of the water. Using your pipet and rubber bulb, pipet exactly 10 mL of water into this flask and weigh the flask with the water in it (gross) to the nearest 0.1 mg. Obtain the mass of the water by subtraction (gross − tare = net). Using the equation below and the data given in Table 1.6, obtain the volume of water delivered and therefore the volume of your pipet.

$$\text{Density} = \frac{\text{mass}}{\text{volume}} \qquad d = \frac{m}{V}$$

Normally, density is given in units of grams per milliliter (g/mL) for liquids, grams per cubic centimeter (g/cm^3) for solids, and grams per liter (g/L) for gases. Repeat this procedure in triplicate—that is, deliver and weigh exactly 10 mL of water three separate times.

EXAMPLE 1.3

Using the procedure given above, a mass of 10.0025 g was obtained for the water delivered by one 10-mL pipet at 22°C. What is the volume delivered by the pipet?

SOLUTION: From the density equation given above, we know that

$$V = \frac{m}{d}$$

For mass we substitute our value of 10.0025 g. For the density, consult Table 1.6. At 22°C, the density is 0.997770 g/mL. The calculation is

$$V = \frac{10.0025 \text{ g}}{0.997770 \text{ g/mL}} = 10.0249 \text{ mL}$$

which must be rounded off to 10.02, because the pipet's volume can be determined only to within a precision of ±0.02 mL.

The *precision* of a measurement is a statement about the internal agreement among repeated results; it is a measure of the reproducibility of a given set of results. The arithmetic mean (average) of the results is usually taken as the "best" value. The simplest measure of precision is the *average deviation from the mean.* The average deviation is calculated by first determining the mean of the measurements, then calculating the deviation of each individual measurement from the mean and, finally, averaging the deviations (treating each as a positive quantity). Study Example 1.4 and then, using your own experimental results, calculate the mean volume delivered by your 10-mL pipet. Also calculate for your three trials the individual deviations from the mean and then state your pipet's volume with its average deviation.

EXAMPLE 1.4

The following volumes were obtained for the calibration of a 10-mL pipet: 10.10, 9.98, and 10.00 mL. Calculate the mean value and the average deviation from the mean.

SOLUTION:

$$\text{Mean} = \frac{10.10 + 9.98 + 10.00}{3} = 10.03$$

Deviations from the mean: $|\text{value} - \text{mean}|$

$$|10.10 - 10.03| = 0.07$$

$$|9.98 - 10.03| = 0.05$$

$$|10.00 - 10.03| = 0.03$$

Average deviation from the mean

$$= \frac{0.07 + 0.05 + 0.03}{3} = 0.05$$

The reported value is therefore 10.03 ± 0.05 mL.

E. Measuring the Density of Antifreeze

Weigh a dry 50-ml flask to the nearest 0.1 mg, and record its mass. Using your pipet, measure a 10-mL sample of antifreeze solution into the 50-mL flask, weigh the flask and its contents, and record this mass. Repeat these measurements two more times to give you an indication of the precision of your measurements. Use the measured mass and volume to calculate the density of the antifreeze for each measurement. Using the three values for the density, calculate the mean density and the average deviation from the mean for your determinations.

REVIEW QUESTIONS

You should be able to answer the following questions before beginning this experiment:

1. What are the basic units of length, mass, volume, and temperature in the SI system?
2. A liquid has a volume of 1.35 liters. What is its volume in mL? In cm^3?
3. If an object weighs 1.47 g, what is its mass in mg?
4. Why should you never weigh a hot object?
5. Why is it necessary to calibrate a thermometer and volumetric glassware?
6. What is precision?
7. Define the term *density*. Can it be determined from a single measurement?
8. What is the density of an object with a mass of 3.66 g and a volume of 0.4018 mL?
9. Weighing an object three times gave the following results: 10.2 g, 10.1 g, and 10.3 g. Find the mean mass and the average deviation from the mean.
10. Normal body temperature is 37.0°C. What is the corresponding Fahrenheit temperature?
11. What is the mass in kilograms of 850 mL of a substance that has a density of 1.174 g/mL?
12. An object weighs exactly 5 g on an analytical balance that has an accuracy of 0.1 mg. To how many significant figures should this mass be recorded?

Name ______________________ Desk ______________________

Date ______________________ Laboratory Instructor ______________________

REPORT SHEET | EXPERIMENT

Basic Laboratory Techniques | 1

A. The Meterstick

Length of this lab book	11.0	in.	27.9	cm	279	mm	0.279	m
Width of this lab book	8.25	in.	21.0	cm	210	mm	0.210	m

Using an equation (including units), show that the above measurements are equivalent.

(11.0 in)(2.54 cm/in) = 27.9 cm
(8.25 in)(2.54 cm/in) = 21.0 cm

Area of this lab book (show calculations) 586 cm^2

(27.9 cm)(21.0 cm) = 586 cm^2

B. The Graduated Cylinder

Volume of water in graduated cylinder 49.8 mL
Volume of water contained in largest test tube ______ mL (depends upon test tube)

C. The Thermometer and Its Calibration

Temperature of water-and-ice mixture 0.1 °C
Temperature of boiling water 96.4 °C
Atmospheric pressure 659 mm Hg
True (corrected) temperature of boiling water 96.3 °C

D. Using the Balance to Calibrate Your 10-mL Pipet

Temperature of water used in pipet 20.0 °C
Corrected temperature 20.1 °C

	Trial 1	*Trial 2*	*Trial 3*	
Mass of Erlenmeyer plus ~10 mL H_2O (gross mass)	46.56	46.68	46.85	g
Mass of Erlenmeyer (tare mass)	36.57	36.74	36.91	g
Mass of ~10 mL of H_2O (net mass)	9.99	9.94	9.94	g

Volume delivered by 10-mL pipet (show calculations)	10.01	9.96	9.96	mL

$$V = \frac{m}{d} = \frac{9.99\text{ g}}{0.9982\text{ g/mL}} = 10.01\text{ mL}$$

Mean volume delivered by 10-mL pipet (show calculations) 9.98 mL

10.01
9.96
9.96
29.93

$$\frac{29.93}{3} = 9.98\text{ mL}$$

	Trial 1	*Trial 2*	*Trial 3*
Individual deviations from the mean	0.03	0.02	0.02

Average deviation from the mean (show calculations) 0.02 mL

$$\frac{0.03 + 0.02 + 0.02}{3} = \frac{0.07}{3} = 0.02$$

Volume delivered by your 10-mL pipet 9.98 mL ± 0.02 mL

E. Measuring the Density of Antifreeze

Temperature of antifreeze 23 °C

	Trial 1	*Trial 2*	*Trial 3*	
Mass of flask + antifreeze	47.64	47.82	47.97	g
Mass of empty flask	36.57	36.74	36.91	g
Mass of antifreeze	11.07	11.08	11.06	g
Density of antifreeze (show calculation below)	1.109	1.110	1.108	g

$$\frac{11.07\text{ g}}{9.98\text{ mL}} = 1.109\text{ g/mL} \qquad \frac{11.08\text{ g}}{9.98\text{ mL}} = 1.110\text{ g/mL} \qquad \frac{11.06\text{ g}}{9.98\text{ mL}} = 1.108\text{ g/mL}$$

Mean (average) density

$$\frac{1.109 + 1.110 + 1.108}{3} = 1.109 = 1.11 \text{ g/mL}$$

Average deviation from the mean
(show calculation below)

$$\frac{(1.109 - 1.109) + (1.110 - 1.109) + (1.108 - 1.109)}{3} = 0.0007 = 0.001$$

QUESTIONS

1. A man who is 5 ft 6 in. tall weighs 140 lb. What is his height in centimeters and his mass in kilograms?

$$5 \text{ ft } 6 \text{ in} = 5 \text{ ft}\left(\frac{12 \text{ in}}{\text{ft}}\right) + 6 \text{ in} = 66 \text{ in} \quad (66 \text{ in})\left(\frac{2.54 \text{ cm}}{\text{in}}\right) = 168 \text{ cm}$$

$$(140 \text{ lb})\left(\frac{1 \text{ kg}}{2.205 \text{ lb}}\right) = 63.5 \text{ kg}$$

2. Determine the boiling point of water at 698.5 mm Hg.

b. p. depression = (760 mm − 698.5 mm)(0.037°C/mm) = 2.3°C
b. p. = 100.0° − 2.3° = 97.7°C

3. A pipet delivers 9.97 g of water at 20°C. What volume does the pipet deliver?

at 19°C: density = 0.998405 g/mL

$$V = \frac{9.97 \text{ g}}{0.998203 \text{ g/mL}} = 9.99 \text{ mL}$$

4. A pipet delivers 10.04, 10.02, 10.08, and 10.06 mL in consecutive trials. Find the mean volume and the average deviation from the mean.

$$\overline{X} = \frac{10.04 + 10.02 + 10.08 + 10.06}{4} = 10.05 \text{ mL}$$

$$\text{average deviation from the mean} = \frac{0.01 + 0.03 + 0.03 + 0.01}{4} = 0.02$$

5. A 141-mg sample was placed on a watch glass that weighed 9.203 g. What is the weight of the watch glass and sample in grams?

$$(141 \text{ mg})\left(\frac{1 \text{ g}}{1000 \text{ mg}}\right) = 0.141 \text{ g}$$

weight = 9.203 g + 0.141 g = 9.344 g

6. (a) Using the defined freezing and boiling points of water, make a plot of degrees Fahrenheit versus degrees Celsius on the graph paper provided.
   (b) Determine the Celsius equivalent of 40°F using your graph. The relationship between these two temperature scales is linear (i.e., it is of the form $y = mx + b$). Consult Appendix B regarding linear relationships and determine the equation that relates degrees Fahrenheit to degrees Celsius.
   (c) Compute the Celsius equivalent of 40°F using this relationship.

   (a) Celsius equivalent of 40°F from graph is 4.5°C.

   (b) °F = m°C + constant. When °C = 0, °F = 32, thus the constant is 32° and

$$m = \frac{\Delta F^\circ}{\Delta C^\circ} = \left(\frac{212 - 32}{100 - 0}\right) = \frac{180}{100} = \frac{9}{5}.\ \text{Thus, } {}^\circ F = 9/5{}^\circ C + 32$$

   (c) °F = 9/5°C + 32 or °C = 5/9(°F − 32). Thus, °C = 5/9°(40 − 32)
   = 5/9(8) = 4°C

Thermometer Calibration Curve

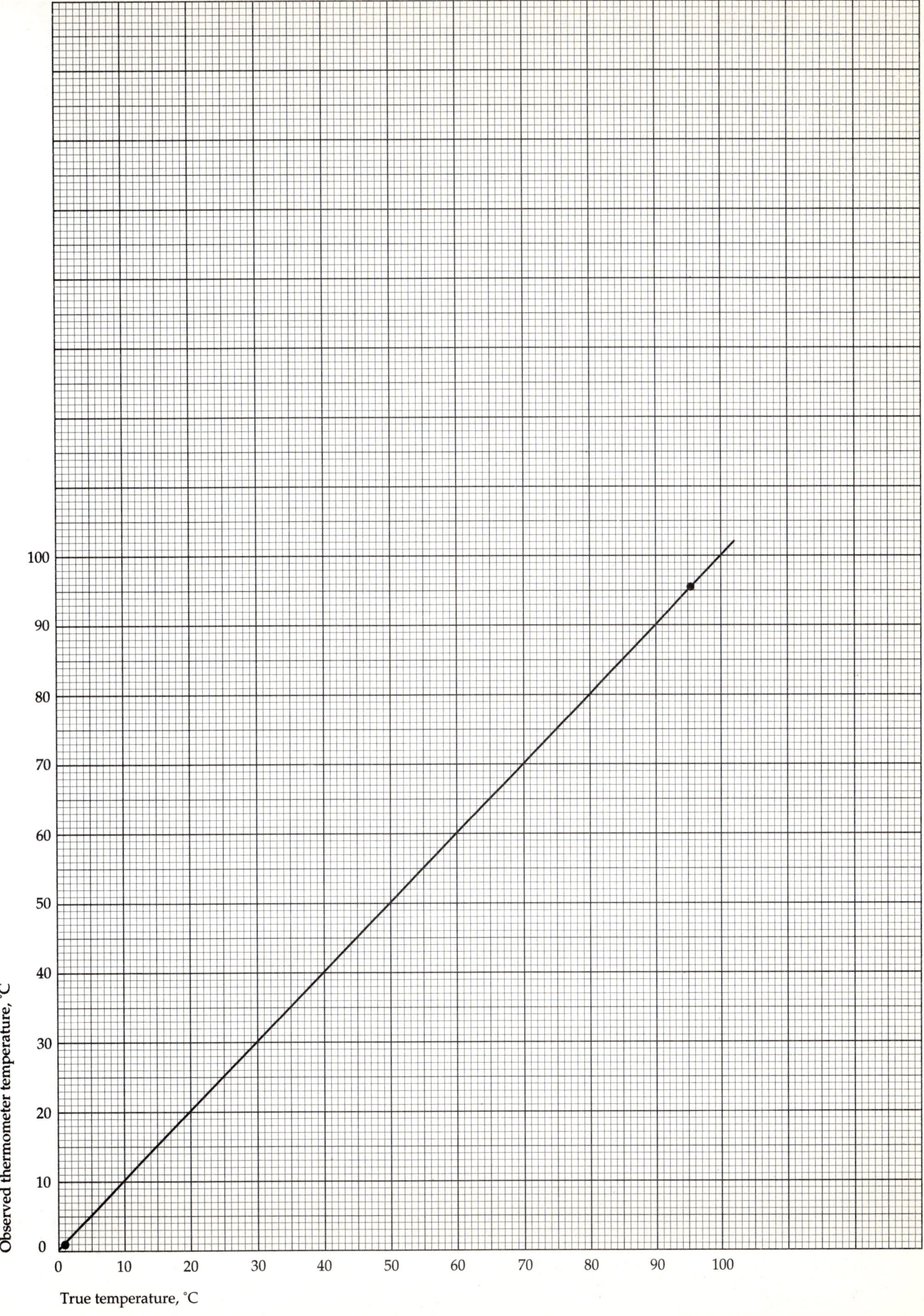

Fahrenheit—Celsius Graph

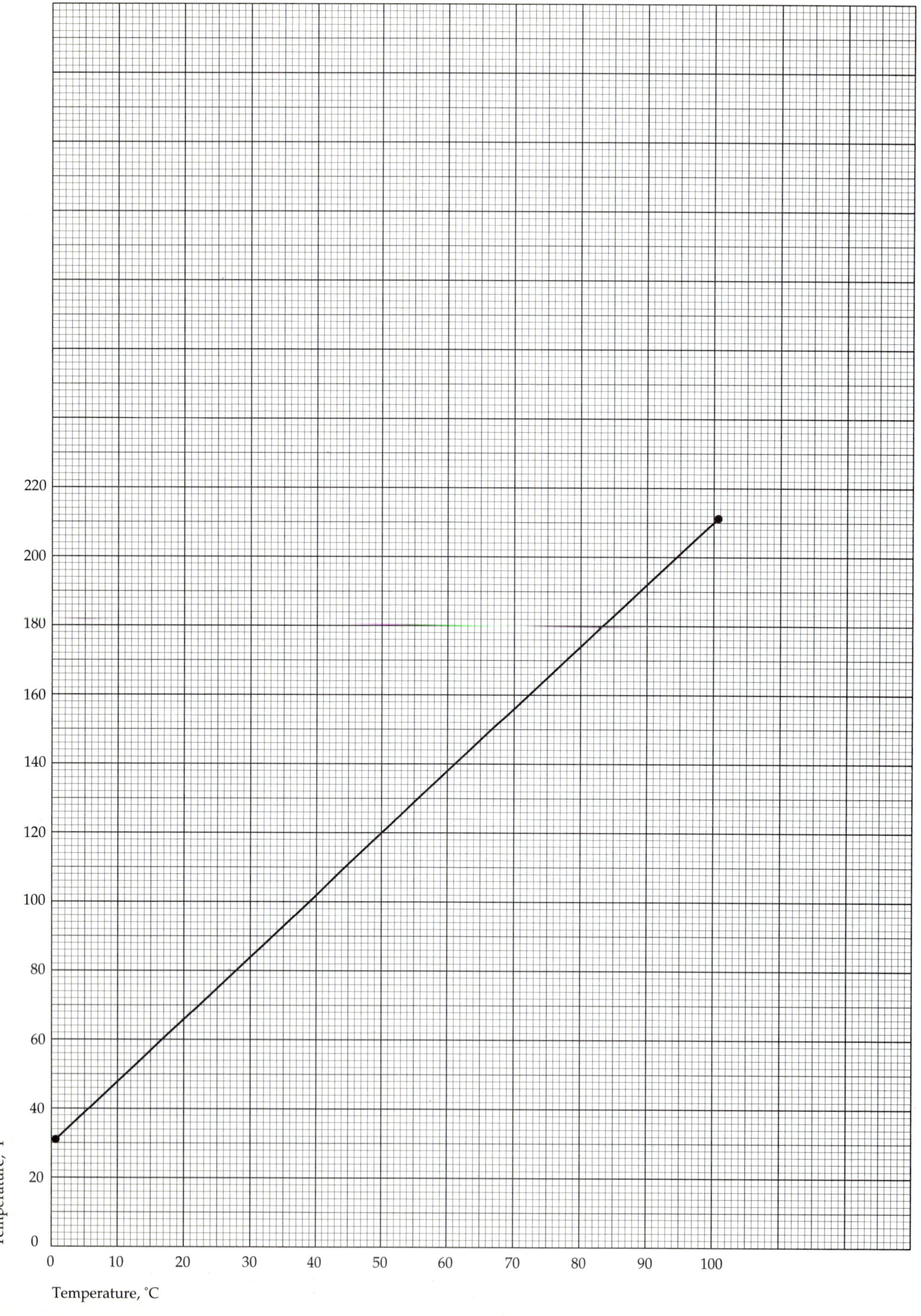

Experiment

Identification of Substances by Physical Properties

OBJECTIVE

To become acquainted with procedures used in evaluating physical properties and the use of these properties to identify substances.

APPARATUS AND CHEMICALS

Apparatus

balance
50-mL beakers (2)
250-mL beaker
50-mL Erlenmeyer flask
10-mL graduated cylinder
large test tubes (2)
small test tubes (6)
test-tube rack
5- and 10-mL pipets
ring stand and ring
wire gauze
no. 3 two-hole stopper with one of the holes slit to the side or a buret clamp
Bunsen burner and hose
stirring rod
dropper
boiling chips
thermometer
spatula
small watch glass
capillary tubes (5)
tubing with right-angle bend
utility clamp
two-hole stopper
small rubber bands (or small sections of $^1/_4$-in. rubber tubing)
rubber bulb for pipet

Chemicals

ethyl alcohol
toluene
soap solution
naphthalene
two unknowns (one liquid, one solid)

DISCUSSION

Properties are those characteristics of a substance that enable us to identify it and to distinguish it from other substances. Direct identification of some substances can readily be made by simply examining them. For example, we see color, size, shape, and texture, and we can smell odors and discern a variety of tastes. Thus, copper can be distinguished from other metals on the basis of its color.

Physical properties are those properties that can be observed without altering the composition of the substance. Whereas it is difficult to assign definitive values to such properties as taste, color, and odor, other physical properties, such as melting point, boiling point, solubility, density, viscosity, and refractive index, can be expressed quantitatively. For example, the melting point of copper is 1087°C and its density is 8.96 g/cm^3. As you probably realize, a specific combination of properties is unique to a given substance, thus making it possible to identify most substances just by careful determination of several properties. This is so important that large books have been compiled listing characteristic properties of many known substances. Many scientists, most notably several German scientists during the latter part of the nineteenth century and earlier part of the twentieth, spent their entire lives gathering data of this sort. Two of the most complete references of this type that are readily available today are The Chemical Rubber Company's *Handbook of Chemistry and Physics* and N. A. Lange's *Handbook of Chemistry.*

In this experiment, you will use the following properties to identify a substance whose identity is unknown to you: solubility, density, melting point, and boiling point. The *solubility* of a substance in a solvent at a specified temperature is the maximum weight of that substance that dissolves in a given volume (usually 100 mL or 1000 mL) of a solvent. It is tabulated in handbooks in terms of grams per 100 mL of solvent; the solvent is usually water.

Density is an important physical property and is defined as the mass per unit volume:

$$d = \frac{m}{V}$$

Melting or freezing points correspond to the temperature at which the liquid and solid states of a substance are in equilibrium. These terms refer to the same temperature but differ slightly in their meaning. The *freezing point* is the equilibrium temperature when approached from the liquid phase—that is, when solid begins to appear in the liquid. The *melting point* is the equilibrium temperature when approached from the solid phase—that is, when liquid begins to appear in the solid.

A liquid is said to boil when bubbles of vapor form within it, rise rapidly to the surface, and burst. Any liquid in contact with the atmosphere will boil when its vapor pressure is equal to atmospheric pressure—that is, the liquid and gaseous states of a substance are in equilibrium. Boiling points of liquids depend upon atmospheric pressure. A liquid will boil at a higher temperature at a higher pressure or at a lower temperature at a lower pressure. The temperature at which a liquid boils at 760 mm Hg is called the *normal* boiling point. To account for these pressure effects on boiling points, people have studied and tabulated data for boiling point versus pressure for a large number of compounds. From these data, nomographs have been constructed. A *nomograph* is a set of scales for connected variables (see Figure 2.5 for an example); these scales are so placed that a straight line connecting the known values on some scales will provide the unknown value at the straight line's intersection with other scales. A nomograph allows you to find the correction necessary to convert the normal boiling point of a substance to its boiling point at any pressure of interest.

PROCEDURE

A. Solubility

Qualitatively determine the solubility of naphthalene (mothballs) in three solvents: water, cyclohexane, and ethyl alcohol. **(CAUTION: *Cyclohexane is highly flammable and must be kept away from open flames.*)** Determine the solubility by adding a few crystals of naphthalene to 2 to 3 mL (it is not necessary to measure either the solute weight or solvent volume) of each of these three solvents in separate, clean, *dry* test tubes. Make an attempt to keep the amount of naphthalene and solvent the same in each case. Place a cork in each test tube and shake briefly. Cloudiness indicates insolubility. Record your conclusions on the report sheet using the abbreviations s (soluble), sp (sparingly soluble), and i (insoluble). Into each of three more clean, *dry* test tubes place 2 or 3 mL of these same solvents and add 4 or 5 drops of

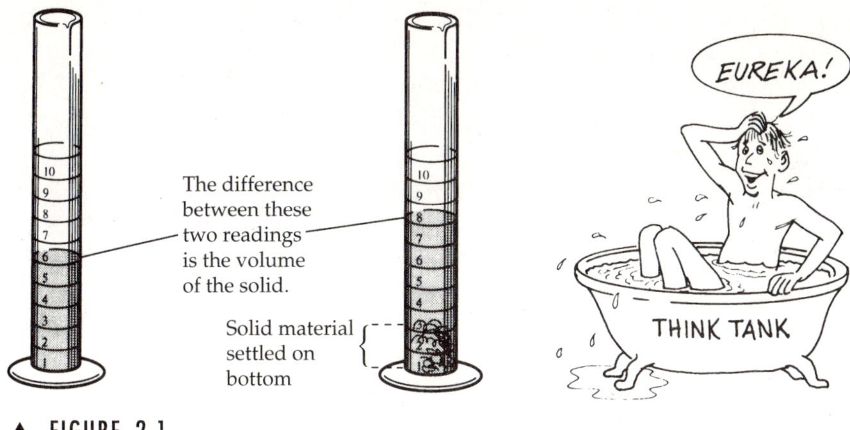

▲ **FIGURE 2.1**

toluene in place of naphthalene. Record your observations. The formation of two layers indicates immiscibility (lack of solubility). Now repeat these experiments using each of the three solvents (water, cyclohexane, and ethyl alcohol) with your solid and liquid unknowns and record your observations.

SAVE your solid and liquid unknowns for Parts B, C, and D, but dispose of the other chemicals in the marked refuse container. Do not dispose of them in the sink.

B. Density

Determine the densities of your two unknowns in the following manner.

The Density of a Solid Weigh about 1.5 g of your solid unknown to the nearest 0.001 g and record the mass. Using a pipet or a wash bottle, half fill a *clean, dry* 10-mL graduated cylinder with a solvent in which your unknown is *insoluble.* Be *careful* not to get the liquid on the inside walls, because you do not want your solid to adhere to the cylinder walls when you add it in a subsequent step. Read and record this volume to the nearest 0.1 mL. Add the weighed solid to the liquid in the cylinder, being careful not to lose any of the material in the transfer process and ensuring that all of the solid is beneath the surface of the liquid. Carefully tapping the sides of the cylinder with your fingers will help settle the material to the bottom. Do not be concerned about a few crystals that do not settle, but if a large quantity of the solid resists settling, add one or two drops of a soap solution and continue tapping the cylinder with your fingers. Now read the new volume to the nearest 0.1 mL. The difference in these two volumes is the volume of your solid (Figure 2.1). Calculate the density of your solid unknown.

You may recall that by measuring the density of metals in this way, Archimedes proved to the king that the charlatan alchemists had in fact not transmuted lead into gold. He did this after observing that he weighed less in the bathtub than he did normally by an amount equal to the weight of the fluid displaced. According to legend, upon making his discovery, Archimedes emerged from his bath and ran naked through the streets shouting *Eureka!* (I have found it).

The Density of a Liquid Weigh a clean, *dry* 50-mL Erlenmeyer flask to the nearest 0.0001 g. Obtain at least 15 mL of the unknown liquid in a clean, *dry* test tube. Using a 10-mL pipet, pipet exactly 10 mL of the unknown

▲ **FIGURE 2.2** Sealing one end of a capillary tube.

liquid into the 50-mL Erlenmeyer flask and quickly weigh the flask containing the 10 mL of unknown to the nearest 0.0001 g. Using the calibration value for your pipet, if you calibrated it, and the weight of this volume of unknown, calculate its density. Record your results and show how (with units) you performed your calculations. SAVE the liquid for your boiling-point determination.

C. Melting Point of Solid Unknown

Obtain a capillary tube and a small rubber band. Seal one end of the capillary tube by carefully heating the end in the edge of the flame of a Bunsen burner until the end *completely* closes. Rotating the tube during heating will help you to avoid burning yourself (Figure 2.2).

Pulverize a small portion of your solid-unknown sample with the end of a test tube on a clean watch glass. Partially fill the capillary with your unknown by gently tapping the pulverized sample with the open end of the capillary to force some of the sample inside. Drop the capillary into a glass tube about 38 to 50 cm in length, with the sealed end down to pack the sample into the bottom of the capillary tube. Repeat this procedure until the sample column is roughly 5 mm in height. Now set up a melting-point apparatus as illustrated in Figure 2.3.

Place the rubber band about 5 cm above the bulb on the thermometer and out of the liquid. Carefully insert the capillary tube under the rubber band with the closed end at the bottom. Place the thermometer with attached capillary into the beaker of water so that the sample is covered by water, the thermometer does not touch the bottom of the beaker, and the open end of the capillary tube is above the surface of the water. Heat the water slowly while gently agitating the water with a stirring rod. Observe the sample in the capillary tube while you are doing this. At the moment that the solid melts, record the temperature. Also record the melting-point range, which is the temperature range between the temperature at which the sample begins

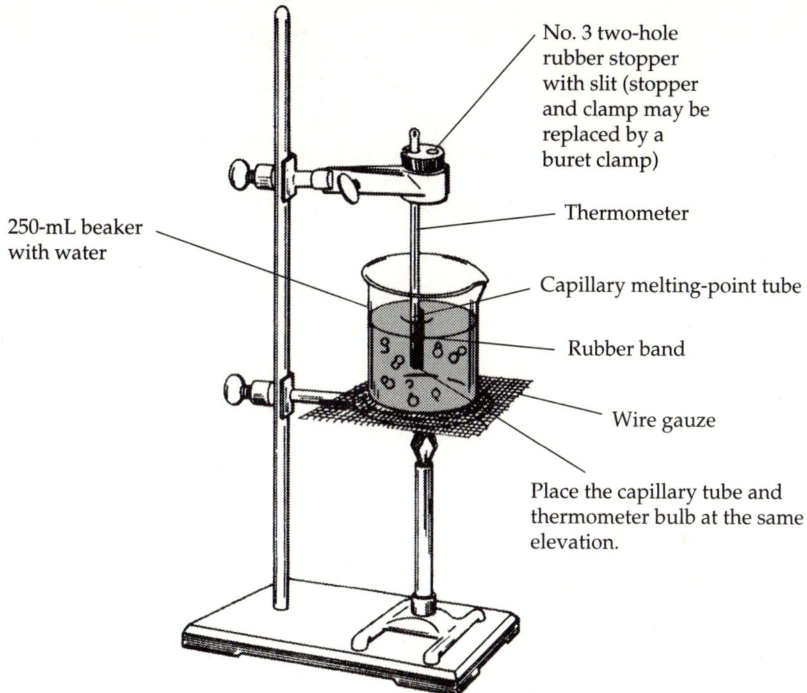

▲ **FIGURE 2.3** Apparatus for melting-point determination.

to melt and the temperature at which all of the sample has melted. Using your thermometer-calibration curve (from Experiment 1), correct these temperatures to the true temperatures and record the melting point and melting-point range. These temperatures may differ by only 1°C or less.

D. Boiling Point of Liquid Unknown

To determine the boiling point of your liquid unknown, put about 3 mL of the material you used to determine the density into a clean, dry test tube. Fit the test tube with a two-hole rubber stopper that has one slit; insert your thermometer into the hole with the slit and one of your right-angle-bend glass tubes into the other hole, as shown in Figure 2.4. Add one or two small boiling chips to the test tube to ensure even boiling of your sample. Position the thermometer so that it is about 1 cm above the surface of the unknown liquid. Clamp the test tube in the ring stand and connect to the right-angle-bend tubing a length of rubber tubing that reaches to the sink. Assemble your apparatus as shown in Figure 2.4. **(CAUTION:** ***Be certain that there are no constrictions in the rubber tubing. Your sample is flammable. Keep it away from open flames.)***

Heat the water gradually and watch for changes in temperature. The temperature will become constant at the boiling point of the liquid. Record the observed boiling point. Correct the observed boiling point to the true boiling point at room atmospheric pressure using your thermometer-calibration curve. The normal boiling point (b.p. at 1 atm = 760 mm Hg) can now be calculated (see Example 2.1, below) using the nomograph provided in Figure 2.5. Your boiling-point correction should not be more than +5°C.

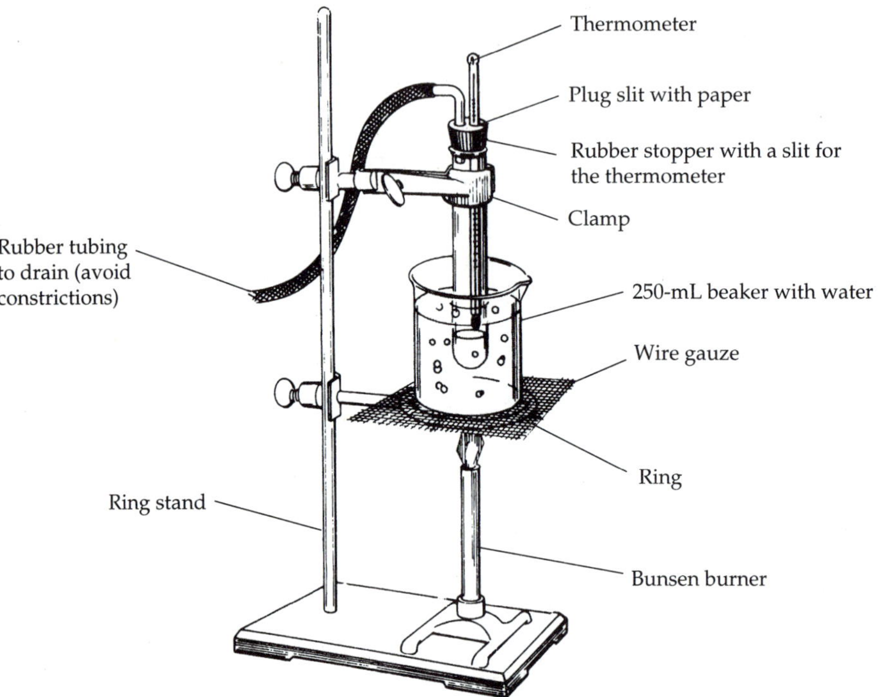

▲ **FIGURE 2.4** Apparatus for boiling-point determination.

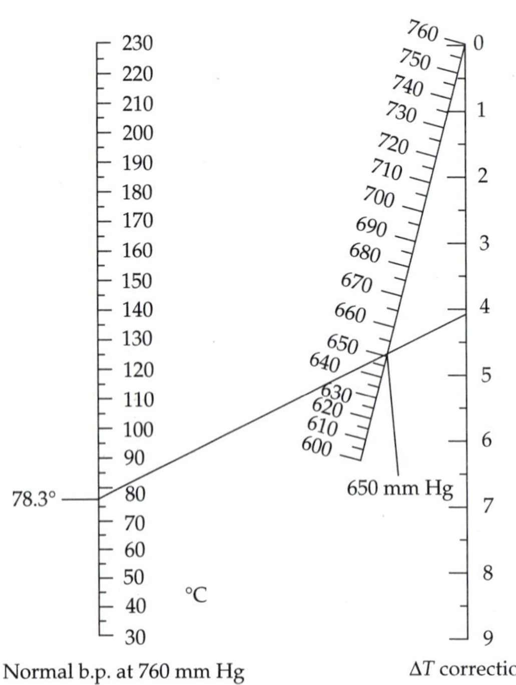

▲ **FIGURE 2.5** Nomograph for boiling-point correction to 760 mm Hg.

TABLE 2.1 Physical Properties of Pure Substances

Substance	Density (g/mL)	Melting point (°C)	Boiling point (°C)	Solubility[a] in		
				Water	Cyclohexane	Alcohol
Acetanilide	1.22	114	304	sp	sp	s
Acetone	0.79	−95	56	s	s	s
Benzophenone	1.15	48	306	i	s	s
Bromoform	2.89	8	150	i	s	s
2,3-Butanedione	0.98	−2.4	88	s	s	s
t-Butyl alcohol	0.79	25	83	s	s	s
Cadmium nitrate • $4H_2O$	2.46	59	132	s	i	s
Chloroform[b]	1.49	−63.5	61	i	s	s
Cyclohexane	0.78	6.5	81.4	i	s	s
p-Dibromobenzene	1.83	86.9	219	i	s	s
p-Dichlorobenzene	1.46	53	174	i	s	s
m-Dinitrobenzene	1.58	90	291	i	s	s
Diphenyl	0.99	70	255	i	s	s
Diphenylamine	1.16	53	302	i	s	s
Diphenylmethane	1.00	27	265	i	s	s
Ether, ethyl propyl	1.37	−79	64	s	s	s
Hexane	0.66	−94	69	i	s	s
Isopropyl alcohol	0.79	−98	83	s	s	s
Lauric acid	0.88	43	225	i	s	s
Magnesium nitrate • $6H_2O$	1.63	89	330[c]	s	i	s
Methyl alcohol	0.79	−98	65	s	sp	s
Methylene chloride[b]	1.34	−97	40.1	i	s	s
Naphthalene	1.15	80	218	i	s	sp
α-Naphthol	1.10	94	288	i	i	s
Phenyl benzoate	1.23	71	314	i	s	s
Propionaldehyde	0.81	−81	48.8	s	i	s
Sodium acetate • $3H_2O$	1.45	58	123	s	i	sp
Stearic acid	0.85	70	291	i	s	sp
Thymol	0.97	52	232	sp	s	s
Toluene	0.87	−95	111	i	s	s
p-Toluidine	0.97	45	200	sp	s	s
Zinc chloride	2.91	283	732	s	i	s

[a]s = soluble; sp = sparingly soluble; i = insoluble.

[b]Toxic. Most organic compounds used in the lab are toxic.

[c]Boils with decomposition.

EXAMPLE 2.1

What will be the boiling point of ethanol at 650 mm Hg when its normal boiling point at 760 mm Hg is known to be 78.3°C?

SOLUTION: The answer is easily found by consulting the nomograph in Figure 2.5. A straight line drawn from 78.3°C on the left scale of normal boiling points through 650 mm Hg on the pressure scale intersects the temperature correction scale at 4°C. Therefore,

$$\text{normal b.p.} - \text{correction} = \text{observed b.p.}$$

$$78.3°C - 4.0°C = 74.3°C$$

Similar calculations could be done for the compounds in Table 2.1 at any pressure listed on the nomograph in Figure 2.5. In this experiment you will observe a boiling point at a pressure other than at 760 mm Hg, and you wish to know its normal boiling point. In order to estimate its normal boiling point, assume that, for example, your observed boiling point is 57.0°C and the observed pressure is 650 mm Hg. Use your observed boiling point of 57.0°C as if it were the normal boiling point and find the correction for a pressure of 650 mm Hg. Using the nomograph, you can see that the correction is 3.8°C. You would then add this correction to your observed boiling point to obtain an approximate normal boiling point:

$$57°C + 3.8°C = 60.8°C, \text{ or } 61°C$$

By consulting Table 2.1, you can find the compound that best fits your data; in this example, the data are for chloroform.

E. Unknown Identification

Your unknowns are substances contained in Table 2.1. Compare the properties that you have determined for your unknowns with those in the table. Identify your unknowns and record your results.

Dispose of your unknowns in the appropriate marked refuse containers.

REVIEW QUESTIONS

Before beginning this experiment in the laboratory, you should be able to answer the following questions:

1. List five physical properties.
2. An 8.192-mL sample of an unknown has a mass of 9.02 g. What is the density of the unknown?
3. Are the substances hexane and naphthalene solids or liquids at room temperature?
4. Could you determine the density of sodium acetate using water? Why or why not?
5. What would be the boiling point of toluene at 670 mm Hg?
6. Why do we calibrate thermometers and pipets?
7. Is chloroform miscible with water? With cyclohexane?
8. When water and bromoform are mixed, two layers form. Is the bottom layer water or bromoform? (See Table 2.1.)
9. What solvent would you use to determine the density of zinc chloride?
10. The density of a solid with a melting point of 47° to 49°C was determined to be 1.13 ± 0.02 g/mL. What is the solid?
11. The density of a liquid whose boiling point is 68° to 69°C was determined to be 0.65 ± 0.05 g/mL. What is the liquid?
12. Which has the greater volume, 10 g of chloroform or 10 g of hexane? What is the volume of each?

Name ______________________ Desk ______________

Date ______________ Laboratory Instructor ______________

Liquid unknown no. ______________

Solid unknown no. ______________

REPORT SHEET | EXPERIMENT 2

Identification of Substances by Physical Properties

A. Solubility

	Water	*Cyclohexane*	*Alcohol*	
Naphthalene	i	s	sp	
Toluene	i	s	s	
Liquid unknown	s	s	i	unknown dependent
Solid unknown	s	i	s	exemplary data only

B. Density

Solid

Final volume of liquid in cylinder 6.8 mL

Initial volume of liquid in cylinder 5.0 mL

Volume of solid 1.8 mL

Mass of solid 2.0 g

Diphenylamine data

Density of solid 1.1 g/mL
(show calculations)

$$d = \frac{2.0\text{ g}}{1.8\text{ mL}} = 1.1\text{ g/mL}$$

Liquid

Volume of liquid 10.0 mL

Volume of liquid corrected for the pipet correction 9.98 mL

Mass of 50-mL Erlenmeyer plus 10 mL of unknown 44.936 g

Mass of 50-mL Erlenmeyer 37.159 g

Mass of liquid 7.777 g

Isopropyl alcohol data

Density 0.779 g/mL
(show calculations)

$$d = \frac{7.777\text{ g}}{9.98\text{ mL}} = 0.779\text{ g/mL}$$

C. Melting Point of Solid Unknown

Observed melting point	52	°C
Corrected (apply thermometer correction to obtain)	52.1	°C
Observed melting-point range	52–53	°C
Corrected (apply thermometer correction to obtain)	52.1–53.1	°C

D. Boiling Point of Liquid Unknown

Observed	79	°C
Corrected (apply thermometer correction to obtain)	79.1	°C
Estimated true (normal) b.p. (apply pressure correction to obtain)	83.1	°C

From Figure 2.5, the boiling point correction is 4°C at 650 mm Hg.

E. Unknown Identification

Solid unknown: diphenylamine

Liquid unknown: isopropyl alcohol

QUESTIONS

1. Is p-toluidine a solid or a liquid at room temperature? Solid.
2. What solvent would you use to measure the density of cadmium nitrate? Cyclohexane.
3. Convert your densities to kg/L and compare these values with those in g/mL.

   since 1 kg = 1000 g and 1 L = 1000 mL

   $$\left(\frac{\text{g}}{\text{mL}}\right)\left(\frac{1000\ \text{mL}}{\text{L}}\right)\left(\frac{1\ \text{kg}}{1000\ \text{g}}\right) = \frac{\text{kg}}{\text{L}}$$

4. If air bubbles were trapped in your solid beneath the liquid level in your density determination, what error would result in the volume measurement, and what would be the effect of this error on the calculated density?

   The volume would be too large; hence, the density would be too small.

5. A liquid unknown was found to be insoluble in water and soluble in cyclohexane and alcohol; the unknown was found to have a boiling point of 57°C at 658 mm Hg. What is the substance? What could you do to confirm your answer?

   Chloroform; determine its density.

6. A liquid that has a density of 0.80 ± 0.01 g/mL is insoluble in cyclohexane. What is the liquid?

   Consult Table 2.1; propionaldehyde

7. What is the boiling point of toluene at 600 mm Hg?

   Its normal boiling point is 111°C (Table 2.1), and from Figure 2.5 its boiling point at 600 mm Hg would be 7°C lower or 111°C − 7°C = 104°C.

   Consult a handbook for the following questions and specify handbook used.
   Handbook: CRC Handbook of Chemistry and Physics

8. Osmium is the densest element known. What are its density and melting point?

   Density*: 22.59 g/mL   Melting point: 3000° ± 10°C

9. What are the colors of $CoCl_2$ and $CoCl_2 \cdot 6H_2O$? What are their solubilities in cold water?

	$CoCl_2$	$CoCl_2 \cdot 6H_2O$
Color:	Blue	Red
Solubility:	97 g/100mL	52.9 g/100mL

10. What are the formula, molar mass, and color of potassium permanganate?

    Formula: $KMnO_4$

    Molar mass: 158.03 g/mol

    Color: dark purple

*Until recently, much confusion existed in the literature as to which is the densest metal, osmium or iridium. The currently accepted densities are 22.588 ± 0.015 g/mL and 22.562 g/mL respectively. See: *Platinum Metal Rev.*, 1995, ***39***, 164.

NOTES AND CALCULATIONS

Experiment 3

Separation of the Components of a Mixture

OBJECTIVE

To become familiar with the methods of separating substances from one another using decantation, extraction, and sublimation techniques.

APPARATUS AND CHEMICALS

Apparatus

balance
Bunsen burner and hose
tongs
evaporating dishes (2)
watch glass
50- or 100-mL graduated cylinder
clay triangles (2) or wire gauze (2)
ring stands (2)
iron rings (2)
glass stirring rods

Chemicals

unknown mixture of sodium chloride, ammonium chloride, and silicon dioxide

DISCUSSION

Materials that are not uniform in composition are said to be impure or heterogeneous and are called *mixtures.* Most of the materials we encounter in everyday life, such as cement, wood, and soil, are mixtures. When two or more substances that do not react chemically are combined, a mixture results. Mixtures are characterized by two fundamental properties:

- Each of the substances in the mixture retains its chemical integrity.
- Mixtures are separable into these components by physical means.

If one of the substances in a mixture is preponderant—that is, if its amount far exceeds the amounts of the other substances in the mixture—then we usually call this mixture an impure substance and speak of the other substances in the mixture as impurities.

The preparation of compounds usually involves their separation or isolation from reactants or other impurities. Thus the separation of mixtures into their components and the purification of impure substances are frequently encountered problems. You are probably aware of everyday problems of this sort. For example, our drinking water usually begins as a mixture of silt, sand, dissolved salts, and water. Since water is by far the largest component in this mixture, we usually call this impure water. How do we purify it? The separation of the components of mixtures is based upon the fact that each component has different physical properties. The components of mixtures are always pure substances, either compounds or elements, and each pure substance possesses a unique set of properties. The properties of every sample of a pure substance are identical under the same conditions of temperature and pressure. This means that once we have determined that a

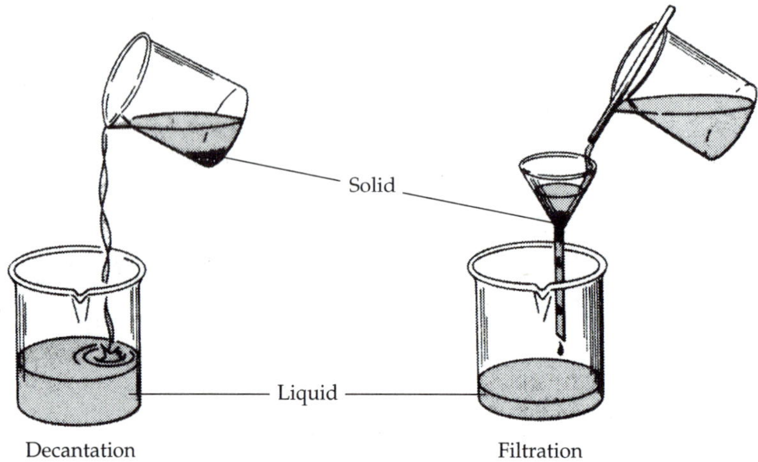

▲ FIGURE 3.1

sample of sodium chloride, NaCl, is water soluble and a sample of silicon dioxide (sand), SiO_2, is not, we realize that all samples of sodium chloride are water soluble and all samples of silicon dioxide are not.

Likewise, every crystal of a pure substance melts at a specific temperature, and at a given pressure, every pure substance boils at a specific temperature.

Although there are numerous physical properties that can be used to identify a particular substance, we will be concerned in this experiment merely with the separation of the components and not with their identification. The methods we will use for the separation depend on differences in physical properties, and they include the following:

1. *Decantation.* This is the process of separating a liquid from a solid (sediment) by gently pouring the liquid from the solid so as not to disturb the solid (Figure 3.1).
2. *Filtration.* This is the process of separating a solid from a liquid by means of a porous substance—a filter—which allows the liquid to pass through but not the solid (see Figure 3.1). Common filter materials are paper, layers of charcoal, and sand. Silt and sand can be removed from our drinking water by this process.
3. *Extraction.* This is the separation of a substance from a mixture by preferentially dissolving that substance in a suitable solvent. By this process a soluble compound is usually separated from an insoluble compound.
4. *Sublimation.* This is the process in which a solid passes directly to the gaseous state and back to the solid state without the appearance of the liquid state. Not all substances possess the ability to be sublimed. Iodine, naphthalene, and ammonium chloride (NH_4Cl) are common substances that easily sublime.

Instructor: The unknown should be shaken to make it as homogeneous as possible.

The mixture that you will separate contains three components: NaCl, NH_4Cl, and SiO_2. Their separation will be accomplished by heating the mixture to sublime the NH_4Cl, extracting the NaCl with water, and finally drying the remaining SiO_2, as illustrated in the scheme shown in Figure 3.2.

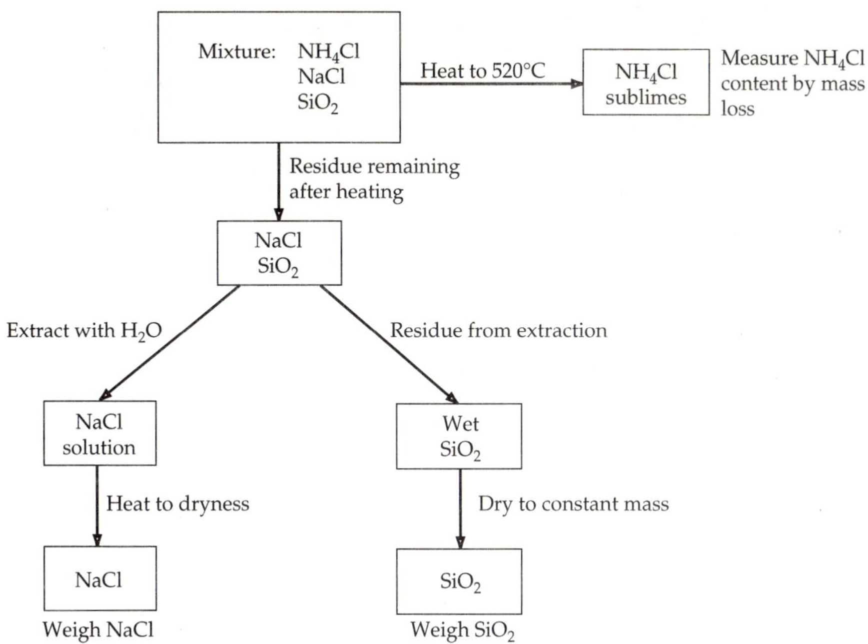

▲ **FIGURE 3.2** Flow diagram for the separation of the components of a mixture.

PROCEDURE

Carefully weigh a clean, dry evaporating dish to the nearest 0.01 g. Then obtain from your instructor a 2- to 3-g sample of the unknown mixture in the evaporating dish. If you obtain your unknown from a bottle, shake the bottle to make the sample mixture as uniform as possible. Weigh the evaporating dish containing the sample and calculate the sample mass.

Place the evaporating dish containing the mixture on a clay triangle (or wire gauze), ring, and ring-stand assembly IN THE HOOD as shown in Figure 3.3. Heat the evaporating dish with a burner until white fumes are no longer formed (a total of about 15 min). Heat carefully to avoid spattering, especially when liquid is present. Occasionally shake the evaporating dish gently, using crucible tongs during the sublimation process.

Allow the evaporating dish to cool until it reaches room temperature and then weigh the evaporating dish with the contained solid. NEVER WEIGH HOT OR WARM OBJECTS! The loss in mass represents the amount of NH_4Cl in your mixture. Calculate this.

Add 25 mL of water to the solid in this evaporating dish and stir gently for 5 min. Next, weigh another clean, dry evaporating dish and watch glass. Decant the liquid carefully into the second evaporating dish, *which you have weighed*, being careful not to transfer any of the solid into the second evaporating dish. Add 10 mL more of water to the solid in the first evaporating dish, stir, and decant this liquid into the second evaporating dish as before. Repeat with still another 10 mL of water. This process extracts the soluble NaCl from the sand. You now have two evaporating dishes—one containing wet sand, and the second a solution of sodium chloride.

Place the evaporating dish containing the sodium chloride solution carefully on the clay triangle on the ring stand. Begin gently heating the solution to evaporate the water. Take care to avoid boiling or spattering, especially

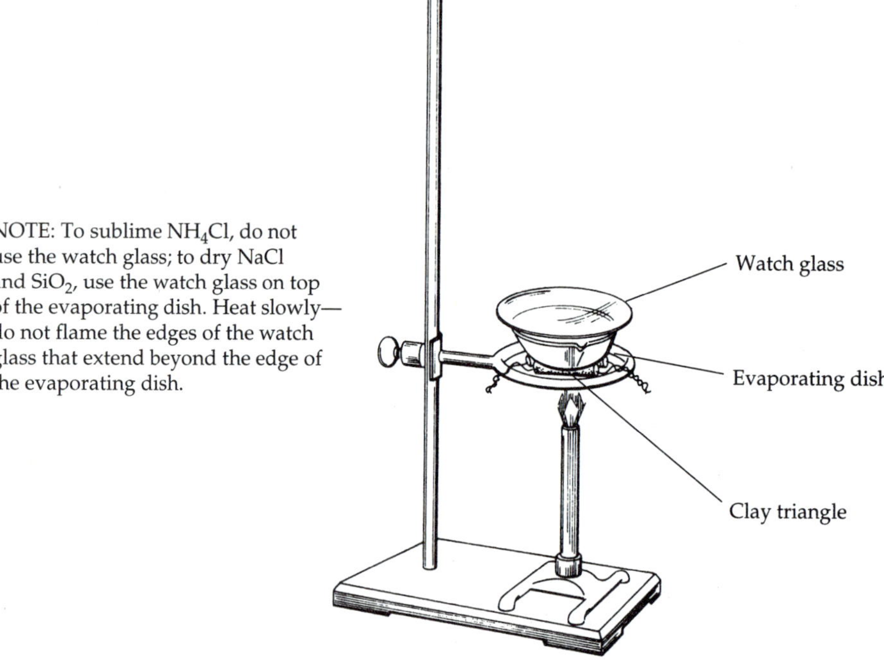

▲ FIGURE 3.3

when liquid is present. Near the end, cover the evaporating dish with the watch glass that was weighed with this evaporating dish, and reduce the heat to prevent spattering. While the water is evaporating, you may proceed to dry the SiO_2 in the other evaporating dish as explained in the next paragraph, if you have another Bunsen burner available. When you have dried the sodium chloride completely, no more water will condense on the watch glass, and it, too, will be dry. Let the evaporating dish and watch glass cool to room temperature and weigh them. The difference between this mass and the mass of the empty evaporating dish and watch glass is the mass of the NaCl. Calculate this mass.

Place the evaporating dish containing the wet sand on the clay triangle on the ring stand and cover the evaporating dish with a clean, dry watch glass. Heat slowly at first until the lumps break up and the sand appears dry. Then heat the evaporating dish to dull redness and maintain this heat for 10 min. Take care not to overheat, or the evaporating dish will crack. When the sand is dry, remove the heat and let the dish cool to room temperature. Weigh the dish after it has cooled to room temperature. The difference between this mass and the mass of the empty dish is the mass of the sand. Calculate this mass. Dispose of the sand in the marked container.

Calculate the percentage of each substance in the mixture using an approach similar to that shown in Example 3.1.

The accuracy of this experiment is such that the combined total of your three components should be in the neighborhood of 99%. If it is less than this, you have been sloppy. If it is more than 100%, you have not sufficiently dried the sand and salt.

EXAMPLE 3.1

What is the percentage of SiO_2 in a 7.69-g sample mixture if 3.76 g of SiO_2 has been recovered?

SOLUTION: The percentage of each substance in such a mixture can be calculated as follows:

$$\% \text{ Component} = \frac{\text{mass component in grams} \times 100}{\text{mass sample in grams}}$$

Therefore, the percentage of SiO_2 in this particular sample mixture is

$$\% \text{ SiO}_2 = \frac{3.76 \text{ g} \times 100}{7.69 \text{ g}} = 48.9\%$$

REVIEW QUESTIONS

Before beginning this experiment in the laboratory, you should be able to answer the following questions:

1. What distinguishes a mixture from an impure substance?
2. Define the process of sublimation.
3. How do decantation and filtration differ? Which should be faster?
4. Why does one never weigh a hot object?
5. How does this experiment illustrate the principle of conservation of matter?
6. A mixture was found to contain 2.10 g of SiO_2, 1.38 g of cellulose, and 6.52 g of calcium carbonate. What is the percentage of SiO_2 in this mixture?
7. How could you separate a mixture of zinc chloride and cyclohexane? Consult Table 2.1 for physical properties.
8. How could you separate zinc chloride from SiO_2?
9. A student found that her mixture was 15% NH_4Cl, 20% NaCl, and 75% SiO_2. Assuming her calculations are correct, what did she most likely do incorrectly in her experiment?
10. Why is the NaCl extracted with water three times as opposed to only once?

NOTES AND CALCULATIONS

Name ______________________ Desk ______________________

Date ______________________ Laboratory Instructor ______________________

Unknown no. ______________________

REPORT SHEET | EXPERIMENT 3

Separation of the Components of a Mixture

A. Mass of evaporating dish and original sample 51.83 g

Mass of evaporating dish 45.91 g

Mass of original sample 5.92 g

Mass of evaporating dish after subliming NH_4Cl 51.27 g

Mass of NH_4Cl 0.56 g

Percent of NH_4Cl (show calculations) 9.5 %

$$\% \ NH_4Cl = \left(\frac{0.56 \text{ g}}{5.92 \text{ g}}\right)(100) = 9.5\%$$

B. Mass of evaporating dish, watch glass, and NaCl 45.38 g

Mass of evaporating dish and watch glass 44.13 g

Mass of NaCl 1.25 g

Percent of NaCl (show calculations) 21.1 %

$$\% \ NaCl = \left(\frac{1.25 \text{ g}}{5.92 \text{ g}}\right)(100) = 21.1\%$$

C. Mass of evaporating dish and SiO_2 49.94 g

Mass of evaporating dish 45.91 g

Mass of SiO_2 4.03 g

Percent of SiO_2 (show calculations) 68.1 %

$$\% \ SiO_2 = \left(\frac{4.03 \text{ g}}{5.92 \text{ g}}\right)(100) = 68.1\%$$

D. Mass of original sample ___5.92___ g

Mass of determined (NH_4Cl + NaCl + SiO_2) ___5.84___ g

Differences in these weights ___0.08___ g

Percent recovery of matter = $\frac{\text{g matter recovered}}{\text{g original sample}} \times 100 =$ ___98.6___ %

Account for your errors

$\left(\frac{5.84}{5.92}\right)(100) = 98.6\%$

Some material was lost on the stirring rod; some was also lost due to spattering while boiling.

QUESTIONS

1. Could the separation in this experiment have been done in a different order? For example, if the mixture were first extracted with water and then the extract and the insoluble residue both heated to dryness, could you determine the amounts of NaCl, NH_4Cl, and SiO_2 originally present? Why or why not?

   Yes. The water would extract both NaCl and NH_4Cl. If you heated the combined NaCl and NH_4Cl sufficiently to sublime the NH_4Cl completely, then by difference you could determine all three components.

Consult a handbook to answer these questions.

2. How could you separate barium sulfate, $BaSO_4$, from NH_4Cl?

   NH_4Cl is much more soluble in water than $BaSO_4$. Extract the NH_4Cl with water and filter off the insoluble $BaSO_4$. Or sublime the NH_4Cl, leaving the $BaSO_4$ behind.

3. How could you separate zinc chloride, $ZnCl_2$, from zinc sulfide, ZnS?

   $ZnCl_2$ is quite soluble in water, whereas ZnS is not. Add water and filter off the ZnS.

4. How could you separate tellurium dioxide, TeO_2, from SiO_2?

   TeO_2 is soluble in HCl, whereas SiO_2 is not.

5. Naphthalene sublimes easily, but potassium bromide does not. How could you separate these two substances?

   By subliming the naphthalene.

Experiment

Chemical Reactions

OBJECTIVE

To observe some typical chemical reactions, identify some of the products and summarize the chemical changes in terms of balanced chemical equations.

APPARATUS AND CHEMICALS

Apparatus

Bunsen burner
6-in. test tube
thistle tube or long-stem funnel
droppers (2)
crucible and cover
glass tubing
ring stand, ring, wire triangle

Chemicals

0.1 *M* sodium oxalate, $Na_2C_2O_4$
10 *M* NaOH
1 *M* K_2CrO_4
mossy zinc
6 *M* NH_3 (*aq*)*
conc. HNO_3
0.1 *M* $NaHSO_3$ (freshly prepared)
2-in. length of copper wire (14, 16, or 18 gauge)
0.1 *M* $BaCl_2$
powdered sulfur
0.1 *M* $KMnO_4$
0.1 *M* $Pb(NO_3)_2$
6 *M* HCl
0.01 *M* $CuSO_4$
6 *M* H_2SO_4
3 *M* $(NH_4)_2CO_3$
$KMnO_4$
Na_2CO_3
Na_2SO_3
ZnS

DISCUSSION

Chemical equations represent what transpires in a chemical reaction. For example, the equation

$$2KClO_3(s) \xrightarrow{\Delta} 2KCl(s) + 3O_2(g)$$

means that potassium chlorate, $KClO_3$, decomposes on heating (Δ is the symbol used for heat) to yield potassium chloride, KCl, and oxygen, O_2. *Before* an equation can be written for a reaction, someone must establish what the products are. How does one decide what these products are? Products are identified by their chemical and physical properties as well as by analyses. That oxygen rather than chlorine gas is produced in the above reaction can be established by the fact that oxygen is a colorless, odorless gas. Chlorine, on the other hand, is a pale yellow-green gas with an irritating odor.

In this experiment, you will observe that in some cases gases are produced, precipitates are formed, or color changes occur during the reactions. These are all indications that a chemical reaction has occurred. To identify some of the products of the reactions, consult Table 4.1, which lists some of the properties of the substances that could be formed in these reactions.

*Reagent bottle may be labeled 6 *M* NH_4OH.

TABLE 4.1 Properties of Reaction Products

Water-soluble solids	Water-insoluble solids	Manganese oxyanions	Gases
KCl: white (colorless solution) NH_4Cl: white (colorless solution) $KMnO_4$: purple $MnCl_2$: pink (very pale) $Cu(NO_3)_2$: blue	CuS: very dark blue or black Cu_2S: black $BaCrO_4$: yellow $BaCO_3$: white $PbCl_2$: white MnO_2: black or brown	MnO_4^-: purple MnO_4^{2-}: dark green MnO_4^{3-}: dark blue	H_2: colorless; odorless NO_2: brown; pungent odor (TOXIC) NO: colorless; slight, pleasant odor CO_2: colorless; odorless Cl_2: pale yellow-green; pungent odor (TOXIC) SO_2: colorless; choking odor (as from matches)(TOXIC) H_2S: colorless; rotten-egg odor (TOXIC)

PROCEDURE

A. A Reaction Between the Elements Copper and Sulfur

PERFORM THIS EXPERIMENT IN THE HOOD WITH A PARTNER. Obtain about a 5-cm (2-in.) length of copper wire and note its properties. Observe that its surface is shiny, that it can be easily bent, and that it has a characteristic color. Make a small coil of the wire by wrapping it around your pencil, and place the wire coil in a crucible. Add sufficient powdered sulfur to the crucible. Cover and place it on a clay triangle on an iron ring for heating. THIS APPARATUS MUST BE SET UP IN THE HOOD, because some sulfur will burn to form noxious sulfur dioxide. Heat the crucible initially with low heat on all sides and then use the hottest flame to heat the bottom of the crucible to red heat. CONTINUE heating until no more smoking occurs, indicating that all the sulfur is burned off. Using the crucible tongs, remove the crucible from the clay triangle without removing the cover, and place it on a heat-resistant pad or wire gauze, *not on the desktop,* to cool. After the crucible has cooled, remove the cover and inspect the substance. Note its properties. Be sure to record your answers on the Report Sheet.

Do this part of Exp. in advance.

1. Does the substance resemble copper?
2. Is it possible to bend the substance without breaking it?
3. What color is it?
4. Has a reaction occurred?

Copper(II) sulfide, CuS, is insoluble in aqueous ammonia, NH_3, (that is, it does not react with NH_3), whereas copper(I) sulfide, Cu_2S, dissolves (that is, reacts) to give a blue solution with NH_3. Place a small portion of your product in a test tube and add 2 mL of 6 *M* NH_3 in the hood. Heat gently with a Bunsen burner.

5. Does your product react with NH_3?
6. Suggest a possible formula for the product.
7. Write a reaction showing the formation of your proposed product:

$$Cu(s) + S_8(s) \longrightarrow ?$$

Waste Disposal Instruction The copper compounds and the acids are toxic and should be handled with care. Avoid spilling any solution, and immediately clean up (using paper towels) any spills that occur. If you spill any solution on your hands, wash them immediately. After completing each series of reactions and before moving on to the next series, dispose of the contents of your test tubes in the designated receptacles. Do not wash the contents down the sink.

B. Oxidation-Reduction Reactions

Many metals react with acids to liberate hydrogen and form the metal salt of the acid. The noble metals do not react with acids to produce hydrogen. Some of the unreactive metals do react with nitric acid, HNO_3; however, in these cases gases that are oxides of nitrogen are formed rather than hydrogen.

Add a small piece of zinc to a test tube containing 2 mL of 6 *M* HCl, and note what happens.

8. Record your observations.
9. Suggest possible products for the observed reaction: Zn(*s*) + HCl(*aq*) ⟶ ?

Place a 1-in. piece of copper wire in a clean test tube. Add 2 mL of 6 *M* HCl, and note if a reaction occurs.

10. Record your observations.
11. Is Cu an active or an inactive metal?

WHILE HOLDING A CLEAN TEST TUBE IN THE HOOD, place a 1-in. piece of copper wire in it and add 1 mL of concentrated nitric acid, HNO_3.

12. Record your observations.
13. Is the gas colored?
14. Suggest a formula for the gas.
15. After the reaction has proceeded for 5 min, carefully add 5 mL of water. Based on the color of the solution, what substance is present in solution?

Potassium permanganate, $KMnO_4$, is an excellent oxidizing agent in acidic media. The permanganate ion is purple and is reduced to the manganous ion, Mn^{2+}, which has a very faint pink color. Place 1 mL of 0.1 *M* sodium oxalate, $Na_2C_2O_4$, in a clean test tube. Add 10 drops of 6 *M* sulfuric acid. Mix thoroughly. To the resulting solution add 1 to 2 drops of 0.1 *M* $KMnO_4$ and stir. If there is no obvious indication that a reaction has occurred, warm the test tube gently in a hot water bath.

16. Record your observations. Was the $KMnO_4$ reduced to Mn^{2+}?

Place 3 mL of 0.1 *M* sodium hydrogen sulfite, $NaHSO_3$, solution in a test tube. Add 1 mL of 10 *M* sodium hydroxide, NaOH, solution and stir. To the mixture in the test tube add 1 drop of 0.1 *M* $KMnO_4$ solution.

17. Record your observations. Was the $KMnO_4$ reduced? Identify the manganese compound formed.

Add additional 0.1 *M* $KMnO_4$ solution, one drop at a time, and observe the effect of each drop until 10 drops have been added.

18. Record your observations.
19. Suggest why the effect of additional potassium permanganate changes as more is added.

WHILE HOLDING A TEST TUBE IN THE FUME HOOD, add one or two crystals of potassium permanganate, $KMnO_4$, to 1 mL of 6 *M* HCl.

20. Record your observations.
21. Note the color of the gas evolved.
22. Based on the color of the gas, what is the gas?

C. Metathesis Reactions

Additional observations are needed before equations can be written for the reaction above, but we see that we can identify some of the products. The remaining reactions are simple, and you will be able, from available information, not only to identify products but also to write equations. A number of reactions may be represented by equations of the following type:

$$AB + CD \longrightarrow AD + CB$$

These are called double-decomposition, or *metathesis,* reactions. This type of reaction involves the exchange of atoms or groups of atoms between interacting substances. The following is a specific example:

$$NaCl(aq) + AgNO_3(aq) \longrightarrow AgCl(s) + NaNO_3(aq)$$

Place a small sample of sodium carbonate, Na_2CO_3, in a test tube and add several drops of 6 *M* HCl.

23. Record your observations.
24. Note the odor and color of the gas that forms.
25. What is the evolved gas?
26. Write an equation for the reaction $HCl(aq) + Na_2CO_3(s) \longrightarrow$? (NOTE: In this reaction the products must have H, Cl, Na, and O atoms in some new combinations, but no other elements can be present.)

Note that H_2CO_3 and H_2SO_3 readily decompose as follows:

$$H_2CO_3(aq) \longrightarrow H_2O(l) + CO_2(g)$$

$$H_2SO_3(aq) \longrightarrow H_2O(l) + SO_2(g)$$

IN THE HOOD, repeat the same test with sodium sulfite, Na_2SO_3.

27. Record your observations.
28. What is the gas?
29. Write an equation for the following reaction (note the similarity to the equation above): $HCl(aq) + Na_2SO_3(s) \longrightarrow$?

IN THE HOOD, repeat this test with zinc sulfide, ZnS.

30. Record your observations.
31. What is the gas?

32. Write an equation for the reaction: $HCl(aq) + ZnS(s) \longrightarrow ?$

To 1 mL of 0.1 *M* lead nitrate, $Pb(NO_3)_2$, solution in a clean test tube add a few drops of 6 *M* HCl.

33. Record your observations.
34. What is the precipitate?
35. Write an equation for the reaction $Pb(NO_3)_2(aq) + HCl(aq) \longrightarrow ?$

To 1 mL of 0.1 *M* barium chloride, $BaCl_2$, solution add 2 drops of 1 *M* potassium chromate, K_2CrO_4, solution.

36. Record your observations.
37. What is the precipitate?
38. Write an equation for the reaction $BaCl_2(aq) + K_2CrO_4(aq) \longrightarrow ?$

To 1 mL of 0.1 *M* barium chloride, $BaCl_2$, solution add several drops of 3 *M* ammonium carbonate, $(NH_4)_2CO_3$, solution in a test tube.

39. What is the precipitate?
40. Write an equation for the reaction $(NH_4)_2CO_3(aq) + BaCl_2(aq) \longrightarrow ?$

After the precipitate has settled somewhat, carefully decant (that is, pour off) the excess liquid. Add 1 mL of water to the test tube, shake it, allow the precipitate to settle, and again carefully pour off the liquid. To the remaining solid, add several drops of 6 *M* HCl.

41. Record your observations.
42. Note the odor.
43. What is the evolved gas? (Recall the reaction in step 26 of this experiment.)

REVIEW QUESTIONS

Before beginning this experiment in the laboratory, you should be able to answer the following questions:

1. Before a chemical equation can be written, what must you know?
2. What observations might you make that suggests that a chemical reaction has occured?
3. How could you distinguish between NO_2 and NO?
4. Define metathesis reactions. Give an example.
5. What is a precipitate?
6. Balance these equations:

$$KBrO_3(s) \xrightarrow{\Delta} KBr(s) + O_2(g)$$

$$MnBr_2(aq) + AgNO_3(aq) \longrightarrow Mn(NO_3)_2(aq) + AgBr(s)$$

7. How could you distinguish between the gases H_2 and H_2S?
8. Using water, how could you distinguish between the white solids KCl and $PbCl_2$?
9. Write equations for the decomposition of $H_2CO_3(aq)$ and $H_2SO_3(aq)$.

NOTES AND CALCULATIONS

Name ______________________ Desk ______________

Date ______________ Laboratory Instructor ______________

Unknown no. ______________

REPORT SHEET | EXPERIMENT

Chemical Reactions | 4

A. A Reaction Between the Elements Copper and Sulfur

1. No.
2. Not without breakage; it's brittle.
3. Blue-black.
4. Yes.
5. Yes, slowly.
6. Cu_2S
7. $Cu(s) + S_8(s) \longrightarrow$ $16Cu(s) + S_8(s) \longrightarrow 8Cu_2S(s)$

B. Oxidation-Reduction Reactions

8. Bubbles are generated and the zinc dissolves.
9. $Zn(s) + HCl(aq) \longrightarrow$ $Zn(s) + 2HCl(aq) \longrightarrow H_2(g) + ZnCl_2(aq)$.
10. No reaction.
11. Copper is an inactive metal.
12. Brown gas is formed, and the copper dissolves.
13. Yes, brown.
14. NO_2
15. $Cu(NO_3)_2$
16. The solution became pale pink; MnO_4^- was reduced to Mn^{2+}.
17. The solution lightened in color, and an odorous gas was produced.

18. The solution becomes purple as additional $KMnO_4$ is added.

19. The HSO_3^- is used up, and there is none left to react with the MnO_4^-.

20. The $KMnO_4$ dissolves, and a gas is evolved.

21. Pungent odor; Cl_2 is produced; pale yellow-green.

22. Cl_2

C. Metathesis Reactions

23. A colorless, odorless gas is produced.

24. Colorless and odorless.

25. CO_2

26. $HCl(aq) + Na_2CO_3(s) \longrightarrow$ $2HCl(aq) + Na_2CO_3(s) \longrightarrow 2NaCl(aq) + H_2O(l) + CO_2(g)$

27. A gas is produced.

28. SO_2

29. $HCl(aq) + Na_2SO_3(s) \longrightarrow$ $2HCl(aq) + Na_2SO_3(s) \longrightarrow 2NaCl(aq) + 2H_2O(l) + SO_2(g)$

30. A gas is produced.

31. H_2S

32. $HCl(aq) + ZnS(s) \longrightarrow$ $2HCl(aq) + ZnS(s) \longrightarrow ZnCl_2(aq) + H_2S(g)$

33. A white precipitate forms.

34. $PbCl_2$

35. $Pb(NO_3)_2(aq) + HCl(aq) \longrightarrow$ $Pb(NO_3)_2(aq) + 2HCl(aq) \longrightarrow PbCl_2(s) + 2HNO_3(aq)$

36. A yellow precipitate forms.

37. $BaCrO_4$

38. $BaCl_2(aq) + K_2CrO_4(aq) \longrightarrow$ $2KCl(aq) + BaCrO_4(s)$

39. $BaCO_3$

40. $(NH_4)_2CO_3(aq) + BaCl_2(aq) \longrightarrow BaCO_3(s) + 2NH_4Cl(aq)$

41. A gas formed that is odorless and colorless.

42. None.

43. CO_2

QUESTIONS: COMPLETE AND BALANCE THE FOLLOWING CHEMICAL REACTIONS

$2HCl(aq) + BaCO_3(s) \longrightarrow BaCl_2(aq) + CO_2(g) + H_2O(l)$

$2HI(aq) + K_2SO_3(s) \longrightarrow 2KI(aq) + SO_2(g) + H_2O(l)$

$Pb(NO_3)_2(aq) + 2KCl(aq) \longrightarrow PbCl_2(s) + 2KNO_3(aq)$

$Ba(NO_3)_2(aq) + Na_2(CrO_4)(aq) \longrightarrow BaCrO_4(s) + 2NaNO_3(aq)$

$K_2CO_3(aq) + Ba(NO_3)_2(aq) \longrightarrow BaCO_3(s) + 2KNO_3(aq)$

$HCl(aq) + AgNO_3(aq) \longrightarrow AgCl(s) + HNO_3(aq)$

NOTES AND CALCULATIONS

Experiment

Chemical Formulas

OBJECTIVE

To become familiar with chemical formulas and how they are obtained.

APPARATUS AND CHEMICALS

Apparatus

balance
Bunsen burner
50-mL graduated cylinder
wire gauze
crucible and cover
carborundum boiling chips or glass beads
250-mL beaker
evaporating dish
ring stand and ring
stirring rod
clay triangle

Chemicals

granular zinc
powdered sulfur
copper wire
6 *M* HCl

DISCUSSION

Chemists use an abbreviated notation to indicate the exact chemical composition of compounds (chemical formulas). We then use these chemical formulas to indicate how new compounds are formed by chemical combinations of other compounds (chemical reactions). However, before we can learn how chemical formulas are written, we must first acquaint ourselves with the symbols used to denote the elements from which these compounds are formed.

Symbols and Formulas

We use one or two letters (with the first letter capitalized) to symbolize a chemical element. These symbols are derived, as a rule, from the first two letters or first syllable or the element's name.

Many elements are found in nature in molecular form; that is, two or more of the same type of atom are tightly bound together. The resultant "package" of atoms, or *molecule,* as it is termed, behaves in many ways as a single distinct object or unit. For example, the oxygen normally found in air consists of molecules that contain two oxygen atoms. We represent this molecular form of oxygen by the chemical formula O_2. The subscript in the formula tells us that two oxygen atoms are present in each oxygen molecule.

Compounds that are composed of molecules are called *molecular compounds,* and they contain more than one type of atom. For example, a molecule of water consists of two hydrogen atoms and one oxygen atom and is represented by the chemical formula H_2O. The absence of a subscript on the O implies there is one oxygen atom per water molecule. Another compound composed of these same elements, but in different proportions, is hydrogen peroxide, H_2O_2. The physical and chemical properties of these two compounds are very different. We shouldn't be surprised, for they are two different substances.

Chemical formulas that indicate the *actual* numbers and types of atoms in a molecule are called *molecular formulas,* whereas chemical formulas that indicate only the *relative* numbers of atoms of a type in a molecule are called *empirical formulas.* The subscripts in an empirical formula are always the smallest whole-number ratios. For example, the molecular formula for hydrogen peroxide is H_2O_2, whereas its empirical formula is HO. The molecular formula for glucose is $C_6H_{12}O_6$; its empirical formula is CH_2O. For many substances the molecular formula and empirical formula are identical, as is the case for water, H_2O, and sulfuric acid, H_2SO_4.

Atom Weights

It is important to know something about masses of atoms and molecules. With a mass spectrometer we can measure the masses of individual atoms with a high degree of accuracy. We know, for example, that the hydrogen-1 atom has a mass of 1.6735×10^{-24} g and the oxygen-16 atom has a mass of 1.674×10^{-24} g. Because it is cumbersome to express such small masses in grams, we use a unit called the *atomic mass unit,* or amu. An amu equals 1.66054×10^{-24} g. Most elements occur as mixtures of isotopes. The average atomic mass of each element expressed in amu is also known as its *atomic weight.* The atomic weights of the elements listed both in the table of elements and in the periodic table inside the front and back covers of this book, respectively, are in amu.

Formula and Molecular Weights

The *formula weight* of a substance is merely the sum of the atomic weights of all atoms in its chemical formula. For example, nitric acid, HNO_3, has a formula weight of 63.0 amu.

$$\begin{aligned} FW &= (\text{AW of H}) + (\text{AW of N}) + 3(\text{AW of O}) \\ &= 1.0 \text{ amu} + 14.0 \text{ amu} + 3(16.0 \text{ amu}) \\ &= 63.0 \text{ amu} \end{aligned}$$

If the chemical formula of a substance is its molecular formula, then the formula weight is also called the *molecular weight.* For example, the molecular formula for formaldehyde is CH_2O. The molecular weight of formaldehyde is therefore

$$\begin{aligned} MW &= 12 \text{ amu} + 2(1.0 \text{ amu}) + 16.0 \text{ amu} \\ &= 30.0 \text{ amu} \end{aligned}$$

For ionic substances such as NaCl that exist as three-dimensional arrays of ions, it is not appropriate to speak of molecules. Similarly, the terms molecular weight and molecular formula are inappropriate for these ionic substances. It is correct to speak of their formula weight, however. Thus, the formula weight of NaCl is

$$\begin{aligned} FW &= 23.0 \text{ amu} + 35.5 \text{ amu} \\ &= 58.5 \text{ amu} \end{aligned}$$

Percentage Composition from Formulas

New compounds are made in laboratories every day, and the formulas of these compounds must be determined. The compounds are often analyzed for their *percentage composition*, i.e., the percentage by mass of each element present in the compound. The percentage composition is useful information in establishing the formula for the substance. If the formula of a compound is known, calculating its percentage composition is a straight-forward matter. In general, the percentage of an element in a compound is given by

$$\frac{(\text{Number of atoms of element})(\text{AW})}{\text{FW of compound}} \times 100$$

If we want to know the percentage composition of formaldehyde, CH_2O, whose formula weight is 30.0 amu, we proceed as follows:

$$\%\text{C} = \frac{12.0 \text{ amu}}{30.0 \text{ amu}} \times 100 = 40.0\%$$

$$\%\text{H} = \frac{2(1.0 \text{ amu})}{30.0 \text{ amu}} \times 100 = 6.7\%$$

$$\%\text{O} = \frac{16.0 \text{ amu}}{30.0 \text{ amu}} \times 100 = 53.3\%$$

The Mole

Even the smallest samples we use in the laboratory contain an enormous number of atoms. A drop of water contains about 2×10^{21} water molecules! The unit that the chemist uses for dealing with such a large number of atoms, ions, or molecules is the *mole*, abbreviated mol. Just as the unit dozen refers to 12 objects, the mole refers to a collection of 6.02×10^{23} objects. This number is called Avogadro's number. Thus, a mole of water molecules contains 6.02×10^{23} H_2O molecules, and a mol of sodium contains 6.02×10^{23} Na atoms. The mass (in grams) of 1 mol of a substance is called its *molar mass*. The molar mass (in grams) of any substance is numerically equal to its formula weight.

Thus:

One CH_2O molecule has a mass of 30.0 amu; 1 mol CH_2O has a mass of 30.0 g and contains 6.02×10^{23} CH_2O molecules

One Na atom has a mass of 23.0 amu; 1 mol Na has a mass of 23.0 g and contains 6.02×10^{23} Na atoms

It is a simple matter to calculate the number of moles of any substance whose mass and formula we know. For example, suppose we have one quart of rubbing alcohol (generally isopropyl alcohol) and know its density to be 0.785 g/mL and we want to know how many moles this is. First we need to convert the volume to our system of units. Since 1 qt is 0.946 L and 1 L contains 1000 mL, our quart of isopropyl alcohol is

$$(1 \text{ qt})\left(\frac{0.946 \text{ L}}{\text{qt}}\right)\left(\frac{1000 \text{ mL}}{\text{L}}\right) = 946 \text{ mL}$$

We can now calculate the mass:

$$946 \text{ mL} \times 0.785 \text{ g/mL} = 743 \text{ g}$$

We next need the chemical formula for isopropyl alcohol. This is C_3H_7OH. The molecular weight is, therefore,

Weight carbon	$3 \times 12.0 = 36.0$ amu
Weight hydrogen	$8 \times 1.0 = 8.0$ amu
Weight oxygen	$1 \times 16.0 = 16.0$ amu
Molecular weight	$C_3H_7OH = 60.0$ amu

Hence,

$$\text{moles of } C_3H_7OH = (743 \text{ g } C_3H_7OH)\left(\frac{1 \text{ mol } C_3H_7OH}{60.0 \text{ g } C_3H_7OH}\right) = 12.4 \text{ mol}$$

Thus our quart of rubbing alcohol contains 12.4 mol of isopropyl alcohol. It should now be apparent to you how much information is contained in a chemical formula.

Empirical Formulas from Analyses

The empirical formula for a substance tells us the relative number of atoms of each element in the substance. Thus, the formula H_2O indicates that water contains two hydrogen atoms for each oxygen atom. This ratio applies on the molar level as well; thus, 1 mol of H_2O contains two mol of H atoms and 1 mol of O atoms. Conversely, the ratio of the number of moles of each element in a compound gives the subscripts in a compound's empirical formula. Thus, the mole concept provides a way of calculating the empirical formula of a chemical substance. This is shown in the following example.

EXAMPLE 5.1

While you are working in a hospital laboratory, a patient complaining of severe stomach cramps and labored respiration dies within minutes of being admitted. Relatives of the patient later tell you that he may have ingested some rat poison. You therefore have his stomach pumped to verify this and also to determine the cause of death. One of the more logical things to do would be to attempt to isolate the agent that caused death and perform chemical analyses on it. Let's suppose that this was done and the analyses showed that the isolated chemical compound contained, by weight, 60.0% potassium, 18.5% carbon, and 21.5% nitrogen. What is the chemical formula for this compound?

SOLUTION: One simple and direct way of making the necessary calculations is as follows. Assume you had 100 g of the compound. This 100 g would contain

$$(100 \text{ g})(0.600) = 60.0 \text{ g potassium}$$
$$(100 \text{ g})(0.185) = 18.5 \text{ g carbon}$$
$$(100 \text{ g})(0.215) = 21.5 \text{ g nitrogen}$$

Divide each of these masses by the appropriate atomic weight to obtain the number of moles of each element in the 100 g:

$$60.0 \text{ g K}\left(\frac{1 \text{ mol K}}{39.0 \text{ g K}}\right) = 1.54 \text{ mol K}$$

$$18.5 \text{ g C}\left(\frac{1 \text{ mol C}}{12.0 \text{ g C}}\right) = 1.54 \text{ mol C}$$

$$21.5 \text{ g N}\left(\frac{1 \text{ mol N}}{14.0 \text{ g N}}\right) = 1.54 \text{ mol N}$$

Then divide each number by 1.54 to determine the simplest whole-number ratio of moles of each element. (In general, after determining the number of moles of each element, we determine the simplest whole-number ratio by dividing each number of moles by the smallest number of moles. In this example all the numbers are the same.)

$$K = \frac{1.54}{1.54} = 1.00$$

$$C = \frac{1.54}{1.54} = 1.00$$

$$N = \frac{1.54}{1.54} = 1.00$$

The ratio obtained in this case is 1.00, and we conclude that the formula is KCN. This is the simplest, or empirical, formula because it uses as subscripts the smallest set of integers to express the correct ratios of atoms present. Because KCN is a common rat poison, we may justifiably conclude that the relatives' suggestion of rat-poison ingestion as the probable cause of death is correct.

Molecular Formulas from Empirical Formulas

The formula obtained from percentage composition is *always* the empirical formula. We can obtain the molecular formula from the empirical formula if we know the molecular weight of the compound. *The subscripts in the molecular formula of a substance are always a whole-number multiple of the corresponding subscripts in its empirical formula.* The multiple is found by comparing the formula weight of the empirical formula with the molecular weight. For example, suppose we determined the empirical formula of a compound to be CH_2O. Its formula weight is

$$\text{FW} = 12.0 \text{ amu} + 2(1.0 \text{ amu}) + 16.0 \text{ amu} = 30.0 \text{ amu}$$

Suppose the experimentally determined molecular weight is 180. Then, the molecule has six times the mass (180/30.0 = 6.00) and must, therefore, have six times as many atoms as the empirical formula. The subscripts in the empirical formula must be multiplied by 6 to obtain the molecular formula: $C_6H_{12}O_6$.

In this experiment, you will determine the empirical formulas of two chemical compounds. One is copper sulfide, which you will prepare according to the following chemical reaction:

$$x\text{Cu}(s) + y\text{S}(s) \longrightarrow \text{Cu}_x\text{S}_y(s)$$

The other is zinc chloride, which you will prepare according to the chemical reaction

$$x\text{Zn}(s) + y\text{HCl}(aq) \longrightarrow \text{Zn}_x\text{Cl}_y(s) + \frac{y}{2}\text{H}_2(g)$$

The objective is to determine the combining ratios of the elements (that is, to determine x and y) and to balance the chemical equations given above.

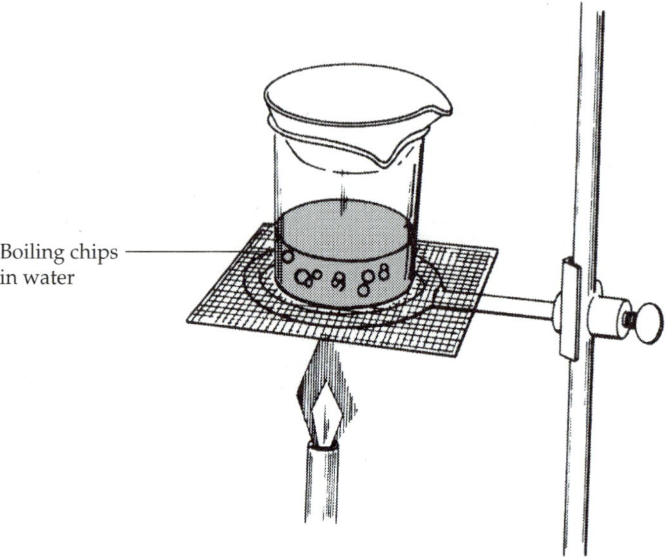

▲ **FIGURE 5.1** Steam bath.

PROCEDURE

Instructor: All students should finish part A before any student begins part B.

A. Zinc Chloride

Clean and dry your evaporating dish and place it on the wire gauze resting on the iron ring. Heat the dish with your Bunsen burner, gently at first, and then more strongly, until all of the condensed moisture has been driven off. This should require heating for about 5 min. Allow the dish to cool to room temperature on a heat-resistant pad (do not place the hot dish on the countertop) and weigh it. Record the mass of the empty evaporating dish to the nearest 0.01 g.

Obtain a sample of granular zinc from your laboratory instructor and add about 0.5 g of it to the weighed evaporating dish. Weigh the evaporating dish containing the zinc and record the total mass to the nearest 0.01 g. Calculate the mass of the zinc.

Slowly, and with constant stirring, add 15 mL of 6 *M* HCl to the evaporating dish containing the zinc. A vigorous reaction will ensue, and hydrogen gas will be produced. **(CAUTION:** ***No flames are permitted in the laboratory while this reaction is taking place, because wet hydrogen gas is very explosive.*****)** If any undissolved zinc remains after the reaction ceases, add an additional 5 mL of acid. Continue to add 5-mL portions of acid as needed until all the zinc has dissolved. **CAUTION:** ***Zinc chloride is caustic and must be handled carefully in order to avoid any contact with your skin. Should you come in contact with it, immediately wash the area with copious amounts of water.***

Instructor: Students have a tendency to overheat the $ZnCl_2$.

Set up a steam bath as illustrated in Figure 5.1 using a 250-mL beaker, and place the evaporating dish on the steam bath. Heat the evaporating dish very carefully on the steam bath until most of the liquid has disappeared. Then remove the steam bath and heat the dish on the wire gauze. During this last stage of heating, the flame must be carefully controlled or there will be spattering, and some loss of product will occur. **(CAUTION:** ***Do not heat to the point that the compound melts, or some will be lost due to sublimation.*****)** Leave the compound looking somewhat pasty while hot.

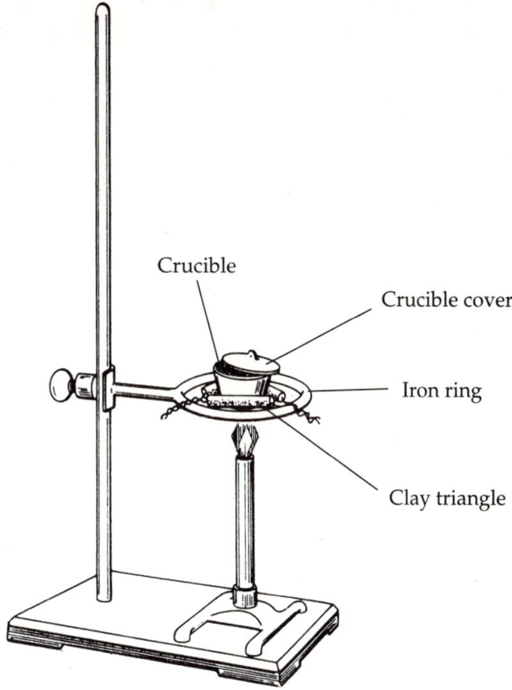

▲ **FIGURE 5.2** Setup for copper sulfide determination.

Allow the dish to cool to room temperature and weigh it. Record the mass. After this first weighing, heat the dish again very gently. Cool it and reweigh it. If these weighings do not agree within 0.02 g, repeat the heating and weighing until two successive weighings agree. This is known as *drying to constant mass* and is the only way to be certain that all the moisture is driven off. Zinc chloride is very deliquescent (rapidly absorbs moisture from the air) and so should be weighed quickly.

Calculate the mass of zinc chloride. The difference in mass between the zinc and zinc chloride is the mass of chlorine. Calculate the mass of chlorine in zinc chloride. From this information, you can readily calculate the empirical formula for zinc chloride and balance the chemical equation for its formation. Perform these operations on the report sheet.

B. Copper Sulfide

Support a clean, dry porcelain crucible and cover on a clay triangle and dry by heating to a dull red in a Bunsen flame, as illustrated in Figure 5.2. Allow the crucible and cover to cool to room temperature and weigh them. Record the mass to the nearest 0.01 g.

Place 1.5 to 2.0 g of tightly wound copper wire or copper turnings in the crucible and weigh the copper, crucible, and lid to the nearest 0.01 g and record your results. Calculate the mass of copper.

In the hood, add sufficient sulfur to cover the copper, place the crucible with cover in place on the triangle, and heat the crucible gently until sulfur ceases to burn (blue flame) at the end of the cover. Do not remove the cover while the crucible is hot. Finally, heat the crucible to dull redness for about 5 min.

Allow the crucible to cool to room temperature. This will take about 10 min. Then weigh with the cover in place. Record the mass. Again cover the contents of the crucible with sulfur and repeat the heating procedure. Allow the crucible to cool and reweigh it. Record the mass. If the last two weighings do not agree to within 0.02 g, the chemical reaction between the copper and sulfur is incomplete. If this is found to be the case, add more sulfur and repeat the heating and weighing until a constant mass is obtained.

Calculate the mass of copper sulfide obtained. The difference in mass between the copper sulfide and copper is the mass of sulfur in copper sulfide. Calculate this mass. From this information the empirical formula for copper sulfide can be obtained, and the chemical equation for its production can be balanced. Perform these operations on your report sheet.

Waste Disposal Instructions All chemicals must be disposed of in the appropriately labeled containers.

REVIEW QUESTIONS

Before beginning this experiment in the laboratory, you should be able to answer the following questions:

1. Define the term *compound*.
2. Why are atomic weights relative weights?
3. How do formula weights and molecular weights differ?
4. What is the percentage composition of Na_2CO_3?
5. A substance was found by analysis to contain 35.74% calcium and 63.36% chlorine. What is the empirical formula for the substance?
6. What is the law of definite proportions?
7. How do empirical and molecular formulas differ?
8. What is the mass in grams of one silver atom?
9. Soda-lime glass is prepared by fusing sodium carbonate, Na_2CO_3; limestone, $CaCO_3$; and sand, SiO_2. The composition of the glass varies, but the commonly accepted reaction for its formation is

   $$Na_2CO_3(s) + CaCO_3(s) + 6SiO_2(s) \longrightarrow Na_2CaSi_6O_{14}(s) + 2CO_2(g)$$

   Using this equation, how many kilograms of sand would be required to produce enough glass to make 5000 400-g wine bottles?
10. Caffeine, a stimulant found in coffee and tea, contains 49.5% carbon, 5.15% hydrogen, 28.9% nitrogen, and 16.5% oxygen by mass. What is the empirical formula of caffeine? If its molar mass is about 195 g, what is its molecular formula?
11. An analysis of an oxide of nitrogen with a molecular weight of 92.02 amu gave 69.57% oxygen and 30.43% nitrogen. What are the empirical and molecular formulas for this nitrogen oxide? Complete and balance the equation for its formation from the elements nitrogen and oxygen.
12. How many potassium atoms are present in 0.01456 g of potassium?

Name ______________________ Desk ______________________

Date ______________________ Laboratory Instructor ______________________

REPORT SHEET | EXPERIMENT

Chemical Formulas | 5

A. Zinc Chloride

1. Mass of evaporating dish and zinc ___41.11___ g
2. Mass of evaporating dish ___40.66___ g
3. Mass of zinc ___0.45___ g

4. Mass of evaporating dish and zinc chloride:
   - first weighing ___41.64___ g
   - second weighing ___41.62___ g
   - third weighing ___41.60___ g
5. Mass of zinc chloride ___0.94___ g

6. Mass of chlorine in zinc chloride ___0.49___ g

$0.94 \text{ g} - 0.45 \text{ g} = 0.49 \text{ g}$

7. Empirical formula for zinc chloride (show calculations) ___$ZnCl_2$___

$$Zn = \frac{0.45 \text{ g}}{65.37 \text{ g/mol}} = 6.9 \times 10^{-3} \text{ mol} \quad Cl = \frac{0.49 \text{ g}}{35.45 \text{ g/mol}} = 0.014 \text{ mol}$$

$$Zn: \frac{0.0069}{0.0069} = 1.0 \qquad Cl: \frac{0.014}{0.0069} = 2.0 \qquad \text{therefore, } ZnCl_2$$

8. Balanced chemical equation for the formation of zinc chloride from zinc and HCl

$Zn(s) + 2HCl(aq) \rightarrow ZnCl_2(aq) + H_2(g)$

B. Copper Sulfide

1. Mass of crucible, cover, and copper ___20.00___ g
2. Mass of crucible and cover ___19.54___ g
3. Mass of copper ___0.46___ g

4. Mass of crucible, cover, and copper sulfide:
   - first weighing ___20.11___ g
   - second weighing ___20.12___ g
   - third weighing ___20.12___ g

5. Mass of copper sulfide 0.58 g

6. Mass of sulfur in copper sulfide 0.12 g

Instructor: Copper (I) sulfide is a non-stoichiometric compound whose real formula is $Cu_{1.9}S$

0.58 g − 0.46 g = 0.12 g

7. Empirical formula for copper sulfide (show calculations) Cu_2S

$$Cu = \frac{0.46\ g}{63.54\ g/mol} = 7.2 \times 10^{-3}\ mol \quad S = \frac{0.12\ g}{32.06\ g/mol} = 3.8 \times 10^{-3}\ mol$$

$$Cu: \frac{0.0072}{0.0038} = 1.9 \quad S: \frac{0.0038}{0.0038} = 1.0 \quad \text{therefore, } Cu_2S$$

8. Balanced chemical equation for the formation of copper sulfide from copper and sulfur

$2Cu(s) + S(s) = Cu_2S(s)$

QUESTIONS

1. Can you determine the molecular formula of a substance from its percent composition?

No

2. Given that zinc chloride has a formula weight of 136.28 amu, what is its formula?

$ZnCl_2$ has a formula weight of 65.37 amu + 2(35.45) amu = 136.27 amu, so $ZnCl_2$ is the formula.

3. Can you determine the atomic weights of zinc or copper by the methods used in this experiment? How? What additional information is necessary in order to do this?

Yes, if the atomic weight of chlorine and sulfur are known and the valences of Cu and Zn in the compounds are known.

4. How many grams of zinc chloride could be formed from the reaction of 7.96 g of zinc with excess HCl?

$Zn(s) + 2HCl(aq) \rightarrow ZnCl_2(aq) + H_2(g)$

$$\text{moles } ZnCl_2 = \text{moles Zn} = (7.96\ g\ Zn)\left(\frac{1\ mol\ Zn}{65.37\ g\ Zn}\right) = 0.122\ mol\ Zn$$

$$\text{mass } ZnCl_2 = (0.122\ mol)\left(\frac{136.28\ g\ ZnCl_2}{mol\ ZnCl_2}\right) = 16.6\ g\ ZnCl_2$$

5. How many kilograms of copper sulfide could be formed from the reaction of 1.80 mol of copper with excess sulfur?

   $16Cu + S_8 \rightarrow 8Cu_2S$

   $$(1.80 \text{ mol Cu})\left(\frac{1 \text{ mol } Cu_2S}{2 \text{ mol Cu}}\right)\left(\frac{159.1 \text{ g}}{1 \text{ mol } Cu_2S}\right)\left(\frac{1 \text{ kg}}{1000 \text{ g}}\right) = 0.143 \text{ kg } Cu_2S$$

6. If copper(I) sulfide is partially roasted in air (reaction with O_2), copper(I) sulfite is first formed. Subsequently, upon heating, the copper sulfite thermally decomposes to copper(I) oxide and sulfur dioxide. Write balanced chemical equations for these two reactions.

   $2Cu_2S + 3O_2 \xrightarrow[\Delta]{} 2Cu_2SO_3$

   $2Cu_2SO_3 \xrightarrow[\Delta]{} 2Cu_2O + 2SO_2$

NOTES AND CALCULATIONS

Experiment

Chemical Reactions of Copper and Percent Yield

OBJECTIVE

To gain some familiarity with basic laboratory procedures, some chemistry of a typical transition element, and the concept of percent yield.

APPARATUS AND CHEMICALS

Apparatus

balance
250-mL beakers (2)
evaporating dish
stirring rod
towel
wire gauze
Bunsen burner and hose
100-mL graduated cylinder
weighing paper
boiling chips
ring stand and iron ring

Chemicals

0.5-g piece of copper wire (16 or 18 gauge)
6 *M* H_2SO_4
methanol
aluminum foil cut in 1-in. squares
conc. HNO_3
3.0 *M* NaOH
granular zinc
acetone
conc. HCl

DISCUSSION

Most chemical syntheses involve separation and purification of the desired product from unwanted side products. Common methods of separation are filtration, sedimentation, decantation, extraction, and sublimation. This experiment is designed to be a quantitative evaluation of your individual laboratory skills in carrying out some of these operations. At the same time, you will become acquainted with two fundamental types of chemical reactions, called redox reactions and metathesis reactions. By means of these reactions, you will carry out several chemical transformations involving copper, and you will finally recover the copper sample with maximum efficiency. The chemical reactions involved are the following:

$$Cu(s) + 4HNO_3(aq) \longrightarrow Cu(NO_3)_2(aq) + 2NO_2(g) + 2H_2O(l) \quad \text{Redox} \quad [1]$$

$$Cu(NO_3)_2(aq) + 2NaOH(aq) \longrightarrow Cu(OH)_2(s) + 2NaNO_3(aq) \quad \text{Metathesis} \quad [2]$$

$$Cu(OH)_2(s) \xrightarrow{\Delta} CuO(s) + H_2O(g) \quad \text{Dehydration} \quad [3]$$

$$CuO(s) + H_2SO_4(aq) \longrightarrow CuSO_4(aq) + H_2O(l) \quad \text{Metathesis} \quad [4]$$

$$CuSO_4(aq) + Zn(s) \longrightarrow ZnSO_4(aq) + Cu(s) \quad \text{Redox} \quad [5]$$

$$3CuSO_4(aq) + 2Al(s) \longrightarrow Al_2(SO_4)_3(aq) + 3Cu(s) \quad \text{Redox} \quad [6]$$

Each of these reactions proceeds to completion. Metathesis reactions proceed to completion whenever one of the components is removed from the solution, such as in the formation of a gas or an insoluble precipitate.

This is the case for reactions [1], [2], and [3], where in reactions [1] and [3] a gas and in reaction [2] an insoluble precipitate are formed. (Reactions [5] and [6] proceed to completion because copper is more difficult to oxidize than either zinc or aluminum.)

The object of this experiment is to recover all of the copper with which you began. This is the test of your laboratory skills.

The percent yield of the copper can be expressed as the ratio of the recovered mass to initial mass, multipled by 100:

$$\%\ \text{yield} = \frac{\text{recovered mass of Cu}}{\text{initial mass of Cu}} \times 100$$

PROCEDURE

Weigh approximately 0.500 g of no. 16 or no. 18 copper wire to the nearest 0.0001 g and record its mass (1). Place it in a 250-mL beaker. IN THE HOOD add 4 or 5 mL of concentrated HNO_3 to the beaker. **(CAUTION: *Be careful not to get any of the nitric acid on yourself. If you do, wash it off immediately with copious amounts of water. The gas produced in this reaction is toxic, and the reaction must be performed in the hood.*)** After the reaction is complete, add 100 mL of distilled H_2O. Describe the reaction as to color change, evolution of a gas, and change in temperature (exothermic or endothermic) on the report sheet (6).

Add 30 mL of 3.0 *M* NaOH to the solution in your beaker and describe the reaction on the report sheet (7). Add two or three boiling chips and carefully heat the solution—while stirring with a stirring rod—just to the boiling point. Describe the reaction on your report sheet (8).

Allow the black CuO to settle, then decant the supernatant liquid. Add about 200 mL of very hot distilled water, stir, and then allow the CuO to settle. Decant once more. What are you removing by the washing and decantation (9)?

Add 15 mL of 6.0 *M* H_2SO_4. What copper compound is present in the beaker now (10)?

Your instructor will tell you whether you should use zinc or aluminum for the reduction of Cu(II) in the following step.

A. Reduction with Zinc

In the hood, add 2.0 g of 30-mesh zinc metal all at once and stir until the supernatant liquid is colorless. Describe the reaction on your report sheet (11). What is present in solution (12)? When gas evolution has become *very* slow, heat the solution gently (but do not boil) and allow it to cool. What gas is formed in this reaction (13)? How do you know (14)?

B. Reduction with Aluminum

In the hood, add several 1-in. squares of aluminum foil and a few drops of concentrated HCl. Continue to add pieces of aluminum until the supernatant liquid is colorless. Describe the reaction on your report sheet (11). What is present in solution (12)? What gas is formed in this reaction (13)? How do you know (14)?

For Either Reduction Method When gas evolution has ceased, decant the solution and transfer the precipitate to a preweighed porcelain evaporating dish and record its mass on the report sheet (3). Wash the precipitated copper with about 5 mL of distilled water, allow it to settle, decant the solution, and

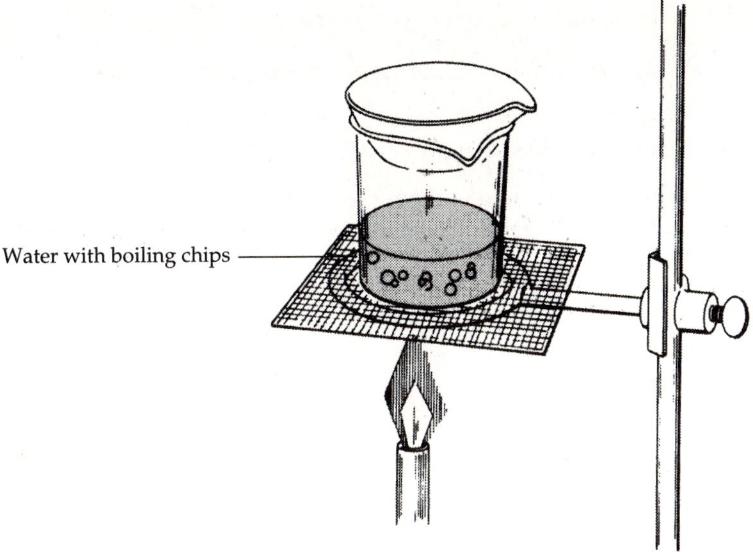

▲ **FIGURE 6.1** Steam bath.

repeat the process. What are you removing by washing (15)? Wash the precipitate with about 5 mL of methanol. **(CAUTION:** ***Keep the methanol away from flames—it is flammable!! Methanol is also extremely toxic. Avoid breathing the vapors as much as possible.*****)** Allow the precipitate to settle, and decant the methanol. Finally, wash the precipitate with about 5 mL of acetone. **(CAUTION:** ***Keep the acetone away from flames—it is extremely flammable!!*****)** Allow the precipitate to settle, and decant the acetone from the precipitate. Prepare a steam bath as illustrated in Figure 6.1 and dry the product on your steam bath for at least 5 min. Wipe the bottom of the evaporating dish with a towel, remove the boiling chips, weigh the evaporating dish plus copper, and record its mass (2). Calculate the final mass of copper (4). Compare the mass with your initial mass and calculate the percent yield (5). What color is your copper sample (16)? Is it uniform in appearance (17)? Suggest possible sources of error in this experiment (18).

Dispose of the chemicals in the designated receptacles.

REVIEW QUESTIONS

Before beginning this experiment in the laboratory, you should be able to answer the following questions:

1. Give an example, other than the ones listed in this experiment, of redox and methathesis reactions.
2. When will reactions proceed to completion?
3. Define *percent yield* in general terms.
4. Name six methods of separating materials.
5. Give criteria in terms of temperature changes for exothermic and endothermic reactions.
6. If 3.40 g of $Cu(NO_3)_2$ are obtained from allowing 2.09 g of Cu to react with excess HNO_3, what is the percent yield of the reaction?

7. What is the maximum percent yield in any reaction?
8. What is meant by the terms *decantation* and *filtration*?
9. When $Cu(OH)_2$*(s)* is heated, copper(II) oxide and water are formed. Write a balanced equation for the reaction.
10. When sulfuric acid and copper(II) oxide are allowed to react, copper(II) sulfate and water are formed. Write a balanced equation for this reaction.

Name ______________________ Desk ______________________

Date ______________________ Laboratory Instructor ______________________

REPORT SHEET | EXPERIMENT 6

Chemical Reactions of Copper and Percent Yield

1. Initial mass of copper ____0.612 g____
2. Mass of copper and evaporating dish ____40.665 g____
3. Mass of evaporating dish ____40.060 g____
4. Mass of recovered copper ____0.605 g____

5. Percent yield (show calculations) ____98.9%____

$$\left(\frac{0.605\text{ g}}{0.612\text{ g}}\right)(100) = 98.9\%$$

6. Describe the reaction $Cu(s) + HNO_3(aq) \longrightarrow$

   Brown fumes are formed, and the copper dissolves to form a blue solution. The reaction is exothermic; heat is produced.

7. Describe the reaction $Cu(NO_3)_2(aq) + NaOH(aq) \longrightarrow$

   A flocculent baby-blue precipitate forms.

8. Describe the reaction $Cu(OH)_2(s) \xrightarrow{\Delta}$

   The baby-blue precipitate turns black.

9. What are you removing by this washing? ____$NaNO_3$____
10. What copper compound is present in the beaker? ____$CuSO_4$____
11. Describe the reaction $CuSO_4(aq) + Zn(s)$, or $CuSO_4(aq) + Al(s)$ ____A brown-black precipitate forms.____
12. What is present in solution? ____$ZnSO_4$ or $Al_2(SO_4)_3$ and the supernatant solution becomes colorless.____
13. What is the gas? ____H_2____
14. How do you know? ____It is colorless, odorless, and flammable.____
15. What are you removing by washing? ____$ZnSO_4$ or $Al_2(SO_4)_3$____
16. What color is your copper sample? ____shiny brown____

17. Is it uniform in appearance? It should be.

18. Suggest possible sources of error in this experiment.

Poor transfer techniques and incomplete reactions from limited amounts of reagents will affect results.

QUESTIONS

1. If your percent yield of copper was greater than 100%, what are two plausible errors you may have made?

The recovered copper was either not washed free of salts or was not completely dried.

2. Consider the combustion of methane, CH_4:

$$CH_4(g) + 2O_2(g) \longrightarrow CO_2(g) + 2H_2O(g)$$

Suppose 2.5 moles of methane are allowed to react with 3 moles of oxygen.
(a) What is the limiting reagent?

Oxygen, because 2.5 moles of CH_4 require 5 moles of O_2 to react completely.

(b) How many moles of CO_2 can be made from this mixture? How many grams of CO_2?

$$3 \text{ mol } O_2 \times \frac{1 \text{ mol } CO_2}{2 \text{ mol } O_2} = 1.5 \text{ mol } CO_2$$

$$1.5 \text{ mol } CO_2 \times \frac{44 \text{ g } CO_2}{\text{mol } CO_2} = 66 \text{ g } CO_2$$

3. Suppose 8.00 g of CH_4 is allowed to burn in the presence of 6.00 g of oxygen. How much (in grams) CH_4, O_2, CO_2, and H_2O remain after the reaction is complete?

$$8.00 \text{ g } CH_4 \times \frac{1 \text{ mol } CH_4}{16.0 \text{ g } CH_4} = 0.500 \text{ mol } CH_4$$

$$6.00 \text{ g } O_2 \times \frac{1 \text{ mol } O_2}{32.0 \text{ g } O_2} = 0.188 \text{ mol } O_2$$

O_2 is the limiting reagent:
0.405 mol CH_4 remain, or 6.48 g CH_4
0.094 mol CO_2 formed, or 4.14 g CO_2
0.188 mol H_2O formed, or 3.38 g H_2O

4. How many milliliters of 3.0 *M* H_2SO_4 are required to react with 0.80 g of CuO according to Equation [4]?

$$0.80 \text{ g CuO} \times \frac{1 \text{ mol CuO}}{79.5 \text{ g CuO}} = 0.010 \text{ mol CuO}$$. Therefore, 0.010 mol H_2SO_4 is required.

$$0.010 \text{ mol } H_2SO_4 \times \frac{1\text{L}}{3 \text{ mol}} = 0.33 \times 10^{-2} \text{ L, or } 3.3 \text{ mL}$$

5. If 3.00 g of Zn is allowed to react with 1.75 g of $CuSO_4$ according to Equation [5], how many grams of Zn will remain after the reaction is complete?

$$3.00 \text{ g Zn} \times \frac{1 \text{ mol Zn}}{65.37 \text{ g Zn}} = 0.0459 \text{ mol Zn}$$

$$1.75 \text{ g } CuSO_4 \times \frac{1 \text{ mol } CuSO_4}{159.5 \text{ g } CuSO_4} = 0.0110 \text{ mol } CuSO_4$$. Hence, Zn is in excess:

$$0.0459 \text{ mol} - 0.0110 \text{ mol} = 0.0349 \text{ mol Zn unreacted}$$

$$0.0349 \text{ mol Zn} \times \frac{65.37 \text{ g Zn}}{1 \text{ mol Zn}} = 2.28 \text{ g Zn}$$

6. What is meant by the term *limiting reagent*?

The substance that is completely consumed in a reaction is called the limiting reagent or limiting reactant because it determines, or limits, the amount of product formed.

NOTES AND CALCULATIONS

Experiment

Chemicals in Everyday Life: What Are They and How Do We Know?

OBJECTIVE

To observe some reactions of common substances found around the home and to learn how to identify them.

APPARATUS AND CHEMICALS

Apparatus

150-mL beaker
medicine droppers (3)
3-in. test tubes (6)
red and blue litmus paper

Chemicals

household ammonia
household bleach (chlorine)
baking soda
chalk
mineral oil
1 *M* NH_4Cl
$(NH_4)_2CO_3$
$Ba(OH)_2$ (sat. soln.)
3 *M* HNO_3
0.2 *M* $BaCl_2$
chemical fertilizer
table salt
vinegar
Epsom salts
NaI
8 *M* NaOH
18 *M* H_2SO_4
2 *M* HCl
0.1 *M* $AgNO_3$
solid unknown containing CO_3^{2-}, Cl^-, SO_4^{2-}, or I^-

DISCUSSION

One important aspect of chemistry is the identification of substances. The identification of minerals—for example, "fool's gold" as opposed to genuine gold—was and still is of great importance to prospectors. The rapid identification of a toxic substance ingested by an infant may expedite the child's rapid recovery, or in fact it may be the determining factor in saving a life. Substances are identified either by the use of instruments or by reactions characteristic of the substance, or both. Reactions that are characteristic of a substance are frequently referred to as a *test*. For example, one may test for oxygen with a glowing splint; if the splint bursts into flame, oxygen is probably present. One tests for chloride ions by adding silver nitrate to an acidified solution. The formation of a white precipitate suggests the presence of chloride ions. Since other substances may yield a white precipitate under these conditions, one "confirms" the presence of chloride ions by observing that this precipitate dissolves in ammonium hydroxide. The area of chemistry concerned with identification of substances is termed *qualitative analysis*.

In this experiment, you will perform tests on or with substances that you are apt to encounter in everyday life, such as table salt, bleach, smelling salts, and baking soda. You probably don't think of these as "chemicals," and yet they are, even though we don't refer to them around the home by their chemical names (which are sodium chloride, chlorine, ammonium carbonate, and sodium bicarbonate, respectively), but rather by a trade name.

After observing some reactions of these household chemicals, you will partially identify an unknown. Your task will be to determine whether the substance contains the carbonate (CO_3^{2-}), chloride (Cl^-), sulfate (SO_4^{2-}), or iodide (I^-) ion.

(CAUTION: ***Even though household chemicals may appear innocuous, NEVER mix them unless you are absolutely certain you know what you are doing. Innocuous chemicals, when combined, can sometimes produce severe explosions or other hazardous reactions.)***

PROCEDURE

A. Household Ammonia

Obtain 1 mL of household ammonia in a 150-mL beaker. Hold a dry piece of red litmus paper over the beaker, being careful not to touch the sides of the beaker or the solution with the paper. Record your observations on the report sheet (1). Repeat the operation using a piece of red litmus paper that has been moistened with tap water. Do you note any difference in the time required for the litmus to change colors or the intensity of the color change? Record your observations on the report sheet (2).

Ammonium salts are converted to ammonia, NH_3, by the action of strong bases. Hence, one can test for the ammonium ion, NH_4^+, by adding sodium hydroxide, NaOH, and noting the familiar odor of NH_3 or by the use of red litmus. The *net* reaction is as follows:

$$NH_4^+(aq) + OH^-(aq) \rightleftharpoons NH_3(aq) + H_2O(l)$$

Place about 1 mL of 1 *M* NH_4Cl, ammonium chloride, in a test tube and hold a moist piece of red litmus in the mouth of the tube. Record your observations (3). Now add about 1 mL of 8 *M* NaOH and repeat the test. (Do not allow the litmus to touch the sides of the tube, because it may come in contact with NaOH, which will turn it blue.) If the litmus does not change color, gently warm the test tube, but do not boil the solution. Record your observations (4).

You may suspect that ordinary garden fertilizer contains ammonium compounds. Confirm your suspicions by placing some solid fertilizer, an amount about the size of a pea, in a test tube; add 1 mL of 8 *M* NaOH and test as above with moist litmus paper. Does the fertilizer contain ammonium salts (5)?

What is the active ingredient in "smelling salts"? Hold a moist piece of red litmus over the mouth of an open jar of ammonium carbonate, $(NH_4)_2CO_3$. Carefully fan your hand over the jar and see if you can detect a familiar odor. Record your observations on the report sheet (6). Most ammonium salts are stable; for example, the ammonium chloride solution that you tested above should not have had any effect on litmus *before* you added the sodium hydroxide. However, $(NH_4)_2CO_3$ is quite unstable and decomposes to ammonia and carbon dioxide:

$$(NH_4)_2CO_3(s) \xrightarrow{\Delta} 2NH_3(g) + CO_2(g) + H_2O(g)$$

Smelling salts contain ammonium carbonate that has been moistened with ammonium hydroxide.

Dispose of the chemicals used in this part, as well as those used in parts B, C, D, and E in the designated receptacles.

B. Baking Soda, $NaHCO_3$

Substances that contain the carbonate ion, CO_3^{2-}, react with acids to liberate carbon dioxide, CO_2, which is a colorless and odorless gas. Carbon dioxide, when released from baking soda by acids (for example, those present in lemon juice or sour milk), helps to "raise" the cake:

$$NaHCO_3(s) + H^+(aq) \longrightarrow CO_2(g) + H_2O(l) + Na^+(aq)$$

Place in a small, dry test tube an amount of solid baking soda about the size of a small pea. **(CAUTION: *Concentrated H_2SO_4 causes severe burns. Do not get it on your skin. If you come in contact with it, immediately wash the area with copious amounts of water.*)** Then add 1 or 2 drops of 18 M H_2SO_4 and notice what happens. Record your observations on the report sheet (7). Repeat this procedure, but use vinegar in place of the sulfuric acid. Record your observations on the report sheet (8).

A confirmatory test for CO_2 is to allow it to react with $Ba(OH)_2$, barium hydroxide, solution. A white precipitate of $BaCO_3$, barium carbonate, is produced:

$$CO_2(g) + Ba(OH)_2(aq) \longrightarrow BaCO_3(s) + H_2O(l)$$

Many substances, such as eggshells, oyster shells, and limestone, contain the carbonate ion. To determine whether common blackboard chalk contains the carbonate ion, place a small piece of chalk in a dry test tube and then add a few drops of 2 M HCl. Test the escaping gas for CO_2 by carefully holding a drop of $Ba(OH)_2$, suspended from the tip of a medicine dropper or a wire loop, a short distance down into the mouth of the test tube. Clouding of the drop is due to the formation of $BaCO_3$ and proves the presence of carbonate. ***(NOTE: Breathing on the drop will cause it to cloud, since your breath contains CO_2!)*** Record your observations on the report sheet (9).

C. Table Salt, NaCl

Chloride salts react with sulfuric acid to liberate hydrogen chloride, which is a pungent and colorless gas that turns moist blue litmus red:

$$2Cl^-(s) + H_2SO_4(aq) \longrightarrow 2HCl(g) + SO_4^{2-}(aq)$$

This reaction will occur independently of whether the substance is $BaCl_2$, KCl, or $ZnCl_2$; the only requirement is that the salt be a chloride. For KCl, the complete equation is

$$2KCl(s) + H_2SO_4(aq) \longrightarrow 2HCl(g) + K_2SO_4(aq)$$

Another reaction characteristic of the chloride ion is its reaction with silver nitrate to form silver chloride, AgCl, a white, insoluble substance:

$$Cl^-(aq) + AgNO_3(aq) \longrightarrow AgCl(s) + NO_3^-(aq)$$

Place in a small, dry test tube an amount of sodium chloride about the size of a small pea and add 1 or 2 drops of 18 M H_2SO_4. Very carefully note the color and odor of the escaping gas by fanning the gas with your hand toward your nose. DO NOT PLACE YOUR NOSE DIRECTLY OVER THE MOUTH OF THE TEST TUBE. Record your observations on the report sheet (10). Complete the equation $NaCl + H_2SO_4 \longrightarrow$? on the report sheet (11).

Place a small amount (about the size of a pea) of NaCl in a small test tube and add 15 drops of distilled water and one drop of 3 M HNO_3. Then add 3 or 4 drops of 0.1 M $AgNO_3$ and mix the contents. Record your observations on the report sheet (12). Why should you use distilled water for this test (13)? Confirm your answer by testing tap water for chloride ions: Add 1 drop of 3 M HNO_3 to about 2 mL of tap water and then add 3 drops of 0.1 M $AgNO_3$.

Does this test indicate the presence of chloride ions in tap water? Record your answer on the report sheet (14).

Sodium ions impart a yellow color to a flame. When potatoes boil over on the gas stove or the campfire, a burst of yellow flames appears because of the presence of sodium ions. Simply handling a utensil will contaminate it sufficiently with sodium ions from the skin so that when the utensil is placed in a hot flame, a yellow color will appear. Obtain a few crystals of table salt on the tip of a clean spatula and place the tip in the flame of your burner for a brief moment. Record your observations on the report sheet (15).

D. Epsom Salts, $MgSO_4 \cdot 7H_2O$

Epsom salts are used as a purgative, and solutions of this salt are used to soak "tired, aching feet." The following tests are characteristic of the sulfate ion, SO_4^{2-}. Place a small quantity of Epsom salts in a small, dry test tube. Add 1 or 2 drops of 18 M H_2SO_4. **(CAUTION:** ***Concentrated H_2SO_4 causes severe burns. Do not get it on your skin. If you come in contact with it, immediately wash the area with copious amounts of water.*****)** Record your observations on the report sheet (16). Note the difference in the behavior of this substance toward sulfuric acid compared with the behavior of baking soda toward sulfuric acid.

Place some Epsom salts (an amount the size of a small pea) in a small test tube and dissolve it in 1 mL of distilled water. Add 1 drop of 3 M HNO_3 and then 1 or 2 drops of 0.2 M $BaCl_2$. Record your observations on the report sheet (17). Barium sulfate is a white, insoluble substance that forms when barium chloride is added to a solution of any soluble sulfate salt, such as Epsom salts, as follows:

$$SO_4^{2-}(aq) + BaCl_2(aq) \longrightarrow BaSO_4(s) + 2Cl^-(aq)$$

E. Bleach, Cl_2 Water

Commercial bleach is usually a 5% solution of sodium hypochlorite, NaOCl. This solution behaves as though only chlorine, Cl_2, were dissolved in it. Since this solution is fairly concentrated, direct contact with the skin or eyes must be avoided! The element chlorine, Cl_2, behaves very differently from the chloride ion. Chlorine is a pale yellow-green gas with an irritating odor, is slightly soluble in water, and is toxic. It is capable of liberating the element iodine, I_2, from iodide salts:

$$Cl_2(aq) + 2I^-(aq) \longrightarrow I_2(aq) + 2Cl^-(aq)$$

Iodine gives a reddish-brown color to water; it is more soluble in mineral oil than in water, and it imparts a violet color to mineral oil. Thus, chlorine can be used to identify iodide salts.

In a small test tube, dissolve a small amount (about the size of a pea) of sodium iodide, NaI, in 1 mL of distilled water; add 5 drops of bleach. Note the color, then add several drops of mineral oil, shake, and allow to separate, which takes about 20 sec. Note that the mineral oil is the top layer. Record your observations on the report sheet (18).

Another reaction characteristic of iodides is that they form a pale yellow precipitate when treated with silver nitrate solution:

$$I^-(aq) + AgNO_3(aq) \longrightarrow AgI(s) + NO_3^-(aq)$$

TABLE 7.1 Reaction of Solid Salts with H_2SO_4

Ion	Reaction
CO_3^{2-}	Colorless, odorless gas, CO_2, evolved
Cl^-	Colorless, pungent gas, HCl, evolved, which turns blue litmus red
SO_4^{2-}	No observable reaction
I^-	Violet vapors of I_2 formed

Dissolve a small amount of sodium iodide in 1 mL of distilled water and add a drop of 3 *M* HNO_3, then add 3 or 4 drops of 0.1 *M* $AgNO_3$ solution. Record your observations on the report sheet (19).

Solid iodide salts react with concentrated sulfuric acid by instantly turning dark brown, with the slight evolution of a gas that fumes in moist air and with the appearance of violet fumes of iodine. **(CAUTION: *Concentrated H_2SO_4 causes severe burns. Do not get it on your skin. If you come in contact with it, immediately wash the area with copious amounts of water.*)** Place a small amount (about the size of a pea) of sodium iodide in a small, dry test tube and add, IN THE HOOD, 1 or 2 drops of 18 *M* H_2SO_4. Record your observations on the report sheet (20).

F. Unknown

Your solid unknown will contain only one of the following ions: carbonate, chloride, sulfate, or iodide. Table 7.1 summarizes the behavior of these ions toward H_2SO_4.

Place a small amount of your unknown (save some for further tests) in a small, dry test tube and add a drop of 18 *M* H_2SO_4. Record your observations and the formula for the unknown ion on the report sheet (21). What additional test might you perform to help identify the ion? Consult with your instructor *before* doing the test. Record your answer on the report sheet.

REVIEW QUESTIONS

Before beginning this experiment in the laboratory, you should be able to answer the following questions:

1. Why is it unwise to haphazardly mix household chemicals or other chemicals?
2. How could you detect the presence of the NH_4^+ ion?
3. How could you detect the presence of the CO_3^{2-} ion?
4. How could you detect the presence of the Cl^- ion?
5. How could you detect the presence of the SO_4^{2-} ion?
6. How could you detect the presence of the I^- ion?
7. How could you detect the presence of the Ag^+ ion?
8. Complete and balance the following equations:

$$LiCl(s) + H_2SO_4(aq) \longrightarrow$$

$$NH_4^+(aq) + OH^-(aq) \rightleftharpoons$$

$$AgNO_3(aq) + I^-(aq) \longrightarrow$$

$$NaHCO_3(s) + H^+(aq) \longrightarrow$$

9. Why should distilled water be used when making chemical tests?

10. Assume you had a mixture of solid Na_2CO_3 and NaCl. Could you use only H_2SO_4 to determine whether or not Na_2CO_3 was present? Explain.

11. Assume you had a mixture of solid Na_2CO_3 and NaCl. How could you show the presence of both carbonate and chloride in this mixture?

12. How could you show the presence of both iodide and sulfate in a mixture? Consult Appendix C for help.

Name ______________________ Desk ______________

Date ______________ Laboratory Instructor ______________________

Unknown no. or letter ______________

REPORT SHEET | EXPERIMENT

Chemicals in Everyday Life: What Are They and How Do We Know? | 7

A. Household Ammonia

1. Effect of household ammonia on dry litmus Very little effect; it might turn light blue.
2. Effect of household ammonia on moist litmus Turns blue.
3. Effect of NH_4Cl on litmus No effect.
4. Effect of NH_4Cl + NaOH on litmus Turns litmus blue.
5. Fertilizers contain ammonium salts: Yes x No ______
6. Smelling salt NH_3 odor

B. Baking Soda, $NaHCO_3$

7. Baking soda + H_2SO_4 A colorless, odorless gas is produced.
8. Baking soda + vinegar A colorless, odorless gas is produced.
9. Chalk contains carbonate ion: Yes x No ______

C. Table Salt, NaCl

10. Effect of H_2SO_4 on table salt A pungent gas (HCl) is formed that turns moist blue litmus red.
11. $H_2SO_4(l) + 2NaCl(s) \longrightarrow$ $Na_2SO_4(s) + 2HCl(g)$
12. Effect of $AgNO_3$ on table salt A white precipitate deposits.
13. Why use distilled water? Because tap water contains Cl^-.
14. Chloride ions in tap water: Yes x No ______
15. Salt in flame A brilliant yellow flame results.

D. Epsom Salts, $MgSO_4 \cdot 7H_2O$

16. Effect of H_2SO_4 on Epsom salts: No reaction, but heat is generated.
17. $BaCl_2$ + Epsom salts: A white precipitate of $BaSO_4$ results.

E. Bleach, Cl_2 Water

18. Bleach + NaI: A brown color results in the water, which becomes purple in mineral oil.
19. Silver nitrate + NaI: A yellow precipitate results.
20. Effect of H_2SO_4 on NaI: The NaI turns brown; HI and I_2 are liberated as a choking purple vapor and rotten egg odor.

F. Unknown

21. Unknown ion: I^-
22. Confirmatory test: Depends upon the unknown (for I^-, react an aqueous solution with Cl_2 water and add mineral oil; a purple layer confirms I^-).

QUESTIONS

1. How could you distinguish sodium chloride (table salt) from sodium iodide (a poison)? Show reactions.

   Use $AgNO_3$ solution: $Ag^+ + Cl^- \longrightarrow AgCl$ (white), turns purple in light; $Ag^+ + I^- \longrightarrow AgI$ (yellow). Use H_2SO_4 on solid: $H_2SO_4 + 2NaCl \longrightarrow 2HCl(g) + Na_2SO_4$ (HCl is a pungent gas). $NaI + H_2SO_4 \longrightarrow NaHSO_4 + HI$; $H_2SO_4 + 8HI \longrightarrow H_2S + 4H_2O + 4I_2$ (solid darkens; purple vapors appear).

2. How could you distinguish solid barium chloride from solid barium sulfate?

   Add concentrated H_2SO_4. No observable reaction occurs with the barium sulfate. Barium chloride will liberate a pungent gas that turns blue litmus red. Also, $BaCl_2$ is soluble in water, whereas $BaSO_4$ is not.

3. Do you think that washing soda, Na_2CO_3, could be used for the same purpose as baking soda, $NaHCO_3$? Would Na_2CO_3 react with HCl? Write the chemical equation. Write the chemical equation for the reaction of $NaHCO_3$ with HCl.

   Yes; yes.

   $2HCl + Na_2CO_3 \longrightarrow 2NaCl + H_2O + CO_2$

   $HCl + NaHCO_3 \longrightarrow NaCl + H_2O + CO_2$

4. Sodium benzoate is a food preservative. What are its formula and its solubility in water? (Consult a handbook.)

   $NaC_7H_5O_2$

   Solubility: 66 g/100 mL at 20°C.

5. Citric acid is often found in soft drinks. What is its melting point? (Consult a handbook.)

   m.p. = 153°C

6. p-Phenylenediamine (also named 1,4-diaminobenzene) dyes hair black. Is this substance a liquid or solid at room temperature? (Consult a handbook.)

   It is a solid, because its melting point is 140°C.

7. Household vinegar is a 5% solution of acetic acid. Consult your textbook or Appendix E and give the formula for acetic acid.

   CH_3CO_2H or $HC_2H_3O_2$

8. Consult the label on a can of Drano (not liquid) and write the chemical formula for the contents. (There are also small pieces of aluminum inside the can.)

   NaOH

NOTES AND CALCULATIONS

Gravimetric Analysis of a Chloride Salt

OBJECTIVE

To illustrate typical techniques used in gravimetric analysis by quantitatively determining the amount of chloride in an unknown.

APPARATUS AND CHEMICALS

Apparatus

balance
250-mL beakers (6)
Bunsen burner
funnels (3)
funnel support
plastic wash bottle
graduated cylinders (2), 10 and 100 mL
ring stand, ring, and wire gauze
stirring rods (3)
rubber policeman (3)
sharkskin filter paper (3)
watch glasses (3)
weighing paper

Chemicals

unknown chloride sample
0.5 *M* $AgNO_3$
6 *M* HNO_3
acetone
distilled water

DISCUSSION

Quantitative analysis is that aspect of analytical chemistry concerned with determining *how much* of one or more constituent is present in a particular sample of material. Information such as percentage composition is essential to establishing formulas for compounds. Two common quantitative methods used in analytical chemistry are gravimetric and volumetric analysis. *Gravimetric analysis* derives its name from the fact that the constituent being determined can be isolated in some weighable form. *Volumetric analysis,* on the other hand, derives its name from the fact that the method used to determine the amount of a constituent involves measuring the volume of a reagent. Usually, gravimetric analyses involve the following steps:

1. Drying and then accurately weighing representative samples of the material to be analyzed.
2. Dissolving the samples.
3. Precipitating the constituent in the form of a substance of known composition by adding a suitable reagent.
4. Isolating the precipitate by filtration.
5. Washing the precipitate to free it of contaminants.
6. Drying the precipitate to a constant mass (to obtain an analytically weighable form of known composition).
7. Calculating the percentage of the desired constituent from the masses of the sample and precipitate.

Although the techniques of gravimetric analysis are applicable to a large variety of substances, we have chosen to illustrate them with an analysis that incorporates a number of other techniques as well. Chloride ion may be quantitatively precipitated from solution by the addition of silver ion according to the following ionic equation:

$$Ag^+(aq) + Cl^-(aq) \longrightarrow AgCl(s) \qquad [1]$$

Silver chloride is quite insoluble (only about 0.0001 g of AgCl dissolves in 100 mL of H_2O at 20°C); hence, the addition of silver nitrate solution to an aqueous solution containing chloride ion precipitates AgCl quantitatively. The precipitate can be collected on a filter paper, dried, and weighed. From the mass of the AgCl obtained, the amount of chloride in the original sample can then be calculated.

This experiment also illustrates the concept of stoichiometry. *Stoichiometry* is the determination of the proportions in which chemical elements combine and the mass relations in any chemical reaction. In this experiment stoichiometry means specifically the mole ratio of the substances entering into and resulting from the combination of Ag^+ and Cl^-. In the reaction of Ag^+ and Cl^- in Equation [1], it can be seen that 1 mol of chloride ions reacts with 1 mol of silver ions to produce 1 mol of silver chloride. Thus,

$$\text{moles } Cl^- = \text{moles AgCl} = \frac{\text{grams AgCl}}{\text{molar mass of AgCl}}$$

$$\begin{aligned}\text{grams Cl in sample} &= (\text{moles } Cl^-)(\text{gram–atomic weight Cl}) \\ &= \frac{(\text{gram–atomic weight Cl})(\text{grams AgCl})}{\text{molar mass of AgCl}} \\ &= \frac{(35.45 \text{ g Cl})(\text{grams AgCl})}{143.32 \text{ g AgCl}} \\ &= (0.2473 \text{ g Cl/g AgCl})(\text{grams AgCl})\end{aligned}$$

The number 0.2473 is called a gravimetric factor. It converts grams of AgCl into grams of Cl. Gravimetric factors are used repeatedly in analytical chemistry and are tabulated in handbooks. The percentage of Cl in the sample can be calculated according to the following formula:

$$\%\text{ Cl in sample} = \frac{(\text{grams Cl in sample})(100)}{\text{gram–weight sample}}$$

EXAMPLE 8.1

In a gravimetric chloride analysis it was found that 0.2516 g AgCl was obtained from an unknown that had a mass of 0.1567 g. What is the percent of chloride in this sample?

SOLUTION: The mass of Cl in the sample is:

$$\begin{aligned}\text{g Cl} &= (0.2473 \text{ g Cl/g AgCl}) \times (0.2516 \text{ g AgCl}) \\ &= 0.06222 \text{ g Cl}\end{aligned}$$

$$\begin{aligned}\%\text{ Cl} &= \frac{(0.06222 \text{ g Cl})(100)}{(0.1567 \text{ g sample})} \\ &= 39.72\%\end{aligned}$$

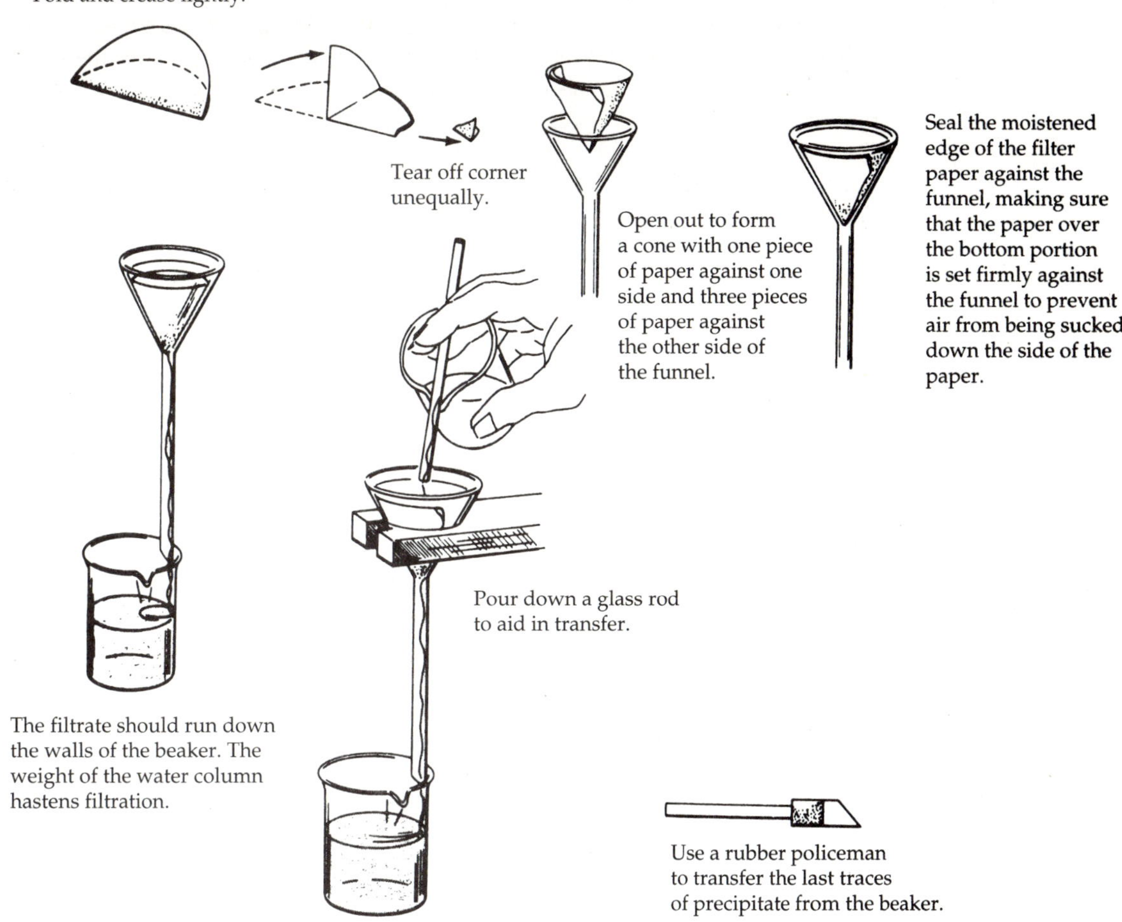

▲ **FIGURE 8.1** Filter paper use.

PROCEDURE

On a piece of weighing paper, weigh to the nearest 0.0001 g about 0.2 to 0.4 g of your unknown sample. Transfer the sample quantitatively to a clean 250-mL beaker (do not weigh the beaker) and label the beaker #1 with a pencil. Record the sample weight. Add 150 mL of distilled water and 1 mL of 6 *M* HNO_3 to the beaker. Repeat with sample numbers 2 and 3, and label the beakers #2 and #3, respectively. Using a different glass rod for each solution, stir until all of the sample has dissolved. Leave the stirring rods in the beakers. Do not place them on the desktop.

Instructor demonstration

While stirring one of the solutions, add to it about 20 mL of 0.5 *M* $AgNO_3$ solution. Place a watch glass over the beaker. Warm the solution gently with your Bunsen burner and keep it warm for 5 to 10 min. Do not boil the solution.

Obtain a filter paper (three of these will be needed) and weigh it accurately. (Be certain that you weigh the paper after it has been folded and torn, not before.) Fold the paper as illustrated in Figure 8.1 and fit it into a glass funnel. Be certain that you open the filter paper in the funnel so that one side has three pieces and one side has one piece of paper against the funnel—not two pieces on each side. *Why?* Your instructor will also demonstrate this for you. Wet the paper with distilled water to hold it in place in the funnel. Completely and quantitatively transfer the precipitate and all the warm solution from the beaker onto the filter, using a rubber policeman (your laboratory instructor will show you how to use a rubber policeman) and a wash bottle to wash out the last traces of precipitate. The level of solution in the filter funnel should always be *below* the top edge of the filter paper. Wash the precipitate on the filter paper with two or three 5-mL portions of water from the wash bottle. Finally, pour three 5-mL portions of acetone through the filter. **(CAUTION: *Acetone is highly flammable! Keep it away from open flames.*)** Remove the filter paper, place it on a numbered watch glass, and store it in your locker until the next period.

Instructor: If time is a problem,students may pool their data.

Repeat the above processes with your other two samples, being sure that you have numbered your watch glasses so that you can identify the samples. The precipitated AgCl must be kept out of bright light, because it is photosensitive and slowly decomposes in the presence of light as follows:

$$2AgCl(s) \xrightarrow{h\nu} 2Ag(s) + Cl_2(g)$$

In this equation $h\nu$ is a symbol for electromagnetic radiation; here it represents radiation in the visible and ultraviolet regions of the spectrum. This is the reaction used by Corning to make photosensitive sunglasses. *In the next period,* when the AgCl is thoroughly dry, weigh the filter papers plus AgCl and calculate the mass of AgCl. From these data calculate the percentage of chloride in your original sample.

Standard Deviation

As a means of estimating the precision of your results, it is desirable to calculate the standard deviation. Before we illustrate how to do this, however, we will define some of the terms above as well as some additional ones that are necessary.

Accuracy: correctness of a measurement, closeness to the true result.

Precision: internal consistency among one's own results—that is, reproducibility.

Error: difference between the true result and the determined result.

Determinate errors: errors in method of performance that can be discovered and eliminated.

Indeterminate errors: random errors, which are incapable of discovery but which can be treated by statistics.

Mean: arithmetic mean or average (μ), where

$$\mu = \frac{\text{sum of results}}{\text{number of results}}$$

For example, if an experiment's results are 1, 3, and 5, then

$$\mu = \frac{1 + 3 + 5}{3} = 3$$

Median: the midpoint of the results for an odd number of results and the average of the two middle results for an even number of results (m). For example, if an experiment's results are 1, 3, and 5, then $m = 3$. If results are 1.0, 3.0, 4.0, and 5.0, then

$$m = \frac{3.0 + 4.0}{2} = 3.5$$

The scatter about the mean or median—that is, the deviations from the mean or median—are measures of precision. Thus the smaller the deviations, the more precise the measurements.

EXAMPLE 8.2

If an experiment's results are 1.0, 2.0, 3.0, and 4.0, calculate the mean, the deviations from the mean, the average deviation from the mean, and the relative average deviation from the mean.

SOLUTION: The mean is calculated as follows:

$$\mu = \frac{1.0 + 2.0 + 3.0 + 4.0}{4} = \frac{10.0}{4} = 2.5$$

The deviations from the mean are

$$|2.5 - 1.0| = 1.5$$

$$|2.5 - 2.0| = 0.5$$

$$|2.5 - 3.0| = 0.5$$

$$|2.5 - 4.0| = 1.5$$

The symbol | | means absolute value, so all differences are positive. The average deviation from the mean is therefore

$$\frac{1.5 + 0.5 + 0.5 + 1.5}{4} = 1.0$$

The relative average deviation from the mean is calculated by dividing the average deviation from the mean by the mean. Thus

$$\text{Relative deviation} = \frac{1.0}{2.5} = 0.40$$

This can be expressed as 40%, 400 parts/thousand (ppt), or 40,000 parts/million (ppm).

EXAMPLE 8.3

If an experiment's results are 1.0, 1.5, 2.0, and 2.5, calculate the mean, the deviations from the mean, the average deviation from the mean, and the relative average deviation from the mean.

SOLUTION: The mean is calculated as follows:

$$\mu = \frac{1.0 + 1.5 + 2.0 + 2.5}{4}$$

$$= \frac{7.0}{4} = 1.75\text{, or 1.8 to two significant figures}$$

The deviations from the mean are

$$|1.8 - 1.0| = 0.8$$

$$|1.8 - 1.5| = 0.3$$

$$|1.8 - 2.0| = 0.2$$

$$|1.8 - 2.5| = 0.7$$

The average deviation from the mean is therefore

$$\frac{0.8 + 0.3 + 0.2 + 0.7}{4} = 0.5$$

The relative average deviation from the mean is

$$\frac{0.5}{1.8} = 0.3$$

$$= 30\%, \text{ or } 300 \text{ ppt, or } 30{,}000 \text{ ppm}$$

Obviously, the data in Example 8.3 are internally more consistent than the data in Example 8.2 and hence are more precise, since the deviations are smaller. Thus, the average deviation and relative average deviation measure precision.

Standard deviation *(s)* is a better measure of precision and is calculated using the formula

$$s = \sqrt{\frac{\text{sum of the squares of the deviations from the mean}}{\text{number of observations} - 1}}$$

$$= \sqrt{\frac{\Sigma_i |\chi_i - \mu|^2}{N - 1}}$$

where s = standard deviation from the mean, χ_i = members of the set, μ = mean, and N = number of members in the set of data. The symbol Σ_i means to sum over the members.

Example 8.4

An experiment's results are 1, 3, and 5. Calculate the mean, the deviations from the mean, the standard deviation, and the relative standard deviation for the data.

SOLUTION: The mean is as follows:

$$\mu = \frac{1 + 3 + 5}{3} = 3$$

The deviations from the mean are

$$|\chi_i - \mu| = \text{deviation}$$

$$|1 - 3| = 2$$

$$|3 - 3| = 0$$

$$|5 - 3| = 2$$

$$s = \sqrt{\frac{2^2 + 0^2 + 2^2}{3 - 1}}$$

$$= \sqrt{\frac{4 + 0 + 4}{2}}$$

$$= \sqrt{\frac{8}{2}}$$

$$= \sqrt{4} = 2$$

The results of this experiment would be reported as 3 ± 2. The relative standard deviation is

$$\frac{2}{3} = 0.7, \text{ or } 70\%$$

EXAMPLE 8.5

The results of an experiment are 2.100, 2.110, and 2.105. Calculate the mean, the deviations from the mean, the standard deviation, and the relative standard deviation.

SOLUTION: The mean is as follows.

$$\mu = \frac{2.100 + 2.110 + 2.105}{3} = 2.105$$

The deviations from the mean are:

$$|2.105 - 2.100| = 0.005$$
$$|2.105 - 2.110| = 0.005$$
$$|2.105 - 2.105| = 0.000$$

The standard deviation is therefore:

$$s = \sqrt{\frac{(0.005)^2 + (0.005)^2 + (0.000)^2}{2}}$$
$$= \sqrt{\frac{5 \times 10^{-5}}{2}}$$
$$= 0.005$$

The results would be reported as 2.105 ± 0.005. The relative standard deviation is

$$\frac{0.005}{2.105} = 0.002 \text{ or } 0.2\%$$

Obviously, the data in Example 8.5 are more precise, though not necessarily more accurate, than the data in Example 8.4, because both the deviations and the standard deviation are smaller in Example 8.5.

Calculate the standard deviation of your data and report the results on your report sheet.

The standard deviation may be used to determine whether a result should be retained or discarded. As a rule of thumb, you should discard any result that is more than two standard deviations from the mean. For example, if you had a result of 49.65% and you had determined that your percentage of chloride was 49.25 ± 0.09%, this result (49.65%) should be discarded. This is because $s = 0.09$ and $|49.25 - 49.65| = 0.40$, which is greater than 2×0.09. This result is more than two standard deviations from the mean.

REVIEW QUESTIONS

Before beginning this experiment in the laboratory, you should be able to answer the following questions:

1. What is the fundamental difference between gravimetric and volumetric analysis?
2. What does *stoichiometry* mean?

3. Why should silver chloride be protected from light? Will your result be high or low if you don't protect your silver chloride from light?
4. Can you eliminate indeterminate errors from your experiment?
5. Does standard deviation give a measure of accuracy or precision?
6. Why don't you open your folded filter paper so that two pieces touch each side of the funnel?
7. If your silver chloride undergoes extensive photodecomposition before you weigh it, will your results be high or low?
8. If an experiment's results are 10.1, 10.4, and 10.6, find the mean, the average deviation from the mean, the standard deviation from the mean, and the relative deviation from the mean.
9. What is meant by the term *gravimetric factor*?

Name ______________________ Desk ______________________

Date ______________ Laboratory Instructor ______________________

Unknown no. or letter ______________________

REPORT SHEET | EXPERIMENT 8

Gravimetric Analysis of a Chloride Salt

	Trial 1	*Trial 2*	*Trial 3*
Mass of sample	0.3126 g	0.4095 g	0.2968 g
Mass of filter paper + AgCl	1.0825	1.3016	1.0492
Mass of filter paper	0.4056	0.4101	0.4059
Mass of AgCl	0.6769	0.8915	0.6433

Mass of Cl in original sample (show calculations)	0.1674	0.2205	0.1591

$$\text{wt. fraction Cl in AgCl} = \frac{35.45}{143.32} = 0.2473$$

$$\text{Mass Cl} = (0.6769)(0.2473) = 0.1674\text{ g}$$

Percent chloride in original sample (show calculations)	53.55%	53.85%	53.61%

$$\left(\frac{0.1674\text{ g}}{0.3126\text{ g}}\right)(100) = 53.55\%$$

Average percent chloride (show calculations) 53.67%

$$\frac{53.55 + 53.85 + 53.61}{3} = 53.67$$

Standard deviation (show calculations) 0.16

$$SD = \sqrt{\frac{(0.12)^2 + (0.18)^2 + (0.06)^2}{2}} = 0.16$$

Relative standard deviation (show calculations) 0.30%

$$\left(\frac{0.16}{53.67}\right)(100) = 0.30\%$$

Do any of your results differ from the mean by more than two standard deviations? No

Reported percent chloride 53.67 ± 0.30 %

QUESTIONS

1. The following percentages of chloride were found: 32.52%, 32.14%, 32.61%, and 32.75%.
   (a) Find the mean, the standard deviation, and the relative standard deviation.
   (b) Can any result be discarded?

$$\overline{x} = \frac{32.52 + 32.14 + 32.61 + 32.72}{4} = 32.50 \qquad \text{SD} = 0.252$$

Relative standard deviation $= \left(\frac{0.252}{32.50}\right)(100) = 0.775\%$; no result differs from the mean by more than 2 SD.

Retain all results.

2. Barium can be analyzed by precipitating it as $BaSO_4$ and weighing the precipitate. When a 0.369-g sample of a barium compound was treated with excess H_2SO_4, 0.153 g of $BaSO_4$ formed. What is the percentage of barium in the compound?

$$(0.153\text{ g BaSO}_4)\left(\frac{137.3\text{ g Ba}}{233.4\text{ g BaSO}_4}\right) = 0.0900\text{ g Ba}$$

$$\%\text{Ba} = \left(\frac{0.0900\text{ g Ba}}{0.369\text{ g}}\right)(100) = 24.4\%$$

3. What is the percentage of sodium in pure table salt?

$$\%\text{Na} = \left(\frac{22.9898}{58.4428}\right)(100) = 39.3373\%$$

4. How many milligrams of sodium are contained in 2.00 g of NaCl?

$$(2.00\text{ g})(0.393373)(1000\text{ mg/g}) = 787\text{ mg Na}$$

5. An impure sample of table salt that weighed 0.4652 g, when dissolved in water and treated with excess $AgNO_3$, formed 1.044 g of AgCl. What is the percentage of NaCl in the impure sample?

$$(1.044\text{ g AgCl})\left(\frac{1\text{ mol AgCl}}{143.3\text{ g AgCl}}\right)\left(\frac{1\text{ mol NaCl}}{1\text{ mol AgCl}}\right)\left(\frac{58.45\text{ g NaCl}}{1\text{ mol NaCl}}\right) = 0.4258\text{ g NaCl}$$

$$\left(\frac{0.4258\text{ g}}{0.4652\text{ g}}\right)(100) = 91.53\%$$

6. List at least three sources of error in this experiment.

Incomplete precipitation, poor transfer, too much washing of the precipitate (loss of precipitate), incomplete drying of the precipitate.

Gravimetric Determination of Phosphorus in Plant Food

Experiment

OBJECTIVE

To illustrate an application of gravimetric analysis to a consumer product.

APPARATUS AND CHEMICALS

Apparatus

balance
beakers (6), any combination, 250 mL or larger
filter paper (Whatman No. 40)
funnels (3)
funnel support
ring stand
stirring rods (3) with rubber policeman

Chemicals

75% aqueous isopropyl alcohol*
10% aqueous $MgSO_4 \cdot 7H_2O$*
2 *M* $NH_3(aq)$*

DISCUSSION

Gravimetric analyses may be difficult and time consuming, but they are inherently quite accurate. The accuracy of an analysis is often directly proportional to the time expended in carrying it out. The ultimate use of the analytical result governs how much time and effort the analytical chemist should expend in obtaining it. For example, before building a mill to process gold ore, an accurate analysis of the ore is required. Mills are very expensive to build and operate, and economic factors determine whether or not construction of the mill is worthwhile. Because of the value of gold, the difference of only a few hundredths of a percent of gold in an ore may be the governing factor as to whether or not to construct a mill. On the other hand, the analysis of an inexpensive commodity chemical, such as a plant food, requires much less accuracy; the economic consequences of giving the consumer an extra 0.2% of an active ingredient are usually small even for a large volume of product. Time is too valuable, whether it be the students' or scientists', to be wasted in the pursuit of the ultimate in accuracy when such is not needed.

Consumer chemicals are subject to quality control by the manufacturer and by various consumer protection agencies. Consumer chemicals are usually analyzed both *qualitatively* to determine what substances they contain and *quantitatively* to determine how much of these substances are present. For example, plant foods are analyzed this way.

Plant foods contain three essential nutrients that are likely to be lacking in soils. These are soluble compounds of nitrogen, phosphorus, and potassium. The labels on the plant food usually have a set of numbers such as 15-30-15. These numbers mean that the plant food is guaranteed to contain at least

*Epsom salts, household ammonia (non-sudsy), and rubbing alcohol, respectively, may be substituted for these reagents.

15% nitrogen, 30% phosphorus (expressed as P_2O_5), and 15% potassium (expressed as K_2O). The rest of the product is other anions or cations necessary to balance charge in the chemical compounds, dyes to provide a pleasing color, and fillers.

EXAMPLE 9.1

What is the minimum percentage of phosphorus in a plant food whose P_2O_5 percentage is guaranteed to be 15%?

SOLUTION: Assuming 100 g of plant food, we would have 15 g of P_2O_5. Using this quantity, we can calculate the amount of P in the sample:

$$(15\text{ g P}_2\text{O}_5)\left(\frac{1\text{ mol P}_2\text{O}_5}{141.9\text{ g P}_2\text{O}_5}\right)\left(\frac{2\text{ mol P}}{1\text{ mol P}_2\text{O}_5}\right)\left(\frac{30.97\text{ g P}}{1\text{ mol P}}\right) = 6.5\text{ g P}$$

Thus,

$$\%\text{P} = \frac{\text{g P}}{\text{g sample}} \times 100 = \frac{6.5\text{ g}}{100\text{ g}} \times 100 = 6.5\%$$

In this experiment we will illustrate one of the quality-control analyses for plant food by gravimetric determination of its phosphorus content. Phosphorus will be determined by precipitation of the sparingly soluble salt magnesium ammonium phosphate hexahydrate according to the reaction

$$5\text{H}_2\text{O}(l) + \text{HPO}_4^{2-}(aq) + \text{NH}_4^+(aq) + \text{Mg}^{2+}(aq) + \text{OH}^-(aq) \rightleftharpoons \text{MgNH}_4\text{PO}_4 \cdot 6\text{H}_2\text{O}(s)$$

EXAMPLE 9.2

If a 10.00-g sample of soluble plant food yields 10.22 g of $MgNH_4PO_4 \cdot 6H_2O$, what are the percentages of P and P_2O_5 in this sample?

SOLUTION: First, solving for the grams of P in the sample

$$(10.22\text{ g MgNH}_4\text{PO}_4 \cdot 6\text{H}_2\text{O})\left(\frac{1\text{ mol MgNH}_4\text{PO}_4 \cdot 6\text{H}_2\text{O}}{245.4\text{ g MgNH}_4\text{PO}_4 \cdot 6\text{H}_2\text{O}}\right)$$

$$\times \left(\frac{1\text{ mol P}}{1\text{ mol MgNH}_4\text{PO}_4 \cdot 6\text{H}_2\text{O}}\right)\left(\frac{30.97\text{ g P}}{1\text{ mol P}}\right) = 1.290\text{ g P}$$

Thus,

$$\%\text{P} = \frac{\text{g P}}{\text{g sample}} \times 100 = \frac{1.290\text{ g P}}{10.00\text{ g sample}} \times 100 = 12.90\%\text{ P}$$

Similarly, solving for grams of P_2O_5:

$$(10.22\text{ g MgNH}_4\text{PO}_4 \cdot 6\text{H}_2\text{O})\left(\frac{1\text{ mol MgNH}_4\text{PO}_4 \cdot 6\text{H}_2\text{O}}{245.4\text{ g MgNH}_4\text{PO}_4 \cdot 6\text{H}_2\text{O}}\right)$$

$$\times \left(\frac{1\text{ mol P}}{1\text{ mol MgNH}_4\text{PO}_4 \cdot 6\text{H}_2\text{O}}\right)\left(\frac{1\text{ mol P}_2\text{O}_5}{2\text{ mol P}}\right)\left(\frac{141.9\text{ g P}_2\text{O}_5}{1\text{ mol P}_2\text{O}_5}\right) = 2.955\text{ g P}_2\text{O}_5$$

$$\text{and } \%\text{P}_2\text{O}_5 = \frac{2.955\text{ g}}{10.00\text{ g}} \times 100 = 29.55\%\text{ P}_2\text{O}_5$$

PROCEDURE

Weigh by difference to the nearest hundredth gram 3.0 to 3.5 g of your unknown sample, using weighing paper. Transfer the sample quantitatively to a 250-mL beaker and record the sample mass. Add 35 to 40 mL of distilled

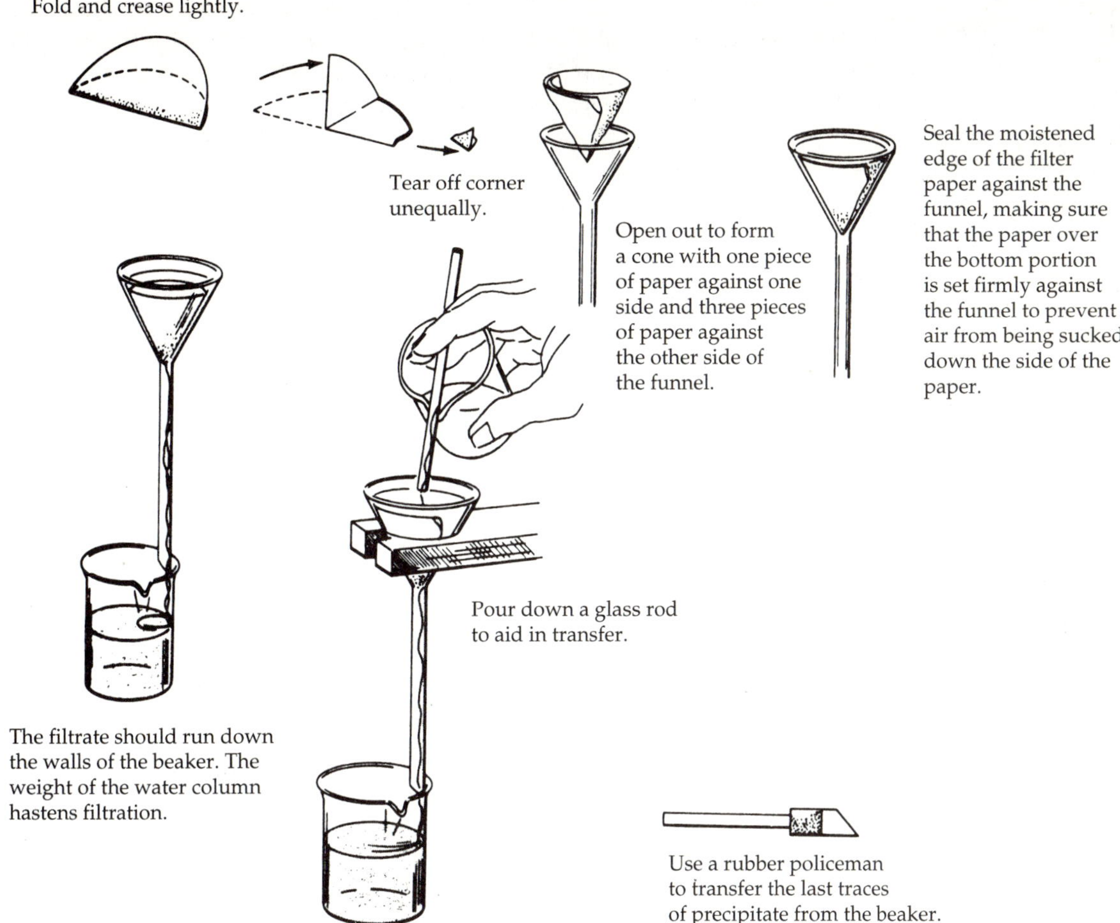

▲ **FIGURE 9.1** Filter paper use. (Filtration of $MgNH_4PO_4 \cdot 6H_2O$ is slow. Time may be saved by filtering by suction with a Büchner funnel.)

water and stir the mixture with a glass stirring rod to dissolve the sample. Although plant foods are all advertised to be water soluble, they may contain a small amount of insoluble residue. If your sample does not completely dissolve, remove the insoluble material by filtration. To the filtrate add about 45 mL of a 10% $MgSO_4 \cdot 7H_2O$ solution. Then add approximately 150 mL of 2 *M* NH_3*(aq)* slowly while stirring. A white precipitate of $MgNH_4PO_4 \cdot 6H_2O$ will form. Allow the mixture to sit at room temperature for 15 minutes to complete the precipitation. Collect the precipitate on a preweighed piece of filter paper (Figure 9.1).

Instructor demonstration

Obtain a filter paper (three of these will be needed) and weigh it accurately. (Be certain that you weigh the paper after it has been folded and torn, not before.) Fold the paper as illustrated in Figure 9.1 and fit it into a glass funnel. Be certain that you open the filter paper in the funnel so that one side has three pieces and one side has one piece of paper against the funnel—not two pieces on each side. *Why?* Your instructor will also demonstrate this for you. Wet the paper with distilled water to hold it in place in the funnel. Completely and quantitatively transfer the precipitate and all of the solution from the beaker onto the filter, using a rubber policeman (your laboratory instructor will show you how to use a rubber policeman). Wash the precipitate with two or three 5-mL portions of distilled water. Do this by adding each portion to the beaker in which you did the precipitation to transfer any remaining precipitate; then pour over the solid on the funnel. Finally, pour two 10 mL portions of 75% isopropyl alcohol through the filter paper. Remove the filter paper, place it on a numbered watch glass, and store it to dry in your locker until the next period.

Repeat the above procedure with two more samples. In the next period, when the $MgNH_4PO_4 \cdot 6H_2O$ is thoroughly dry, weigh the filter papers plus $MgNH_4PO_4 \cdot 6H_2O$. Record the mass and calculate the percentages of phosphorus and P_2O_5 in your original samples.

Standard Deviation

As a means of estimating the precision of your results, it is desirable to calculate the standard deviation. Before we illustrate how to do this, however, we will define some of the terms above as well as some additional ones that are necessary.

Accuracy: correctness of a measurement, closeness to the true result.
Precision: internal consistency among one's own results—that is, reproducibility.
Error: difference between the true result and the determined result.
Determinate errors: errors in method of performance that can be discovered and eliminated.
Indeterminate errors: random errors that are incapable of discovery but which can be treated by statistics.
Mean: arithmetic mean or average (μ), where

$$\mu = \frac{\text{sum of results}}{\text{number of results}}$$

For example, if an experiment's results are 1, 3, and 5, then

$$\mu = \frac{1 + 3 + 5}{3} = 3$$

Median: the midpoint of the results for an odd number of results and the average of the two middle results for an even number of results (m). For example, if an experiment's results are 1, 3, and 5, then $m = 3$. If results are 1.0, 3.0, 4.0, and 5.0, then

$$m = \frac{3.0 + 4.0}{2} = 3.5$$

The scatter about the mean or median—that is, the deviations from the mean or median—are measures of precision. Thus the smaller the deviations, the more precise the measurements.

Example 9.3

If an experiment's results are 1.0, 2.0, 3.0, and 4.0, calculate the mean, the deviations from the mean, the average deviation from the mean, and the relative average deviation from the mean.

SOLUTION: The mean is calculated as follows:

$$\mu = \frac{1.0 + 2.0 + 3.0 + 4.0}{4} = \frac{10.0}{4} = 2.5$$

The deviations from the mean are

$$|2.5 - 1.0| = 1.5$$
$$|2.5 - 2.0| = 0.5$$
$$|2.5 - 3.0| = 0.5$$
$$|2.5 - 4.0| = 1.5$$

The symbol | | means absolute value, so all differences are positive. The average deviation from the mean is therefore

$$\frac{1.5 + 0.5 + 0.5 + 1.5}{4} = 1.0$$

The relative average deviation from the mean is calculated by dividing the average deviation from the mean by the mean. Thus,

$$\text{Relative deviation} = \frac{1.0}{2.5} = 0.40$$

This can be expressed as 40%, 400 parts/thousand (ppt), or 40,000 parts/million (ppm).

Standard deviation *(s)* is a better measure of precision and is calculated using the formula

$$s = \sqrt{\frac{\text{sum of the squares of the deviations from the mean}}{\text{number of observations} - 1}}$$

$$= \sqrt{\frac{\Sigma_i |\chi_i - \mu|^2}{N - 1}}$$

where s = standard deviation from the mean, χ_i = members of the set, μ = mean, and N = number of members in the set of data. The symbol Σ_i means to sum over the members.

Example 9.4

An experiment's results are 1, 3, and 5. Calculate the mean, the deviations from the mean, the standard deviation, and the relative standard deviation for the data.

SOLUTION: The mean is as follows:

$$\mu = \frac{1 + 3 + 5}{3} = 3$$

The deviations from the mean are

$$|\chi_i - \mu| = \text{deviation}$$
$$|1 - 3| = 2$$
$$|3 - 3| = 0$$
$$|5 - 3| = 2$$

$$s = \sqrt{\frac{2^2 + 0^2 + 2^2}{3 - 1}}$$

$$= \sqrt{\frac{4 + 0 + 4}{2}}$$

$$= \sqrt{\frac{8}{2}}$$

$$= \sqrt{4} = 2$$

The results of this experiment would be reported as 3 ± 2. The relative standard deviation is

$$\frac{2}{3} = 0.7, \text{ or } 70\%$$

Calculate the standard deviation of your data and report the results on your report sheet.

The standard deviation may be used to determine whether a result should be retained or discarded. As a rule of thumb, you should discard any result that is more than two standard deviations from the mean. For example, if you had a result of 49.65% and you had determined that your percentage of phosphorus was 49.25 ± 0.09%, this result (49.65%) should be discarded. This is because $s = 0.09$ and $|49.25 - 49.65| = 0.40$, which is greater than 2×0.09. This result is more than two standard deviations from the mean.

REVIEW QUESTIONS

Before beginning this experiment in the laboratory, you should be able to answer the following questions:

1. Why would only three significant figures be required for the analysis of a consumer chemical such as P_2O_5 in plant food?
2. Epsom salts, household ammonia, and rubbing alcohol are also consumer chemicals that could be used in this experiment. What are the chemical formulas for these substances?
3. The label on a plant food reads 23-19-17. What does this mean?
4. What is the minimum percentage of phosphorus in the plant food in question 3?
5. What is the minimum percentage of potassium in the plant food in question 3?
6. What is the formula weight of $CuSO_4 \cdot 5H_2O$?
7. Does standard deviation give a measure of accuracy or precision?
8. If an experiment's results are 12.1, 12.4 and 12.6, find the mean, the average deviation from the mean, the standard deviation from the mean, and the relative average deviation from the mean.
9. What is meant by the term *indeterminate errors*?
10. Differentiate between qualitative and quantitative analysis.
11. What is meant by the term *accuracy*?

Name ______________________ Desk ______________

Date ______________________ Laboratory Instructor ______________

Unknown no. ______________

REPORT SHEET | EXPERIMENT 9

Gravimetric Determination of Phosphorus in Plant Food

Plant Food Name RAP-ID-GROW Label Percentage 23-19-17

	Trial 1	Trial 2	Trial 3
Mass of sample	5.03 g	5.20 g	5.37 g
Mass of filter paper and $MgNH_4PO_4 \cdot 6H_2O$	3.71 g	3.82 g	3.92 g
Mass of filter paper	0.40 g	0.41 g	0.44 g
Mass of $MgNH_4PO_4 \cdot 6H_2O$	3.31 g	3.41 g	3.48 g
Mass of P in original	0.418 g	0.430 g	0.439 g

Sample (show calculations)

(1) $\text{Mass P} = \dfrac{(30.97)(3.31\text{ g})}{245.4} = 0.418\text{ g}$ (2) $\dfrac{(30.97)(3.41\text{ g})}{245.4} = 0.430\text{ g}$

(3) $\dfrac{(30.97)(3.48\text{ g})}{245.4} = 0.439\text{ g}$

	Trial 1	Trial 2	Trial 3
Percent phosphorus (show calculations)	8.31%	8.27%	8.18%

(1) $\%P = \left(\dfrac{0.418\text{ g P}}{5.03\text{ g sample}}\right)(100) = 8.31\%$ (2) $\left(\dfrac{0.430}{5.20}\right)(100) = 8.27\%$

(3) $\left(\dfrac{0.439}{5.37}\right)(100) = 8.18\%$

Average percent phosphorus (show calculations) 8.25%

$$\text{ave.} = \frac{8.31 + 8.27 + 8.18}{3} = 8.25$$

Standard deviation (show calculations) 0.07

$$\text{SD} = \sqrt{\frac{(0.06)^2 + (0.02)^2 + (0.07)^2}{2}} = 0.07$$

Do any of your results differ from the mean by more than two standard deviations? No

Reported percent phosphorus 8.25% ± 0.07

Percent P_2O_5 (show calculations) 18.92% ± 0.10

$$\%P_2O_5 = (2.29)(\%P) = (2.29)(8.25) = 18.9$$

Guaranteed minimum percent P_2O_5 from the bottle label 19%

Do the guaranteed and experimental values agree within two standard deviations? Yes

QUESTIONS

1. What is the percent phosphorus in $MgKPO_4 \cdot 6H_2O$?

$$\frac{(30.97 \text{ g/mol P})}{(266.47 \text{ g/mol } MgKPO_4 \cdot 6H_2O)} = 11.62\%$$

2. $MgNH_4PO_4 \cdot 6H_2O$ has a solubility of 0.023 g/100 mL in water. Suppose a 5.02-g sample were washed with 20 mL of water. What fraction of the $MgNH_4PO_4 \cdot 6H_2O$ would be lost?

$$\left(\frac{0.023 \text{ g}}{100 \text{ mL}}\right)(20 \text{ mL}) = 0.0046 \text{ g}$$

$$\left(\frac{0.0046 \text{ g}}{5.02 \text{ g}}\right) = 0.00092 \text{ or } 0.092\%$$

3. $MgNH_4PO_4 \cdot 6H_2O$ loses H_2O stepwise as it is heated. Between 40° and 60°C the monohydrate is formed, and above 100°C the anhydrous material is formed. What are the phosphorus percentages of the monohydrate and of the anhydrous material?

For $MgNH_4PO_4 \cdot H_2O$:

$$\%P = \frac{(30.97 \text{ g P})(100)}{155.28 \text{ g } MgNH_4PO_4 \cdot H_2O} = 19.94\%$$

For $MgNH_4PO_4$:

$$\%P = \frac{(30.97 \text{ g P})(100)}{137.28 \text{ g } MgNH_4PO_4} = 22.56\%$$

4. Ignition of $MgNH_4PO_4 \cdot 6H_2O$ produces NH_3, H_2O, and magnesium pyrophosphate, $Mg_2P_2O_7$. Complete and balance the equation for this reaction. If 5.00 g of $MgNH_4PO_4 \cdot 6H_2O$ are ignited, how many grams of $Mg_2P_2O_7$ would be formed?

$$2MgNH_4PO_4 \cdot 6H_2O(g) \xrightarrow{\Delta} Mg_2P_2O_7(s) + 13H_2O(g) + 2NH_3(g)$$

$$\left(\frac{5.00 \text{ g } MgNH_4PO_4 \cdot 6H_2O}{245.28 \text{ g/mol } MgNH_4PO_4 \cdot 6H_2O}\right)\left(\frac{1 \text{ mol } Mg_2P_2O_7}{2 \text{ mol } MgNH_4PO_4 \cdot 6H_2O}\right) \times (222.56 \text{ g/mol } Mg_2P_2O_7) = 2.27 \text{ g}$$

5. What is the percentage of P_2O_5 in $Mg_2P_2O_7$?

$$\left(\frac{141.94 \text{ g/mol } P_2O_5}{222.56 \text{ g/mol } Mg_2P_2O_7}\right)\left(\frac{1 \text{ mol } P_2O_5}{1 \text{ mol } Mg_2P_2O_7}\right)(100) = 63.78\%$$

NOTES AND CALCULATIONS

Experiment

Paper Chromatography: Separation of Cations and Dyes

OBJECTIVE

To become acquainted with chromatographic techniques as a method of separation (purification) and identification of substances.

APPARATUS AND CHEMICALS

Apparatus

petri or evaporating dishes (2)
paper chromatography strips
600-mL beaker
capillary pipets (6)
6-in. glass stirring rod
metric ruler
4-in. watch glasses (2)
12.5-cm Whatman #1 filter paper
scissors
50- or 100-mL graduated cylinder
sample vials (3)
paper clips (3)

Chemicals

0.5 *M* $Cu(NO_3)_2$
0.5 *M* $Ni(NO_3)_2$
15 *M* NH_3
1% dimethylglyoxime in ethanol
0.5 *M* $Fe(NO_3)_3$
solvent (90% acetone, 10% 6 *M* HCl, freshly prepared)
isopropyl alcohol
unknown solutions

DISCUSSION

Chromatography (from the Greek *chrōma*, for color, and *graphein*, to write) is a technique often used by chemists to separate components of a mixture. In 1906, the Russian botanist Mikhail Tsvett separated color pigments present in leaves by allowing a solution of these pigments to flow down a column packed with an insoluble material such as starch, alumina (Al_2O_3), or silica (SiO_2). Because different color bands appeared along the column, he called the procedure chromatography. Color is not a requisite property to achieve separation of compounds by this procedure. Colorless compounds can be made visible by being allowed to react with other reagents, or they can be detected by physical means. Consequently, because of its simplicity and efficiency, this technique has wide applicability for separating and identifying compounds such as drugs and natural products.

The basis of chromatography is the *partitioning* (that is, separation arising from differences in solubility) of compounds between a stationary phase and a moving phase. Stationary phases such as alumina (Al_2O_3), silica (SiO_2), and paper (cellulose) have enormous surface areas. The molecules or ions of the substances to be separated are continuously being adsorbed and then released (desorbed) into the solvent flowing over the surface of the stationary phase. This brings about a separation of the components—that is, they will travel with different speeds in the moving solvent because there are generally different attractions between these components and the stationary phase.

In paper chromatography, a small spot of the mixture to be separated is placed at one end of a strip of paper, and solvent is allowed to move up the paper, through the spot, by capillary action. The solvent and various components of the mixture each travel at different speeds along the paper. In this experiment, Fe^{3+}, Cu^{2+}, and Ni^{2+} will be separated with a solvent that consists of a mixture of acetone, water, and hydrochloric acid. A diagram of a portion of a typical chromatogram strip is shown in Figure 10.1.

The identity of components in a mixture can be deduced by comparing a chromatogram of the unknown with chromatograms of mixtures of components suspected to be present in the unknown. An additional aid in identification of a compound is its R_f value, which is defined as the ratio of the distance traveled by a compound to the distance traveled by the solvent. For example, for Cu^{2+} (see Figure 10.1),

$$R_f(\mathrm{Cu}) = \frac{d_{\mathrm{Cu}}}{d_s} \qquad [1]$$

The R_f value of a compound is a characteristic of the compound, the support, and the solvent used, and it serves to identify the constitutents of a mixture. An estimate of the relative amount of each of the constituents in a mixture can be made from the relative intensities and sizes of the various bands in a chromatogram.

The three ions studied in this experiment will be discerned in the following manner: Fe^{3+} in water imparts a red-brown or rust color and thus will produce a rust-colored band on the paper. Although Cu^{2+} is blue, the color is faint and not easily detected, especially if copper is present in small amounts. In aqueous solution, however, Cu^{2+} reacts with NH_3 (from ammonium hydroxide) to form a complex ion, $[Cu(NH_3)_4]^{2+}$, which is deep blue and therefore readily observed. Finally, Ni^{2+} reacts with an organic reagent, dimethylglyoxime, to produce a strawberry-red color. The following reactions will occur during the development process on the paper:

$$Fe^{3+} + 3OH^- \longrightarrow Fe(OH)_3$$

Faint yellow → Rust colored

$$Cu^{2+} + 4NH_3 \longrightarrow [Cu(NH_3)_4]^{2+}$$

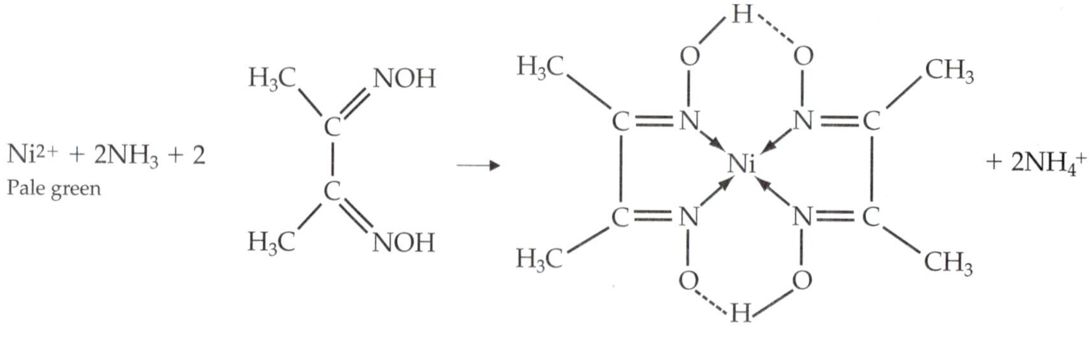

Pale blue → Deep blue

$Ni^{2+} + 2NH_3 + 2$ (H3C)(NOH)C–C(NOH)(CH3) → [Ni(dimethylglyoximate)2, with O–H···O bridges] $+ 2NH_4^+$

Pale green → Strawberry red

PROCEDURE

There are two simple ways of performing this experiment: either by circular horizontal chromatography or by ascending chromatography using paper strips. One half of the class should perform this experiment using one

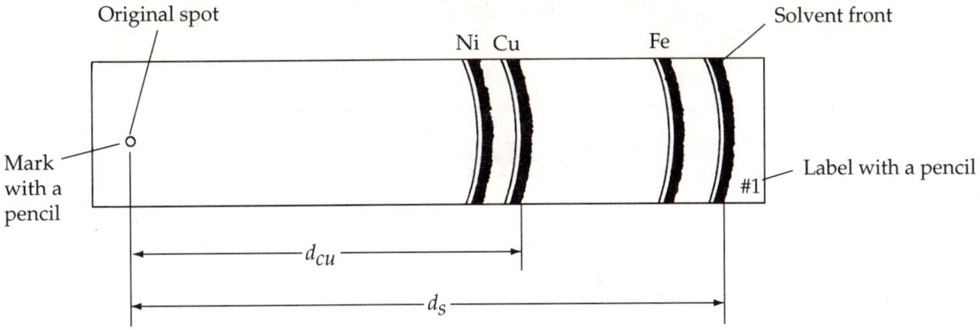

▲ **FIGURE 10.1**

technique, and the other half should use the other technique. The two techniques can then be compared in terms of time required, separation, and ease of identification.

Make six capillary pipets (2 in. long) by drawing out 6-mm glass tubing. Your instructor will demonstrate how this is done, or you can obtain the capillary tubes from your instructor.

Instructor: If capillary pipets are not provided, demonstrate how to make them.

Obtain an unknown; you will analyze this at the same time you are conducting the experiment on the known solutions. Your unknown may contain one, two, or three of the ions being studied. Record the colors of the *known solutions* on your report sheet (1).

Do not pour the chemicals down the drain. Dispose of them in the designated receptacles.

A. Horizontal Circular Technique

Cut a wick about 1 cm wide on a piece of 12.5-cm Whatman # 1 filter paper, as shown in Figure 10.2. Avoid touching the filter paper with oily hands. Trim about 1 cm from the end of the wick so that the filter paper will fit into the petri dish. Mark the center of the paper with a pencil dot (do not use a pen). Place the cut filter paper on top of another piece of filter paper, which will serve as an absorbing pad. Using the following technique, "spot" the filter paper at the center directly on the pencil dot sequentially with 1 drop of each of the known solutions of Cu^{2+}, Ni^{2+}, and Fe^{3+} (let the spot dry com-

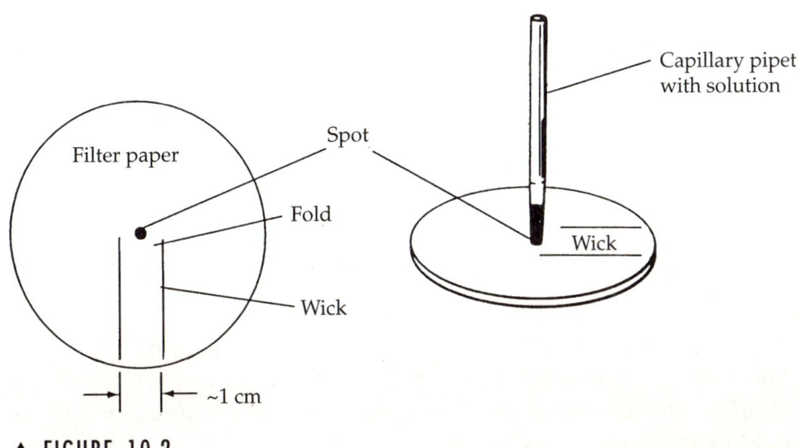

▲ **FIGURE 10.2**

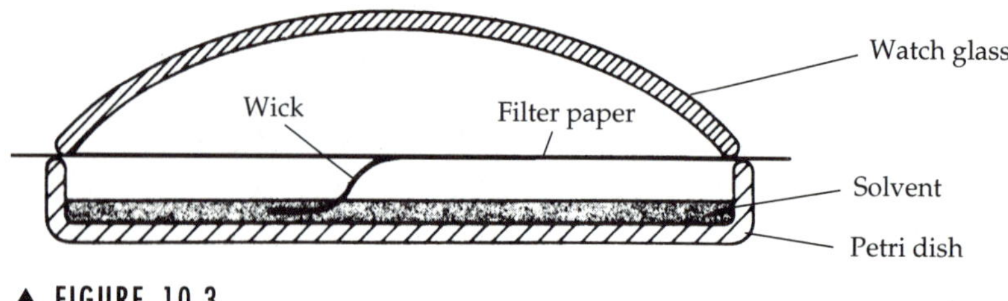

▲ FIGURE 10.3

pletely between applications). Pour a small quantity (only 1 to 2 mL) of Cu^{2+} solution into a clean sample vial and fill your capillary pipet by dipping the end into the solution. Allow the solution to rise by capillary action. Withdraw the pipet and touch the inside of the vial with the tip of the pipet to remove the hanging drop. Spot the filter paper on the pencil dot by touching it with the capillary held perpendicular to the paper. Allow the solution to flow out of the capillary until a spot having about a 5-to 7-mm diameter is obtained. Dry the filter paper completely by waving it in the air. In the same way, apply a single drop of the Ni^{2+} and then the Fe^{3+} solution to the same spot; be certain that the paper is dry before making each application. Spot a second piece of filter paper that you have prepared for chromatography with 1 drop of your unknown solution in the same manner.

Place two petri (or evaporating) dishes on your desktop away from direct sunlight or heat. Fill the dishes with the solvent mixture to a depth of 5 to 7 mm. **(CAUTION:** ***Acetone is highly flammable! Keep it away from open flames.*****)** Carefully place the (dry) spotted filter papers on the rims of the petri dishes with the wicks bent down into the solvent (Figure 10.3). Cautiously place a 4-in. watch glass on top of each filter paper; be careful not to push the papers into the solvent. The purpose of the cover is to prevent uneven evaporation of the solvent from the paper by providing an enclosed atmosphere that is saturated with solvent vapor. These conditions permit the solvent to travel across the paper in a more uniform manner; this is necessary for effective separation of the ions. Do not disturb the systems while the chromatograms are developing. When the solvent front has nearly reached the edge of each dish (15 to 25 min), carefully remove the watch glass and filter paper. Immediately mark the position of the solvent front with a pencil. Because the solvent evaporates quickly, this marking must be made as soon as possible. Allow each paper to dry by fanning the air with it.

Which of the known ions do you detect without resorting to the use of any other reagents (or development) (2)? Is there any difference between the ring front of this ion and the solvent front (3)? Mark the ring front of this ion.

Developing Reactions IN THE HOOD, pour about 5 mL of 15 *M* ammonium hydroxide into a clean, shallow dish and rest the filter paper containing the knowns on top of the dish. Do not permit the paper to dip into the solution. What color develops (4)? What ion does this indicate (5)? Mark the ring front.

To detect the third ion, dip a new piece of filter paper into a 1% solution of dimethylglyoxime; then, using the new piece as a brush, paint the test filter paper. As an alternative procedure, you may spray the paper with a dimethylglyoxime-containing aerosol. What color develops (6)? What ion does this indicate (7)? Mark the ring front.

Measure the distance to each of the ring fronts in millimeters (8). Calculate the R_f values for the three known ions (9) using Equation [1].

Repeat the above sequence of developing reactions with paper containing your unknown. What ions are present in your unknown (10)? Record the distances in millimeters to the ring fronts (11) and calculate the R_f values (12).

Relative amounts of the components of a mixture can be determined by using the methods employed in this experiment. Prepare a mixture of these ions in the following manner: Mix together 5 mL of the Fe^{3+} and 5 mL of the Ni^{2+} solutions, and then add 2 or 3 drops of the Cu^{2+} solution as a trace contaminant. Apply 1 drop of this mixture to a prepared piece of filter paper. Run and develop the chromatogram. Record your observations (13). Compare your results with those of someone who performed the experiment using the other technique (14).

Dyes You may believe that food coloring consists of one substance or that the black ink in felt-tip pens is a single substance. Spot a piece of filter paper heavily with a black felt-tip pen (such as a Flair pen). Make the spot quite dark to be sure that enough ink has been transferred to the paper. Obtain a chromatogram, using as the solvent a solution prepared by diluting 10 mL of isopropyl alcohol with 5 mL of water.

Attach all the dry, developed chromatograms to your report sheet.

B. Ascending Strip Technique

Obtain a piece of chromatography paper strip about 50 cm long. Cut this into 15-cm strips and mark each strip with a pencil dot. Label as diagrammed in Figure 10.4. You will be running three chromatograms simultaneously by this technique: the known, the unknown, and the trace Cu^{2+}. Place the developing solvent to a depth of 10 to 12 mm in the bottom of a 600-mL beaker (Figure 10.5). Spot the strips as described above for the circles: one strip with the three knowns, one strip with the unknown solution, and one strip with the solution containing a trace of Cu^{2+}. Attach the labeled ends of the strips to a 6-in. glass rod by folding the ends over the rod and clipping the paper together using paper clips (Figure 10.5). Place the rod with the three strips attached across the top of the beaker. Be certain that the spots on the strips are completely dry before you place the strips in the beaker. Make sure that the strips touch neither one another nor the walls of the beaker and that the bottom of each strip is resting in the solution. Cover the beaker carefully with a watch glass. Do not disturb the beaker while the chromatograms are developing. When the solvent front has nearly reached the union of the folded part of the paper, carefully remove the watch glass and the glass rod with the strips. Immediately mark the solvent fronts with a pencil. Allow the

Instructor: Aluminum foil may be used to cover the beaker instead of a watch glass.

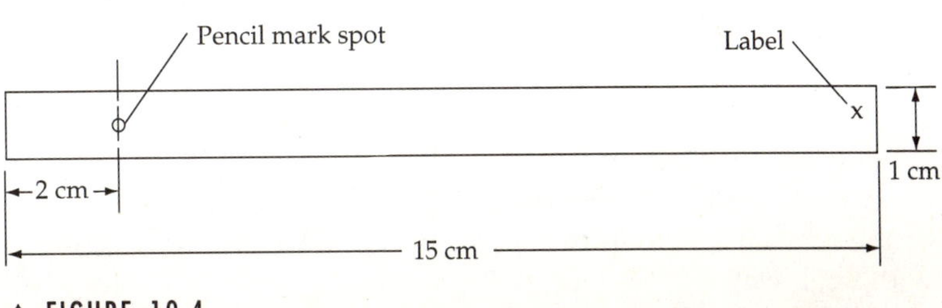

▲ FIGURE 10.4

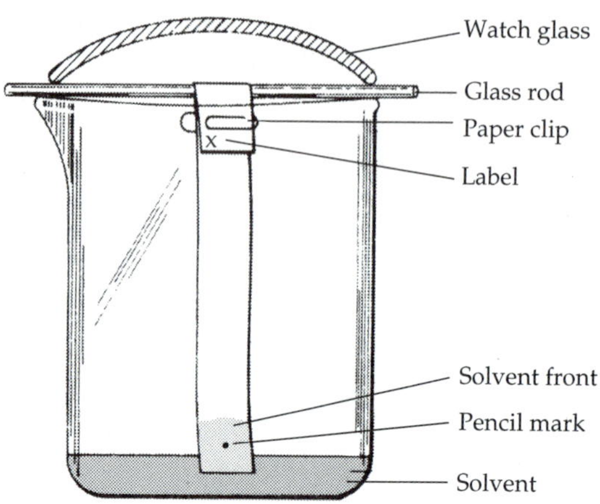

▲ FIGURE 10.5

strips to dry by fanning the air with them, and proceed as directed below with the developing solutions.

Which of the known ions do you detect without resorting to the use of any other reagents (or development) (2)? Is there any difference between the ring front of this ion and the solvent front (3)? Mark the ring front of this ion.

Developing Reactions IN THE HOOD, pour about 5 mL of 15 *M* ammonium hydroxide into a clean, shallow dish and rest the filter paper containing the knowns on top of the dish. Do not permit the paper to dip into the solution. What color develops (4)? What ion does this indicate (5)? Mark the ring front.

To detect the third ion, dip a new piece of filter paper into a 1% solution of dimethylglyoxime; then, using the new piece as a brush, paint the test filter paper. As an alternate procedure, you may spray the paper with a dimethylglyoxime-containing aerosol. What color develops (6)? What ion does this indicate (7)? Mark the ring front.

Measure the distance to each of the ring fronts in millimeters (8). Calculate the R_f values for the three known ions (9) using Equation [1].

Repeat the above sequence of developing reactions with paper containing your unknown. What ions are present in your unknown (10)? Record the distances in millimeters to the ring fronts (11) and calculate the R_f values (12).

Relative amounts of the components of a mixture can be determined by using the methods employed in this experiment. Prepare a mixture of these ions in the following manner: Mix together 5 mL of the Fe^{3+} solution and 5 mL of the Ni^{2+} solution, and then add 2 or 3 drops of the Cu^{2+} solution as a trace contaminant. Apply 1 drop of this mixture to a prepared piece of filter paper. Run and develop the chromatogram. Record your observations (13). Compare your results with those of someone who performed the experiment using the other technique (14).

Dyes You may believe that food coloring consists of one substance or that the black ink in felt-tip pens is a single substance. Spot a piece of filter paper heavily with a black felt-tip pen. Make the spot quite dark to be sure that enough ink has been transferred to the paper. Obtain a chromatogram, using as the solvent a solution prepared by diluting 10 mL of isopropyl alcohol with 5 mL water.

Attach all the dry, developed chromatograms to your report sheet.

REVIEW QUESTIONS

Before beginning this experiment in the laboratory, you should be able to answer the following questions:

1. What does the technique of chromatography allow us to do?
2. What is the meaning and utility of an R_f value? What would you expect to influence an R_f value?
3. What are the developing reactions that allow the identification of Ni^{2+} and Cu^{2+}?
4. What forces cause the eluting solution to move along the chromatographic support material?
5. Why should the solvent front be marked immediately?
6. If the solvent front is 76 mm and the ring front of an unknown is 48 mm, what is the R_f value?
7. Would you expect changing the solvent to change the R_f value? Why?
8. Suggest simple reasons why the Ni^{2+} and Fe^{3+} ions have different R_f values.
9. Is it necessary that compounds be colored to be separated by chromatography?
10. Why should the petri dish (or beaker) be covered during the development of the chromatogram?

NOTES AND CALCULATIONS

Name ______________________ Desk ______________________

Date ______________ Laboratory Instructor ______________________

Unknown no. or letter ______________________

REPORT SHEET | EXPERIMENT

Paper Chromatography: Separation of Cations and Dyes | 10

1. Color of solutions: Cu^{2+} blue Fe^{3+} pale yellow Ni^{2+} green
2. Ion requiring no development iron, rust color
3. Yes X No ______
4. Ammonia develops a deep blue color.
5. The ion that causes the color of (4) is $Cu(NH_3)_4^{2+}$
6. Dimethylglyoxime develops a strawberry red color with (7) Ni^{2+} ion.
8. Ring front distances for knowns in mm are: solvent 88 mm; Fe^{3+} 88 mm; Cu^{2+} 51 mm; Ni^{2+} 46 mm.
9. R_f values (show calculations at end of report): Fe^{3+} 1.0 ; Cu^{2+} 0.58 ; Ni^{2+} 0.52 .
10. Ions present in unknown Fe^{3+} = Cu^{2+}, Ni^{2+}
11. Ring front distances for unknown Solvent 77 mm, Fe^{3+} 77 mm, Cu^{2+} 43 mm, Ni^{2+} 39 mm
12. R_f values for unknown Fe^{3+}: 77/77 = 1.0, Cu^{2+}: 43/77 = 0.56, Ni^{2+}: 39/77 = 0.51
13. Observation on relative amounts By their color intensities, Fe^{3+} and Ni^{2+} are about the same. Cu^{2+} is only a very faint, trace amount.
14. Comment on the relative merits of the two techniques.

    The ascending strip method is slower, but it gives better results (better separation or resolution).

Calculations for R_f's (9) and R_f's (12):

$$R_f\text{Fe} = \frac{88\text{ mm}}{88\text{ mm}} = 1.0; \qquad R_f\text{Cu} = \frac{51\text{ mm}}{88\text{ mm}} = 0.58; \quad R_f\text{Ni} = \frac{46\text{ mm}}{88\text{ mm}} = 0.52$$

ATTACH CHROMATOGRAMS

NOTES AND CALCULATIONS

Experiment

Molecular Geometries of Covalent Molecules: Lewis Structures and VSEPR Theory

OBJECTIVE

To become familiar with Lewis structures, the principles of VSEPR theory, and the three-dimensional structures of covalent molecules.

APPARATUS

Prentice-Hall Molecular Model Set for General and Organic Chemistry or Styrofoam balls* and pipe cleaners

DISCUSSION

Types of Bonding Interactions

Whenever atoms or ions are strongly attached to one another, we say that there is a *chemical bond* between them. There are three types of chemical bonds: ionic, covalent, and metallic. The term *ionic bond* refers to electrostatic forces that exist between ions of opposite charge. Ions may be formed by the transfer of one or more electrons from an atom with a low ionization energy to an atom with a high electron affinity. Thus, ionic substances generally result from the interaction of metals on the far left side of the periodic table with nonmetals on the far right side of the periodic table (excluding the noble gases, group 8A). A *covalent bond* results from the sharing of electrons between two atoms. The more familiar examples of covalent bonding are found among nonmetallic elements interacting with one another. This experiment illustrates the geometric (three-dimensional) shapes of molecules and ions resulting from covalent bonding among various numbers of elements, and two of the consequences of geometric structure—isomers and polarity. *Metallic bonds* are found in metals such as gold, iron, and magnesium. In the metals, each atom is bonded to several neighboring atoms. The bonding electrons are relatively free to move throughout the three-dimensional structure of the metal. Metallic bonds give rise to such typical metallic properties as high electrical and thermal conductivity and luster.

Lewis Symbols

The electrons involved in chemical bonding are the *valence electrons,* those residing in the incomplete outer shell of an atom. The American chemist G. N. Lewis suggested a simple way of showing these valence electrons, which we now call Lewis electron-dot symbols or simply Lewis symbols. The **Lewis symbol** for an element consists of the chemical abbreviation for the element plus a dot for each valence electron. For example, oxygen has the electron configuration $[He]s^2p^4$; its Lewis symbol therefore shows six valence electrons. The dots are placed on the four sides of the atomic abbreviation.

*Available from Snow Foam Products, Inc., 9917 W. Gidley St., El Monte, CA 91731.

Each side can accommodate up to two electrons. All four sides are equivalent; the placement of two electrons versus one electron is arbitrary. The number of valence electrons of any representative element is the same as the group number of the element in the periodic table.

The Octet Rule

Atoms often gain, lose, or share electrons so as to achieve the same number of electrons as the noble gas closest to them in the periodic table. Because all noble gases (except He) have eight valence electrons, many atoms undergoing reactions also end up with eight valence electrons. This observation has led to the **octet rule:** *Atoms tend to gain, lose, or share electrons until they are surrounded by eight valence electrons.* An octet of electrons consists of full *s* and *p* subshells on an atom. In terms of Lewis symbols, an octet can be thought of as four pairs of valence electrons arranged around the atom as in the configuration for Ne, which is :N̈e:. There are many exceptions to the octet rule, but it provides a useful framework for many important concepts of bonding.

Covalent Bonding

Ionic substances have several characteristic properties. They are usually brittle, have high melting points, and are crystalline solids with well-formed faces that can often be cleaved along smooth, flat surfaces. These characteristics result from electrostatic forces that maintain the ions in a rigid, well-defined, three-dimensional arrangement.

The vast majority of chemical substances do not have the characteristics of ionic materials. Most of the substances with which we come in daily contact, such as water, gasoline, oxygen, and sugar, tend to be either liquids, gases, or low-melting solids. Many of them, such as charcoal lighter fluid, vaporize readily. Many are pliable in their solid form, for example, plastic bags and paraffin wax.

For the very large class of substances that do not behave like ionic substances, we need a different model for the bonding between atoms. Lewis reasoned that atoms might acquire a noble-gas electron configuration by sharing electrons with other atoms to form *covalent bonds.* The hydrogen molecule, H_2, provides the simplest possible example of a covalent bond. When two hydrogen atoms are close to each other, electrostatic interactions occur between them. The two positively charged nuclei and the two negatively charged electrons repel each other, whereas the nuclei and electrons attract each other. The attraction between the nuclei and the electrons cause electron density to concentrate between the nuclei. As a result, the overall electrostatic interactions are attractive in nature.

In essence, the shared pair of electrons in any covalent bond acts as a kind of "glue" to bind the two hydrogen atoms together in the H_2 molecule.

Lewis Structures

The formation of covalent bonds can be represented using Lewis symbols as shown below for H_2:

H• + •H → H : H

The formation of a bond between two F atoms to give a F_2 molecule can be represented in a similar way.

$$:\ddot{\underset{..}{F}}\cdot + \cdot\ddot{\underset{..}{F}}: \rightarrow :\ddot{\underset{..}{F}}:\ddot{\underset{..}{F}}:$$

By sharing the bonding electron pair, each fluorine atom acquires eight electrons (an octet) in its valence shell. It thus achieves the noble-gas electron configuration of neon. The structures shown here for H_2 and F_2 are called *Lewis structures* (or Lewis electron-dot structures). In writing Lewis structures, we usually show each electron pair shared between atoms as a line, to emphasize it is a bond, and the unshared electron pairs as dots. Writing them this way, the Lewis structures for H_2 and F_2 are shown as follows:

$$H-H \qquad :\ddot{\underset{..}{F}}-\ddot{\underset{..}{F}}:$$

The number of valence electrons for the nonmetal is the same as the group number. Therefore, one might predict that 7A elements, such as F, would form one covalent bond to achieve an octet; 6A elements, such as O, would form two covalent bonds; 5A elements, such as N, would form three covalent bonds; and 4A elements, such as C, would form four covalent bonds. For example, consider the simple hydrogen compounds of the nonmetals of the second row of the periodic table.

$$H-\ddot{\underset{..}{F}}: \qquad H-\underset{\underset{H}{|}}{\ddot{O}}: \qquad H-\underset{\underset{H}{|}}{\ddot{N}}-H \qquad H-\underset{\underset{H}{|}}{\overset{\overset{H}{|}}{C}}-H$$

Thus, the Lewis model succeeds in accounting for the compounds of nonmetals, in which covalent bonding predominates.

Multiple Bonds

The sharing of a pair of electrons constitutes a single covalent bond, generally referred to simply as a *single bond*. In many molecules, atoms attain complete octets by sharing more than one pair of electrons between them. When two electron pairs are shared, two lines (representing a *double bond*) are drawn. A *triple bond* corresponds to the sharing of three pairs of electrons. Such multiple bonding is found in CO_2 and N_2.

$$\ddot{\underset{..}{O}}=C=\ddot{\underset{..}{O}} \qquad :N\equiv N:$$

Drawing Lewis Structures

Lewis structures are useful in understanding the bonding in many compounds and are frequently used when discussing the properties of molecules. To draw Lewis structures, we follow a regular procedure.

1. *Sum the valence electrons from all atoms.* Use the periodic table as necessary to help determine the number of valence electrons on each atom. For an anion, add an electron to the total for each negative charge. For a cation, subtract an electron for each positive charge.

2. *Write the symbols for the atoms to show which atoms are attached to which, and connect them with a single bond* (a dash, representing two electrons). Chemical formulas are often written in the order in which the atoms are connected in the molecule or ion, as in HCN. When a central atom has a group of other atoms bonded to it, the central atom is usually written first, as in CO_3^{2-} or BF_3. In other cases you may need more information before you can draw the Lewis structure.
3. *Complete the octets of the atoms bonded to the central atom.* (Remember, however, that hydrogen can have only two electrons.)
4. *Place any leftover electrons on the central atom,* even if doing so results in more than an octet.
5. *If there are not enough electrons to give the central atom an octet, try multiple bonds.* Use one or more of the unshared pairs of electrons on the atoms bonded to the central atom to form double or triple bonds.

Formal Charge

When we draw a Lewis structure, we are describing how the electrons are distributed in a molecule (or ion). In some instances we can draw several different Lewis structures that all obey the octet rule. How do we decide which one is the most reasonable? One approach is to do some "bookkeeping" of the valence electrons to determine the *formal charge* of each atom in each Lewis structure. The formal charge of an atom is the charge that an atom in a molecule or ion would have if all atoms had the same electronegativity. To calculate the formal charge on any atom in a Lewis structure, we assign the electrons to the atoms as follows:

1. All of the unshared (nonbonding) electrons are assigned to the atom on which they are found.
2. Half of the bonding electrons are assigned to each atom in the bond.
3. Check atom formal charges for reasonableness (zero formal charges are preferred, otherwise more electronegative atoms should have negative formal charges).

EXAMPLE 11.1

Draw the Lewis structures for SCN^-. Calculate formal charges, and designate the preferred structure.

SOLUTION: The total number of valence electrons is 6(S) + 4(C) + 5(N) + 1 (minus charge) = 16 electrons. There are three possible Lewis structures:

$$[:\overset{\cdot\cdot}{\underset{\cdot\cdot}{S}}-C\equiv N:]^- \qquad [:\overset{\cdot\cdot}{S}=C=N:]^- \qquad [:S\equiv C-\overset{\cdot\cdot}{\underset{\cdot\cdot}{N}}:]^-$$

I	II	III
−1 0 0	0 0 −1	+1 0 −2

All the atoms have octets in each structure. Because nitrogen is more electronegative than sulfur, structure II is preferred on the basis of formal charges. Structure III is not favored, because of the large charge separation.

Bond Polarity and Dipole Moments

A covalent bond between two different kinds of atoms is usually a polar bond. This is because the two different atoms have different electronegativities and the electrons in the bond are then not shared equally by the two

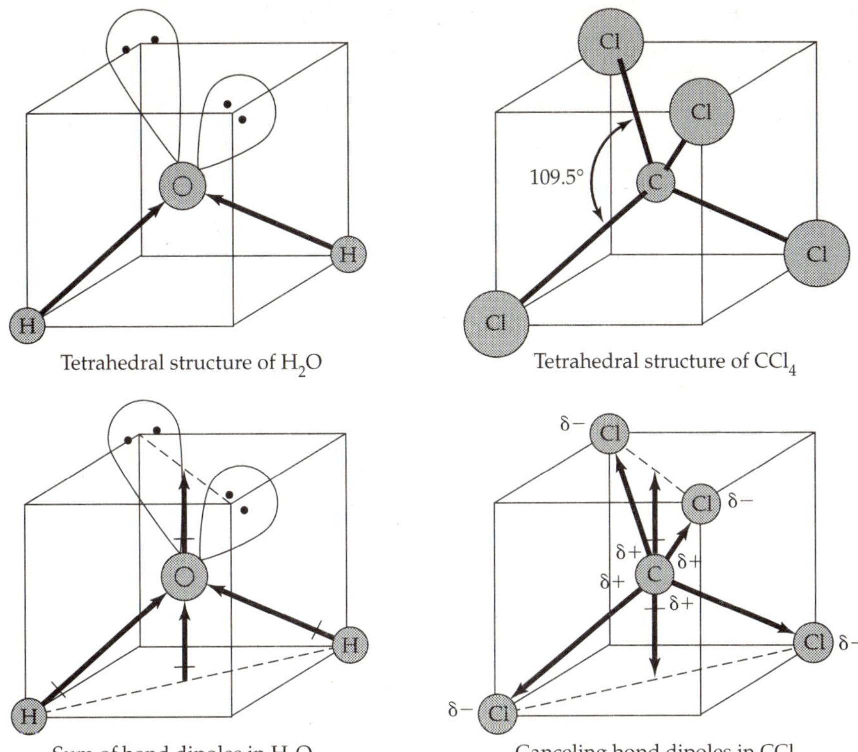

▲ **FIGURE 11.1** The bond dipoles in H_2O add to give a net molecular dipole that bisects the HOH angle and points upward in this drawing. In contrast, because of the symmetry of CCl_4, the bond dipoles cancel and the CCl_4 molecule has no molecular dipole moment.

bound atoms. This creates a bond dipole moment, μ, which is a charge separation, Q, over a distance, r:

$$\mu = Qr$$

The bonding electrons have an increased attraction to the more electronegative atom, thus creating an excess of electron density (δ^-) near it and a deficiency of electron density (δ^+) near the less electronegative atom. We symbolize the bond dipole (a vector) with an arrow and a cross, $\mapsto$, with the point of the arrow representing the negative and the cross the positive end of the dipole. Thus, in the polar covalent molecule H—Cl, because chlorine is more electronegative than hydrogen, the bond dipole is as illustrated below:

$$\overset{\delta^+}{\mathrm{H}} \mapsto \overset{\delta^-}{\mathrm{Cl}}$$

The bond dipole is a vector (it has magnitude and direction), and the dipole moments of polyatomic molecules are the vector sums of the individual bond dipoles as illustrated in Figure 11.1.

Because H_2O has a molecular dipole and CCl_4 does not, H_2O is a polar molecule and CCl_4 is a nonpolar molecule.

VSEPR Theory

In covalent molecules, atoms are bonded together by sharing pairs of valence-shell electrons. Electron pairs repel one another and try to stay out of each other's way. *The best arrangement of a given number of electron pairs is the one that minimizes the repulsions among them.* This simple idea is the basis of valence-shell electron pair repulsion theory, or the **VSEPR** model. Thus, as illustrated in Table 11.1, two electron pairs are arranged *linearly*, three pairs are arranged in a *trigonal planar* fashion, four are arranged *tetrahedrally*, five are arranged in a *trigonal bipyramidal* geometry, and six are arranged *octahedrally*. The shape of a molecule or ion can be related to these five basic arrangements of electron pairs.

Predicting Molecular Geometries When we draw Lewis structures, we encounter two types of valence-shell electron pairs: **bonding pairs,** which

TABLE 11.1 Electron-Pair Geometries as a Function of the Number of Electron Pairs

Number of electron pairs	Arrangement of electron pairs	Electron-pair geometry	Predicted bond angles
2	 180°	Linear	180°
3	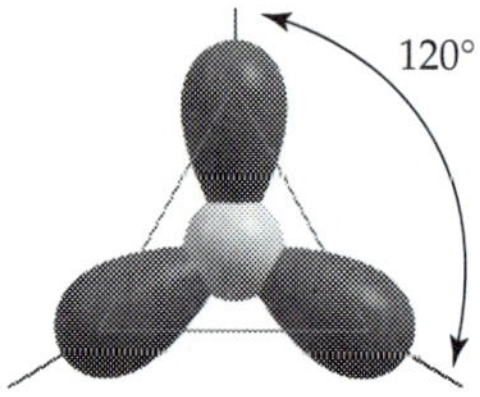 120°	Trigonal planar	120°
4	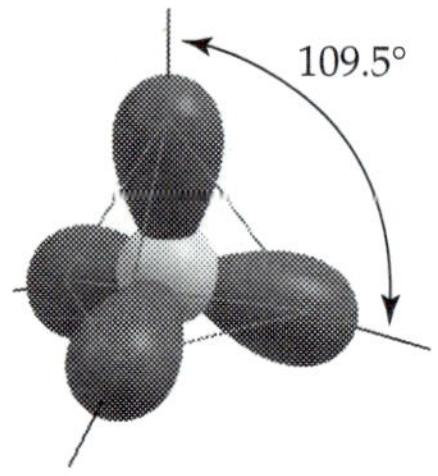 109.5°	Tetrahedral	109.5°
5	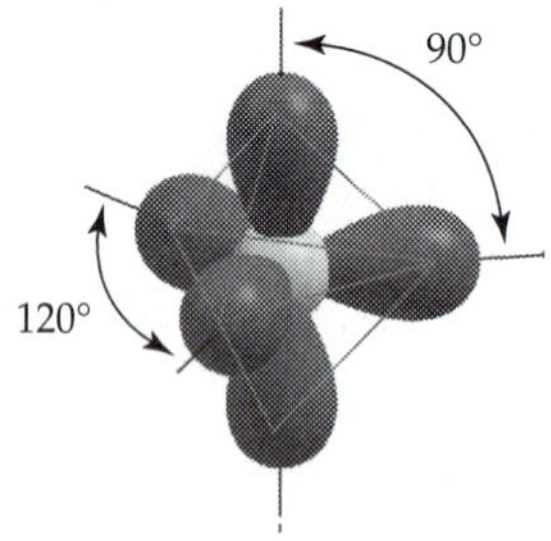 90° 120°	Trigonal bipyramidal	120° 90°
6	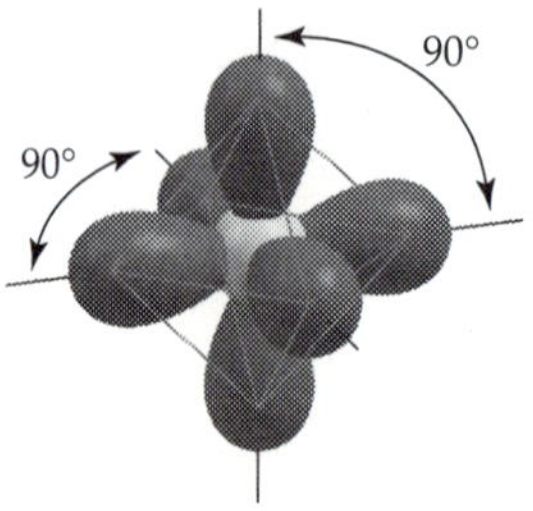 90° 90°	Octahedral	90° 180°

are shared by atoms in bonds, and **nonbonding pairs** (or *lone pairs*) such as in the Lewis structure for NH_3.

Nonbonding pair

H—N̈—H
|
H

Bonding pairs

Because there are four electron pairs around the N atom, the electron-pair repulsions will be minimized when the electron pairs point toward the vertices of a tetrahedron (see Table 11.1). The arrangement of electron pairs about the central atom of an AB_n molecule is called its **electron-pair geometry.** However, when we use experiments to determine the structure of a molecule, we locate atoms, not electron pairs. The **molecular geometry** of a molecule (or ion) is the arrangement of the *atoms* in space. We can predict the molecular geometry of a molecule from its electron-pair geometry. In NH_3, the three bonding pairs point toward three vertices of a tetrahedron. The hydrogen atoms are therefore located at the three vertices of a tetrahedron and the lone pair is located at a fourth vertex. The arrangement of the atoms in NH_3 is thus a *trigonal pyramid,* and the sequence of steps to arrive at this prediction are illustrated in Figure 11.2. We see that the trigonal-pyramidal molecular geometry of NH_3 is a consequence of its tetrahedral electron-pair geometry. When describing the shapes of molecules, we always give the molecular geometry rather than the electron-pair geometry.

Following are the steps used to predict molecular geometries with the VSEPR model:

1. Sketch the Lewis structure of the molecule or ion.
2. Count the total number of electron pairs around the central atom, and arrange them in a way that minimizes electron-pair repulsions (see Table 11.1)
3. Describe the molecular geometry in terms of the angular arrangement of the bonding pairs, which corresponds with the arrangement of bound atoms.

Application of the VSEPR model to molecules that contain multiple bonds reveals that a double or triple bond has essentially the same effect on bond angles as does a single bond. This observation leads to one further rule:

4. A double or triple bond is counted as one bonding pair when predicting geometry.

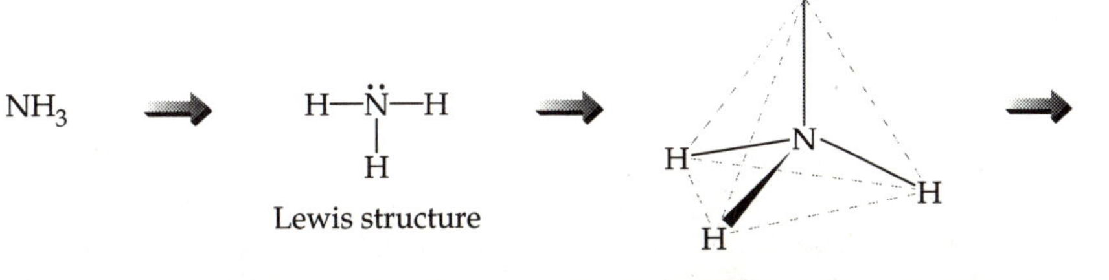

Electron-domain geometry
(tetrahedral)

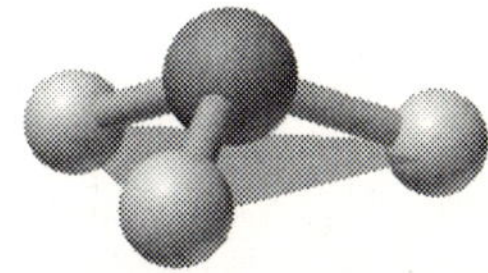

Molecular geometry
(trigonal pyramidal)

▲ **FIGURE 11.2**

EXAMPLE 11.2

Using the VSEPR model, predict the molecular geometries of (a) SCl_2 and (b) NO_2^-.

SOLUTION:

a. The Lewis structure for SCl_2 is: :Cl—S—Cl: The central S atom is surrounded by two nonbonding electron pairs and two single bonds. Thus, the electron pair geometry is tetrahedral, and the molecular geometry is bent.

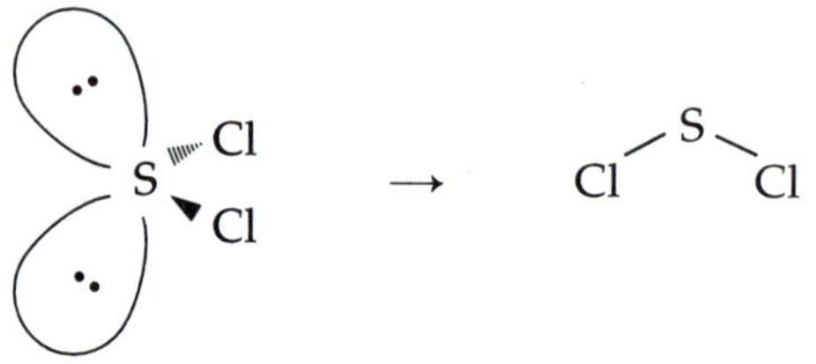

b. For NO_2^- we can draw two equivalent resonance structures.

Because of resonance, the bonds between N and the two O atoms are of equal length. Both resonance structures show one nonbonding electron pair, one single bond, and one double bond about the central N atom. When we predict geometry, a double bond is counted as one electron pair (rule 4). Thus, the arrangement of valence-shell electrons is trigonal planar. Two of these positions are occupied by O atoms, so the molecule has a bent shape.

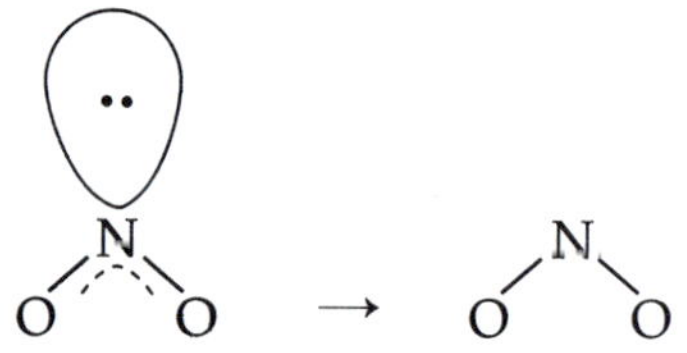

Valence-Bond (VB) Theory: Covalent Bonding and Orbital Overlap

Although the VSEPR model provides a simple means for predicting molecular shapes, it does not explain why bonds exist between atoms. In 1931 Linus Pauling developed a bonding model called **valence-bond theory,** based on the marriage of Lewis's notion of electron-pair bonds to the idea of atomic orbitals. By extending this approach to include the ways in which atomic orbitals can mix with one another, we obtain a picture that corresponds nicely to the VSEPR model.

In the Lewis theory, covalent bonding occurs when atoms share electrons. Such sharing concentrates electron density between the nuclei. In the valence-bond theory, the buildup of electron density between two nuclei is visualized as occurring when a valence atomic orbital of one atom merges with that of another atom. The orbitals are then said to share a region of space, or to overlap. The overlap of orbitals allows two electrons of opposite spin to share the common space between the nuclei, forming a covalent bond as shown for the H_2, HCl, and Cl_2 molecules in Figure 11.3.

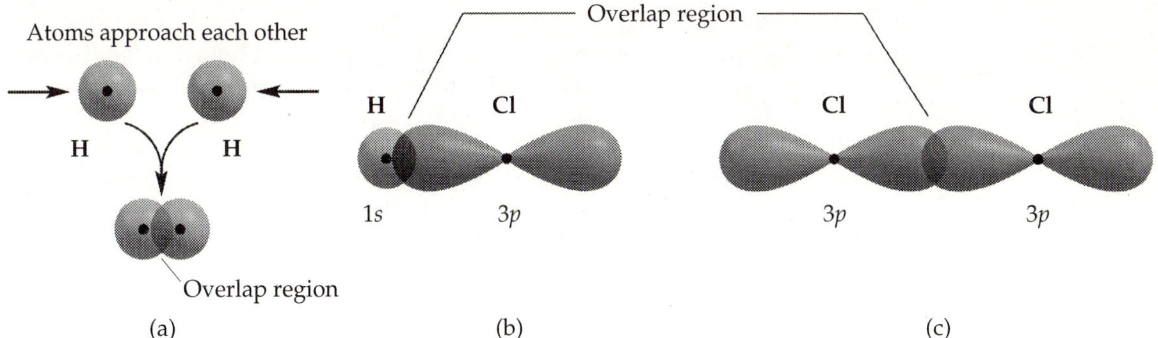

▲ **FIGURE 11.3** The overlap of orbitals to form covalent bonds. (a) The bond in H_2 results from the overlap of two 1*s* orbitals from two H atoms. (b) The bond in HCl results from the overlap of a 1*s* orbital of H and one of the lobes of a 3*p* orbital of Cl. (c) The bond in Cl_2 results from the overlap of two 3*p* orbitals from two Cl atoms.

Hybrid Orbitals

Although the idea of orbital overlap allows us to understand the formation of covalent bonds, it is not always easy to extend these ideas to polyatomic molecules. We need to explain both the formation of electron-pair bonds and the observed geometry. Consider the molecule BeF_2:

$$:\ddot{\underset{..}{F}}-Be-\ddot{\underset{..}{F}}:$$

The VSEPR model predicts, correctly, that this molecule is linear. Beryllium has a $1s^22s^2$ electron configuration with no unpaired electrons. Which orbitals on Be overlap with orbitals on F to form bonds? If the electron configuration of Be were promoted to $1s^22s^12p^1$, then a Be *s* and a *p* orbital could be used for bonding. We can also answer the former question by "mixing" the 2*s* orbital with one of the 2*p* orbitals on Be to generate two new orbitals as shown in Figure 11.4.

Like *p* orbitals, each of the new *sp* orbitals has two lobes. The two new orbitals are identical in shape, but their large lobes point in opposite direc-

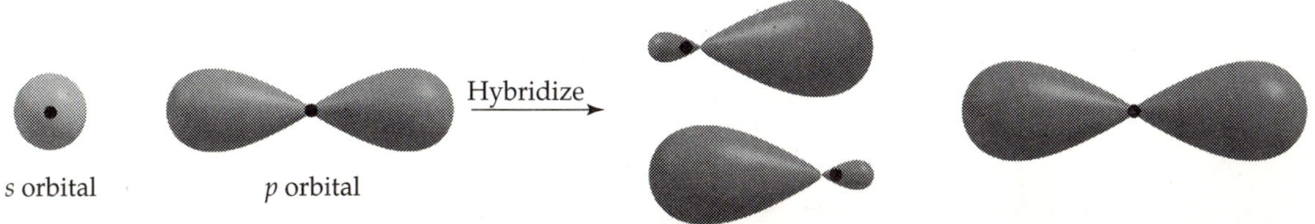

▲ **FIGURE 11.4** One *s* orbital and one *p* orbital can hybridize to form two equivalent *sp* hybrid orbitals. The two hybrid orbitals have their large lobes pointing in opposite directions, 180° apart.

tions. We have created two **hybrid orbitals,** orbitals formed by mixing two or more atomic orbitals on an atom, a procedure called hybridization. Other types of hybrid orbitals and their geometric arrangements are shown in Table 11.2.

In this experiment, you will draw Lewis structures for molecules and ions, predict their geometric structures from VSEPR and valence-bond theories, and predict some of their properties, based on molecular models.

TABLE 11.2 Geometrical Arrangements Characteristic of Hybrid Orbital Sets

Atomic orbital set	Hybrid orbital set	Geometry	Examples
s,p	Two *sp*	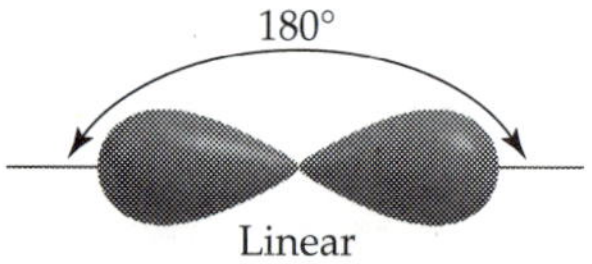 180°; Linear	BeF_2, $HgCl_2$
s,p,p	Three sp^2	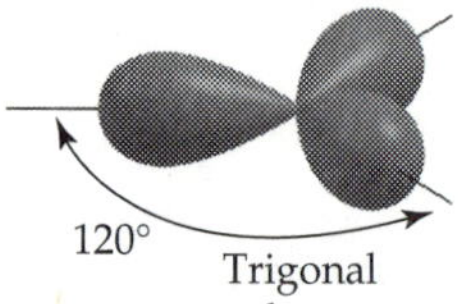 120°; Trigonal planar	BF_3, SO_3
s,p,p,p	Four sp^3	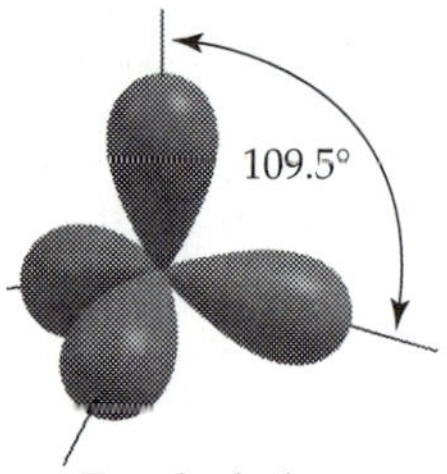 109.5°; Tetrahedral	CH_4, NH_3, H_2O, NH_4^+
s,p,p,p,d	Five sp^3d	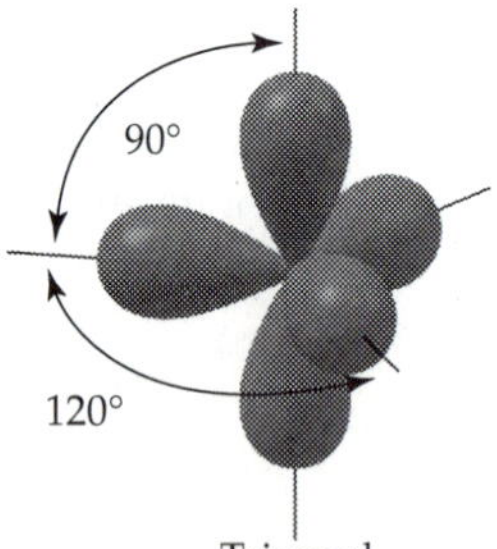 90°; 120°; Trigonal bipyramidal	PF_5, SF_4, BrF_3, $SbCl_5$
s,p,p,p,d,d	Six sp^3d^2	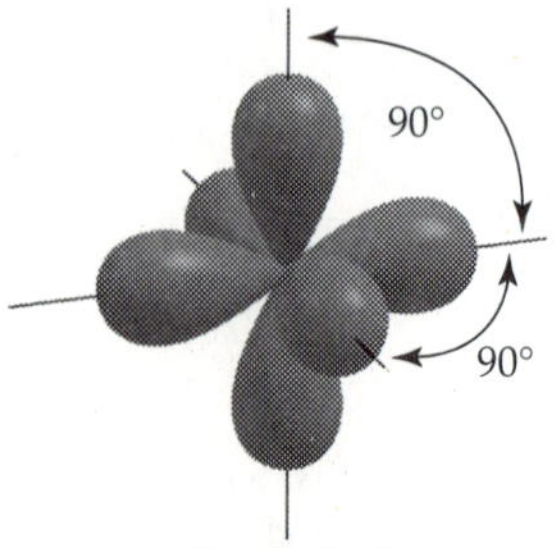 90°; 90°; Octahedral	SF_6, ClF_5, XeF_4, PF_6^-

REVIEW QUESTIONS

Before beginning this experiment in the laboratory, you should be able to answer the following questions:

1. Distinguish among ionic, covalent, and metallic bonding.
2. Which of the following molecules possess polar covalent bonds: H_2, N_2, HCl, HCN, CO_2?
3. Which of the molecules in question 2 have molecular dipole moments?
4. What are the favored geometrical arrangements for AB_n molecules for which the A atom has 2, 3, 4, 5, and 6 pairs of electrons in its valence shell?
5. How many equivalent orbitals are involved in each of the following sets of hybrid orbitals: sp, sp^2, sp^3d, sp^3d^2?
6. Define the term *formal charge*.
7. Calculate the formal charges of the atoms in CO, CO_2, and CO_3^{2-}.

NOTES AND CALCULATIONS

Name ______________________________ Desk ______________

Date ______________ Laboratory Instructor ______________

REPORT SHEET | EXPERIMENT

Molecular Geometries of Covalent Molecules: Lewis Structures and VSEPR Theory

11

1. Using an appropriate set of models, make molecular models of the compounds listed below and complete the table.

Molecular formula	No. of bond pairs	No. of lone pairs	Hybridization of central atoms	Molecular geometry	Bond angle(s)	Dipole moment (yes or no)
$BeCl_2$	2	0	*sp*	linear	180°	no
BF_3	3	0	sp^2	trigonal planar	120°	no
$SnCl_2$	2	1	sp^2	bent	120°	yes
CH_4	4	0	sp^3	tetrahedral	109.5°	no
NH_3	3	1	sp^3	pyramidal	109.5°	yes
H_2O	2	2	sp^3	bent	109.5°	yes
PCl_5	5	0	sp^3d	trigonal pyramidal	120°/90°	no
SF_4	4	1	sp^3d	saw horse	120°/90°	yes
BrF_3	3	2	sp^3d	tetrahedral	90°/180°	yes
XeF_2	2	4	sp^3d^2	linear	180°	no
SF_6	6	0	sp^3d^2	octahedral	90°	no
IF_5	5	1	sp^3d^2	square based pyramidal	90°	yes
XeF_4	4	2	sp^3d^2	square planar	90°	no

2. From your models of SF_4, BrF_3, and XeF_4, deduce whether different atom arrangements, called geometrical isomers, are possible, and, if so, sketch them below. Indicate the preferred geometry for each case and suggest a reason for your choice. Indicate which structures have dipole moments, and show their direction.

	Molecule	*Dipole moment*	*Preferred geometry*	*Reason*
(a) SF_4	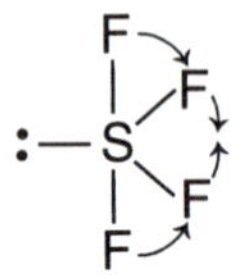 :—S(F)(F)(F)(F)	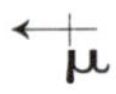 Yes (μ ←)	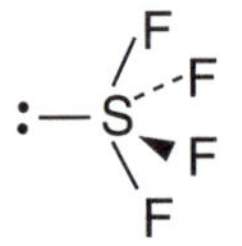 :—S(F)(F)(F)(F) sawhorse (or seesaw)	lp – bp repulsion> bp – bp
	or F—S(F)(F)(F) with lone pair 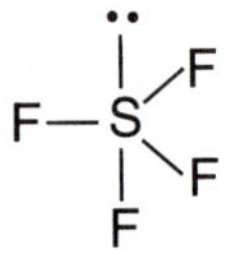			
(b) BrF_3	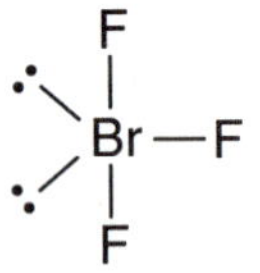 F, Br—F, F with two lone pairs	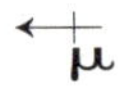 Yes (μ ←)	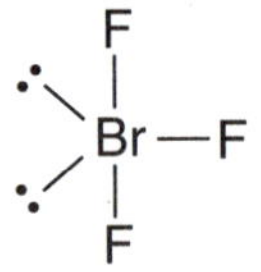 F, Br—F, F with two lone pairs	lp – lp repulsion> bp – bp
	or F, F, Br—F with axial lone pairs	No		
	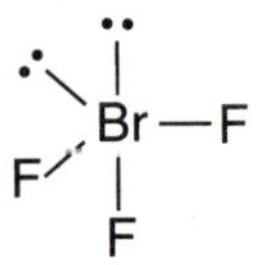 or F, F, Br—F with lone pairs	Yes  (μ)		
(c) XeF_4	 F, F, Xe, F, F with axial lone pairs	No	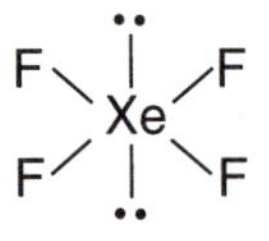 F, F, Xe, F, F with axial lone pairs	lp – lp repulsion> bp – bp
	or F, F, F, Xe, F with lone pairs	Yes		

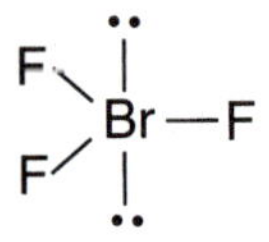

3. Predict the structures of the following ions, and state the hybridization of the central atom.

Ion	*Structure*	*Central atom hybridization*
N_3^-	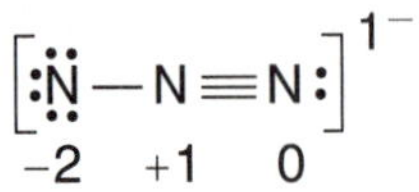 [:N̈—N≡N:]$^{1-}$ (formal charges −2, +1, 0) or [N=N=N]$^{1-}$ (formal charges −1, +1, −1) 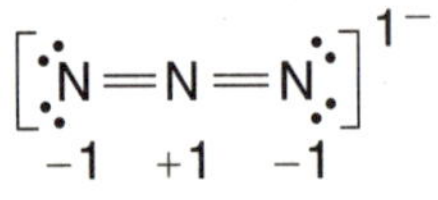	*sp*

Ion	*Structure*	*Central atom hybridization*
CO_3^{2-}	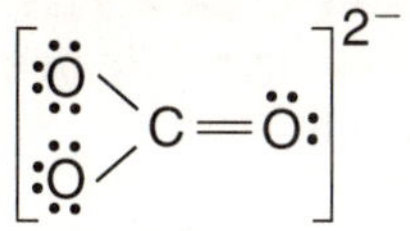	sp^2
NO_3^-	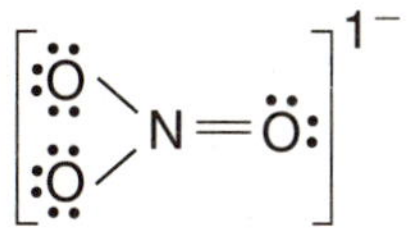	sp^2
BF_4^-	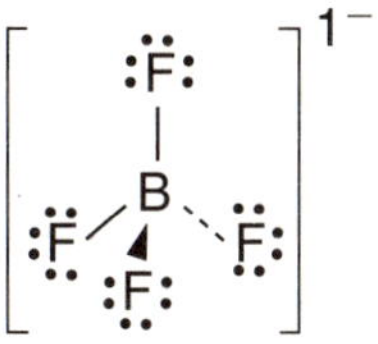	sp^3

4. Because lone pairs are larger than bonding pairs, lone pair–lone pair interactions are greater than lone pair–bonding pair interactions, which are in turn larger than bonding pair–bonding pair interactions. Using this notion, suggest how the following species would distort from regular geometries.

(a) OF_2

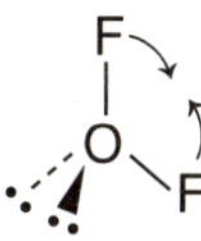

F—O—F angle less than 109.5°; actually is 103.1°

(b) SCl_2

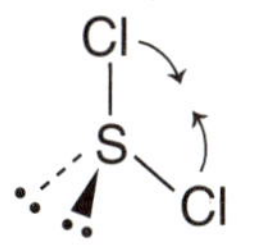

Cl—S—Cl angle less than 109.5°; actually is 102.7°

(c) PF_3

F—P—F angle less than 109.5°; actually is 97.7°

5. There are several families of hydrocarbons, among which are alkanes, alkenes, and alkynes. The parents of each family are CH_4 (methane), $H_2C=CH_2$ (ethene), and $H-C\equiv C-H$ (ethyne), respectively. Predict the geometries of these molecules, give the hybridization of carbon in each molecule, and suggest whether they are polar or nonpolar molecules.

Molecule		*C-hybridization*	*Polar (yes or no)*
CH_4	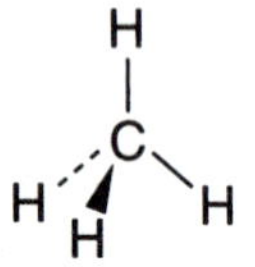 tetrahedral	sp^3	no
C_2H_4	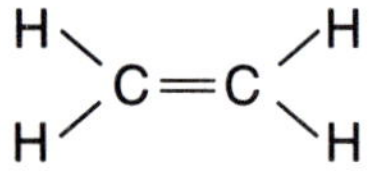 planar	sp^2	no
C_2H_2	$H-C\equiv C-H$ linear	sp	no

6. Calculate the formal charges of all atoms in NH_3, NO_2^-, and NO_3^-.

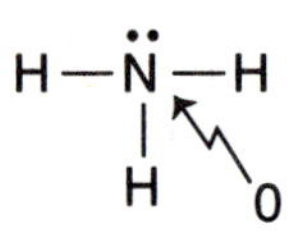

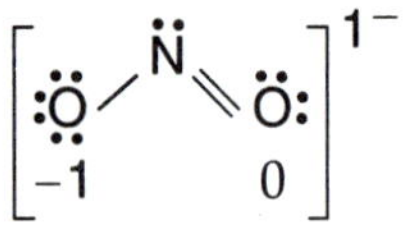

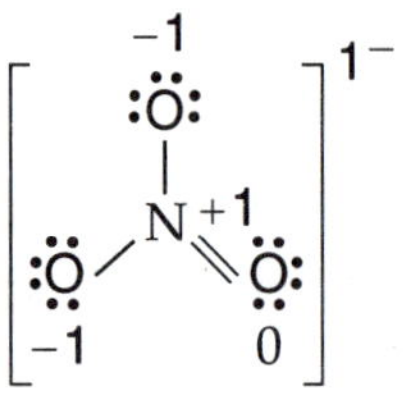

NH_3: H—N—H, H; N: 0

$[NO_2]^{1-}$: O: −1, N, O: 0

$[NO_3]^{1-}$: O (top): −1, N: +1, O (left): −1, O (right): 0

Experiment

12

Atomic Spectra and Atomic Structure

OBJECTIVE

To gain some understanding of the relationship between emission (line) spectra and atomic structure.

APPARATUS AND CHEMICALS

Apparatus

spectroscope with illuminated scale or diffraction grating
hydrogen lamp
mercury-vapor lamp
high-voltage power supply with lamp holder
Nichrome wire loop
Bunsen burner and hose

Chemicals

6 *M* HCl
0.1 *M* $CaCl_2$
0.1 *M* LiCl
unknown solution (mixture containing two or more cations from above)
0.1 *M* NaCl
0.1 *M* $SrCl_2$
0.1 *M* KCl
0.1 *M* $BaCl_2$
unknown solution (one cation from above)

INTRODUCTION

Our present understanding of atomic structure has come from studies of the properties of light or radiant energy and emission and absorption spectra. Radiant energy is characterized by two variables: its wavelength, λ, and its frequency, v. The wavelength and frequency are related to one another by the relation

$$v\lambda = c \qquad [1]$$

where c is the speed of light, 3.00×10^8 meters per second, or m/s; wavelength is usually expressed in nanometers, or nm (10^{-9} m); and frequency is given in cycles per second, or hertz (Hz).

EXAMPLE 12.1

What is the frequency that corresponds to a wavelength of 500 nm?

SOLUTION:

$$v = \frac{c}{\lambda} = \left(\frac{3.00 \times 10^8 \text{ m/s}}{500 \text{ nm}}\right)\left(\frac{10^9 \text{ nm}}{1 \text{ m}}\right)$$
$$= 6.00 \times 10^{14} \text{ s}^{-1} \text{ or } 6.00 \times 10^{14} \text{ Hz}$$

Radiation of different wavelengths affects matter differently. Infrared radiation may cause a "heat burn"; visible and near-ultraviolet light may cause a sunburn or suntan; and X rays may cause tissue damage or even cancer. Some wavelength units for various types of radiation are given in Table 12.1.

TABLE 12.1 Wavelength Units for Electromagnetic Radiation

Unit	Symbol	Length (m)	Type of radiation
Ångstrom	Å	10^{-10}	X ray
Nanometer	nm	10^{-9}	ultraviolet, visible light
Micrometer	μm	10^{-6}	infrared
Millimeter	mm	10^{-3}	infrared
Centimeter	cm	10^{-2}	microwaves
Meter	m	1	TV, radio

A particular source of radiant energy may emit a single wavelength, as in the light from a laser, or many different wavelengths, as in the radiations from an incandescent lightbulb or a star. Radiation composed of a single wavelength is termed *monochromatic.* When the radiation from a source such as the sun, or other stars, is separated into its components, a *spectrum* is produced. This separation of radiations of differing wavelengths can be achieved by passing the radiation through a prism. Each component of the polychromatic radiation is bent to a different extent by the prism, as shown in Figure 12.1. This rainbow of colors, containing light of all wavelengths, is called a *continuous spectrum.* The most familiar example of a continuous spectrum is the rainbow, produced by the dispersion of sunlight by raindrops or mist.

It was found that by placing a narrow slit between the source of radiation and the prism, the quality of the spectrum was improved and the monochromatic components were more sharply resolved. An instrument used for studying line spectra is called a *spectroscope.* A spectroscope (Figure 12.2) contains the following: a slit for admitting a narrow, collimated beam of light; a prism for dispersing the light into its components; an eyepiece for viewing the spectrum; and an illuminated scale against which the spectrum may be viewed.

Most substances will emit light energy if heated to a high enough temperature. For example, a fireplace poker will glow red if left in the fireplace flame for several minutes. Similarly, neon gas will emit bright red light when excited with a sufficiently high electrical voltage. When energy is absorbed by a substance, electrons within the atoms of the substance may be excited

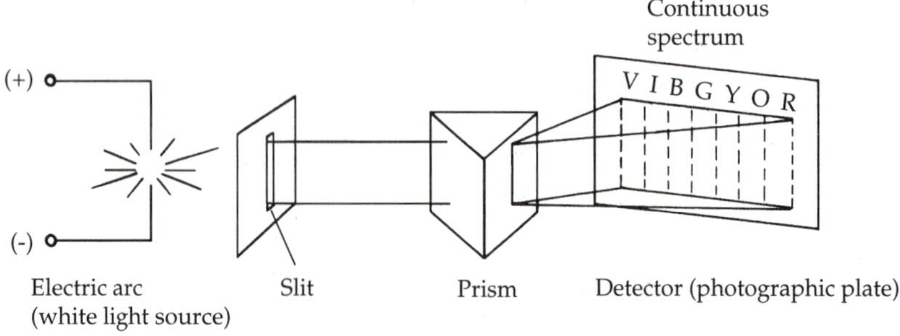

▲ **FIGURE 12.1** The spectrum of "white" light. When light from the sun or from a high-intensity incandescent bulb is passed through a prism, the component wavelengths are spread out into a continuous rainbow spectrum.

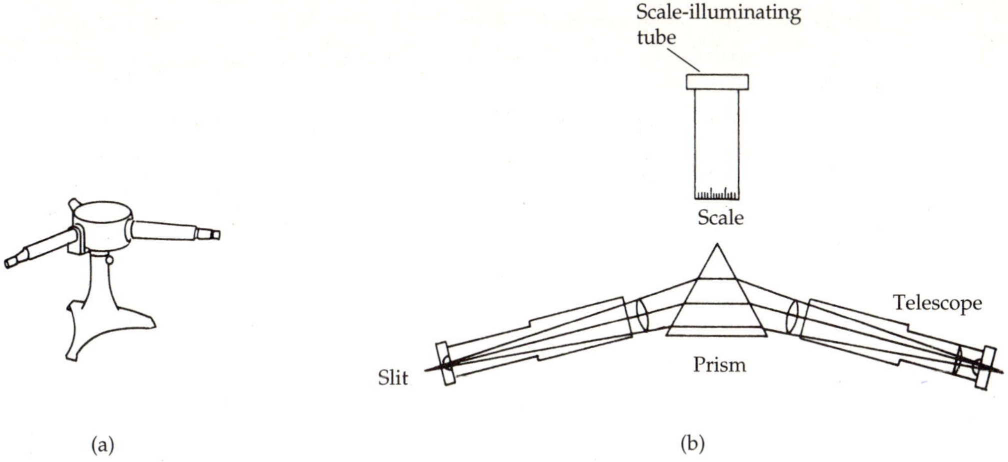

▲ **FIGURE 12.2** (a) A sketch of the spectroscope to be used. (b) A diagram showing its component parts. When viewed through the telescope, the spectrum will appear superimposed on the numerical scale.

to positions of higher potential energy, farther away from the nucleus. When these electrons return to their normal, or ground-state, position, energy will be emitted. Usually some of the energy emitted occurs in the visible region of the electromagnetic spectrum (4000 to 7000 Å, or 400 to 700 nm).

Excited atoms do not emit a continuous spectrum, but rather emit radiation at only certain discrete, well-defined, fixed wavelengths. For example, if you have ever spilled table salt (NaCl) into a flame, you have seen the characteristic yellow emission of excited sodium atoms. If the light emitted by the atoms of a particular element is viewed in a spectroscope, only certain bright-colored lines are seen in the spectrum. See Figure 12.3, which illustrates the emission spectrum of hydrogen.

The observation that a given excited atom emits radiation at only certain fixed wavelengths indicates that the atom can undergo energy changes only of certain fixed, definite amounts. An atom does not emit continuous radia-

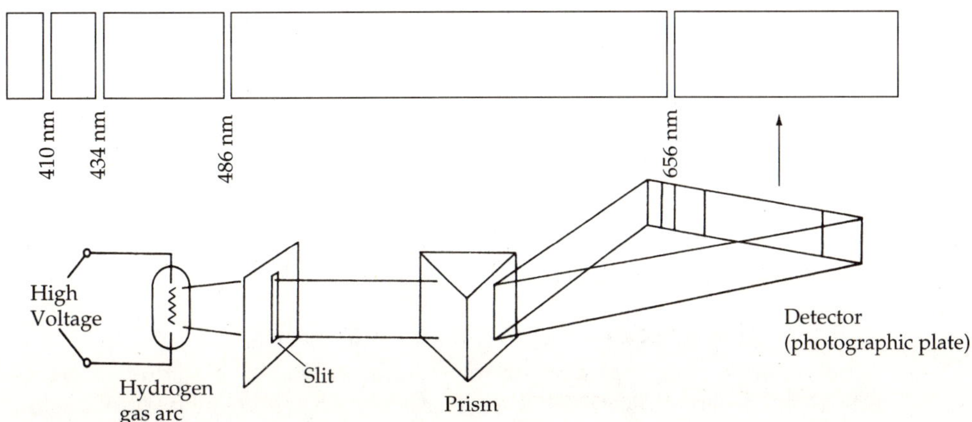

▲ **FIGURE 12.3** The emission spectrum of hydrogen.

tion, but rather emits energy corresponding to definite regular changes in the energies of its component electrons. The experimental demonstration of bright-line atomic emission spectra implied a regular, fixed electronic microstructure for the atom and led to the Bohr model for the hydrogen atom.

In this experiment, you will obtain emission spectra for a few elements and explain your results in terms of the Bohr model. You will use a spectroscope like the one illustrated in Figure 12.2 to obtain your data.

CALIBRATION OF THE SPECTROSCOPE

The scale of the spectroscope has arbitrary divisions and must be calibrated with a known spectrum before the spectrum of an unknown may be obtained. Calibration is accomplished by viewing the emission spectrum of mercury because the emission wavelengths for mercury are very precisely known. You will record the positions on the spectroscope scale of the mercury emission lines and prepare a calibration curve by plotting these positions against the known wavelengths of the lines, which are:

Violet:	404.7 nm
Blue:	435.8 nm
Green:	546.1 nm
Yellow:	579.0 nm

Instructor: Check the electrical connections before the students turn on the power supply.

Turn on the illuminated scale of the spectroscope, and look through the eyepiece to make sure that the scale is visible but not so brightly lighted that the mercury spectral lines will be obscured when the mercury lamp is turned on. With the power supply unplugged, position the power supply and mercury lamp directly in front of the slit opening of the spectroscope. **(CAUTION:** ***The power supply develops several thousand volts. Do not touch any portion of the power supply, wire leads, or lamps unless the power supply is unplugged from the wall outlet. In addition to visible light, the lamps may emit ultraviolet radiation. Ultraviolet radiation is damaging to your eyes. Wear your safety glasses at all times during this experiment, since they will absorb some of the ultraviolet radiation. Do not look directly at any of the lamps while they are illuminated.*** **DO NOT LET THE POWER SUPPLY OR LAMP TOUCH THE SPECTROSCOPE.)** With your Instructor's permission, turn on the power supply and then turn on the power-supply switch to illuminate the mercury lamp. Look into the eyepiece and adjust the slit opening so as to maximize the brightness and sharpness of the emission lines on the scale. If necessary, adjust the position of the illuminated scale so that the numbered divisions are easily read but they do not obscure the mercury spectral lines. Once the slit and scale have been adjusted, do not move them during the course of the rest of the experiment. Record on your report sheet the color and location of the mercury lines on the numbered scale for each line in the visible spectrum of mercury. On the graph paper provided, plot the observed scale reading versus the known wavelength for each line. You will then use this calibration curve to determine the wavelengths of the emission lines for some other atoms.

DISCUSSION

A. Emission Spectrum of Atomic Hydrogen

Atoms absorb and emit radiation with characteristic wavelengths. This was one of the observations that led the Danish physicist Niels Bohr to develop a model for the structure of the hydrogen atom. Within this model the electron of the hydrogen atom moves about the central proton in a circular orbit. Only orbits of certain radii and having certain energies are allowed. In the absence of radiant energy, an electron in an atom remains indefinitely in one of the allowed energy states or orbits. When electromagnetic energy impinges upon the atom, the atom may absorb energy, and in the process an electron will be promoted from one energy state to another. The frequency of energy absorbed is related to the energy difference:

$$\Delta E = hv = \frac{hc}{\lambda} \qquad [2]$$

In the Bohr model, the radius of the orbit is related to the principal quantum number, n:

$$\text{radius} = n^2(5.3 \times 10^{-11}\ \text{m}) \qquad [3]$$

and the energy of the electron is also related to n:

$$E_n = -R_H\left(\frac{1}{n^2}\right) \qquad [4]$$

Thus, as n increases, the electron moves farther from the nucleus and its energy increases. The constant R_H in Equation [4] is called the *Rydberg constant;* it has the value 2.18×10^{-18} J.

Example 12.2

What is the energy of a hydrogen electron when $n = 3$? When $n = 2$?

SOLUTION:

$$E_3 = (-2.18 \times 10^{-18}\ \text{J})\left(\frac{1}{3^2}\right)$$

$$= -2.42 \times 10^{-19}\ \text{J}$$

$$E_2 = (-2.18 \times 10^{-18}\ \text{J})\left(\frac{1}{2^2}\right)$$

$$= -5.45 \times 10^{-19}\ \text{J}$$

The orbital radii and energies are illustrated for $n = 1, 2,$ and 3 in Figure 12.4.

According to Bohr's theory, if an electron were to move from an outer orbit to an inner orbit, a photon of light should be emitted having the energy

$$\Delta E = E_{\text{inner}} - E_{\text{outer}} = -R_H\left(\frac{1}{n^2_{\text{inner}}} - \frac{1}{n^2_{\text{outer}}}\right) \qquad [5]$$

Thus, from Example 12.2, an electron moving from $n = 3$ to $n = 2$ would emit light of energy

$$(5.45 - 2.42)(10^{-19}\ \text{J}) = 3.03 \times 10^{-19}\ \text{J}$$

The wavelength of this photon is given by the Planck relation

$$\lambda = \frac{hc}{\Delta E} \qquad [6]$$

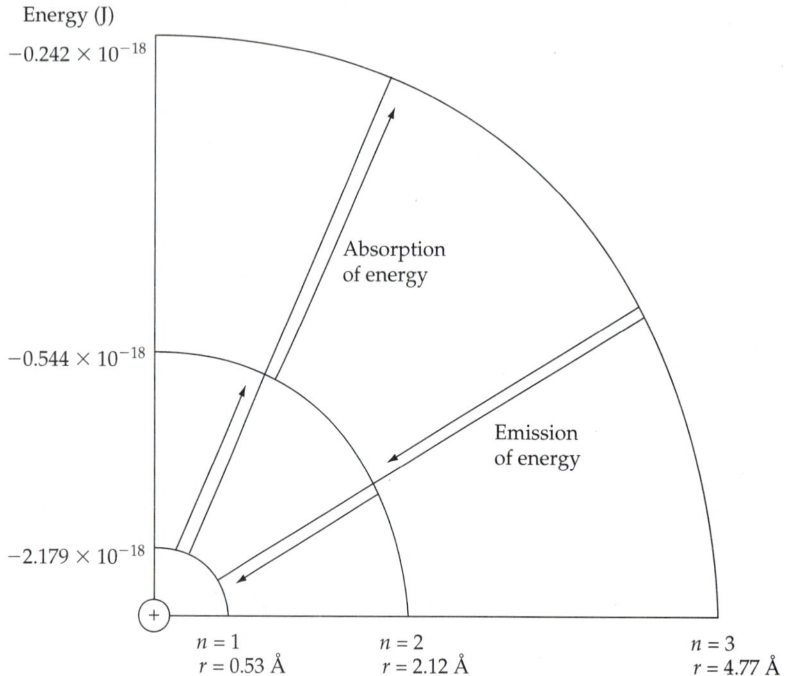

▲ **FIGURE 12.4** Radii and energies of the three lowest energy orbits in the Bohr model of hydrogen. The arrows refer to transitions of the electron from one allowed energy state to another. When the transition takes the electron from a lower- to a higher-energy state, absorption occurs. When the transition is from a higher- to a lower-energy state, emission occurs.

EXAMPLE 12.3

Calculate the wavelength of light emitted for the $n = 3 \longrightarrow n = 2$ transition.

SOLUTION:

$$\lambda = \frac{(6.63 \times 10^{-34}\,\text{J} - \text{s})(3.00 \times 10^{8}\,\text{m/s})}{3.03 \times 10^{-19}\,\text{J}}$$

$$= 6.56 \times 10^{-7}\,\text{m}$$

$$= 656\,\text{nm}$$

By similar calculations Bohr predicted wavelengths for the hydrogen emission spectrum that agreed exactly with the experimental values. He even predicted emission wavelengths in the infrared and ultraviolet regions of the spectrum that had not yet been measured but were later confirmed.

You will measure the emission wavelengths in the visible region for hydrogen with the spectroscope and assign these wavelengths to their corresponding transitions by calculations similar to those in Examples 12.2 and 12.3.

PROCEDURE

Turn on the illuminated scale of the spectroscope and look through the eyepiece to make sure that the scale is visible but not so brightly lighted that the hydrogen spectral lines will be obscured when the hydrogen lamp is turned on. With the power supply unplugged, position the power supply and hydrogen lamp directly in front of the slit opening of the spectroscope. **(CAUTION:** ***The power supply develops several thousand volts. Do not touch any portion of the power supply, wire leads, or lamps unless the power supply is unplugged from the wall outlet. In addition to visible light, the lamps may emit ultraviolet radiation. Ultraviolet radiation is damaging to your eyes. Wear your safety glasses at all times during this experiment, since they will absorb some of the ultraviolet radiation. Do not look directly at any of the lamps while they are illuminated.*** **DO NOT LET THE POWER SUPPLY OR LAMP TOUCH THE SPECTROSCOPE.)** With your instructor's permission, turn on the power supply and then turn on the power-supply switch to illuminate the hydrogen lamp. You should have adjusted the slit and the illuminated scale in the calibration step. Should they require further adjustment, adjust the slit so as to maximize the line intensity and sharpness. On your report sheet, record the color and location of the hydrogen lines on the numbered scale for each line in the visible spectrum of hydrogen. You should easily observe the red, blue-green, and violet lines. A second, faint violet line may also be visible if the room and scale illumination are not too bright.

Instructor: Check the set up before the power supply is turned on.

Use the calibration curve that you constructed to determine the wavelengths of the lines in the hydrogen emission spectrum. Find the true wavelengths of these lines in your textbook or a handbook and calculate the percent error in your determination of the wavelength of each line:

$$\%\ \text{error} = \frac{(\text{true value} - \text{experimental value})}{(\text{true value})} \times 100 \qquad [7]$$

Use Equations [5] and [6] to calculate the wavelengths in nanometers for the $n = 3 \longrightarrow n = 2$, $n = 4 \longrightarrow n = 2$, $n = 5 \longrightarrow n = 2$, and $n = 6 \longrightarrow n = 2$ transitions. How do these calculated values compare with your experimental values? Assign the transitions.

B. Emission Spectra of Group 1A and Group 2A Elements

DISCUSSION

Since the energies of the electrons in the atoms of different elements are different, the emission spectrum of each element is unique. The emission spectrum may be used to detect the presence of an element in both a qualitative and a quantitative way. A number of common metallic elements emit light strongly in the visible region, allowing their detection with a spectroscope. For these elements, the emissions are so intense that the elements may often be recognized by the gross color that they impart to a flame. For example, lithium ions impart a red color to a flame; sodium ions, a yellow color; potassium ions, a violet color; calcium ions, a brick-red color; strontium ions, a bright red color; and barium ions, a green color. If we examine the emission spectra of these ions with a spectroscope, we find that as with mercury and hydrogen, the emission spectra are composed of a series of lines. The series is unique for each metal. Consequently, a flame into which both lithium and strontium, for example, had been placed would be red, and we could not tell with our naked eye that both these ions were present. However, with the aid of a spectroscope, we could detect the presence of both ions.

PROCEDURE

You will obtain the emission spectra of the ions of each of the elements listed in the preceding discussion. You will use a Bunsen burner as an excitation source and observe the gross color imparted to the flame by these ions. With this information, you will determine the contents of two unknown solutions—one containing only one of these metal ions, and the other a mixture containing two or more of these ions.

There are two ways to introduce the metal ions into the flame. First, you can use a wire loop to pick up a drop of the metal-ion solution and then place the drop into the flame for vaporization. Although this method is simple and inexpensive, it produces only a brief burst of color before the sample evaporates completely. Second, you can introduce a fine mist of sample into the flame by using a spray bottle. This method produces a longer-lived emission that is therefore easier to see. Your instructor will tell you which method to use.

Instructor: Tell the students which method to use.

WORK IN PAIRS FOR THIS PART OF THE EXPERIMENT

If you use the wire-loop method, obtain several 6-in. lengths of Nichrome wire and about 10 mL of 6 *M* HCl. Bend the last 1/4-in. of each wire into a small circular loop for picking up the sample solutions. Dip the loop into the 6 *M* HCl solution to remove any oxides that are present, rinse the loop in distilled water, and then heat the loop in the hottest part of the flame until no color is imparted to the flame by the wires.

If the sprayer is used, check to ensure that the sprayer produces a *fine* mist. If the sprayer nozzle is adjustable, try adjusting it to make a very fine mist. If the nozzle cannot be adjusted, ask your instructor how to clean it to improve the mist that it produces.

Instructor: Check the position of the Bunsen burner.

Set up a Bunsen burner directly in front of the slit of the spectroscope but at a sufficient distance to avoid damage to the spectroscope. Have your instructor check the burner placement before you ignite the burner. Ignite the burner and adjust the flame so that it is as hot as possible. Adjust the illuminated scale of the spectroscope so that approximate positions of the emission lines may be determined. It is not necessary to make exact measurements of the emission wavelengths.

With either method of introduction (loop or mist), introduce the metal-ion solution into the flame and note the gross color of the flame. Record your observations on your report sheet. Then, while looking through the eyepiece of the spectroscope, have your partner introduce the metal-ion solution into the flame, noting the color, intensity, and approximate scale position of the brightest lines in the emission spectrum of the metal ions. Repeat the above with each of the other metal-ion solutions. If you use the wire-loop method, use a new loop for each solution, or clean the wire loop with HCl and then distilled water and place the wire loop in the flame until it imparts no color to the flame.

After obtaining the emission spectrum of each of the known metal-ion solutions, obtain the emission spectrum of a single metal unknown, and by matching the colors, intensities, and positions of the lines in the emission spectrum to those of a known metal, identify your unknown. Record your results on your report sheet.

Obtain an unknown mixture and its emission spectrum as above. Then compare the color, intensity, and positions of the brightest lines in this spectrum with those of the knowns. In this way, *determine which metal ions are present in the unknown mixture.* Record your results on your report sheet.

REVIEW QUESTIONS

Before beginning this experiment in the laboratory, you should be able to answer the following questions:

1. Name the colors of visible light, beginning with that of lowest energy (longest wavelength).
2. Distinguish between absorption and emission of energy.
3. A system proposed by the U.S. Navy for underwater submarine communication, called ELF (for "extremely low frequency"), operates with a frequency of 76 Hz. What is the wavelength of this radiation in meters? In miles? (1 mile = 1.61 km).
4. What is the energy in joules of the frequency given in question 3?
5. Red and green light have wavelengths of about 650 nm and 490 nm, respectively. Which light has the higher frequency—red or green? Which light has the higher energy—red or green?
6. Mg emits radiation at 285 nm. Could a spectroscope be used to detect this emission?
7. If boron emits radiation at 518 nm, what color will boron impart to a flame?
8. From the wavelengths and colors given for the mercury emission spectrum in this experiment, construct a graphical representation of the mercury emission spectrum as it would appear on the scale of a spectroscope.

NOTES AND CALCULATIONS

Name ______________________ Desk ______________________

Date ______________ Laboratory Instructor ______________________

Single-ion unknown no. ____________ Unknown mixture no. ______________________

REPORT SHEET | EXPERIMENT

Atomic Spectra and Atomic Structure | 12

Calibration of Spectroscope

Lines observed in emission spectrum of mercury

Color	*Position on scale*	*Wavelength*
Violet	1.71	404.7 nm
Blue	3.62	435.8 nm
Green	7.46	546.1 nm
Yellow	8.00	579.0 nm

A. Emission Spectrum of Hydrogen

Lines observed in emission spectrum of hydrogen

Color	*Position on scale*	*Wavelength from calibration curve*	*Assignment*
Violet	2.1	405 nm	$n = 6 \longrightarrow n = 2$
Blue	3.1	433 nm	$n = 5 \longrightarrow n = 2$
Green	4.9	483 nm	$n = 4 \longrightarrow n = 2$
Red	10.9	650 nm	$n = 3 \longrightarrow n = 2$

In order to make the assignments of the observed transitions, use equations [5] and [6] to calculate the wavelengths of the

$n = 6 \longrightarrow n \longrightarrow 2$ transition 410 nm

$n = 5 \longrightarrow n = 2$ transition 434 nm

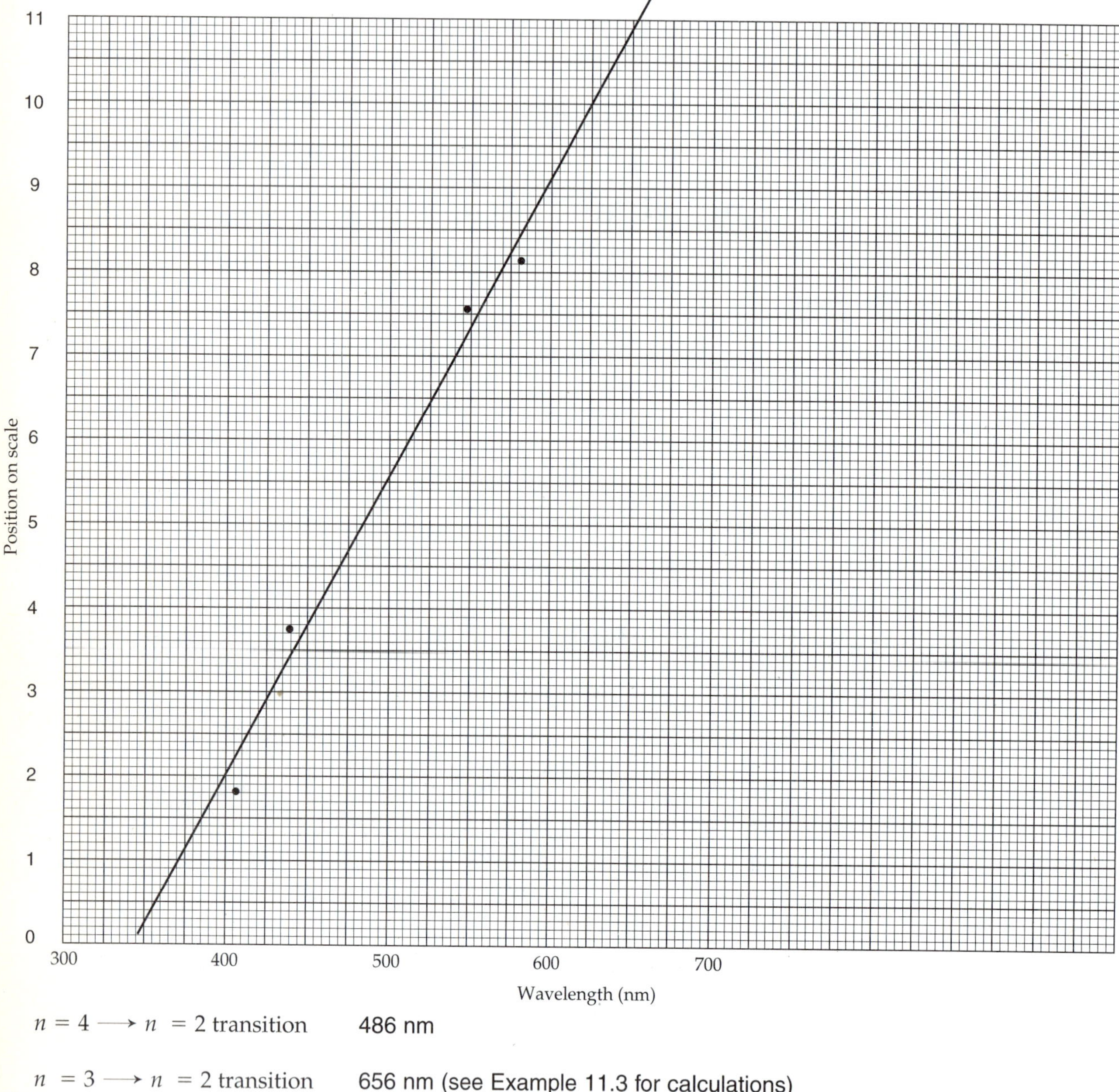

$n = 4 \longrightarrow n = 2$ transition 486 nm

$n = 3 \longrightarrow n = 2$ transition 656 nm (see Example 11.3 for calculations)

Consult your text or a handbook to find the accepted values for the emission spectrum of hydrogen, and calculate the percent errors of your experimental results.

Accepted wavelength	*Observed wavelength*	*Percent error*
365 nm	not observed	
410 nm	405 nm	1
434 nm	433 nm	0.2
486 nm	483 nm	0.6
656 nm	650 nm	0.9

Percent error calculations

line 1

$$\frac{(410 - 405)}{410}(100) = 1.2\%$$

line 2

$$\frac{(434 - 433)}{434}(100) = 0.2\%$$

line 3

$$\frac{(486 - 483)}{486}(100) = 0.6\%$$

line 4

$$\frac{(656 - 650)}{656}(100) = 0.9\%$$

B. Emission Spectra of Group 1A and Group 2A Elements

Which method did you use to introduce the metal ion into the Bunsen burner flame? Aspiration

1. *Known metal ions*	*Gross flame color*	*Position and colors of the lines observed on the spectroscope scale*
Li	red	8.7-9.4 (red)
Na	yellow	7.6-8.3 (yellow)
K	red (pink)	6-7.6 (green), 7.6-8.2 (yellow)
Ca	red-orange	7.0-7.6 (green), 7.6-8.1 (yellow), 8.1-8.8 (red)
Sr	red	8.0-9.0 (red)
Ba	pale green	6-7.5 (green)

unknown dependent typical results

2. *Unknowns*	*Gross flame color*	*Position and colors of the lines observed on the spectroscope scale*
Single-ion unknown	bright red	8.7-9.4 (red)
Mixture	yellow	8.7-9.4 (red) and 7.6-8.3 (yellow)

What is the identity of the single-ion unknown? Li

What ions are present in the mixture? Li , Na

QUESTIONS

1. What is the purpose of the slit in the spectroscope?
   To focus the emitted light.

2. Why is the spectroscope scale illuminated?
   So that you may see it.

3. Why was the emission spectrum of mercury used to calibrate the spectroscope?
   Because its emission spectrum is very accurately known.

4. Could the emission spectrum of some other element be used to calibrate the spectroscope?
   Yes, if its emission spectrum is precisely known. H_2 would be suitable.

5. In addition to the spectral lines that you observed in the emission spectrum of hydrogen, several other lines are also present in other regions of the spectrum. Calculate the wavelengths of the $n = 4 \longrightarrow n = 1$ and $n = 4 \longrightarrow n = 3$ transitions and indicate in which regions of the spectrum these transitions would occur.

$n = 4 \rightarrow n = 1$:

$$E_4 = (-2.18 \times 10^{-18}\ \text{J})\left(\frac{1}{4^2}\right) = -1.36 \times 10^{-19}\ \text{J}$$

$$E_1 = -2.18 \times 10^{-18}\ \text{J}$$

$$\Delta E = (21.8 - 1.36)(10^{-19}\ \text{J}) = 20.44 \times 10^{-19}\ \text{J}$$

$$\lambda = \frac{(6.63 \times 10^{-34}\ \text{J-s})(3.00 \times 10^{8}\ \text{m/s})}{20.44 \times 10^{-19}\ \text{J}} = 9.73 \times 10^{-8}\ \text{m}$$

$$= 97.3\ \text{nm}$$ vacuum ultraviolet

$n = 4 \rightarrow n = 3$

$$E_4 = -1.36 \times 10^{-18}\ \text{J};\ E_3 = -0.242 \times 10^{-18}\ \text{J}$$

$$\Delta E = (1.36 - 0.242)(10^{-18}\ \text{J}) = 1.118 \times 10^{-18}\ \text{J}$$

$$\lambda = \frac{(6.63 \times 10^{-34}\ \text{J-s})(3.00 \times 10^{8}\ \text{m/s})}{1.118 \times 10^{-18}\ \text{J}} = 1.78 \times 10^{-7}\ \text{M}$$

$$= 1.78 \times 10^{3}\ \text{nm}$$ near infrared

6. Of the metal ions tested, sodium gives the brightest and most persistent color in the flame. Do you think that potassium could be detected visually in the presence of sodium by burning this mixture in a flame? Could you detect both with a spectroscope?
   No. Technically, yes; however, with this spectrometer it is very difficult.

7. The minimum energy required to break the oxygen–oxygen bond in O_2 is 495 kJ/mol. What is the longest wavelength of radiation that possesses the necessary energy to break the O—O bond? What type of electromagnetic radiation is this?

$$\lambda = \frac{hc}{E} = \frac{(6.63 \times 10^{-34}\ \text{J-s})(3.00 \times 10^{8}\ \text{m/s})}{(495 \times 10^{3}\ \text{J/mol})(1\ \text{mol}/6.02 \times 10^{23}\ \text{molecules})}$$

$$= 2.42 \times 10^{-7}\ \text{m}$$

$= 242$ nm   This is ultraviolet radiation

Experiment

Behavior of Gases: Molar Mass of a Vapor

OBJECTIVE

To observe how changes in temperature and pressure affect the volume of a fixed amount of a gas; to determine the molar mass of a gas from a knowledge of its mass, temperature, pressure, and volume.

APPARATUS AND CHEMICALS

Apparatus

gas-law demonstration apparatus
balance
125-mL Erlenmeyer flask
250-mL graduated cylinder (one per class)
rubber band
wire gauze
boiling chips
barometer
Bunsen burner and hose
600-mL beaker
2-in. square of aluminum foil
pins
ring stand and iron ring
utility clamp
thermometer

Chemicals

volatile unknown liquid

DISCUSSION

The Effect of Pressure on the Volume of a Gas

The effect of pressure on the volume of a gas can be determined by using a gas buret, as shown in Figure 13.1. The volume of the buret is graduated in terms of cubic centimeters (cm^3) or milliliters (mL). When the stopcock is opened, air can enter the buret, and the level of mercury will be equal in both tubes. If the stopcock is then closed, a fixed volume of air is trapped in the buret at prevailing atmospheric pressure. Raising the leveling bulb increases the pressure on the gas; the new pressure on the gas corresponds to the prevailing atmospheric pressure plus the height that the mercury in the leveling bulb is above the level of mercury in the buret. It is found that when the pressure is doubled, the volume is halved; and when the pressure is halved, the volume is doubled. From such experiments we conclude that at constant temperature, the volume of a given amount of gas is *inversely* proportional to the pressure. This is *Boyle's law,* which Boyle enunciated in 1662. This may be expressed mathematically as

$$\frac{V_1}{V_2} = \frac{P_2}{P_1}$$

or

$$V = \frac{k}{P}$$

where V_1 is the volume at pressure P_1, V_2 is the volume at pressure P_2, and k is a proportionality constant.

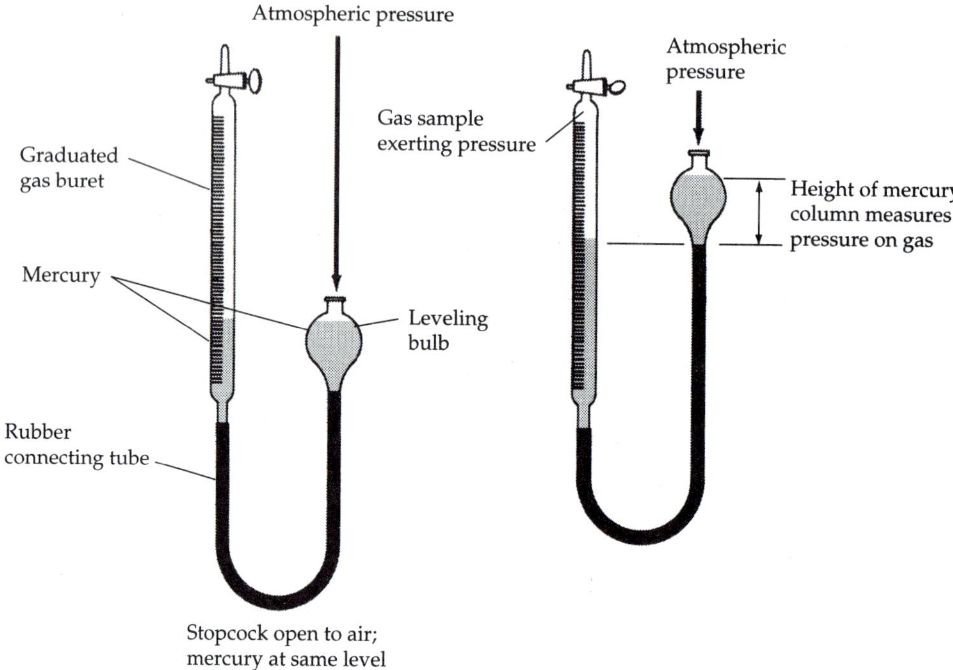

▲ **FIGURE 13.1** A gas buret.

By means of this law, we can calculate the volume of a gas at any pressure, provided we know the volume at a given pressure. Example 13.1 is illustrative.

Example 13.1

If 100 mL of a gas is enclosed in the buret at 760 mm Hg, what volume would the gas occupy at 1520 mm Hg at the same temperature?

SOLUTION: We could let $V_1 = 100$ mL, $P_1 = 760$ mm Hg, and $P_2 = 1520$ mm Hg, substitute these values into Equation [1], and solve for V_2. A better approach is to reason through the problem. Since the pressure increases from 760 mm Hg to 1520 mm Hg, the new volume must be *less* than 100 mL. The volume decreases in the same ratio that the pressure increases. The new volume is, therefore,

$$V_2 = 100 \text{ mL} \times \frac{760 \text{ mm Hg}}{1520 \text{ mm Hg}} = 50.0 \text{ mL}$$

If the volume had been multiplied by the fraction 1520 mm Hg/760 mm Hg, a fraction larger than 1, the answer would have been larger than 100 mL. And this, we know, must be incorrect.

The Effect of Temperature on the Volume of a Gas

The relation between temperature and volume of a given sample of gas can be studied in a gas buret while the pressure is held constant. We are all familiar with the fact that gases expand when heated. It has been observed that at constant pressure the volume of a given mass of a gas is directly proportional to the absolute (or Kelvin) temperature. This is known as *Charles's law.*

To convert a Celsius temperature to the Kelvin (or absolute) scale, add 273°. Thus 20°C equals (20° + 273°) = 293 K.

Charles's law may be expressed mathematically as follows:

$$\frac{V_1}{V_2} = \frac{T_1}{T_2}$$

or

$$V = k'T$$

where V_1 is the volume at temperature T_1, V_2 is the volume at T_2, and k' is a proportionality constant. An application of this law is illustrated in Example 13.2.

EXAMPLE 13.2

Suppose a sample of oxygen has a volume of 100 mL at a temperature of 20°C. What will be its volume at 100°C at the same pressure?

SOLUTION: Since the temperature increases, the volume must increase. The volume at 100°C equals the original volume multiplied by a fraction made up of the Kelvin temperatures, which is larger than 1:

$$V \text{ at } 100°\text{C} = 100 \text{ mL} \times \frac{373 \text{ K}}{293 \text{ K}} = 127 \text{ mL}$$

Since gases can occupy any volume, it is most informative to compare them under the same conditions. Standard conditions of temperature and pressure, designated as STP, are often used for this purpose. Standard temperature is 0°C, and standard pressure is 760 mm Hg, or one atmosphere (1 atm).

Problems involving changes both in temperature and pressure are worked by combining the effects due to each of these changes. Example 13.3 is illustrative.

EXAMPLE 13.3

Suppose a sample of oxygen at 100°C occupies a volume of 100 mL at 700 mm Hg. What will be its volume at STP?

SOLUTION:

$$V_1 = 100 \text{ mL}$$

$$V_2 = ?$$

$$T_1 = 100° + 273° = 373 \text{ K}$$

$$T_2 = 0° + 273° = 273 \text{ K}$$

$$P_1 = 700 \text{ mm Hg}$$

$$P_2 = 760 \text{ mm Hg}$$

The volume at STP will be equal to the original volume multiplied by a fraction made up of the two temperatures and also another fraction made up of the two pressures. The temperature fraction will be less than 1, because the temperature decreases, and this results in a decrease in volume. The pressure fraction will be less than 1, because the pressure increases, and an increase in pressure also causes the volume to decrease:

$$\text{Volume at STP} = 100 \text{ mL} \times \frac{273 \text{ K}}{373 \text{ K}} \times \frac{700 \text{ mm Hg}}{760 \text{ mm Hg}}$$

$$= 67.4 \text{ mL}$$

Ideal-Gas Law

It is possible to relate the four variables of a gas—pressure, volume, temperature, and number of moles—in one equation, which is referred to as the *ideal-gas law.* Before doing this, we must recognize *Avogadro's law,* which states that *equal volumes of all gases, at the same conditions of temperature and pressure, contain the same number of molecules.* Thus, 6.022×10^{23} molecules (Avogadro's number) of any gaseous substance should occupy the same volume under the same conditions. One mole of an ideal gas at STP occupies 22.4 L, a value known as the *molar volume.*

According to Charles's law, the volume of a fixed number of moles, n, of a gas at a constant pressure is directly proportional to the Kelvin temperature:

$$V \propto T \quad \text{at constant } P \text{ and } n$$

According to Boyle's law, the volume of a fixed number of moles of a gas at constant temperature is inversely proportional to the pressure:

$$V \propto \frac{1}{P} \quad \text{at constant } T \text{ and } n$$

And according to Avogadro's law,

$$\mathrm{V} \propto n \quad \text{at constant } P \text{ and } T$$

If a quantity is proportional to two or more variables, it is proportional to the product of those variables:

$$V \propto T \times \frac{1}{P} \times n \qquad [3]$$

The proportionality symbol, $\propto$, in Equation [3] can be replaced by an equals sign by introducing a proportionality constant, R:

$$V = R \times T \times \frac{1}{P} \times n$$

or

$$PV = nRT \qquad [4]$$

Equation [4] is the ideal-gas law. R, the ideal-gas constant, can be calculated by considering 1 mol of an ideal gas at STP:

$$R = \frac{P \times V}{n \times T} = \frac{760 \text{ mm} \times 22{,}400 \text{ mL}}{1 \text{ mol} \times 273 \text{ K}} = 62{,}400 \text{ mL–mm Hg/mol–K}$$

or, in other units,

$$R = \frac{1 \text{ atm} \times 22.4 \text{ L}}{1 \text{ mol} \times 273 \text{ K}} = 0.0821 \text{ L–atm/mol–K}$$

The number of moles of a substance (n) equals its mass in grams, m, divided by the number of grams per mole (that is, its molar mass, $\mathcal{M}$): $n = m/\mathcal{M}$. Making this substitution into Equation [4] gives

$$PV = \left(\frac{m}{\mathcal{M}}\right)RT \qquad [5]$$

The units of P, V, T, and m must, of course, be expressed in units consistent with the value of R. Example 13.4 illustrates how this equation may be used to calculate the molar mass of a gas.

EXAMPLE 13.4

A gaseous sample weighing 0.896 g was found to occupy a volume of 524 mL at 730 mm Hg and 28°C. What is the molar mass of the gas?

SOLUTION: Solving Equation [5] for molar mass and substituting the appropriate values, we find

$$\mathcal{M} = \frac{mRT}{PV} \qquad [6]$$

$$= \frac{0.896 \text{ g} \times 62{,}400 \text{ mL–mm Hg/mol–K} \times 301\text{K}}{730 \text{ mm Hg} \times 524 \text{ mL}}$$

$$= 44.0 \text{ g/mol}$$

The first portion of this experiment will be a demonstration of Charles's and Boyle's laws. The laboratory instructor will vary the pressure, temperature, and volume of a sample of air using a gas buret.

In the second portion of this experiment, you will determine the molar mass of a volatile liquid using Equation [6]. A small quantity of liquid sample is placed in a pre-weighed flask and vaporized so as to expel all the air from the flask, leaving it filled with the vapor at a known temperature (temperature of boiling water) and atmospheric pressure. The flask plus vapor is then cooled so that the vapor condenses. The flask plus condensed vapor plus air is then weighed. The mass of air, being nearly identical before and after, cancels out and allows one to determine the mass of the vapor. The above data, in conjunction with the volume of the flask, permit the calculation of the molar mass.

PROCEDURE

Instructor: See page xix for a description of a convenient demonstration apparatus.

A. Demonstration Experiment

Record the data as collected by the laboratory instructor on the report sheet. Calculate the volumes by means of the appropriate gas laws and compare these with the observed volumes. Express the differences as percent error.

B. Molar Mass of a Vapor

Record the number of your unknown liquid (1). Place a small square of aluminum foil over the mouth of a clean, dry 125-mL Erlenmeyer flask. Fold the foil loosely around the neck and secure with a rubber band (Figure 13.2). Make a very small hole in the center of the foil with a pin. Weigh the flask, foil cap, and rubber band (2).

Place about 350 mL of water and some boiling chips in a 600-mL beaker and begin heating it to bring the water to a boil. While the water is heating, remove the foil from the Erlenmeyer flask and place about 2 mL of your unknown liquid in the flask; then replace the foil and rubber band. Clamp the flask at the top of the neck. After the water has been brought to a boil, record its temperature (3) and barometric pressure (4). If you calibrated your thermometer in Experiment 1, apply its correction. Insert the Erlenmeyer flask as far as possible into the boiling water, holding it by hand using the clamp. It is not necessary to clamp the flask to the ring stand. After about 4 minutes remove the flask from the boiling water and examine it to see if all of the unknown liquid has vaporized (including any that condensed in the neck). If it has not, reinsert the flask into the boiling water for a few minutes. After all of the liquid has vaporized, remove the flask by means of the clamp and set it aside to cool.

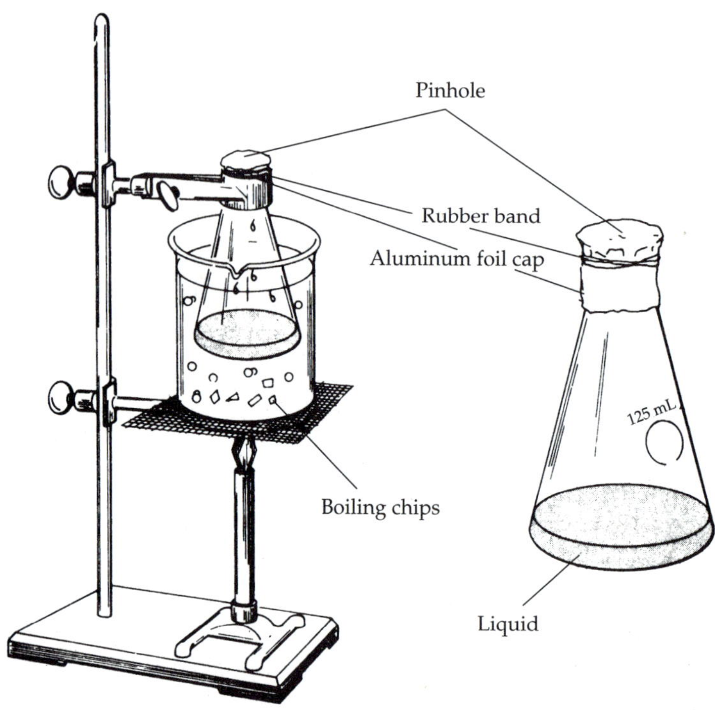

▲ FIGURE 13.2

After the flask has cooled to room temperature, wipe it dry and remove any water that may adhere to the aluminum. Weigh the flask, cap, rubber band, and condensed unknown liquid (5). Calculate the mass of the condensed liquid (6).

Remove the cap and fill the flask completely with water. Measure the volume by pouring the water into a large graduated cylinder (7).

Calculate the molar mass of the unknown using Equation [6].

REVIEW QUESTIONS

Before beginning this experiment in the laboratory, you should be able to answer the following questions:

1. How does the pressure of an ideal gas at constant volume change as the temperature increases?
2. How does the volume of an ideal gas at constant temperature change as the pressure increases?
3. How does the volume of an ideal gas at constant temperature and pressure change as the number of molecules increases?
4. Write the ideal-gas equation and give the units for each term when $R = 0.0821$ L−atm/K−mol.
5. Show by mathematical equations how one can determine the molar mass of a volatile liquid by measurement of the pressure, volume, temperature, and mass of the liquid.

6. If 0.80 g of a gas occupies 295 mL at 25°C and 680 mm Hg of pressure, what is the molar mass of the gas?

7. A sample of nitrogen occupies a volume of 600 mL at 60°C and 650 mm Hg of pressure. What will be its volume at STP?

8. Consider Figure 13.1. If the height of the mercury column in the leveling bulb is 40 mm greater than that in the gas buret and atmospheric pressure is 650 mm, what is the pressure on the gas trapped in the buret?

9. Consider Figure 13.1. If the level of the mercury in the leveling bulb is lowered, what happens to the volume of the gas in the gas buret?

10. Show that Boyle's law, Charles's law, and Avogadro's law can be derived from the ideal-gas law.

11. Methane burns in oxygen to produce CO_2 and H_2O

$$CH_4(g) + 2O_2(g) \rightarrow 2H_2O(l) + CO_2(g)$$

If 2.5 L of gaseous CH_4 is burned at STP, what volume of O_2 is required for complete combustion? What volume of CO_2 is produced?

12. Calculate the density of O_2 at STP (a) using the ideal-gas law and (b) using the molar volume and molar mass of O_2. How do the densities compare?

NOTES AND CALCULATIONS

Name ______________________ Desk ______________________

Date ______________ Laboratory Instructor ______________________

Unknown no. ______________________

REPORT SHEET | EXPERIMENT

Behavior of Gases: Molar Mass of a Vapor | 13

A. Demonstration Experiment

1. Boyle's law—effect of pressure at constant temperature

	Trial 1	*Trial 2*
First pressure, barometer reading, mm Hg	650 mm	650 mm
Difference in mercury levels (+ or −), mm Hg	+126 mm	−112 mm
Second pressure, mm Hg	776 mm	538 mm
First volume, mL	46.5 mL	29.3 mL
Second volume, measured, mL	39.2 mL	35.2 mL
Second volume, calculated, mL	38.9 mL	35.4 mL
Percent error	0.6%	0.6%

(show calculations)

Trial 1

$$V_2 = V_1\left(\frac{P_1}{P_2}\right) = 46.5\text{ mL}\left(\frac{650\text{ mm}}{760\text{ mm}}\right) = 38.9\text{ mL}$$

$$\%\text{ error} = \left(\frac{39.2 - 38.9}{38.9}\right)(100) = 0.6\%$$

Trial 2

$$V_2 = 29.3\text{ mL}\left(\frac{650\text{ mm}}{538\text{ mm}}\right) = 35.4\text{ mL}$$

$$\%\text{ error} = \left(\frac{35.2 - 35.4}{35.4}\right)(100) = 0.6\%$$

2. Charles's law—effect of temperature at constant pressure

First temperature	22.1 °C =	295.25 K
Second temperature	95.2 °C =	368.35 K
First volume, mL		29.4 mL
Second volume, measured, mL		36.1 mL
Second volume, calculated, mL		36.7 mL
Percent error		2%

(show calculations)

$$V_2 = \frac{V_1 T_2}{T_1} = \frac{(29.4 \text{ mL})(368.35 \text{ K})}{(295.25 \text{ K})} = 36.7 \text{ mL}$$

$$\% \text{ error} = \left(\frac{36.7 - 36.1}{36.7}\right)(100) = 2\%$$

B. Molar Mass of a Vapor

unknown dependent—typical data

1. Unknown liquid number 1
2. Mass of flask + cap + rubber band 66.4456 g
3. Temperature of boiling water 95.2 °C
   Correction +0.1 °C
   Corrected temperature 95.3 °C
4. Barometric pressure 650 mm Hg
5. Mass of flask + rubber band + cap + condensed vapor 66.7559 g
6. Mass of condensed liquid 0.3103 g
7. Volume of flask 152.5 mL
8. Molar mass of vapor 72.0 g/mol
   (show calculations)
   PV = *n*RT   *n* = PV/RT   W = $\mathcal{M} \times n$ so $\mathcal{M}$ = W/*n*

$$\mathcal{M} = \frac{(0.3103 \text{ g})(62{,}400 \text{ mL} \cdot \text{mm Hg/mol-K})(368.45 \text{ K})}{(650 \text{ mm Hg})(152.5 \text{ mL})} = 72.0 \text{ g/mol}$$

Data are for tetrahydrofuran C_4H_8O MW = 72.10 g/mole

QUESTIONS

1. If an insufficient amount of liquid unknown had been used, how would this have affected the value of the experimental molar mass?
   It would be too low because the flask would not be filled with vapor.
2. What are the major sources of error in your determination of the molar mass?
   Volume measurement and ensuring that the flask is filled with vapor and no vapor escapes from the flask. Removing the water from the outside of the flask is very important.

3. If the flask was not thoroughly wiped dry, how would this affect the molar mass?

   The molar mass would be too high because the mass would be too high.

4. Isobutyl alcohol has a boiling point of 108°C. How would you modify the procedure used in this experiment to determine its molar mass?

   One would need to use a liquid in the bath that has a higher boiling point than 108°C.

GAS-LAW PROBLEMS

1. What volume will 300 mL of gas at 20°C and a pressure of 355 mm Hg occupy if the temperature is reduced to −80°C and the pressure increased to 760 mm Hg?

$$V_2 = \frac{(355\text{ mm})(300\text{ mL})(193\text{ K})}{(293\text{ K})(760\text{ mm})} = 92.3\text{ mL}$$

2. A sample of gas of mass 2.82 g occupies a volume of 639 mL at 27°C and 1.00 atm pressure. What is the molar mass of the gas?

$$\mathcal{M} = \frac{(2.82\text{ g})(300\text{ K})(0.0821\text{ L–atm/mol–K})}{(0.639\text{ L})(1.00\text{ atm})} = 147\text{ g/mol}$$

3. A gas is placed in a storage tank at a pressure of 20.0 atm at 20.3°C. As a safety device, there is a small metal plug in the tank made of a metal alloy that melts at 125°C. If the tank is heated, what is the maximum pressure that will be attained in the tank before the plug will melt and release gas?

$$P_2 = \frac{P_1T_2}{T_1} = \frac{(20.0\text{ atm})(398\text{ K})}{(293.3\text{ K})} = 27.1\text{ atm}$$

4. Sixty liters of a gas were collected over water when the barometer read 670 mm Hg, and the temperature was 20°C. What volume would the dry gas occupy at standard conditions? (HINT: Consider Dalton's law of partial pressures.)

   Total pressure = vapor pressure (H_2O) + pressure (gas)
   vapor pressure of H_2O at 20°C is 17.5 mm Hg
   hence, gas pressure = 670 − 17.5 = 652.5 mm Hg

$$V_2 = \frac{P_1V_1T_2}{T_1P_2} = \frac{(652.5\text{ mm Hg})(60\text{ L})(273\text{ K})}{(293\text{ K})(760\text{ mm Hg})} = 48\text{ L}$$

5. Eight moles of hydrogen gas at 0°C are forced into a steel cylinder with a volume of 300 mL. What is the pressure of the gas in the cylinder?

$$P = \frac{(8.0\text{ mol})(62{,}400\text{ mL–mm Hg/mol–K})(273\text{ K})}{(300\text{ mL})} = 4.5 \times 10^5\text{ mm Hg}$$

6. What is the density of He at STP? Why do helium-filled balloons rise in air?

$\mathcal{M}$ He = 4.00 g, and the molar volume of an ideal gas is 22.4 L.

$$\text{Density} = \frac{4.00\text{ g}}{22.4\text{ L}} = 0.179\text{ g/L}$$

Helium is much less dense than air (about 1.2 g/L)

7. What volume in milliliters will 6.5 g of CO_2 occupy at STP?

$$PV = nRT \qquad V = \frac{nRT}{P} = \frac{\left(\frac{6.5\text{ g}}{44\text{ g/mol}}\right)(0.082\text{ L–atm/mol–K})(273\text{ K})}{1\text{ atm}} = 3.3\text{ L or } 3.3 \times 10^3\text{ mL}$$

$$\text{also} \left(\frac{6.5\text{ g}}{44\text{ g/mol}}\right)\left(\frac{22.4\text{ L}}{\text{mol}}\right) = 3.3\text{ L}$$

8. If 20.0 g of O_2 and 4.4 g of CO_2 are placed in a 5.00-L container at 21°C, what is the pressure of this mixture of gases?

$$\frac{20.0\text{ g }O_2}{32.0\text{ g/mol}} = 0.625\text{ mol }O_2 \qquad \frac{4.4\text{ g }CO_2}{44\text{ g/mol}} = 0.100\text{ mol }CO_2$$

total moles gas = 0.725 mol

$$P = \frac{nRT}{V} = \frac{0.725\text{ mol}(0.0821\text{ L–atm/mol–K})(294\text{ K})}{(5.00\text{ L})} = 3.50\text{ atm}$$

9. A mixture of cyclopropane gas and oxygen is used as an anesthetic. Cyclopropane contains 85.7% C and 14.3% H by mass. At 50.0°C and 0.984 atm pressure, 1.56 g cyclopropane has a volume of 1.00 L. What is the molecular formula of cyclopropane?

In 100 g of cyclopropane there are 85.7 g C and 14.3 g H.

$$\text{mol C} = 85.7\text{ g} \times \frac{1\text{ mol C}}{12.0\text{ g C}} = 7.14\text{ mol C}$$

$$\text{mol H} = 14.3\text{ g} \times \frac{1\text{ mol H}}{1.01\text{ g}} = 14.2\text{ mol H}$$ ∴ empirical formula is CH_2, and its formula weight is 14.0 amu

$$\mathcal{M}\text{ of cyclopropane} = \frac{(1.56\text{ g})(323\text{ K})(0.0821\text{ L–atm/mol–K})}{(1.00\text{ L})(0.984\text{ atm})} = 42.0\text{ g/mol}$$

$$\frac{42.0}{14.0} = 3.00,$$ Therefore molecular formula is $(CH_2)_3$ or C_3H_6.

Experiment

Determination of *R*: The Gas-Law Constant

OBJECTIVE

To gain a feeling for how well real gases obey the ideal-gas law and to determine the ideal-gas-law constant, *R*.

APPARATUS AND CHEMICALS

Apparatus

balance	barometer
Bunsen burner and hose	glass tubing with 60-degree bends (2) and straight pieces (2)
test tube	125-mL Erlenmeyer flask
250-mL beaker	rubber tubing
8-oz wide-mouth bottle	thermometer
rubber stoppers (2)	Styrofoam cups
pinch clamp	ring stand
clamp	

Chemicals

$KClO_3$ $\qquad$ MnO_2

DISCUSSION

Most gases obey the ideal-gas equation, $PV = nRT$, quite well under ordinary conditions, that is, at room temperature and atmospheric pressure. Small deviations from this law are observed, however, because real-gas molecules are finite in size and exhibit mutual attractive forces. The van der Waals equation,

$$\left(P + \frac{n^2a}{V^2}\right)(V - nb) = nRT$$

where a and b are constants characteristic of a given gas, takes into account these two causes for deviation and is applicable over a much wider range of temperatures and pressures than the ideal-gas equation. The term nb in the expression $(V - nb)$ is a correction for the finite volume of the molecules; the correction to the pressure by the term n^2a/V^2 takes into account the intermolecular attractions.

In this experiment you will determine the numerical value of the gas-law constant *R*, in its common units of L−atm/mol−K. This will be done using both the ideal-gas law and the van der Waals equation together with measured values of pressure, *P*, temperature, *T*, volume, *V*, and number of moles, *n*, of an enclosed sample of oxygen. Then you will perform an error analysis on the experimentally determined constant.

The oxygen will be prepared by the decomposition of potassium chlorate, using manganese dioxide as a catalyst:

$$2KClO_3(s) \xrightarrow[\Delta]{MnO_2(s)} 2KCl(s) + 3O_2(g)$$

If the $KClO_3$ is accurately weighed before and after the oxygen has been driven off, the weight of the oxygen can be obtained by difference. The oxygen can be collected by displacing water from a bottle, and the volume of gas can be determined from the volume of water displaced. The pressure of the gas may be obtained through use of Dalton's law of partial pressures, the vapor pressure of water, and atmospheric pressure. Dalton's law states that the pressure of a mixture of gases in a container is equal to the sum of the pressures that each gas would exert if it were present alone:

$$P_{\text{total}} = \sum_i P_i$$

Because this experiment is conducted at atmospheric pressure, $P_{\text{total}} = P_{\text{atmospheric}}$. Hence,

$$P_{\text{atmospheric}} = P_{O_2} + P_{H_2O\ \text{vapor}}$$

PROCEDURE

Instructor: Demonstate the safe way to insert the glass tubing into rubber stoppers.

Add a small amount of MnO_2 (about 0.02 g) and approximately 0.3 g of $KClO_3$ to a test tube and accurately weigh to the nearest 0.001 g. Your instructor will demonstrate how to insert the glass tubing into the rubber stoppers. Be extremely careful to follow his or her instructions. Assemble the apparatus illustrated in Figure 14.1, but do not attach the test tube. Be sure that tube B does not extend below the water level in the bottle. Fill glass tube A and the rubber tubing with water by loosening the pinch clamp and attaching a rubber bulb to tube B and applying pressure through it. Close the clamp when the tube is filled.

Mix the solids in the test tube by rotating the tube, being certain that none of the mixture is lost from the tube, and attach tube B as shown in Figure 14.1. **(CAUTION:** ***When you attach the test tube, be certain that none of the*** $KClO_3$ ***and*** MnO_2 ***comes into contact with the rubber stopper, or a severe explosion may result. Make certain that the clamp holding the test tube is secure so that the test tube cannot move*****).**

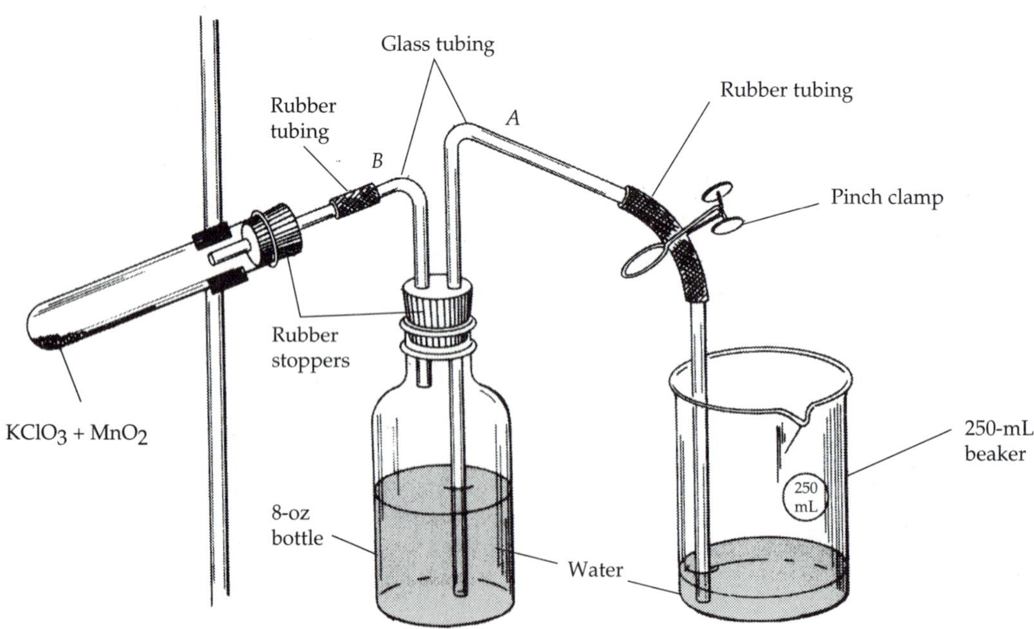

▲ **FIGURE 14.1** Apparatus for determination of R.

TABLE 14.1 Density and Vapor Pressure of Pure Water at Various Temperatures

Temperature (°C)	Density (*d*) (g/mL)	Temperature (°C)	H_2O vapor pressure (mm Hg)
15	0.999099	15	12.8
16	0.998943	16	13.6
17	0.998774	17	14.5
18	0.998595	18	15.5
19	0.998405	19	16.5
20	0.998203	20	17.5
21	0.997992	21	18.6
22	0.997770	22	19.8
23	0.997538	23	21.1
24	0.997296	24	22.4
25	0.997044	25	23.8
26	0.996783		
27	0.996512		
28	0.996232		

Fill the beaker about half full of water, insert glass tube *A* in it, open the pinch clamp, and lift the beaker until the levels of water in the bottle and beaker are identical. Then close the clamp, discard the water in the beaker, and dry the beaker. The purpose of equalizing the levels is to produce atmospheric pressure inside the bottle and test tube.

Set the beaker with tube *A* in it on the desk and open the pinch clamp. A little water will flow into the beaker, but if the system is airtight and has no leaks, the flow will soon stop and tube *A* will remain filled with water. If this is not the case, check the apparatus for leaks and start over. Leave in the beaker the water that has flowed into it; at the end of the experiment, the water levels will be adjusted and this water will flow back into the bottle.

Heat the lower part of the test tube gently (be certain that the pinch clamp is open) so that a slow but steady stream of gas is produced, as evidenced by the flow of water into the beaker. When the rate of gas evolution slows considerably, increase the rate of heating, and heat until no more oxygen is evolved. Allow the apparatus to cool to room temperature, making certain that the end of the glass tube in the beaker is always below the surface of the water. Equalize the water levels in the beaker and the bottle as before and close the clamp. Weigh a 125-mL Erlenmeyer flask* to the nearest 0.01 g and empty the water from the beaker into the flask.* Weigh the flask* with the water in it. Measure the temperature of the water and, using the density of water in Table 14.1, calculate the volume of the water displaced. This is equal to the volume of oxygen produced. Remove the test tube from the apparatus and accurately weigh the tube plus the contents. The difference in mass between this and the original mass of the tube plus MnO_2 and $KClO_3$ is the mass of the oxygen produced.

Record the barometric pressure. The vapor pressure of water at various temperatures is also given in Table 14.1.

*Or Styrofoam cup. The volume of water may also be measured directly, but less accurately, with a graduated cylinder.

Waste Disposal Instructions $KClO_3$ is a powerful oxidizing agent and must not be disposed of in a waste basket! Do not attempt to clean out the residue that remains in the test tube. Return the test tube to the instructor or follow his instructions for disposal of its contents.

Calculate the gas-law constant, R, from your data, using the ideal-gas equation. Calculate R using the van der Waals equation $(P + n^2a/V^2)(V - nb) = nRT$ (for O_2, $a = 1.360$ L^{2}atm/mol^2), and $b = 31.83$ cm^3/mol). Be sure to keep your units straight.

Error Analysis

Determine the maximum and the minimum value of R consistent with the experimental reliability of your data from the ideal-gas law:

$$R = \frac{PV}{nT} = \frac{(32.00\ \text{g/mol})PV}{mT}$$

Assume that the reliabilities for the various measured quantities in this experiment are as follows:

$$P = \pm 0.1\ \text{mm Hg} \qquad T = \pm 1°\text{C}$$

$$V = \pm 0.0001\ \text{L} \qquad m = \pm 0.0001\ \text{g}$$

To determine the maximum value of R, use the maximum values that the pressure and volume may have and the minimum values that the mass and temperature may have. Similarly, calculate the minimum value of R from the minimum values of P and V and the maximum values for m and T. Determine the average value of R and assign an uncertainty range to this average value.

EXAMPLE 14.1

Assume that the measured quantities were as follows: $P = 705.5$ mm Hg; $T = 20°$C; $V = 242.9$ mL; and $m = 0.3002$ g. What would be the maximum and minimum values of R, the average value of R, and the uncertainty range to be assigned to this average value?

SOLUTION: First put the measured quantities into proper units as follows:

$$P = \frac{705.5\ \text{mm Hg}}{760\ \text{mm Hg/atm}} = 0.928\ \text{atm}$$

$$V = 242.9\ \text{mL} = 0.2429\ \text{L}$$

$$m = 0.3002\ \text{g}$$

$$T = (20°\text{C} + 273)\ \text{K} = 293\ \text{K}$$

Therefore,

$$\text{Maximum } R = \frac{[705.6\ \text{mm Hg}/(760\ \text{mm Hg/atm})](0.2430\ \text{L})(32.00\ \text{g/mol})}{(0.3001\ \text{g})(292\ \text{K})}$$

$$= 0.0823\ \text{L–atm/mol–K}$$

$$\text{Minimum } R = \frac{[705.4\ \text{mm Hg}/(760\ \text{mm Hg/atm})](0.2428\ \text{L})(32.00\ \text{g/mol})}{(0.3003\ \text{g})(294\ \text{K})}$$

$$= 0.0816\ \text{L–atm/mol–K}$$

The average value for R is, therefore,

$$\text{Average } R = \frac{0.0823 + 0.0816}{2} \text{ L-atm/mol-K}$$
$$= 0.0820 \text{ L-atm/mol-K}$$

Note that the minimum and maximum values of R differ from the average by 0.0004. Consequently, the uncertainty in R can be written as ± 0.0004 L–atm/ mol–k and the data would be reported as

$$R = 0.0820 \pm 0.0004 \text{ L-atm/mol-K}$$

REVIEW QUESTIONS

Before beginning this experiment in the laboratory, you should be able to answer the following questions:

1. Under what conditions of temperature and pressure would you expect gases to obey the ideal-gas equation?
2. Calculate the value of R in L–atm/mol–K by assuming that an ideal gas occupies 22.4 L/mol at STP.
3. Why do you equalize the water levels in the bottle and the beaker?
4. Why does the vapor pressure of water contribute to the total pressure in the bottle?
5. What is the value of an error analysis?
6. Suggest reasons why real gases might deviate from the ideal-gas law on the molecular level.
7. At present, automobile batteries are sealed. When lead storage batteries discharge, they produce hydrogen. Suppose the void volume in the battery is 100 mL at 1 atm of pressure and 25°C. What would be the pressure increase if 0.05 g H_2 were produced by the discharge of the battery? Does this present a problem? Do you know why sealed lead storage batteries were not used in the past?
8. Why is the corrective term to the volume subtracted and not added to the volume in the van der Waals equation?
9. A sample of pure gas at 20°C and 670 mm Hg occupied a volume of 562 cm^3. How many moles of gas does this represent? (HINT: Use the value of R that you found in question 2.)
10. A certain compound containing only carbon and hydrogen was found to have a vapor density of 2.550 g/L at 100°C and 760 mm Hg. If the empirical formula of this compound is CH, what is the molecular formula of this compound?
11. Which gas would you expect to behave more like an ideal gas, Ne or HBr? Why?

NOTES AND CALCULATIONS

Name ______________________ Desk ______________________

Date ______________ Laboratory Instructor ______________________

REPORT SHEET | EXPERIMENT

Determination of *R*: The Gas-Law Constant | 14

1. Mass of test tube + $KClO_3$ + MnO_2 ____21.5392____ g
2. Mass of test tube + contents after reaction ____21.3964____ g
3. Mass of oxygen produced ____0.1428____ g
4. Mass of 125-mL flask* + water ____230.320____ g
5. Mass of 125-mL flask* ____96.625____ g
6. Mass of water ____133.695____ g
7. Temperature of water ____25.0° C____
8. Density of water ____0.99707 g/ml____
9. Volume of water ____134.1 mL____ = volume of O_2 gas 134.1 mL = 0.1341 L
10. Barometric pressure ____645.0____ mm Hg
11. Vapor pressure of water ____23.8 mm Hg____
12. Pressure of O_2 gas (show calculations) ____621.2 mm Hg____

$$P_{total} = P_{O_2} + P_{H_2O}$$

$$645.0 \text{ mm Hg} - 23.8 \text{ mm Hg} = 621.2 \text{ mm Hg} = P_{O_2}$$

13. Gas-law constant, *R*, from ideal-gas law (show calculations) ____0.0824 L−atm/mol−K____

$$R = \frac{(621.2 \text{ mm Hg}/760 \text{ mm Hg/atm})(0.1341 \text{ L})}{(0.1428 \text{ g}/32.0 \text{ g/mol})(298 \text{ K})} = 0.0824 \text{ L−atm/mol−K}$$

14. *R* from the van der Waals equation (show calculations) ____0.0826 L−atm/mol−K____

$$R = \frac{\left[(0.817358 \text{ atm} + \frac{(0.00446 \text{ mol})^2(1.360 \text{ L}^2 - \text{atm/mol}^2)}{(0.13409 \text{ L})^2}\right][(0.13409 \text{ L}) - (0.00466 \text{ mol})(0.03183 \text{ L/mol})]}{(0.00446 \text{ mol})(298 \text{ K})}$$

$$= 0.0826 \text{ L–atm/mol–K}$$

15. Accepted value of *R* ____0.08206 L–atm/mol–K____ (source of *R* value) ____Brown, LeMay, and Bursten____

16. Uncertainty in *R* (show calculations) ____0.0008 ±0.0004____

$$\text{Max:} \quad R = \frac{PV}{nT} = \frac{(621.3 \text{ mm Hg}/760 \text{ mm Hg/atm})(0.1342 \text{ L})(32.0 \text{ g/mol})}{(297 \text{ K})(0.1427 \text{ g})} = 0.0828 \text{ L–atm/mol–K}$$

$$\text{Min:} \quad R = \frac{PV}{nT} = \frac{(621.1 \text{ mm Hg}/760 \text{ mm Hg/atm})(0.1342 \text{ L})(32.0 \text{ g/mol})}{(299 \text{ K})(0.1429 \text{ g})} = 0.0820 \text{ L–atm/mol–K}$$

*Or Styrofoam cup.

QUESTIONS

1. Does your value of R agree with the accepted value within your uncertainty limits?
   Yes

2. Discuss possible sources of error in the experiment. Indicate the ones that you feel are most important.
   Measurement uncertainties

3. Which gas would you expect to deviate more from ideality, H_2 or HBr? Explain your answer.
   HBr, because it is more polar than H_2, and hence the intermolecular forces are greater for HBr than H_2.

4. How does the solubility of oxygen in water affect the value of R you determined? Explain your answer.
   It will lower it somewhat because the pressure will be lowered.

5a. Use the van der Waals equation to calculate the pressure exerted by 1.000 mol of Cl_2 in 22.41L at 0.0°C. The van der Waals constants for Cl_2 are: $a = 6.49\ L^2\ atm/mol^2$ and $b = 0.0562\ L/mol$.

a. If we solve van der Waals equation for pressure, we have

$$P = \frac{nRT}{V - nb} - \frac{n^2a}{V^2}$$

Substituting $n = 1.000$ mol, $R = 0.08206$ L–atm/K–mol, $T = 273.2$ K, $V = 22.41$ L, $a = 6.49\ L^2\ atm/mol^2$, and $b = 0.0562$ L/mol:

$$P = \frac{(1.000\ \text{mol})(0.08206\ \text{L–atm/K–mol})(273.1\ \text{K})}{22.41\ \text{L} - (1.000\ \text{mol})(0.0562\ \text{L/mol})} - \frac{(1.000\ \text{mol})^2(6.49\ \text{L}^2\ \text{atm/mol}^2)}{(22.41\ \text{L})^2}$$

$$= 1.003\ \text{atm} - 0.013\ \text{atm} = 0.990\ \text{atm}$$

5b. Which factor is the major cause for deviation from ideal behavior, the volume of the Cl_2 molecules or the attractive forces between them?

b. The first term in the equation above (1.003) corrects for the molecular volume. The second term (−0.013) corrects for intermolecular attraction and is the major reason for deviation from ideal behavior.

6. How much potassium chlorate is needed to produce 20.0 mL of oxygen gas at 670 mm Hg and 20°C?

$$\text{mol } O_2 = \frac{PV}{RT} = \frac{(670\ \text{mm Hg}/760\ \text{mm Hg/atm})(20.0\ \text{mL}/1000\ \text{mL/L})}{(0.0821\ \text{L–atm/mol–K})(293\ \text{K})} = 7.34 \times 10^{-4}\ \text{mol}$$

$$\text{g } KClO_3 = \left(\frac{2\ \text{mol } KClO_3}{3\ \text{mol } O_2}\right)\left(7.34 \times 10^{-4}\ \text{mol } O_2\right)\left(\frac{122.45\ \text{g } KClO_3}{\text{mol } KClO_3}\right) = 0.0599\ \text{g } KClO_3$$

7. If oxygen gas were collected over water at 20°C and the total pressure of the wet gas were 670 mm Hg, what would be the partial pressure of the oxygen?
   $P_{O_2} = 670 - 17.5 = 653$ mm Hg

8. An oxide of nitrogen was found by elemental analysis to contain 30.4% nitrogen and 69.6% oxygen. If 23.0 g of this gas were found to occupy 5.6 L at STP, what are the empirical and molecular formulas for this oxide of nitrogen?

   22.4 L or 1 mol weighs 92.0 g and contains 92.0 g × 0.304 = 28.0 g nitrogen, which is 2 mol of N, and 92.0 g × 0.696 = 64.0 g oxygen, which is 4 mol of O. Thus, the formula is N_2O_4.

9. The gauge pressure in an automobile tire reads 32 pounds per square inch (psi) in the winter at 32°F. The gauge reads the difference between the tire pressure and the atmospheric pressure (14.7 psi). In other words, the tire pressure is the gauge reading plus 14.7 psi. If the same tire were used in the summer at 110°F and no air had leaked from the tire, what would be the tire gauge reading in the summer? (HINT: Recall that °C = $\frac{5}{9}$ (°F − 32).)

   $$P = 32 \text{ psi} + 14.7 \text{ psi} = 46.7 \text{ psi}$$

   In the summer $P = 46.7\left(\frac{316.5 \text{ K}}{273 \text{ K}}\right) = 54.1$ psi

   Gauge would read 54.1 − 14.7 = 39.4 psi

NOTES AND CALCULATIONS

Experiment

Activity Series

OBJECTIVE

To become familiar with the relative activities of metals in chemical reactions.

APPARATUS AND CHEMICALS

Apparatus

small test tubes* (13)

test-tube rack

Chemicals

0.2 *M* $Ca(NO_3)_2$

0.2 *M* $Zn(NO_3)_2$

0.2 *M* $FeSO_4$

0.2 *M* $Cu(NO_3)_2$

7 small pieces each of calcium, magnesium, zinc, iron wool, tin, copper

0.2 *M* $Mg(NO_3)_2$

0.2 *M* $Fe(NO_3)_3$

0.2 *M* $SnCl_4$

6 *M* HCl

DISCUSSION

Chemical elements are usually classified by their properties into three groups: metals, nonmetals, and metalloids. Most of the known elements are metals. Their physical properties include high thermal and electrical conductivity, high luster, malleability (ability to be pounded flat without shattering), and ductility (ability to be drawn out into a fine wire). All common metals are solids at room temperature except mercury, which is a liquid. The periodic table illustrated in Figure 15.1 shows the three classifications of the elements.

All elements to the left of the shaded area are metals except hydrogen. Those to the right are nonmetals. Those in the shaded area have intermediate properties and are called semimetals or metalloids. Families or groups of

*A spot plate may be used in place of test tubes.

1A																	8A
H	2A											3A	4A	5A	6A	7A	He
Li	Be											B	C	N	O	F	Ne
Na	Mg	Transition elements										Al	Si	P	S	Cl	Ar
K	Ca	Sc	Ti	V	Cr	Mn	Fe	Co	Ni	Cu	Zn	Ga	Ge	As	Se	Br	Kr
Rb	Sr	Y	Zr	Nb	Mo	Tc	Ru	Rh	Pd	Ag	Cd	In	Sn	Sb	Te	L	Xe
Cs	Ba	La	Hf	Ta	W	Re	Os	Ir	Pt	Au	Hg	Tl	Pb	Bi	Po	At	Rn

▲ **FIGURE 15.1** Partial periodic table of the elements.

elements consist of elements in vertical columns in the periodic table. Elements within a group or family (called congeners) have similar chemical properties because they have similar valence electronic structures; that is, the number of valence electrons (electrons in the outermost shell) is the same for all members of a family or group. For historical reasons, most of the groups have names, and some are often referred to by them. These are

1. Group 1, called *alkali metals* because they react with oxygen to form bases
2. Group 2, called *alkaline earth metals* because their presence makes soils alkaline
3. Group 3, no common name
4. Group 4, no common name
5. Group 5, called *pnictides,* from the Greek word meaning choking suffocation
6. Group 6, called *chalcogens,* from Greek roots meaning ore former
7. Group 7, called *halogens,* from Greek roots meaning salt former
8. Group 8, called *rare, noble,* or *inert gases* because they are rare and were thought to be unreactive

Those most frequently referred to by group name are the alkali metals, the alkaline earth metals, the halogens, and the rare gases.

The three broad categories of the elements also have somewhat similar chemical properties. For example, metals, as compared with the other elements, all have relatively low ionization potentials and enter into chemical combination with nonmetals by *losing* electrons to become cations. This can be symbolized by the following equation:

$$M \longrightarrow M^{n+} + ne^-$$

Nonmetals, as compared with metals, have relatively high electron affinities and enter into chemical combination with metals by *gaining* electrons to become anions. This can be symbolized by the following equation:

$$X + ne^- \longrightarrow X^{n-}$$

Specific examples of these types of reactions can be divided into several useful categories, which will be illustrated by the following examples.

Electron-Transfer Reactions

1 REACTIONS WITH OXYGEN

$$2Mg(s) + O_2(g) \longrightarrow 2MgO(s)$$

In this reaction, magnesium is oxidized by oxygen, which is reduced by magnesium. This can be better illustrated by breaking down the reaction into fictitious although helpful steps:

$$2(Mg \longrightarrow Mg^{2+} + 2e^-) \quad \text{oxidation}$$

$$O_2 + 4e^- \longrightarrow 2O^{2-} \quad \text{reduction}$$

$$2Mg + O_2 \longrightarrow 2MgO \quad \text{oxidation-reduction (redox) reaction}$$

In *oxidation,* the oxidized element loses electrons and becomes more positive. In *reduction,* the reduced element gains electrons and becomes more negative. Oxidation is always associated with a concomitant reduction.

2 REACTIONS WITH WATER

$$2Na(s) + 2H_2O(l) \longrightarrow 2NaOH(aq) + H_2(g)$$
$$Ca(s) + 2H_2O(l) \longrightarrow Ca(OH)_2(aq) + H_2(g)$$

The *ionic* equations for these reactions better illustrate the electron-transfer process:

$$2Na(s) + 2H_2O(l) \longrightarrow 2Na^+(aq) + 2OH^-(aq) + H_2(g)$$
$$Ca(s) + 2H_2O(l) \longrightarrow Ca^{2+}(aq) + 2OH^-(aq) + H_2(g)$$

3 REACTIONS WITH ACIDS

$$Zn(s) + 2HCl(aq) \longrightarrow ZnCl_2(aq) + H_2(g)$$

or

$$Zn(s) + 2H^+(aq) + 2Cl^-(aq) \longrightarrow Zn^{2+}(aq) + 2Cl^-(aq) + H_2(g)$$

Since the chloride ion is merely a spectator—that is, it does not participate in the reaction—it may be omitted, yielding the *net ionic equation:*

$$Zn(s) + 2H^+(aq) \longrightarrow Zn^{2+}(aq) + H_2(g)$$

or simply

$$Zn + 2H^+ \longrightarrow Zn^{2+} + H_2$$

4 ELECTRON TRANSFER AMONG METALS

$$Zn(s) + Cu(NO_3)_2(aq) \longrightarrow Zn(NO_3)_2(aq) + Cu(s)$$

or

$$Zn(s) + Cu^{2+}(aq) \longrightarrow Zn^{2+}(aq) + Cu(s)$$

or simply

$$Zn + Cu^{2+} \longrightarrow Zn^{2+} + Cu$$

Note once again that in the ionic equation the spectator ion (NO_3^-) has been omitted because it takes no active part in the reaction and serves only to provide electrical neutrality. Therefore, any other soluble salt of copper(II), such as chloride, sulfate, or acetate, could perform the same function.

In this game of "musical electrons," there are only enough electrons for one atom of the system. In order to achieve the lowest energy level for the system, the more active metal of a pair will lose electrons to the more passive metal or will react more vigorously with water, acids, or oxygen. In some cases no reaction at all will occur. Without prior knowledge we have no way of predicting these events.

PROCEDURE

A. Reactions of Metals with Acid

To each of six test tubes containing 0.5 mL of dilute 6 *M* HCl, add a small piece of the metals Ca, Cu, Fe, Mg, Sn, and Zn. Observe the test tubes and note any changes that occur (such as the evolution of a gas, whether it is vigorous or not, and any color changes). Enter your observations on the report sheet, and write both complete and ionic equations for each reaction noted. After completing each series of reactions, dispose of the contents of your test tubes in the designated containers.

B. Reactions of Metals with Solutions of Metal Ions

(WORK IN PAIRS FOR THIS STEP.) Add a small piece of calcium metal to each of seven test tubes containing, respectively, about 0.5 mL of $Ca(NO_3)_2$, $Cu(NO_3)_2$, $FeSO_4$, $Fe(NO_3)_3$, $Mg(NO_3)_2$, $SnCl_4$, and $Zn(NO_3)_2$ solutions. Note any reaction that occurs by observing whether a color change occurs on the surface of the metal or in the solution or whether a gas is evolved. Record your observations on the report sheet. Write both complete and ionic equations for any reaction that occurs. After completing each series of reactions, dispose of the contents of your test tubes in the designated containers.

Repeat the preceding process by adding a small piece of copper to another 0.5 mL of each of the metal-cation solutions. Do the same for iron, magnesium, tin, and zinc, and record all your observations. Write both complete molecular and ionic equations for all reactions.

C. Relative-Activity Series

From the information contained in the table you constructed in Part B, you can rank these six metals according to their relative chemical reactivities. This can be done in the following manner: One of the metals will replace all others in solution. For example, if calcium metal is oxidized by solutions containing cations of each of the other metals, then it is the most reactive. One of the other metals that will replace all but calcium is the next most reactive. Finally, one of the metals will not replace any of the other metal cations from solution. Therefore, it is the least reactive. List the metals on your report sheet in terms of decreasing reactivity, starting with the most reactive (1) and ending with the least reactive (6).

REVIEW QUESTIONS

Before beginning this experiment in the laboratory, you should be able to answer the following questions:

1. What distinguishes a metal from a nonmetal?
2. What does ionization potential measure?
3. What does electron affinity measure?
4. Why must oxidation be accompanied by a reduction?
5. How does one determine the relative reactivities of metals?
6. Complete and balance the following:

$$Mg + O_2 \longrightarrow$$

$$Zn + HCl \longrightarrow$$

$$Zn + Cu^{2+} \longrightarrow$$

7. Balance the following reactions and identify the species that have been oxidized and the species that have been reduced.

Reaction	Species oxidized	Species reduced
$Cl_2 + I^- \longrightarrow I_2 + Cl^-$		
$WO_2 + H_2 \longrightarrow W + H_2O$		
$Ca + H_2O \longrightarrow H_2 + Ca(OH)_2$		
$Al + O_2 \longrightarrow Al_2O_3$		

8. If the following redox reactions are found to occur spontaneously, identify the more active metal in each reaction.

Reaction	More-active metal
$2Li + Cu^{2+} \longrightarrow 2Li^+ + Cu$	
$Cr + 3V^{3+} \longrightarrow 3V^{2+} + Cr^{3+}$	
$Cd + 2Ti^{3+} \longrightarrow 2Ti^{2+} + Cd^{2+}$	

NOTES AND CALCULATIONS

Name ______________________ Desk ______________________

Date ______________ Laboratory Instructor ______________________

REPORT SHEET | EXPERIMENT

Activity Series | 15

A. Reactions of Metals with Acid

Metal	Reaction with HCl	Observation	Equations
Ca	Yes	Ca all dissolved with very vigorous evolution of H_2 gas; colorless solution	$Ca(s) + 2HCl \longrightarrow CaCl_2(aq) + H_2(g)$ $Ca(s) + 2H^+ \longrightarrow Ca^{2+} + H_2(g)$
Cu	No	No reaction	————
Mg	Yes	Mg dissolved, but more slowly than Ca, to produce a colorless solution with H_2 gas evolution	$Mg(s) + 2HCl \longrightarrow MgCl_2 + H_2(g)$ $Mg(s) + 2H^+ \longrightarrow Mg^{2+} + H_2(g)$
Fe	Yes	Very, very slow reaction with evolution of some H_2 gas	$Fe(s) + 2HCl \longrightarrow FeCl_2 + H_2(g)$ $Fe(s) + 2H^+ \longrightarrow Fe^{2+} + H_2(g)$
Sn	Yes	Extremely slow evolution of H_2	$Sn(s) + 2HCl \longrightarrow SnCl_2 + H_2(g)$ $Sn(s) + 2H^+ \longrightarrow Sn^{2+} + H_2(g)$
Zn	Yes	Zn dissolved to produce colorless solution with H_2 gas evolution	$Zn(s) + 2HCl \longrightarrow ZnCl_2 + H_2(g)$ $Zn(s) + 2H^+ \longrightarrow Zn^{2+} + H_2(g)$
Example: Co	Yes, slowly	Gas evolved; solution turned blue	$Co + 2HCl \longrightarrow CoCl_2 + H_2$ $Co + 2H^+ \longrightarrow Co^{2+} + H_2$

B. Reactions of Metals with Solutions of Metal Ions

Metal \ Metal ions	Ca^{2+}	Cu^{2+}	Fe^{3+}	Fe^{2+}	Mg^{2+}	Sn^{4+}	Zn^{2+}	Al^{3+}
Ca		Solution changed from blue to green as Ca dissolved	Ca dissolved to give dirty green solution	Ca dissolved to give a dirty green solution & a red-brn ppt	Ca dissolved; milky white suspension formed	Ca dissolved to form milky solution	Ca dissolved; white ppt. formed	Yes, calcium dissolves
Cu	N.R.[a]		N.R.	N.R.	N.R.	N.R.	N.R.	N.R.
Fe	N.R.	Very slow reaction			N.R.	Slow reaction	N.R.	N.R.
Mg	N.R.	Solution became darker & a black ppt. formed on Mg	Mg slowly dissolved as rust ppts	Mg dissolved very slowly		Mg dissolved and dark gray ppt. formed	Mg dissolved	Yes, magnesium dissolves
Sn	N.R.	Very slow reaction	N.R.	N.R.	N.R.		Very little reaction	N.R.
Zn	N.R.	Zn turned black	Zn dissolved slowly and a red-brown ppt. formed	Almost no reaction	N.R.	Zn dissolved slowly		N.R.
Example: Al	N.R.	Yes, aluminum turns brown	Yes, turns aluminum dark	Yes, aluminum turns dark; solution colorless	N.R.	Yes, aluminum turns dark; solution colorless	Yes, aluminum turns dark	

[a]N.R. means no reaction occurred.

Example:

$$2Al + 3Zn(NO_3)_2 \longrightarrow 2Al(NO_3)_3 + 3Zn$$
$$2Al + 3Zn^{2+} \longrightarrow 2Al^{3+} + 3Zn$$

Complete equation	*Net ionic equation*
$Ca + Mg(NO_3)_2 \longrightarrow Ca(NO_3)_2 + Mg$	$Ca + Mg^{2+} \longrightarrow Ca^{2+} + Mg$
$Ca + Zn(NO_3)_2 \longrightarrow Ca(NO_3)_2 + Zn$	$Ca + Zn^{2+} \longrightarrow Ca^{2+} + Zn$
$Ca + Fe(NO_3)_2 \longrightarrow Ca(NO_3)_2 + Fe$	$Ca + Fe^{2+} \longrightarrow Ca^{2+} + Fe$
$3Ca + 2Fe(NO_3)_3 \longrightarrow 3Ca(NO_3)_2 + 2Fe$	$3Ca + 2Fe^{3+} \longrightarrow 3Ca^{2+} + 2Fe$
$2Ca + SnCl_4 \longrightarrow 2CaCl_2 + Sn$	$2Ca + Sn^{4+} \longrightarrow 2Ca^{2+} + Sn$
$Ca + Cu(NO_3)_2 \longrightarrow Ca(NO_3)_2 + Cu$	$Ca + Cu^{2+} \longrightarrow Ca^{2+} + Cu$
$Mg + Zn(NO_3)_2 \longrightarrow Mg(NO_3)_2 + Zn$	$Mg + Zn^{2+} \longrightarrow Mg^{2+} + Zn$
$Mg + Fe(NO_3)_2 \longrightarrow Mg(NO_3)_2 + Fe$	$Mg + Fe^{2+} \longrightarrow Mg^{2+} + Fe$
$3Mg + 2Fe(NO_3)_2 \longrightarrow 3Mg(NO_3)_2 + 2Fe$	$3Mg + 2Fe^{3+} \longrightarrow 3Mg^{2+} + 2Fe$
$2Mg + SnCl_4 \longrightarrow 2MgCl_2 + Sn$	$2Mg + Sn^{4+} \longrightarrow 2Mg^{2+} + Sn$
$Mg + Cu(NO_3)_2 \longrightarrow Mg(NO_3)_2 + Cu$	$Mg + Cu^{2+} \longrightarrow Mg^{2+} + Cu$
$Zn + Fe(NO_3)_2 \longrightarrow Zn(NO_3)_2 + Fe$	$Zn + Fe^{2+} \longrightarrow Zn^{2+} + Fe$
$3Zn + 2Fe(NO_3)_3 \longrightarrow 3Zn(NO_3)_2 + 2Fe$	$3Zn + 2Fe^{3+} \longrightarrow 3Zn^{2+} + 2Fe$
$2Zn + SnCl_4 \longrightarrow 2ZnCl_2 + Sn$	$2Zn + Sn^{4+} \longrightarrow 2Zn^{2+} + Sn$
$Zn + Cu(NO_3)_2 \longrightarrow Zn(NO_3)_2 + Cu$	$Zn + Cu^{2+} \longrightarrow Zn^{2+} + Cu$
$2Fe + SnCl_4 \longrightarrow 2FeCl_2 + Sn$	$2Fe + Sn^{4+} \longrightarrow 2Fe^{2+} + Sn$
$Fe + Cu(NO_3)_2 \longrightarrow Fe(NO_3)_2 + Cu$	$Fe + Cu^{2+} \longrightarrow Fe^{2+} + Cu$
$Sn + Cu(NO_3)_2 \longrightarrow Sn(NO_3)_2 + Cu$	$Sn + Cu^{2+} \longrightarrow Sn^{2+} + Cu$

C. Relative-Activity Series

Most reactive — Least reactive

1. Ca 2. Mg 3. Zn 4. Fe 5. Sn 6. Cu

QUESTIONS

1. Which of these six metals should be the most reactive toward oxygen? Ca

2. Which of the oxides would be expected to be thermally unstable and decompose according to

   $2MO \xrightarrow{\Delta} 2M + O_2$ Sn

3. Sodium is slightly less reactive than calcium. Predict the outcome of the following reactions:

   $Na + H_2O \longrightarrow$ $2Na + 2H_2O \longrightarrow 2NaOH + H_2$

   $Na + O_2 \longrightarrow$ $4Na + O_2 \longrightarrow 2Na_2O$ (actually $2Na + O_2 \longrightarrow Na_2O_2$ occurs)

   $Na + HCl \longrightarrow$ $2Na + 2HCl \longrightarrow 2NaCl + H_2$

   $Na + Ca^{2+} \longrightarrow$ No reaction

4. Which is more reactive, Fe^{2+} or Fe^{3+}, and why?

   Fe^{3+} has more chemical potential than Fe^{2+}, and it reacted faster and to a greater extent with Zn. The emf for the half reaction $Fe^{3+} + 3e^- \longrightarrow Fe$ is greater than the emf for the half reaction $Fe^{2+} + 2e^- \longrightarrow Fe$. Therefore, Fe^{3+} is a stronger oxidizing agent than Fe^{2+}.

5. From the data in Table B, rank the activity of aluminum.

   Because aluminum does not replace magnesium ion, it is less reactive than magnesium. It does replace zinc ion and is therefore more reactive than zinc. Hence, the order of reactivity is Mg > Al > Zn.

6. For each of the following reactions, indicate which substance is oxidized and which is reduced. Which substance is the oxidizing agent and which is the reducing agent?

	Substance oxidized	*Substance reduced*	*Oxidizing agent*	*Reducing agent*
$2Al(s) + 3Cl_2(g) \longrightarrow 2AlCl_3(s)$	Al	Cl_2	Cl_2	Al
$8H^+(aq) + MnO_4^-(aq) + 5Fe^{2+}(aq) \longrightarrow 5Fe^{3+}(aq) + Mn^{2+}(aq) + 4H_2O(l)$	Fe^{2+}	MnO_4^-	MnO_4^-	Fe^{2+}
$FeS(s) + 3NO_3^-(aq) + 4H^+(aq) \longrightarrow 3NO(g) + SO_4^{2-}(aq) + Fe^{3+}(aq) + 2H_2O(l)$	FeS	NO_3^-	NO_3^-	FeS

Experiment

Electrolysis, the Faraday, and Avogadro's Number

OBJECTIVE

To determine the values for the faraday and Avogadro's number by electrolysis.

APPARATUS AND CHEMICALS

Apparatus

DC source of electricity
insulated copper wires (2)
50-mL buret
250-mL beaker
barometer
ring stand
ammeter
timer or watch
clamp
thermometer
glass stirring rod
millimeter ruler or meter stick

Chemicals

3 *M* H_2SO_4

DISCUSSION

The passage of an electric current through a solution is accompanied by chemical reactions at the electrodes. Oxidation (loss of electrons) occurs at the anode; reduction (gain of electrons) occurs at the cathode. The amount of reaction that occurs at the electrodes is directly proportional to the number of electrons transferred. A *faraday* is defined as the total charge carried by Avogadro's number of electrons; or, in other words, 1 faraday represents the charge on 1 mol of electrons.

In this experiment you will determine the value of the faraday by measuring the amount of charge required to reduce 1 mol of H^+ ions according to the reaction

$$2H^+(aq) + 2e^- \longrightarrow H_2(g) \quad \text{or} \quad 1H^+(aq) + 1e^- \longrightarrow \tfrac{1}{2}H_2(g) \qquad [1]$$

Electric charge is conveniently measured in coulombs. A coulomb, C, is the amount of electrical charge that passes a point in a circuit when a current of 1 ampere, *A*, flows for 1 second, *s*:

$$1\ \text{C} = 1\ \text{A} \times 1\ \text{s}$$

Therefore, the number of coulombs passing through a solution in a cell can be obtained by multiplying the current in amperes by the time in seconds for which it flows. The charge on the electron can also be measured in coulombs and is equal to 1.60×10^{-19} C.

EXAMPLE 16.1

A current of 3.00 A was passed through a solution of sulfuric acid for exactly 20.0 min. How many electrons and how many coulombs were passed through the solution?

SOLUTION:

$$\text{coulombs} = 3.00\text{ A} \times 20.0\text{ min} \times \frac{60\text{ s}}{\text{min}} \times \frac{1\text{ C}}{\text{A–s}}$$

$$= 3.60 \times 10^3\text{ C}$$

$$\text{electrons} = \frac{3.60 \times 10^3\text{ C}}{1.60 \times 10^{-19}\text{ C/electron}}$$

$$= 2.25 \times 10^{22}\text{ electrons}$$

From Equation [1] we note that one hydrogen ion is reduced for every electron passed through the solution and that one molecule of H_2 is produced for every two electrons. Thus, in Example 16.1 the 2.25×10^{22} electrons would produce 1.125×10^{22} molecules of H_2.

If we were to measure the volume, pressure, and temperature of the hydrogen gas associated with the electrolysis in Example 16.1, we could calculate the number of moles of H_2 produced using the relation $n = PV/RT$. Example 16.2 illustrates how this information can be used to calculate Avogadro's number.

EXAMPLE 16.2

A current of 3.00 A was passed through a solution of sulfuric acid for 20 min. The hydrogen produced was collected over water at 20°C and 657.5 mm Hg and was found to occupy a volume of 534 mL. How many moles of H_2 were produced, and what is the value of Avogadro's number, N?

SOLUTION: At 20°C the vapor pressure of H_2O is 17.5 mm Hg. Therefore,

$$P_{\text{total}} = P_{H_2O} + P_{H_2}$$

$$P_{H_2} = 657.5\text{ mm Hg} - 17.5\text{ mm Hg}$$

$$= 640.0\text{ mm Hg}$$

Solving the ideal-gas equation for n gives:

$$n = \frac{PV}{RT} = \frac{[640.0\text{ mm Hg}/(760\text{ mm Hg/atm})](0.534\text{ L})}{(0.0821\text{ L–atm/mol–K})(293\text{ K})}$$

$$= 0.0187\text{ mol }H_2$$

Because two moles of electrons are required for each mole of H_2 produced,

$$\text{mol electrons} = (0.0187\text{ mol }H_2) \times (2\text{ mol electrons/mol }H_2)$$

$$= 0.0374\text{ mol electrons}$$

Avogadro's number is the number of electrons in 1 mol of electrons. From the previous example,

$$2.25 \times 10^{22}\text{ electrons} = 0.0374\text{ mol electrons}$$

Therefore,

$$N = \frac{2.25 \times 10^{22}\text{ electrons}}{0.0374\text{ mol}}$$

$$= 60.2 \times 10^{22}\text{ electrons/mol}$$

$$= 6.02 \times 10^{23}\text{ electrons/mol}$$

The faraday, also known as Faraday's constant (abbreviated $\mathcal{F}$), is defined as the number of coulombs that is equivalent to 1 mol of electrons. Thus, in the examples above, the 3600 C corresponds to the charge associated with 0.0374 mol electrons, or

$$\frac{3600\ \text{C}}{0.0374\ \text{mol}} = 96{,}300\ \text{C/mol electrons}$$

This corresponds very closely to the accepted value for the faraday, which is

$$1\ \mathcal{F} = 96{,}500\ \text{C/mol electrons}$$

We should also note from Example 16.2 above that $(2.25 \times 10^{22}/2)$ molecules of H_2 were produced and represent 0.0187 mol H_2:

$$0.0187\ \text{mol}\ H_2 = (2.25 \times 10^{22}/2)\ \text{molecules}\ H_2$$

or

$$1\ \text{mol}\ H_2 = \frac{2.25 \times 10^{22}\ \text{molecules}\ H_2}{2 \times 0.0187\ \text{mol}\ H_2} = 6.02 \times 10^{23}\ \text{molecules}\ H_2/\text{mol}\ H_2$$
$$= N$$

PROCEDURE

Obtain a DC source with an attached ammeter. In a 250-mL beaker, add 100 mL of distilled water and then 50 mL of dilute (3 *M*) sulfuric acid. Stir with a glass rod to mix well. Fill a 50-mL buret with this solution and invert it in the solution, holding it in place with a ring stand and clamp. Attach the copper wire cathode to the negative terminal of the DC source and place the other end uninsulated into the inverted mouth of the buret. Be certain that all of the bare part of the wire is wholly inside the buret—otherwise some of the H_2 generated will not be collected in the buret. The anode electrode should be hung over the edge of the beaker and immersed in the acid solution, with the other end attached to the positive electrode of the DC source. The top of the solution in the buret should be within the graduated region of the buret so that the volume may be accurately measured, as illustrated in Figure 16.1.

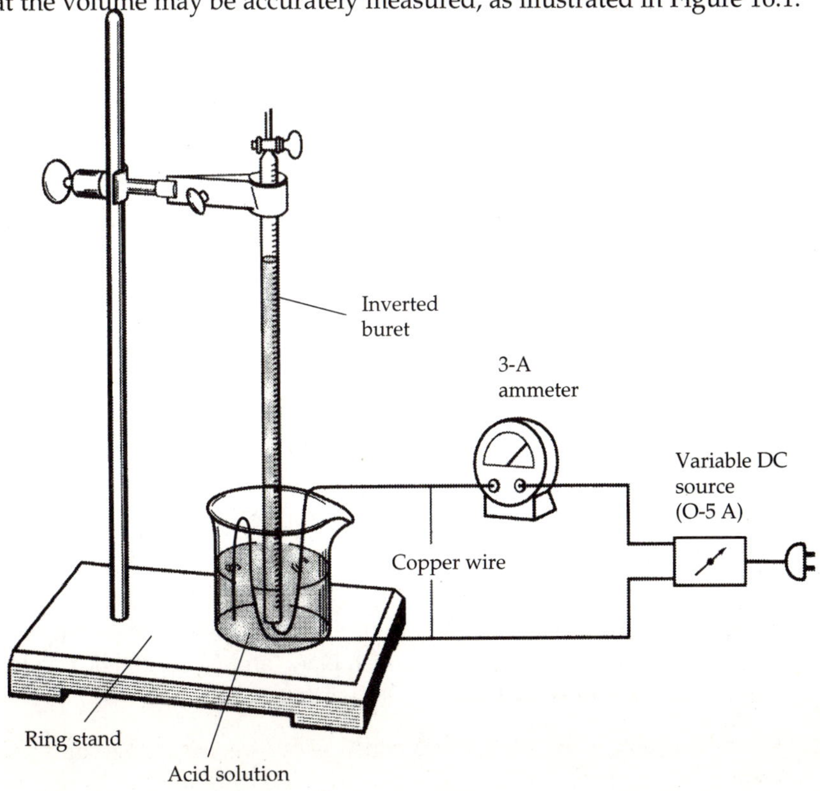

▲ **FIGURE 16.1**

Record the time, turn on the DC source, and record the ammeter reading. During electrolysis, be careful not to move the electrodes, because this may change the current. It is important to maintain a steady current throughout the duration of the electrolysis. If current does fluctuate, it may be necessary to use an average value. Continue the electrolysis until at least 20 mL of hydrogen has been collected. Note the time when the electrolysis is stopped. Measure the volume of H_2 collected.

Measure the height of the water column in the buret above the solution in your beaker. Also measure the temperature of the acid solution and obtain the barometric pressure and the vapor pressure of water at the solution temperature. Note that

$$P_{H_2} = P_{\text{barometric}} - P_{H_2O\text{ column}} - P_{H_2O\text{ vapor}}$$

$$P_{H_2O\text{ column}} = \frac{\text{height of } H_2O \text{ in mm}}{13.6 \text{ mm } H_2O/\text{mm Hg}}$$

If time permits, repeat the experiment.

Dispose of the sulfuric acid in the designated containers.

REVIEW QUESTIONS

Before beginning this experiment in the laboratory, you should be able to answer the following questions:

1. What type of chemical reaction is occurring to produce H_2 at the cathode in this experiment?
2. Define the term *coulomb.*
3. What process is occurring at the anode?
4. Why do you include the height of the water column in the buret in your calculation of the pressure?
5. Why is H_2SO_4 present in the electrolysis solution?
6. If a current of 40.0 μA (40.0 × 10^{-6} A) is drawn from a solar cell for six months, how many faradays are involved?
7. When an aqueous NaCl solution is electrolyzed, how many faradays need to be transferred at the anode to release 0.100 mol of Cl_2 gas?

$$2Cl^-(aq) \longrightarrow Cl_2(g) + 2e^-$$

8. How long must a current of 0.50 A pass through a sulfuric acid solution in order to liberate 0.200 L of H_2 gas at STP?
9. How much silver was there in the solution if all the silver was removed as Ag metal by electrolysis for 0.50 hr with a current of 1.00 mA (1 mA = 10^{-3} A)?
10. Electrolysis of molten NaCl is done in a Downs cell operating at 7.0 volts (V) and 4.0 × 10^4 A. How much Na(*s*) and $Cl_2(g)$ can be produced in eight hours in such a cell?

Name ______________________________ Desk ______________________________

Date ______________________ Laboratory Instructor ______________________________

REPORT SHEET | EXPERIMENT

Electrolysis, the Faraday, and Avogadro's Number | 16

1. Final volume in buret 29.4 mL
2. Initial volume in buret 49.7 mL
3. Volume of hydrogen 20.3 mL

4. Temperature of solution 21°C
5. Height of water column 350 mm
6. Barometric pressure 648.7 mm Hg
7. Vapor pressure of H_2O (see Appendix K) 18.7 mm Hg
8. Pressure of H_2 604.3 mm Hg 0.795 atm (show calculations)

$$P_{H_2} = 648.7 \text{ mm} - 18.7 \text{ mm} - \frac{350 \text{ mm } H_2O}{13.6 \text{ mm } H_2O/\text{mm Hg}} = 604.3 \text{ mm Hg}$$

$$\frac{604.3 \text{ mm Hg}}{760 \text{ mm Hg/atm}} = 0.795 \text{ atm}$$

9. Moles of H_2 produced (show calculations) 6.69×10^{-4} mol

$$n = \frac{PV}{RT} = \frac{(0.795 \text{ atm})(0.0203 \text{ L})}{(0.0821 \text{ L-atm/mol-K})(294 \text{ K})} = 6.69 \times 10^{-4} \text{ mol}$$

10. Time reaction started 2:58.05
11. Time reaction terminated 3:01.25
12. Elapsed time 3 min 20 s = 200 s
13. Current 0.71 A
14. Number of coulombs passed 1.4×10^2 C = (0.71 C/s)(200 s)
15. Value for faraday 100,000 C Accepted value 96,500 C (show calculations)

$$\mathcal{F} = \left(\frac{1.4 \times 10^2 \text{ C}}{6.69 \times 10^{-4} \text{ mol } H_2}\right)\left(\frac{1 \text{ mol } H_2}{2 \text{ mol } e^-}\right) = 1.0 \times 10^5 \text{ C}$$

16. Value for Avogadro's number 6.63×10^{23} Accepted value 6.02×10^{23} (show calculations)

$$N = \left(\frac{1.4 \times 10^2 \text{ C}}{1.60 \times 10^{-19} \text{ C}/e^-}\right)\left(\frac{1 \text{ mol } H_2}{2 \text{ mol } e^-}\right)\left(\frac{1}{6.69 \times 10^{-4} \text{ mol } H_2}\right) = 6.5 \times 10^{23}$$

QUESTIONS

1. Discuss the major sources of error in this experiment.

   (1) Overvoltage required for electrolysis of $H_2O \longrightarrow 2H_2 + O_2$

   (2) Resistance in wire

2. Calculate the percentage error in $\mathcal{F}$ and N.

$$\mathcal{F}: \left(\frac{100{,}000 - 96{,}500}{96{,}500}\right)(100) = 3.6\%$$

$$N: \left(\frac{6.5 \times 10^{23} - 6.02 \times 10^{23}}{6.02 \times 10^{23}}\right)(100) = 7.9\%$$

3. If the amperage in your electrolysis cell were increased by a factor of 2, what effect would this have on the time required to produce the same amount of hydrogen?

   Because C = amperes × seconds, increasing the amperage by a factor of two would decrease the time required to produce the same amount of hydrogen by a factor of two.

4. Would an increase in the concentration of the sulfuric acid increase the rate of hydrogen production? Explain.

   No. The resistance of the solution would decrease, but this would have no effect on the rate of hydrogen production.

5. Electrolysis of an NaCl solution with a current of 2.00 A for a period of 200 s produced 59.6 mL of Cl_2 at 650 mm pressure and 27°C. Calculate the value of the faraday from these data.

$$2Cl^- \longrightarrow Cl_2 + 2e^- \qquad \text{coulombs} = (2.00\ \text{A})(200\ \text{s}) = 400\ \text{C}$$

$$PV = nRT;\ n = \frac{PV}{RT} = \frac{(650\ \text{mm}/760\ \text{mm/atm})(0.0596\ \text{L})}{(0.0821\ \text{L–atm/mol–K})(300\ \text{K})} = 2.07 \times 10^{-3}\ \text{mol}$$

$$\mathcal{F} = \left(\frac{400\ \text{C}}{2.07 \times 10^{-3}\ \text{mol}}\right)\left(\frac{1\ \text{mol}\ Cl_2}{2\ \text{mol}\ e^-}\right) = 9.66 \times 10^5\ \text{C}$$

6. Why are different products obtained when molten and aqueous NaCl are electrolyzed? Predict the products in each case.

   Electrolysis of molten NaCl produces Na and Cl_2, whereas electrolysis of aqueous NaCl produces H_2 and Cl_2. Water is easier to reduce than Na^+. Thus, H_2 is formed in the aqueous medium.

Experiment

Electrochemical Cells and Thermodynamics

OBJECTIVE

To become familiar with some fundamentals of electrochemistry, including the Nernst equation, by constructing electrochemical (voltaic) cells and measuring their potentials at various temperatures. The quantities ΔG, ΔH, and ΔS are calculated from the temperature variation of the measured emf.

APPARATUS AND CHEMICALS

Apparatus

- DC voltmeter, or potentiometer (to measure mV)
- emery cloth
- 600-mL beaker
- thermometer
- glass stirring rods (3)
- Bunsen burner and hose
- clamps (2)
- alligator clips and lead wires (2 sets)
- 50-mL test tubes (3)
- glass U-tubes (to fit large test tubes) (3)
- cotton
- ring stand, iron ring, and wire gauze

Chemicals

- 1 *M* $Pb(NO_3)_2$
- 1 *M* $SnCl_2$
- lead, tin, and copper strips or wire
- 1 *M* $Cu(NO_3)_2$
- 0.1 *M* KNO_3
- ice
- agar

DISCUSSION

BACKGROUND

Electrochemistry is that area of chemistry that deals with the relations between chemical changes and electrical energy. It is primarily concerned with oxidation-reduction phenomena. Chemical reactions can be used to produce electrical energy in cells that are referred to as *voltaic,* or galvanic, cells. Electrical energy, on the other hand, can be used to bring about chemical changes in what are termed *electrolytic* cells. In this experiment you will investigate some of the properties of voltaic cells.

In principle, any spontaneous redox reaction can be used to produce electrical energy. This task can be accomplished by means of a voltaic cell, a device in which electron transfer takes place through an external circuit or pathway rather than directly between reactants. One such spontaneous reaction occurs when a strip of zinc is immersed in a solution containing Cu^{2+}. As the reaction proceeds, the blue color of the Cu^{2+} (*aq*) ions begins to fade and metallic copper deposits on the zinc strip. At the same time, the zinc begins to dissolve. The redox reaction that occurs is given in Equation [1].

$$Zn(s) + Cu^{2+}(aq) \longrightarrow Zn^{2+}(aq) + Cu(s) \qquad [1]$$

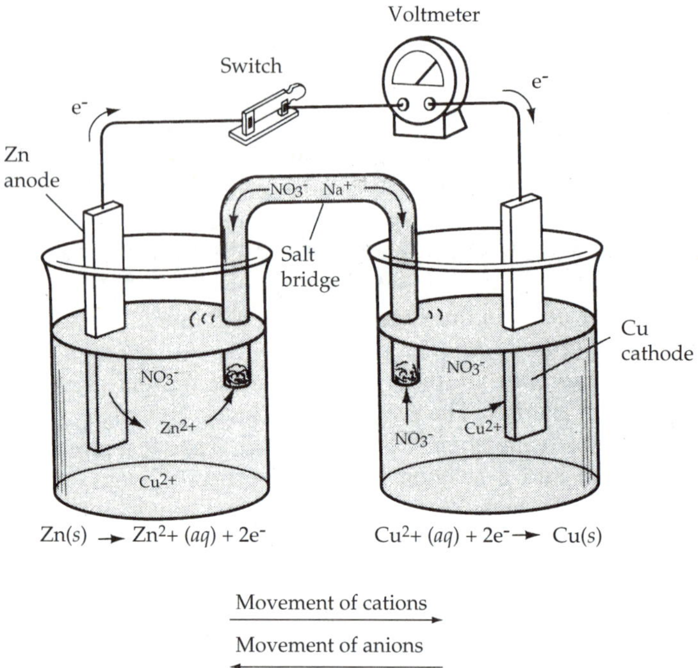

▲ **FIGURE 17.1** A complete and functioning voltaic cell using a salt bridge to complete the electrical circuit.

Figure 17.1 shows a voltaic cell that utilizes the same reaction. The two solid metal strips connected by the external circuit are called electrodes. The electrode at which oxidation occurs is called the *anode,* and the electrode at which reduction occurs is called the *cathode.* The voltaic cell may be regarded as two "half-cells," one corresponding to the oxidation half-reaction and the other to the reduction half-reaction. Recall that a substance that loses electrons is said to be oxidized, whereas a substance that gains electrons is said to be reduced. In our example, Zn is oxidized and Cu^{2+} is reduced:

Anode (oxidation half-reaction) $\quad Zn(s) \longrightarrow Zn^{2+}(aq) + 2e^-$

Cathode (reduction half-reaction) $\quad Cu^{2+}(aq) + 2e^- \longrightarrow Cu(s)$

Because Zn^{2+} ions are formed in one compartment and Cu^{2+} ions are depleted in the other, a salt bridge is used to maintain electrical neutrality by allowing the migration of ions between these compartments.

The cell voltage, or electromotive force (*emf*), is indicated on the voltmeter in units of volts. The cell emf is also called the cell potential. The magnitude of the emf is a quantitative measure of the driving force or thermodynamic tendency for the reaction to occur. In general, the emf of a voltaic cell depends on the substances that make up the cell as well as on their concentration and temperature. Hence, it is a common practice to compare *standard cell potentials,* symbolized by E°_{cell}. These potentials correspond to cell voltages under standard conditions: gases at 1 atm pressure, solutions at 1 *M* concentration and at 25°C. For the Zn/Cu voltaic cell in Figure 17.1, the standard cell potential at 25°C is 1.10 V:

$$Zn(s) + Cu^{2+}(aq, 1\,M) \longrightarrow Zn^{2+}(aq, 1\,M) + Cu(s) \qquad E^\circ_{cell} = 1.10\text{ V}$$

Recall that the superscript ° denotes standard state conditions.

The cell potential is the difference between two electrode potentials, one associated with the cathode and the other associated with the anode. By convention, the potential associated with each electrode is chosen to be the potential for reduction to occur at that electrode. Thus, standard electrode potentials are tabulated for reduction reactions, and they are denoted by the symbol, E°_{red}. The cell potential is given by the standard reduction potential of the cathode reaction, E°_{red} (cathode), *minus* the standard reduction potential of the anode reaction, E°_{red} (anode):

$$E^\circ_{cell} = E^\circ_{red}(\text{cathode}) - E^\circ_{red}(\text{anode}) \qquad [2]$$

Because it is not possible to directly measure the potential of an isolated half-cell reaction, the standard hydrogen half-reduction reaction, in which H^+ (aq) is reduced to $H_2(g)$ under standard conditions, has been selected as a reference. It has been assigned a standard reduction potential of exactly 0 V:

$$2H^+(aq, 1\,M) + 2e^- \longrightarrow H_2(g, 1\text{ atm}) \qquad E^\circ_{red} = 0\text{ V}$$

An electrode designed to produce this half-reaction is called the standard hydrogen electrode (SHE). Figure 17.2 shows a voltaic cell using a SHE and a standard Zn^{2+}/Zn electrode. The spontaneous reaction occurring in this cell is the oxidation of Zn and the reduction of H^+:

$$Zn(s) + 2H^+(aq) \longrightarrow Zn^{2+}(aq) + H_2(g)$$

The standard cell potential for this cell is 0.76 V. By using the defined standard reduction potential of H^+ ($E^\circ_{red} = 0$ V) and Equation [2], we can determine the standard reduction potential for the Zn^{2+}/Zn half-reaction:

$$E^\circ_{cell} = E^\circ_{red}(\text{cathode}) - E^\circ_{red}(\text{anode})$$

$$0.76\text{ V} = 0\text{ V} - E^\circ_{red}(\text{anode})$$

$$E^\circ_{red}(\text{anode}) = -0.76\text{ V}$$

Thus, a standard reduction potential of -0.76 V can be assigned to the reduction of Zn^{2+} to Zn:

$$Zn^{2+}(aq, 1\,M) + 2e^- \longrightarrow Zn(s) \qquad E^\circ_{red} = -0.76\text{ V}$$

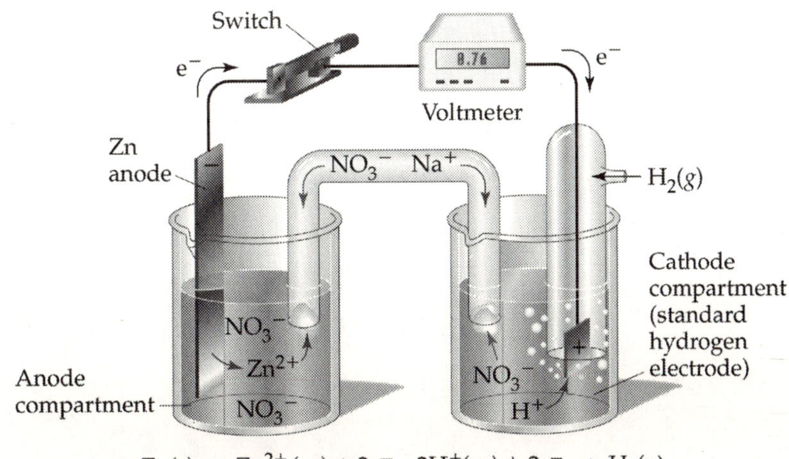

▲ **FIGURE 17.2** Voltaic cell using a standard hydrogen electrode.

Notice that the reaction is written as a reduction even though it is "running in reverse," as an oxidation in the cell in Figure 17.2. Whenever we assign a potential to a half-cell reaction, we always write the reaction as a reduction reaction.

Standard reduction potentials for other half-reactions can be established in a manner similar to that used for the Zn^{2+}/Zn half-reaction. The table in Appendix G lists some standard reduction potentials. Example 17.1 illustrates how this method can be used to determine the standard reduction potential for the Cu^{2+}/Cu half-reaction:

EXAMPLE 17.1

The cell in Figure 17.1 may be represented by the following notation:

$$\mathrm{Zn} \mid \mathrm{Zn}^{2+} \parallel \mathrm{Cu}^{2+} \mid \mathrm{Cu}$$

The double bar represents the salt bridge. Given that E°_{cell} for this cell is 1.10 V, the Zn electrode is the anode, and the standard reduction potential of Zn^{2+} is −0.76 V, calculate the E°_{red} for the reduction of Cu^{2+} to Cu:

$$\mathrm{Cu}^{2+}(aq,\ 1\ M) + 2\mathrm{e}^- \longrightarrow \mathrm{Cu}(s)$$

SOLUTION: We use Equation [2] and the information we are given:

$$E^\circ_{\text{cell}} = E^\circ_{\text{red}}(\text{cathode}) - E^\circ_{\text{red}}(\text{anode})$$
$$1.10\ \text{V} = E^\circ_{\text{red}}(\text{cathode}) - (-0.76\ \text{V})$$
$$E^\circ_{\text{red}}(\text{cathode}) = 1.10\ \text{V} - 0.76\ \text{V} = 0.34\ \text{V}$$

The free-energy change, ΔG, associated with a chemical reaction is a measure of the driving force or spontaneity of the process. If the free-energy change of a process is negative, the reaction will occur spontaneously in the direction indicated by the chemical equation.

The cell potential of a redox process is related to the free-energy change as follows:

$$\Delta G = -n\mathcal{F}E \qquad [3]$$

In this equation, $\mathcal{F}$ is Faraday's constant, the electrical charge on 1 mol of electrons is

$$1\mathcal{F} = 96{,}500 \frac{\text{C}}{\text{mol e}^-} = 96{,}500 \frac{\text{J}}{\text{V–mol e}^-}$$

and n represents the number of moles of electrons transferred in the reaction. For the case when both reactants and products are in their standard states, Equation [3] takes the following form:

$$\Delta G^\circ = -n\mathcal{F}E^\circ \qquad [4]$$

EXAMPLE 17.2

Calculate the standard free-energy change associated with the redox reaction $2Ce^{4+} + Tl^{+} \longrightarrow 2Ce^{3+} + Tl^{3+}$ ($E^\circ = 0.450$ V). Would this reaction occur spontaneously under standard conditions?

SOLUTION:

$$\Delta G^\circ = -n\mathcal{F}E^\circ$$
$$= -(2\ \text{mol e}^-)\left(\frac{96{,}500\ \text{J}}{\text{V–mol e}^-}\right)(0.450\ \text{V})$$

$$= -86.9 \times 10^3 \text{ J}$$

$$= -86.9 \text{ kJ}$$

Since $\Delta G^\circ < 0$, this reaction would occur spontaneously.

The standard free-energy change of a chemical reaction is also related to the equilibrium constant for the reaction as follows:

$$\Delta G^\circ = -RT \ln K \qquad [5]$$

where R is the gas-law constant (8.314 J/K-mol) and T is the temperature in Kelvin. Consequently, E° is also related to the equilibrium constant. From Equations [4] and [5] it follows that

$$-n\mathcal{F}E^\circ = -RT\, ln\, K$$

$$E^\circ = \frac{RT}{n\mathcal{F}} \ln K \qquad [6]$$

When $T = 298$ K, $\ln K$ is converted to $\log K$, and the appropriate values of R and $\mathcal{F}$ are substituted, Equation [6] becomes

$$E^\circ = \frac{0.0592}{n} \log K \qquad [7]$$

We can see from this relation that the larger K is, the larger the standard-cell potential will be.

In practice, most voltaic cells are not likely to be operating under standard-state conditions. It is possible, however, to calculate the cell emf, E, under nonstandard-state conditions from a knowledge of E°, temperature, and concentrations of reactants and products:

$$E = E^\circ - \frac{0.0592}{n} \log Q \qquad [8]$$

Q is called the reaction quotient; it has the form of an equilibrium-constant expression, but the concentrations used to calculate Q are not necessarily equilibrium concentrations. The relationship given in Equation [8] is referred to as the Nernst equation (see Example 17.3).

Let us consider the operation of the cell shown in Figure 17.1 in more detail. Earlier we saw that the reaction

$$Cu^{2+} + Zn \longrightarrow Zn^{2+} + Cu$$

is spontaneous. Consequently, it has a positive electrochemical potential ($E^\circ = 1.10$ V) and a negative free energy ($\Delta G^\circ = -n\mathcal{F}E^\circ$). As this reaction occurs, Cu^{2+} will be reduced and deposited as copper metal onto the copper electrode. The electrode at which reduction occurs is called the cathode. Simultaneously, zinc metal from the zinc electrode will be oxidized and go into solution as Zn^{2+}. The electrode at which oxidation occurs is called the anode. Effectively, then, electrons will flow in the external wire from the zinc electrode through the voltmeter to the copper electrode and be given up to copper ions in solution. These copper ions will be reduced to copper metal and plate out on the copper electrode. Concurrently, zinc metal will give up electrons to become Zn^{2+} ions in solution. These Zn^{2+} ions will diffuse through the salt bridge into the copper solution and replace the Cu^{2+} ions that are being removed.

EXAMPLE 17.3

Calculate the cell potential for the following cell,

$$Zn \mid Zn^{2+}(0.60\ M) \parallel Cu^{2+}(0.20\ M) \mid Cu$$

given the following:

$$Cu^{2+} + Zn \longrightarrow Cu + Zn^{2+} \qquad E^\circ = 1.10\ V$$

(HINT: Recall that Q includes expressions for species in solution but not for pure solids.)

SOLUTION:

$$E = E^\circ - \frac{0.0592}{n} \log \frac{[Zn^{2+}]}{[Cu^{2+}]}$$

$$= 1.10\ V - \frac{0.0592}{2} \log \frac{[0.60]}{[0.20]}$$

$$= 1.10 - 0.014$$

$$= 1.086$$

$$= 1.09\ V$$

You can see that small changes in concentrations have small effects on the cell emf.

A list of the properties of electrochemical cells and some definitions of related terms are given in Table 17.1.

Chemists have developed a shorthand notation for electrochemical cells, as seen in Example 17.1. The notation for the Cu-Zn cell that explicitly shows concentrations is as follows:

$$Zn \mid Zn^{2+}(xM) \parallel Cu^{2+}(yM) \mid Cu$$

Anode (oxidation)   Cathode (reduction)

In this notation, the anode (oxidation half-cell) is written on the left and the cathode (reduction half-cell) is written on the right.

Your objective in this experiment is to construct a set of three electrochemical cells and to measure their cell potentials. With a knowledge of two half-cell potentials and the cell potentials obtained from your measurements, you will calculate the other half-cell potentials and the equilibrium constants for the reactions. By measuring the cell potential as a function of tempera-

TABLE 17.1 Summary of Properties of Electrochemical Cells and Some Definitions

Voltaic cells: $E > 0$, $\Delta G < 0$: reaction spontaneous, K large (greater than 1)
Electrolytic cells: $E < 0$, $\Delta G > 0$; reaction nonspontaneous, K small (less than 1)
Anode: electrode at which oxidation occurs
Cathode: electrode at which reduction occurs
Oxidizing agent: species accepting electrons to become reduced
Reducing agent: species donating electrons to become oxidized

ture, you may also determine the thermodynamic constants, ΔG, ΔH, and ΔS, for the reactions. This can be done with the aid of Equation [9]:

$$\Delta G = \Delta H - T\Delta S \qquad [9]$$

ΔG may be obtained directly from measurements of the cell potential using the relationship:

$$\Delta G = -n\mathcal{F}E$$

A plot of ΔG versus temperature in Kelvin will give $-\Delta S$ as the slope and ΔH as the intercept. A more accurate measure of ΔH can be obtained, however, by substituting ΔG and ΔS back into Equation [9] and calculating ΔH.

EXAMPLE 17.4

For the voltaic cell

$$\text{Pb} \mid \text{Pb}^{2+}(1\ M) \parallel \text{Cu}^{2+}(1\ M) \mid \text{Cu}$$

the following data were obtained:

$$E = 0.464 \text{ V} \qquad T = 298 \text{ K}$$
$$E = 0.468 \text{ V} \qquad T = 308 \text{ K}$$
$$E = 0.473 \text{ V} \qquad T = 318 \text{ K}$$

Calculate ΔG, ΔH, and ΔS for this cell.

SOLUTION:

$$\Delta G = -n\mathcal{F}E$$

At 298 K,

$$\begin{aligned}\Delta G &= -(2 \text{ mol e}^-)(96{,}500 \text{ J/V-mol e}^-)(0.464 \text{ V})(1 \text{ kJ}/1000 \text{ J})\\ &= -89.6 \text{ kJ}\end{aligned}$$

At 308 K,

$$\begin{aligned}\Delta G &= -(2 \text{ mol e}^-)(96{,}500 \text{ J/V-mol e}^-)(0.468 \text{ V})(1 \text{ kJ}/1000 \text{ J})\\ &= -90.3 \text{ kJ}\end{aligned}$$

At 318 K,

$$\begin{aligned}\Delta G &= -(2 \text{ mol e}^-)(96{,}500 \text{ J/V-mol e}^-)(0.473 \text{ V})(1 \text{ kJ}/1000 \text{ J})\\ &= -91.3 \text{ kJ}\end{aligned}$$

Since $\Delta G = \Delta H - T\Delta S$, a plot of ΔG versus T in Kelvin gives $\Delta H = -64.2$ kJ/mol and $\Delta S = 85.0$ J/mol-K.

PROCEDURE

Before setting up pairs of half-cells, make a complete cell of the type shown in Figure 17.1 in the following manner: Place 30 mL of 1 M $Pb(NO_3)_2$ and 30 mL of 1 M $Cu(NO_3)_2$ in separate large test tubes. Obtain a lead strip and a copper strip and clean the surfaces of each with emery cloth or sandpaper. Boil 100 mL of 0.1 M KNO_3. Remove this solution from the heat and add to the boiling solution 1 g of agar, stirring constantly until all the agar dissolves. Invert a U-tube and fill the U-tube with this solution before it cools, leaving about a half inch of air space at each end of the U-tube as shown in Figure 17.3. The cotton plugs must protrude from the ends of the U-tube.

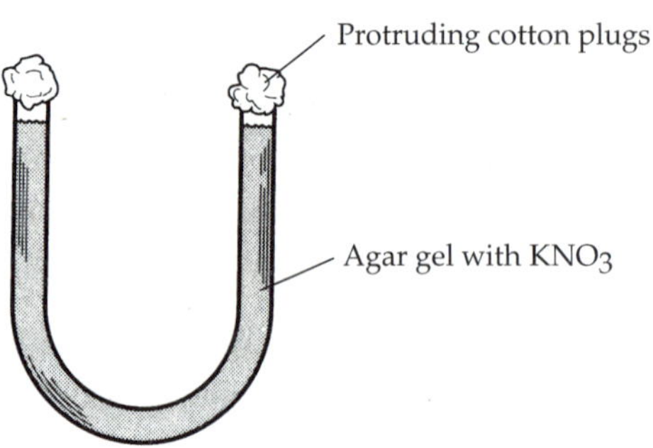

▲ FIGURE 17.3

Construct two additional agar-filled U-tubes in the same manner. Place a U-tube into the test tubes as a salt bridge, as shown in Figure 17.4.

Insert the lead strip into the $Pb(NO_3)_2$ solution and the copper strip into the $Cu(NO_3)_2$ solution. Obtain a voltmeter and attach the positive lead to the copper strip and the negative lead to the lead strip using alligator clips. Read the voltage. Be certain that the alligator clips make good contact with the metal strips. Record this voltage and the temperature of the cells on your report sheet. If your measured potential is negative, reverse the wire connection. Now construct the following cells and measure their voltages in the same manner:

$$Sn \mid Sn^{2+}(1\ M) \parallel Cu^{2+}(1\ M) \mid Cu$$

$$Pb \mid Pb^{2+}(1\ M) \parallel Sn^{2+}(1\ M) \mid Sn$$

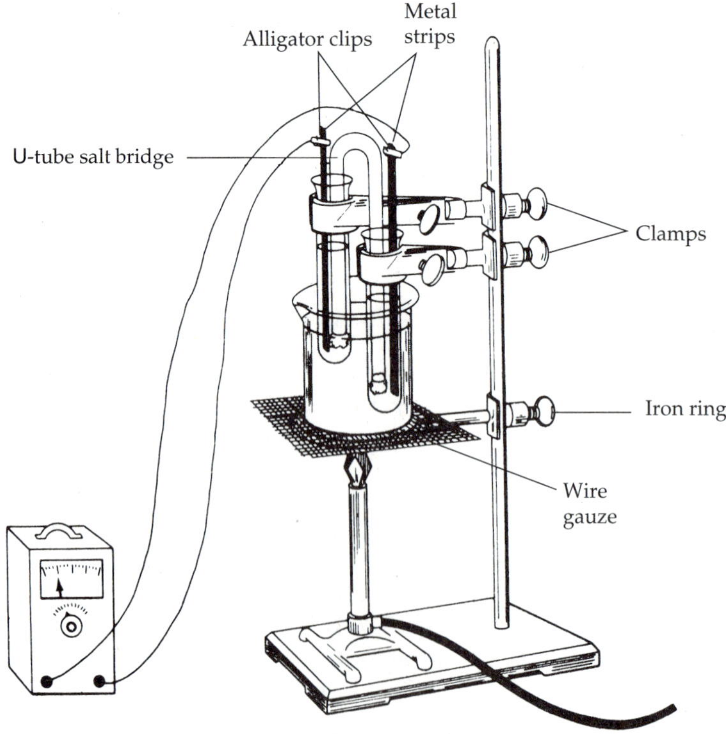

▲ FIGURE 17.4

Record the voltage and temperature of each cell on your report sheet. From the measured voltages, calculate the half-cell potentials for the lead and tin half-cells and the equilibrium constants for these two reactions. In these calculations use $E° = 0.34$ V for the Cu^{2+}/Cu couple.

Now choose any one of the three cells and measure the cell potential as a function of temperature as follows: Insert the metal strips into each of two 50-mL test tubes containing their respective 1 *M* cation solutions and place these test tubes in a 600-mL beaker containing about 200 mL of distilled water.

Place the U-tube into the two test tubes and connect the metallic strips to the voltmeter as before (see Figure 17.3). Because the voltage changes are of the order of 30 mV for the temperature range you will study, be certain that the voltmeter you use is sensitive enough to detect these small changes.

Begin heating the water in the 600-mL beaker, using your Bunsen burner. Be certain that the test tubes are firmly clamped in place. DO NOT MOVE any part of the cell, because the voltage will fluctuate if you do. Heat the cell to approximately 70°C. Measure the temperature and record it on your report sheet. Determine the cell potential at this temperature and record it on your report sheet. Remove the Bunsen burner and record the temperature and voltage at 15° intervals as the cell cools to room temperature. Finally, replace the beaker of water with a beaker containing an ice-water mixture; be careful not to move the test tubes and their contents. After the cell has been in the ice-water mixture for about 10 min, measure the temperature of the ice-water mixture in the 600-mL beaker and record it and the cell potential. You now have determined the cell potential at various temperatures. Calculate ΔG for the cell at each of these temperatures and plot ΔG versus temperature on the graph paper provided. The slope of the plot is $-\Delta S$. From the values of ΔG and ΔS, calculate ΔH at 298 K. If time permits, determine the temperature dependence of E for another cell.

Dispose of the solutions in the test tubes in the designated receptacles.

REVIEW QUESTIONS

Before beginning this experiment in the laboratory, you should be able to answer the following questions:

1. Define the following terms: *faraday, salt bridge, anode, cathode, voltaic cell, electrolytic cell.*
2. Write a chemical equation for the reaction that occurs in the following cell: $Ag \mid Ag^+ \parallel Cu^{2+} \mid Cu$.
3. Given the following $E°$'s, calculate the standard-cell potential for the cell in question 2.

$$Cu^{2+}(aq) + 2e^- \longrightarrow Cu(s) \qquad E° = +0.34 \text{ V}$$

$$Ag^+(aq) + e^- \longrightarrow Ag(s) \qquad E° = +0.80 \text{ V}$$

4. Calculate the voltage of the following cell:

$$Zn \mid Zn^{2+}(0.10\ M) \parallel Cu^{2+}(0.40\ M) \mid Cu$$

5. Calculate the cell potential, the equilibrium constant, and the free-energy change for

$$\mathrm{Ba}(s) + \mathrm{Mn}^{2+}(aq)(1\ M) \rightleftharpoons \mathrm{Ba}^{2+}(aq)(1\ M) + \mathrm{Mn}(s)$$

given the following $E°$ values:

$$\mathrm{Ba}^{2+}(aq) + 2e^- \longrightarrow \mathrm{Ba}(s) \qquad E° = -2.90\ \mathrm{V}$$

$$\mathrm{Mn}^{2+}(aq) + 2e^- \longrightarrow \mathrm{Mn}(s) \qquad E° = -1.18\ \mathrm{V}$$

6. Would you normally expect $\Delta H°$ to be positive or negative for a voltaic cell? Justify your answer.

7. Predict whether the following reactions are spontaneous or not.

$$\mathrm{Pd}^{2+} + \mathrm{H}_2 \longrightarrow \mathrm{Pd} + 2\mathrm{H}^+ \qquad \mathrm{Pd}^{2+} + 2e^- \longrightarrow \mathrm{Pd} \qquad E° = 0.987\ \mathrm{V}$$

$$\mathrm{Sn}^{4+} + \mathrm{H}_2 \longrightarrow \mathrm{Sn}^{2+} + 2\mathrm{H}^+ \qquad \mathrm{Sn}^{4+} + 2e^- \longrightarrow \mathrm{Sn}^{2+} \qquad E° = 0.154\ \mathrm{V}$$

$$\mathrm{Ni}^{2+} + \mathrm{H}_2 \longrightarrow \mathrm{Ni} + 2\mathrm{H}^+ \qquad \mathrm{Ni}^{2+} + 2e^- \longrightarrow \mathrm{Ni} \qquad E° = -0.250\ \mathrm{V}$$

$$\mathrm{Cd}^{2+} + \mathrm{H}_2 \longrightarrow \mathrm{Cd} + 2\mathrm{H}^+ \qquad \mathrm{Cd}^{2+} + 2e^- \longrightarrow \mathrm{Cd} \qquad E° = -0.403\ \mathrm{V}$$

From your answers, decide which of the above metals could be reduced by hydrogen.

8. Identify the oxidizing agents and reducing agents in the reactions in question 7.

Name ______________________________ Desk ______________

Date ______________ Laboratory Instructor ______________

REPORT SHEET | EXPERIMENT

Electrochemical Cells and Thermodynamics | 17

	Shorthand cell designation	*Temperature (°C)*	*E cell (measured)*	*ΔG (calculated)*	*K_{eq} (calculated)*
1.	$Pb\|Pb^{2+}\|\|Cu^{2+}\|Cu$	26	0.451 V	−87.0 kJ	1.57×10^{15}
2.	$Sn\|Sn^{2+}\|\|Cu^{2+}\|Cu$	26	0.562 V	−109 kJ	1.10×10^{19}
3.	$Pb\|Pb^{2+}\|\|Sn^{2+}\|Sn$	26	0.125 V	−24.1 kJ	1.64×10^{4}

Show calculations for $\Delta G°$ and K_{eq} for an exemplary pair.

$\Delta G = -n\mathcal{F}E = -(2 \text{ mol e}^-)(96{,}500 \text{ J/V-mol e}^-)(0.125 \text{ V}) = -24{,}125 \text{ J or } -24.1 \text{ kJ}$

$$\Delta G = -RT \ln K_{eq} \text{ or } \ln K_{eq} = \frac{\Delta G}{-RT} = \frac{-24{,}125 \text{ J}}{-(8.3144 \text{ J/K})(299 \text{ K})} = 9.704;\ K_{eq} = 1.64 \times 10^4$$

	Half-cell equation	*E half-cell (calculated)*
4.	$Pb^{2+} + 2e^- \longrightarrow Pb$	−0.11 V
5.	$Sn^{2+} + 2e^- \longrightarrow Sn$	−0.22 V
6.	$Cu^{2+} + 2e^- \longrightarrow Cu$	+0.34 V (given)

Effect of Temperature on Cell Potential

Cell designation: E (measured)	*Temperature (°C)*	*Temperature (K)*	*ΔG (calculated)*
$Pb\|Pb^{2+}\|\|Cu^{2+}\|Cu$; 0.446 V	6	279	−86.1 kJ
0.451 V	26	299	−87.0 kJ
0.468 V	68	341	−90.3 kJ
$Sn\|Sn^{2+}\|\|Cu^{2+}\|Cu$; 0.548 V	6	279	−105.7 kJ
0.562 V	26	299	−108.5 kJ
0.581 V	68	341	−112.1 kJ

ΔS determined from the slope of a plot of ΔG versus T ______ $\Delta S°$ Pb|Cu = −67.74 J/mol–K; $\Delta S°$ Sn|Cu = 103.22 J/mol–K

$\Delta H°$ calculated at 298 K ______ (show calculations) $\Delta H°$ Pb|Cu = −66.8 kJ/mV; $\Delta H°$ Sn|Cu = −77.6 kJ/mol

from $\Delta G = \Delta H - T\Delta S$ $\quad \Delta H = \Delta G + T\Delta S$

$= -87.0 \text{ kJ/mol} + (299 \text{ K})(67.74 \text{ J/mol-K})(1 \text{ kJ/1000 J})$

$= -66.8 \text{ kJ/mol}$

Is the cell reaction endothermic or exothermic? $\Delta H < 0$ for all three cells; the cell reactions are all exothermic.

QUESTIONS

1. Write the net ionic equations that occur in the following cells:

$Pb\|Pb(NO_3)_2\|\|AgNO_3\|Ag$	$Pb + 2\,Ag^+ \longrightarrow Pb^{2+} + 2\,Ag$
$Zn\|ZnCl_2\|\|Pb(NO_3)_2\|Pb$	$Zn + Pb^{2+} \longrightarrow Zn^{2+} + Pb$
$Pb\|Pb(NO_3)_2\|\|NiCl_2\|Ni$	$Pb + Ni^{2+} \longrightarrow Pb^{2+} + Ni$

2. Which of the following reactions should have the larger emf under standard conditions? Why?

$$CuSO_4(aq) + Pb(s) \rightleftharpoons PbSO_4(s) + Cu(s)$$

$$Cu(NO_3)_2(aq) + Pb(s) \rightleftharpoons Pb(NO_3)_2(aq) + Cu(s)$$

The reaction $CuSO_4 + Pb \longrightarrow PbSO_4(s) + Cu(s)$ will have the larger emf because the insolubility of $PbSO_4$ will drive the reaction to the right as it removes Pb^{2+} from solution.

3. Calculate ΔG for the reaction in Example 17.3.

$\Delta G = -n\mathcal{F}E$

$= -(2)(96{,}500)(1.09) = -210$ kJ

4. Voltages listed in textbooks and handbooks are given as *standard-cell potentials* (voltages). What is meant by a standard cell? Were the cells constructed in this experiment standard cells? Why or why not?
A standard cell has all concentrations equal to 1 molar, temperatures at 298 K, and gases at 1 atmosphere pressure. No, the temperature was not 298 K.

5. As a standard voltaic cell runs, it "discharges," and the cell potential decreases with time. Explain.
The cell potential is concentration dependent, and as the concentrations decrease, the potential decreases.

6. Using standard potentials given in the appendices, calculate the standard cell potentials and the equilibrium constants for the following reactions:

$$Cu(s) + 2Ag^+(aq) \rightleftharpoons Cu^{2+}(aq) + 2Ag(s)$$

$$Zn(s) + Fe^{2+}(aq) \rightleftharpoons Zn^{2+}(aq) + Fe(s)$$

(1) $E° = 0.799\ V - 0.337\ V = 0.462$ V; $\log K_{eq} = \dfrac{nE°}{0.0592} = \dfrac{2(0.462)}{0.0592} = 15.6$

$K_{eq} = 4 \times 10^{15}$

(2) $E° = 0.763\ V - 0.440\ V = 0.323\ V$ $\quad \log K_{eq} = \dfrac{nE°}{0.0592} = \dfrac{2(0.323)}{0.0592} = 10.9$

$K_{eq} = 8 \times 10^{10}$

Free Energy vs. Temperature

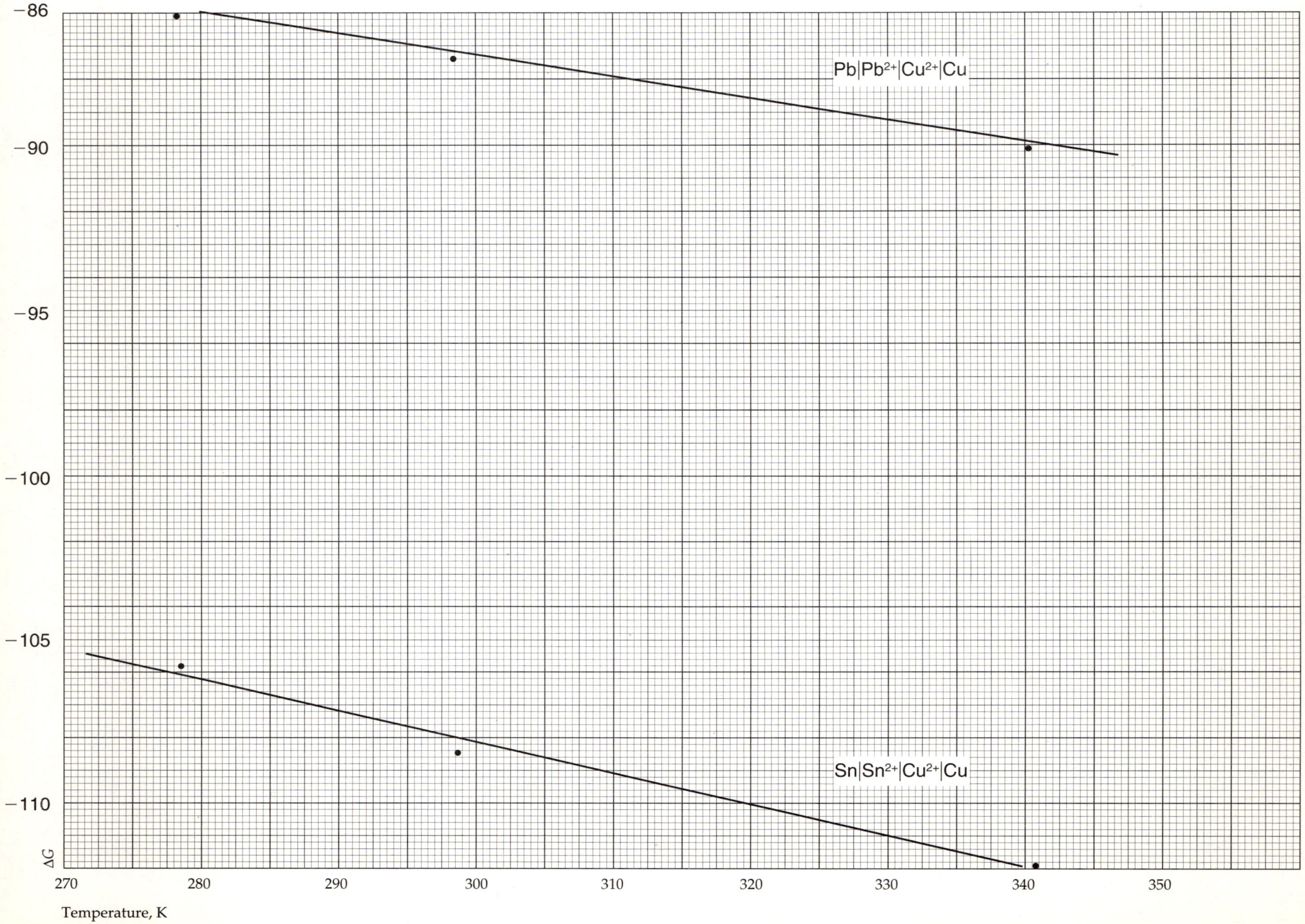

ΔG: −86, −90, −95, −100, −105, −110

Temperature, K: 270, 280, 290, 300, 310, 320, 330, 340, 350

$Pb|Pb^{2+}|Cu^{2+}|Cu$

$Sn|Sn^{2+}|Cu^{2+}|Cu$

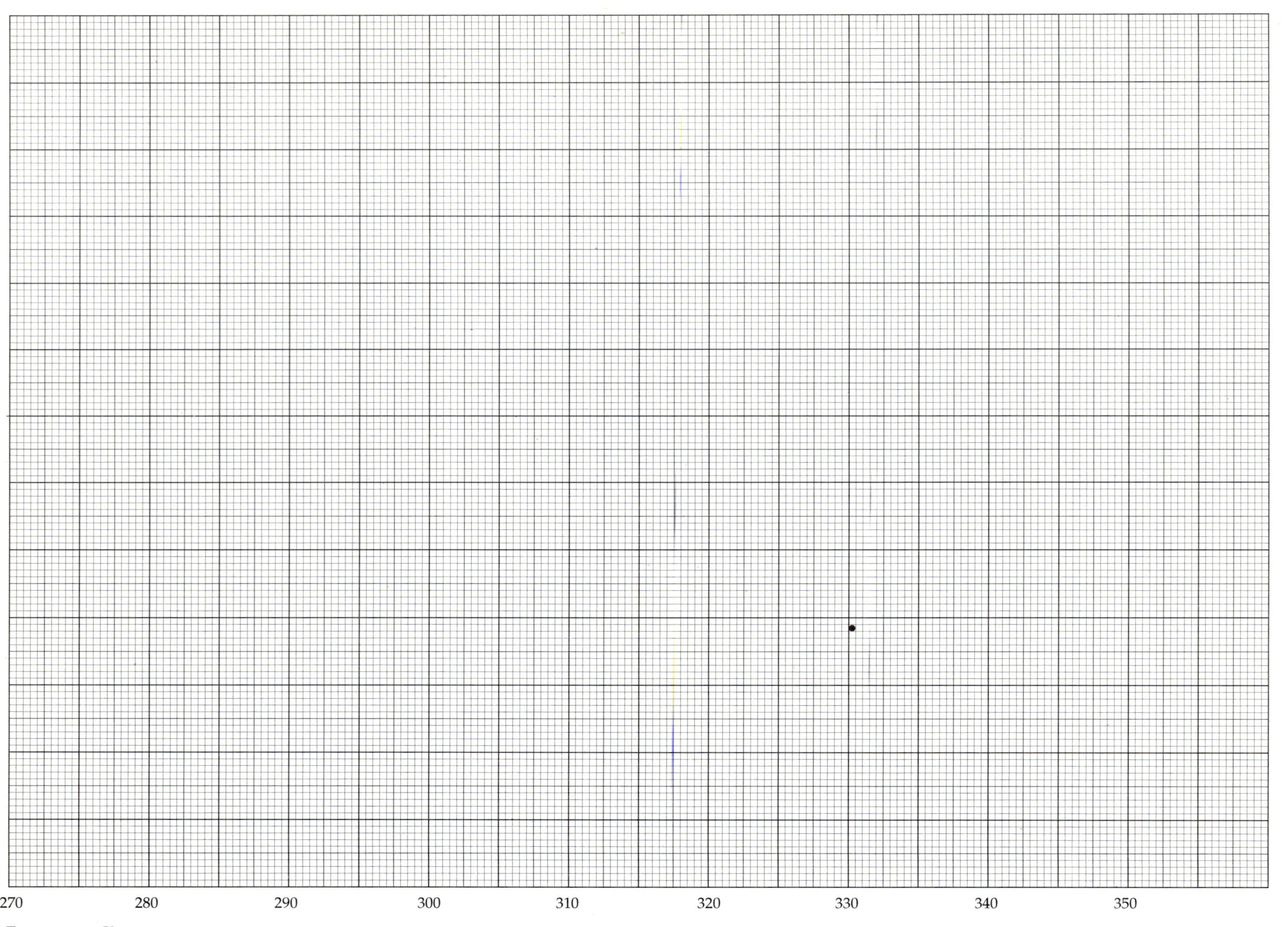
ΔG
270
280
290
300
310
320
330
340
350
Temperature, K

Experiment

The Chemistry of Oxygen: Basic and Acidic Oxides and the Periodic Table

OBJECTIVE

To illustrate the chemistry of oxides and to become familiar with acids and bases and the concept of pH.

APPARATUS AND CHEMICALS

Apparatus

400-mL beaker
150-mL beakers (5)
6-in. test tube
deflagrating spoon
pneumatic trough
rubber tubing
clay triangle
no. 2 one-hole stopper
250-mL wide-mouth bottles (6)
glass squares (6)
crucible and lid
pH paper
Bunsen burner and hose
crucible tongs
ring stand and clamp

Chemicals

$KClO_3$ (solid)
H_3BO_3 (solid)
$HClO_4$ (10% solution)
ZnO (solid)
calcium metal shavings
steel wool
sulfur lumps
MnO_2 (powdered)
Na_2O_2 (solid)
6 *M* HCl
19% NaOH (50% by weight)
magnesium ribbon, 2-in.
red phosphorus
charcoal pieces

WORK IN PAIRS, BUT EVALUATE YOUR DATA INDEPENDENTLY.

DISCUSSION

Most elements react with molecular oxygen under various conditions to produce oxides. These reactions are oxidation-reduction reactions and are illustrated by the examples below. The symbol Ae stands for any element; the formula for the oxide in general depends on the number of valence electrons of the element.

$$Ae + O_2 \longrightarrow AeO_2$$

$$2Ae + O_2 \longrightarrow 2AeO$$

$$4Ae + 3O_2 \longrightarrow 2Ae_2O_3$$

Oxidation is always accompanied by a reduction reaction. In the examples above, the element Ae is oxidized while oxygen (O_2) is reduced.

Oxides of elements may be either ionic or covalent in character. In general, the more ionic oxides are formed from the elements at the left of the periodic table, whereas the elements at the right form covalent oxides. The ionic oxides that dissolve to any extent in water react with water to form basic solutions; for example:

$$AeO(s) + H_2O(l) \longrightarrow Ae^{2+}(aq) + 2OH^-(aq)$$

$$O^{2-}(aq) + H_2O(l) \longrightarrow 2OH^-(aq)$$

The oxide ion reacts with water to form hydroxide. If the water were evaporated from the above solution, we could obtain the base $Ae(OH)_2$. These ionic oxides are called *basic oxides* or *basic anhydrides* (base without water). For example:

$$Ba(OH)_2 - H_2O = BaO \qquad \text{(basic anhydride)}$$

The covalent oxides react with water to form acidic solutions. For example:

$$AeO(g) + H_2O(l) \longrightarrow H_2AeO_2(aq)$$

$$AeO_2(g) + H_2O(l) \longrightarrow H_2AeO_3(aq)$$

In water these acids ionize to various degrees to furnish hydrogen ions:

$$H_2AeO_3(aq) + H_2O(l) \rightleftharpoons H_3O^+(aq) + HAeO_3^-(aq)$$

$$HAeO_3^-(aq) + H_2O(l) \rightleftharpoons H_3O^+(aq) + AeO_3^{2-}(aq)$$

Thus, the covalent oxides are termed *acidic oxides* or *acid anhydrides.* We can determine the corresponding formula for the acidic or basic anhydride from the formula of a given acid or base by subtracting water from the formula to eliminate *all* hydrogen atoms. For example:

$$2HNO_2 - H_2O \longrightarrow N_2O_3$$

$$2H_3BO_3 - 3H_2O \longrightarrow B_2O_3$$

$$2NaOH - H_2O \longrightarrow Na_2O$$

To determine whether an oxide is acidic or basic, the water solution of the oxide can be checked with litmus paper. However, not all of the oxides are appreciably soluble. In those cases in which the oxides are not water-soluble, one determines whether the oxides are acidic or basic by noting whether the oxide reacts with bases or acids. Acidic oxides react with bases, and basic oxides with acids.

It is not possible to classify all oxides either as acidic or basic. Some behave as both and are called *amphoteric.* Aluminum oxide is amphoteric, for it reacts both with acids and with strong bases:

$$Al_2O_3(s) + 6H^+(aq) \longrightarrow 2Al^{3+}(aq) + 3H_2O(l)$$

$$Al_2O_3(s) + 3H_2O(l) + 2OH^-(aq) \longrightarrow 2Al(OH)_4^-(aq)$$

Note that the formulas $Ae(OH)_2$ and H_2AeO_2 represent the same chemical composition. By convention, we indicate bases by placing the hydroxyl group at the end of the formula. Acids are written with the acidic hydrogens at the front of the formula.

Although there are several different definitions of acids and bases, a particularly useful one for aqueous solutions was proposed in 1923 by J.N. Brønsted and T.M. Lowry. In this scheme, *an acid is defined as a proton donor and a base as a proton acceptor.*

Consider the reaction of a strong acid, such as HCl, with water:

$$HCl(aq) + H_2O(l) \longrightarrow H_3O^+(aq) + Cl^-(aq) \qquad [1]$$

The HCl acts as a proton donor (acid), and the H_2O acts as a proton acceptor (base). Because HCl is a good donor of protons, the reaction essentially goes to completion in dilute solution—that is, all the HCl is present as H_3O^+ and Cl^-. These reactions are then merely proton-transfer reactions in which the stronger base (in this case, H_2O compared with Cl^-) competes for the protons. Recall that oxidation-reduction reactions are electron-transfer reactions in which a competition for electrons is developed. The more electronegative element gains the electrons from the less electronegative element.

A weak acid, such as acetic acid, is a poor donor of protons. Such a weak acid can donate protons to water only to a limited extent, and, in the resulting equilibrium, the concentrations of the undissociated reactants are much greater than the concentrations of the dissociated products. The reaction of acetic acid with water can be expressed as

$$HC_2H_3O_2(aq) + H_2O(l) \rightleftharpoons H_3O^+(aq) + C_2H_3O_2^-(aq) \qquad [2]$$

This is an equilibrium, and we can write the *equilibrium-constant (K_c)* expression for it. Thus,

$$K_c = \frac{[H_3O^+][C_2H_3O_2^-]}{[HC_2H_3O_2][H_2O]} \qquad [3]$$

Because the concentration of molecular water is large in comparison with the concentration of all other species present, the concentration of water changes relatively little in the course of the reaction and remains essentially constant. Thus, the concentration of water can be combined with the equilibrium constant, K_c, and we obtain what is called an *acid-ionization constant* or *acid dissociation constant, K_a*:*

$$K_a = K_c[H_2O] = \frac{[H_3O^+][C_2H_3O_2^-]}{[HC_2H_3O_2]} \text{ or } K_a = \frac{[H^+][C_2H_3O_2^-]}{[HC_2H_3O_3]} \qquad [4]$$

If the concentration of all the species in Equation [4] are known, the dissociation constant, K_a, of acetic acid can be calculated.

Similarly, the weak base ammonia only partially reacts according to

$$NH_3(aq) + H_2O(l) \rightleftharpoons NH_4^+(aq) + OH^-(aq) \qquad [5]$$

for which the equilibrium constant expression is

$$K_c = \frac{[NH_4^+][OH^-]}{[NH_3][H_2O]} \qquad [6]$$

and the base-dissociation-constant expression is

$$K_b = K_c[H_2O] = \frac{[NH_4^+][OH^-]}{[NH_3]}$$

where K_b is called the *base-dissociation constant.*

The reaction of HCl, an acid, with NaOH, a base, is also an acid-base reaction. Remembering that an aqueous solution of HCl consists of H_3O^+ and Cl^- ions and that NaOH is a strong electrolyte that dissociates to produce Na^+ and OH^- ions when dissolved in water, we can write

$$H_3O^+(aq) + Cl^-(aq) + Na^+(aq) + OH^-(aq) \longrightarrow 2H_2O(l) + Na^+(aq) + Cl^-(aq)$$

Here the proton is transferred from the H_3O^+ ion to the OH^- ion. Noting that the Na^+ and Cl^- ions appear as both reactants and products (that is, they are

*Note that both H^+ and H_3O^+ are often used interchangeably.

spectators and can be canceled out of the equation), we can write the following net ionic equation:

$$H_3O^+(aq) + OH^-(aq) \longrightarrow 2H_2O(l) \quad [7]$$

Equation [7] is the net ionic equation for the reaction of any strong acid with any strong OH^--containing base. The reaction is referred to as *neutralization.*

Water is an amphoteric or amphiprotic substance—that is, it can act as either a Brønsted acid (proton donor) or a Brønsted base (proton acceptor). This is evidenced by the reaction of water with itself,

$$H_2O(l) + H_2O(l) \rightleftharpoons H_3O^+(aq) + OH^-(aq) \quad [8]$$

where one molecule of water donates a proton to the other. Since Equation [8] is also an equilibrium, we can write the equilibrium-constant expression for this reaction as

$$K_c = \frac{[H_3O^+][OH^-]}{[H_2O]^2} \quad [9]$$

Again, because the concentration of molecular water is large in aqueous solutions when compared with the concentration of the other species present, its concentration changes very little during the course of the reaction and can be incorporated into the dissociation constant for water, known as K_w:

$$K_w = K_c[H_2O]^2 = [H_3O^+][OH^-] \text{ or } K_w = [H^+][OH^-] \quad [10]$$

K_w is often called the ion-product constant for water. At 25°C, $K_w = 1.0 \times 10^{-14}$. This relationship is essentially true for any aqueous solution at 25°C, irrespective of the presence of other ions in solution. Hence, if we know the concentration of H_3O^+ ions in an aqueous solution, we can always use Equation [10] to calculate the concentration of the OH^- ions present in this same solution. In fact, since $K_w = [H_3O^+][OH^-] = 1.0 \times 10^{-14}$, if either $[OH^-]$ or $[H_3O^+]$ is known the other may be calculated. As can be seen from Equation [10], since $K_w = 1.0 \times 10^{-14}$, the value of the H_3O^+ or OH^- concentration can be small but can *never* be equal to zero!

Example 18.1

Calculate the $[OH^-]$ concentration in an aqueous solution where $[H_3O^+] = 3.0 \times 10^{-5}$ *M*.

SOLUTION:

$$[H_3O^+][OH^-] = 1.0 \times 10^{-14}$$

$$[OH^-] = \frac{1.0 \times 10^{-14}}{[H_3O^+]} = \frac{1.0 \times 10^{-14}}{3.0 \times 10^{-5}}$$

$$= 3.3 \times 10^{-10}\ M$$

In a neutral solution, one where $[OH^-] = [H_3O^+]$, both concentrations are equal to 1.0×10^{-7}, or 0.00000010. This latter number and others like it are very inconvenient to write. Because of this, the concept of pH was developed as a convenience in expressing the concentration of the hydronium ion in dilute solutions of acids and bases. The pH of a solution is defined as the negative logarithm of the hydronium ion concentration or hydrogen ion concentration:

$$pH = -\log [H_3O^+] \text{ or } pH = -\log [H^+] \quad [11]$$

And consequently,

$$[H_3O^+] = 10^{-pH}$$

For example, a solution containing 1.0×10^{-3} mol of HCl in 1 L of aqueous solution (10^{-3} *M*) contains 0.0010 *M* $[H_3O^+]$ and has a pH of 3.00. This pH is calculated as follows:

$$pH = -\log [H_3O^+] = -\log [1.0 \times 10^{-3}] = -[0 - 3.0] = 3.00$$

A review on the use of logs can be found in Appendix A.

EXAMPLE 18.2

What is the pH of a 0.033 *M* HCl solution?

SOLUTION:

$$\begin{aligned} pH &= -\log [H_3O^+] \\ &= -\log [3.3 \times 10^{-2}] \\ &= -(\log 3.3 + \log 10^{-2}) \\ &= -(0.52 - 2.00) \\ &= 1.48 \end{aligned}$$

From Equation [8] we can see that in a neutral solution, the hydronium ion concentration must equal the hydroxide ion concentration. These concentrations can be calculated by letting

$$[H_3O^+] = [OH^-] = x$$

Substituting into Equation [10], we have

$$(x)(x) = x^2 = 1.0 \times 10^{-14}$$

$$x = 1.0 \times 10^{-7}\ M$$

Then the pH of a neutral solution can be calculated:

$$pH = -\log (1.0 \times 10^{-7}) = 7.00$$

Since an acid solution is one in which $[H_3O^+] > [OH^-]$, then $[H_3O^+] > 10^{-7}$ and $[OH^-] < 10^{-7}$.* Likewise, since a basic solution is one in which $[OH^-] > [H_3O^+]$, then $[OH^-] > 10^{-7}$ and $[H_3O^+] < 10^{-7}$. These results are summarized in Table 18.1.

The hydronium ion concentration can also be calculated from the pH. For example, if a solution has a pH of 4.0 or 5.3, the hydronium ion concentration can be calculated:

$$\begin{aligned} -\log [H_3O^+] &= 4.0 \\ \log [H_3O^+] &= -4.0 \\ [H_3O^+] &= 1 \times 10^{-4}\ M \end{aligned}$$

or

$$\begin{aligned} -\log [H_3O^+] &= 5.3 \\ \log [H_3O^+] &= -5.3 = -6.0 + 0.7 \\ [H_3O^+] &= 5 \times 10^{-6}\ M \end{aligned}$$

*The symbols > and < mean "greater than" and "less than," respectively.

TABLE 18.1 Important Relations in Acidic and Basic Aqueous Solutions

	$[H_3O^+]$	pH	$[OH^-]$
Acidic	$>10^{-7}$	<7	$<10^{-7}$
Neutral	10^{-7}	7	10^{-7}
Basic	$<10^{-7}$	>7	$>10^{-7}$

The pH of aqueous solutions may be determined in essentially two ways. One is by use of an electronic instrument having an electrode that is sensitive to hydronium ion concentration. These instruments are called "pH meters." The other way is by the use of organic dyes that possess two or more different pH-dependent colored forms. These dyes are called *acid-base indicators.*

Acid-base indicators derive their ability to act as pH indicators from the fact that they can exist in two forms. The acid form possesses one color, which by loss of a proton is converted to a differently colored base form. The pertinent equilibrium for the indicator is

$$\underset{\text{(Color A)}}{HIn_A} \rightleftharpoons H^+ + \underset{\text{(Color B)}}{In_B^-} \qquad [12]$$

where HIn_A is the acid form of the indicator having color A and In_B^- is the base form of the indicator having color B. As an acid is added to a solution containing an indicator, Le Châtelier's principle tells us that the equilibrium in Equation [12] will shift to the left, the indicator will be present predominantly as HIn_A, and the solution will have color A. Likewise, if a base is added, the indicator will be present predominantly in the form of In_B^-, and the solution will have color B. Figure 18.1 shows some common indicators, with their respective colors and pH-transition regions.

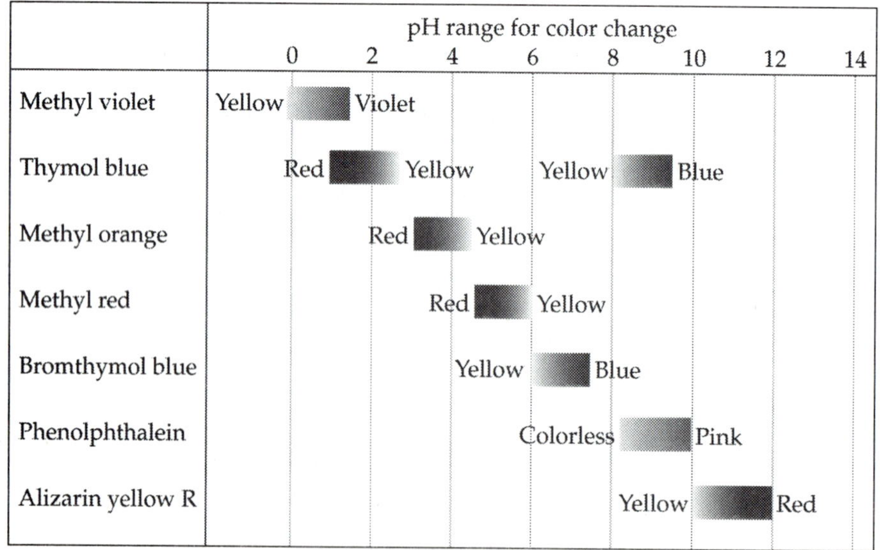

▲ **FIGURE 18.1** The pH ranges for the color changes of some common acid-base indicators. Most indicators have a useful range of about 2 pH units.

The pH paper is paper that has been impregnated with a series of organic dyes similar to those listed in Figure 18.1. The paper changes various colors according to the pH of the solution in which it is immersed. Using pH paper, you should be able to estimate the pH of a solution to the nearest pH unit in the region pH 1 to 11.

In this experiment, you will prepare oxygen gas and a number of typical oxides of various elements. You will then dissolve these oxides in water and test the acidity or basicity of the resulting solution. The following key points are to be especially noted:

1. Is there any correlation between the locations of the elements in the periodic table and the acidic and basic properties of their oxides?
2. Is there any correlation between the location of the element in the periodic table and its reactivity toward oxygen?
3. How does a catalyst affect a reaction?

PROCEDURE

A. Preparation of Oxygen

Dispose of all chemicals used in Parts A through D in the appropriate receptacles. Do not dispose of the $KClO_3$ in waste baskets, because it may start a fire.

Assemble the apparatus as illustrated in Figure 18.2, but do not yet connect the test tube to the delivery tube. **(CAUTION:** ***Be certain that the test tube is securely clamped. It* must *NOT be free to wiggle or tilt accidentally so that the contents [$KClO_3$ and MnO_2] come into contact with the rubber stopper, or a SEVERE EXPLOSION may result!*)**

Instructor: Tell students about the hazards of $KClO_3$-MnO_2 and how to dispose of them.

First heat about 0.25 g of potassium chlorate ($KClO_3$) in a clean, dry test tube and compare the vigor and rate of the reaction with the following mixture: carefully mix about 10 g of $KClO_3$ with 0.2 g of manganese dioxide (MnO_2). Place about 0.25 g of this mixture in a clean, dry test tube and heat it gently—just enough to melt the $KClO_3$. If it decomposes readily

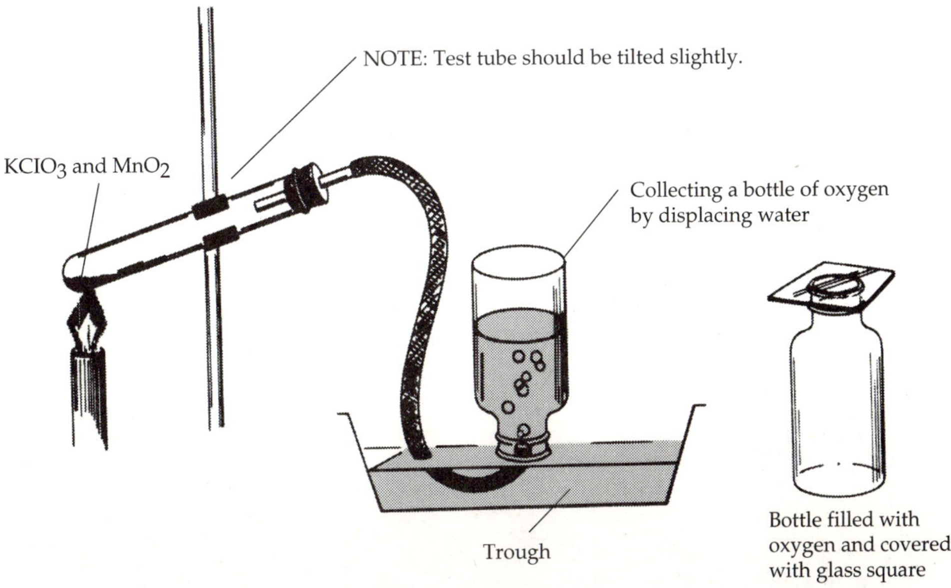

▲ **FIGURE 18.2** Apparatus for oxygen production.

without sparks or combusting, the mixture is safe to use in your apparatus. **(CAUTION: *Potassium chlorate is a powerful oxidizing agent and must be kept away from material that can be oxidized easily, or explosions may occur.*)** Allow the test tube to cool to room temperature. Then add the remainder of the mixture and connect the delivery tube.

Completely fill six 250-mL wide-mouth bottles with tap water; cover the bottles with a glass square and invert them in the trough as required to collect the oxygen. Gently heat the mixture of $KClO_3$ and MnO_2 so as to maintain a moderate rate of oxygen formation. Prior to collecting any oxygen, allow some of the gas to escape into the atmosphere. This will sweep air from the generator and allow you to collect pure oxygen. Collect six bottles of oxygen by displacing water, and immediately cover the bottles with glass squares as they are filled. Remove the delivery tube from the water before you remove the flame or before the formation of oxygen stops. Why?

B. Preparation of Oxides

Prepare oxides of the elements as directed below by burning them in the oxygen you collected. Label each bottle so that you can identify it and avoid confusion. Keep the bottles covered with your glass square as much as possible during the combustion (Figure 18.3). Immediately after each combustion, add about 30 to 50 mL of water, then cover the bottle with the glass square, shake the bottle to dissolve the oxide, and set it aside for later tests.

Magnesium Hold a 10-cm piece of magnesium with crucible tongs, ignite it, and quickly place it into a bottle of oxygen. **(CAUTION: *To avoid injury to your eyes, do not look directly at the brilliant light.*)**

Calcium Calcium burns brilliantly but is difficult to ignite. Place a shaving of calcium in a crucible and ignite it in air over a Bunsen burner for 15 min using the maximum temperature possible. After the crucible has cooled, wash out the contents with 50 mL of water into a 150-mL beaker.

Iron Pour a little water into a bottle of oxygen—enough to cover the bottom. The water layer will help prevent cracking of the bottle. Ignite a small, loosely packed wad of steel wool by holding it with tongs in a Bunsen burner flame until it ignites. Quickly thrust it into the bottle.

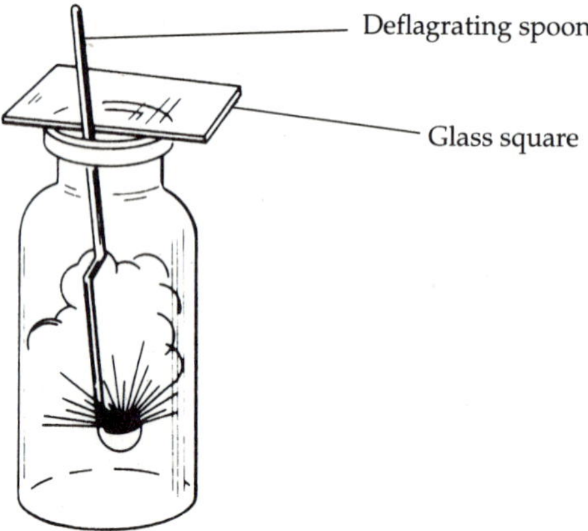

▲ **FIGURE 18.3** Cover bottles with glass square as much as possible during combustion.

Carbon Ignite a small piece of charcoal, holding it either with tongs or in a clean deflagrating spoon. Thrust the glowing charcoal into a bottle of oxygen.

Phosphorus and Sulfur DO THESE COMBUSTIONS IN THE HOOD. Clean the deflagrating spoon in the dilute HCl and rinse with water. Then heat the spoon to remove any combustible material before it is used for the combustion of each of these elements. After the spoon has cooled, add a bit (no more in size than half a pea) of sulfur or red phosphorus. **(CAUTION:** ***Do not touch the phosphorus with your hands, because it will cause burns.)*** Ignite each with your burner and thrust the spoon into separate bottles of oxygen. After the combustion subsides, heat the spoon to burn off all remaining traces of phosphorus or sulfur; then clean the spoon by dipping it in dilute HCl.

On your report sheet, record your observations on each of the combustions shown above.

C. Reactions of Oxides with Water

Measure the pH of the water you used to make the above solutions with a strip of pH paper and record your results. Then dip a 2-in. strip of pH paper into the solution in each of the above bottles. Estimate the pH to the nearest unit and record your results on your report sheet.

The oxides of elements from groups 1, 3, and 7 of the periodic table are difficult to prepare or are extremely insoluble or unstable. Therefore, you will be provided with either the oxide or a solution of the oxide.

Sodium When sodium burns in air, the peroxide, Na_2O_2, forms rather than the oxide, Na_2O. Because the peroxide is readily available and because its reaction with water is similar to that of sodium oxide, we shall use it to be indicative of the group 1 elements. In a 6-in. test tube, carefully boil a very small amount (no more than the size of half a pea, since larger quantities react violently!) of sodium peroxide in 5 mL of water for a few seconds. Be careful not to point the test tube toward yourself or a neighbor. **(CAUTION:** ***Do not touch the*** Na_2O_2 ***with your hands, because it will cause severe burns.)*** Cool the solution and test it with pH paper. Record your results.

Boron When the oxide of boron, B_2O_3, reacts with water, it forms a substance whose composition may be represented either as $B(OH)_3$ or H_3BO_3. Decide which of these formulas is preferred by dissolving a small amount of a substance, which we shall arbitrarily call "boron hydroxide," in about 5 mL of warm water. Cool the solution and test it with pH paper. Record your results. Which formula do you prefer, H_3BO_3 or $B(OH)_3$?

Chlorine The oxides of chlorine—Cl_2O, ClO_2, and Cl_2O_7—are gases and are unstable. An aqueous solution of one of these, labeled $HClO_4$, is available for you to test. Obtain 1 mL of this solution, check its pH, and record your results. **(CAUTION:** ***Do not let the*** $HClO_4$ ***solution come into contact with any organic material, because it is a very powerful oxidizing agent and reacts violently.)***

Neutralization Pour the sulfur oxide solution into the calcium oxide solution and test it with pH paper. How does this mixture differ from the individual solutions? Did a chemical reaction occur? What is it? Now mix the phosphorus oxide and magnesium oxide solutions and test the mixture with pH paper.

Record your results on your report sheet.

D. Insoluble Oxide

Begin heating about 200 mL of water in a 400-mL beaker. Place about 0.25 g of zinc oxide, ZnO, in a 6-in. test tube and add 5 mL of water. Test the pH of the mixture and record your results.

Place about 0.25 g of ZnO in another test tube and *carefully* add 5 mL of 14 *M* NaOH. **(CAUTION:** ***Do not get any of the*** **14 *M* NaOH** ***on yourself. If you do, wash it off with water immediately!*****)** Place this test tube in the beaker of hot water for several minutes and then allow it to cool. Does the ZnO appear to dissolve in the NaOH solution?

Place another 0.25 g of ZnO in a test tube and add about 5 mL of 6 *M* HCl. Does the ZnO react with the acid? What kind of oxide is ZnO—acidic, basic, or amphoteric?

REVIEW QUESTIONS

Before beginning this experiment in the laboratory, you should be able to answer the following questions:

1. Distinguish between ionic and covalent bonding.
2. What is the anhydride of sulfuric acid?
3. Define an acid and a base according to a Brønsted-Lowry definition.
4. How do K_w and K_c differ for the dissociation of water?
5. Why do we use the concept of pH?
6. State when solutions are acidic, neutral, and basic in terms of $[OH^-]$, $[H_3O^+]$, and pH.
7. If the $[H^+]$ in water is 10^{-4} *M*, what is the pH and what is the $[OH^-]$?
8. Complete and balance the following:

$$MgO(s) + H_2O(l) \longrightarrow$$

$$SO_3(g) + H_2O(l) \longrightarrow$$

9. If the pH of a solution is 7.4, what are the hydrogen- and hydroxide-ion concentrations?
10. What are the hydroxide- and hydrogen-ion concentrations of 0.024 *M* $Ca(OH)_2$?
11. What is the pH of a solution if the indicators methyl red and bromthymol blue impart a yellow color to the solution? Consult Figure 18.2.
12. Which solution would have a higher pH, 0.01 *M* HCl or 0.02 *M* NaOH? Which solution is acidic and which is basic?

Name ______________________ Desk ______________________

Date ______________ Laboratory Instructor ______________________

REPORT SHEET | EXPERIMENT

The Chemistry of Oxygen: Basic and Acidic Oxides and the Periodic Table | 18

A. Preparation of Oxygen

1. Write the equation for the reaction by which you prepared oxygen. $2KClO_3 \xrightarrow[\Delta]{MnO2} 2KCl + 3O_2$
2. Compare the thermal decomposition of $KClO_3$ alone with the decomposition of a mixture of $KClO_3$ and MnO_2 as to the following:
   (a) The relative ease or the temperature at which decomposition occurs:
   Higher temperatures were needed in the absence of MnO_2 to produce O_2.

   (b) The amount of oxygen that may be produced from each sample:
   the same
3. The term used to refer to substances that alter the rate of a reaction is a catalyst.

B. Preparation of Oxides

Describe any changes occurring during the reaction of each element with oxygen, and the properties of the products formed. Write the equation for each reaction.

	Observations	Equations	Oxidation state of element in the oxide
Mg	burned brightly and vigorously; made a pop sound	$2Mg + O_2 \longrightarrow 2MgO$	+2
Ca	difficult to combust it; oxide is white	$2Ca + O_2 \longrightarrow 2CaO$	+2
Fe	the iron sparked and some dark solid formed	$4Fe + 3O_2 \longrightarrow 2Fe_2O_3$	+3
C	the carbon glowed as it burned	$C + O_2 \longrightarrow CO_2$	+4
P	glowed brightly in O_2; a lot of white fumes formed	$4P + 5O_2 \longrightarrow 2P_2O_5$	+5
S	burned with a blue flame; a choking gas formed	$S + O_2 \longrightarrow SO_2$	+4

C. Reactions of Oxides with Water

pH of water 6.5

On the line opposite its periodic group, write the formula for each oxide (or hydroxide) studied in this experiment. Indicate the pH of its solution, the $[H_3O^+]$ and $[OH^-]$ concentrations, and the equations for the formation of the acid or base.

Group	Formula of oxide	pH	$[H_3O^+]$	$[OH^-]$	Equation for reaction
1 Na	Na_2O_2	12	1×10^{-12}	1×10^{-2}	$2Na_2O_2 + 2H_2O \longrightarrow 4NaOH + O_2$
2 Mg	MgO	11	1×10^{-11}	1×10^{-3}	$MgO + H_2O \longrightarrow Mg(OH)_2$
2Ca	CaO	9	1×10^{-9}	1×10^{-5}	$CaO + H_2O \longrightarrow Ca(OH)_2$
3 B	B_2O_3	6	1×10^{-6}	1×10^{-8}	$B_2O_3 + 3H_2O \longrightarrow 2H_3BO_3$
4 C	CO_2	6	1×10^{-6}	1×10^{-8}	$CO_2 + H_2O \longrightarrow H_2CO_3$
5 P	P_2O_5	2	1×10^{-2}	1×10^{-12}	$P_2O_5 + 3H_2O \longrightarrow 2H_3PO_4$
6 S	SO_2	2	1×10^{-2}	1×10^{-12}	$SO_2 + H_2O \longrightarrow H_2SO_3$
7 Cl	Cl_2O_7	1	1×10^{-1}	1×10^{-13}	$Cl_2O_7 + H_2O \longrightarrow 2HClO_4$
Transition Fe	Fe_2O_3	8	1×10^{-8}	1×10^{-6}	$Fe_2O_3 + 3H_2O \longrightarrow 2Fe(OH)_3$

Write equations for the chemical reactions that occurred between the aqueous solutions of the oxides of sulfur and calcium, and between the aqueous solutions of the oxides of phosphorus and magnesium. Indicate the acids and bases in each reaction.

ACID BASE

$H_2SO_3 + Ca(OH)_2 \longrightarrow CaSO_3 + 2H_2O$

$2H_3PO_4 + 3Mg(OH)_2 \longrightarrow Mg_3(PO_4)_2 + 6H_2O$

D. Insoluble Oxide

Compare the solubility of ZnO in H_2O, NaOH, and HCl.

Insoluble in H_2O, but soluble in both NaOH and HCl.

What kind of oxide is ZnO?

amphoteric

QUESTIONS

1. Comment on the positions of the elements in the periodic table and on whether their oxides are acid or base producers.

   Elements on the left side of the periodic table (metals) produce basic oxides.

   Elements on the right side of the periodic table (nonmetals) produce acidic oxides.

2. Where in the periodic table would you expect most amphoteric oxides to be located?

   At the end of the transition series and the beginning of the representitive elements. Near the step-line.

3. Deduce and write formulas for the anhydrides of the following:

$HClO_2$	Cl_2O_3	KOH	K_2O	H_7SbO_6	Sb_2O_5
H_2SO_4	SO_3	$Ba(OH)_2$	BaO	HNO_3	N_2O_5
H_3AsO_4	As_2O_5	$Al(OH)_3$	Al_2O_3	H_2CO_3	CO_2

4. If the pH of an aqueous solution is 4.72, what are the hydrogen- and hydroxide-ion concentrations of this solution?

$$4.72 = -\log[H_3O^+];\ \log[H_3O^+] = -5 + 0.28 \text{ and } [H_3O^+] = 1.9 \times 10^{-5}\ M$$

$$[OH^-] = \frac{1.0 \times 10^{-14}}{[H_3O^+]} = \frac{1.0 \times 10^{-14}}{1.9 \times 10^{-5}} = 5.3 \times 10^{-10}\ M$$

5. If 7.40 g of $Ca(OH)_2$ is dissolved in sufficient water to make 100 mL of solution, what is the hydroxide-ion concentration of this solution?

$$M = \left(\frac{7.40\ \text{g}}{74.0\ \text{g/mol}}\right)\left(\frac{1}{0.1\ \text{L}}\right) = 1.00\ M \qquad Ca(OH)_2 \longrightarrow Ca^{2+} + 2\ OH^-$$

$[OH^-] = 2[Ca(OH)_2] = 2(1.00\ M) = 2.00\ M$

6. Determine the hydroxide-ion concentration, the pH, and the hydrogen-ion concentrations of solutions containing 15.0 g of the following substances in 1.00 L of solution: (HINT: First write balanced equations for the reactions that occur when the oxides are dissolved in water.)

Substance	Molarity of substance	$[OH^-]$	pH	$[H_3O^+]$
CaO	0.268	0.536	13.73	1.87×10^{-14}
Na_2O	0.242	0.484	13.68	2.07×10^{-14}

$Na_2O + H_2O \longrightarrow 2NaOH \qquad [OH^-] = 2 \times [Na_2O]$

$$M \text{ of } Na_2O = \left(\frac{15.0\ \text{g}}{62.0\ \text{g/mol}}\right)\left(\frac{1}{1.00\ \text{L}}\right) = 0.242\ M \text{ and } [OH\ \ ] = 0.484\ M$$

$$[H^+] = \frac{1.0 \times 10^{-14}}{0.484} = 2.07 \times 10^{-14}\ M \text{ and pH} = -\log[2.07 \times 10^{-14}] = 13.68$$

7. If 50 mL of 0.50 *M* KOH solution is added to 75 mL of 0.20 *M* H_2SO_4 solution, will the solution be neutral, acidic, or basic? Write a balanced chemical equation for the reaction and justify your answer.

$2\ KOH + H_2SO_4 \longrightarrow K_2SO_4 + 2H_2O$

$$50\ \text{mL} \times \frac{0.50\ \text{mmol KOH}}{\text{mL}} = 25\ \text{mmol KOH}$$

$$75\ \text{mL} \times \frac{0.20\ \text{mmol}\ H_2SO_4}{\text{mL}} = 15\ \text{mmol}\ H_2SO_2$$

Because two moles of KOH react with one mole H_2SO_4, this solution contains 2.5 mmol H_2SO and no base. It is acidic.

8. Complete and balance the following equations and tell whether you predict the oxide to be acidic, basic, or amphoteric.

$Zn + O_2$	$\longrightarrow$	$Zn + O_2$	$\longrightarrow$	2ZnO	amphoteric
$Ga + O_2$	$\longrightarrow$	$4Ga + 3O_2$	$\longrightarrow$	$2Ga_2O_3$	amphoteric
$As_4 + O_2$	$\longrightarrow$	$As_4 + 5O_2$	$\longrightarrow$	As_4O_{10}	acidic
$Li + O_2$	$\longrightarrow$	$4Li + O_2$	$\longrightarrow$	$2Li_2O$	basic

Experiment

Colligative Properties: Freezing-Point Depression and Molar Mass

OBJECTIVE

To become familiar with colligative properties and to use them to determine the molar mass of a substance.

APPARATUS AND CHEMICALS

Apparatus

600-mL beaker
thermometer
large test tube
250 mL wide-mouth glass bottle
towel
wire gauze
clamp
balance
Bunsen burner and hose
wire stirrer
weighing paper
ring stand and ring
two-hole rubber stopper with slit

Chemicals

sulfur, "roll" or precipitated, or unknown solid
naphthalene

WORK IN PAIRS, BUT EVALUATE YOUR DATA INDEPENDENTLY.

DISCUSSION

Solutions are homogeneous mixtures that contain two or more substances. The major component is called the *solvent*, and the minor component is called the *solute.* Since the solution is primarily composed of solvent, physical properties of a solution resemble those of the solvent. Some of these physical properties, called *colligative properties,* are independent of the nature of the solute and depend only upon the solute concentration. The colligative properties include vapor-pressure lowering, boiling-point elevation, freezing-point lowering, and osmotic pressure. The *vapor pressure* is just the escaping tendency of the solvent molecules. When the vapor pressure of a solvent is equal to atmospheric pressure, the solvent boils. At this temperature the gaseous and liquid states of the solvent are in dynamic equilibrium, and the rate of molecules going from the liquid to the gaseous state is equal to the rate of molecules going from the gaseous state to the liquid state. It has been found experimentally that the dissolution of a nonvolatile solute (one with very low vapor pressure) in a solvent lowers the vapor pressure of the solvent, which in turn raises the boiling point and lowers the freezing point. This is shown graphically by the phase diagram given in Figure 19.1

You are probably familiar with some common uses of these effects: Antifreeze is used to lower the freezing point and raise the boiling point of coolant (water) in an automobile radiator, and salt is used to melt ice. These effects are expressed quantitatively by the *colligative-property law,* which states that the freezing point and boiling point of a solution differ from those

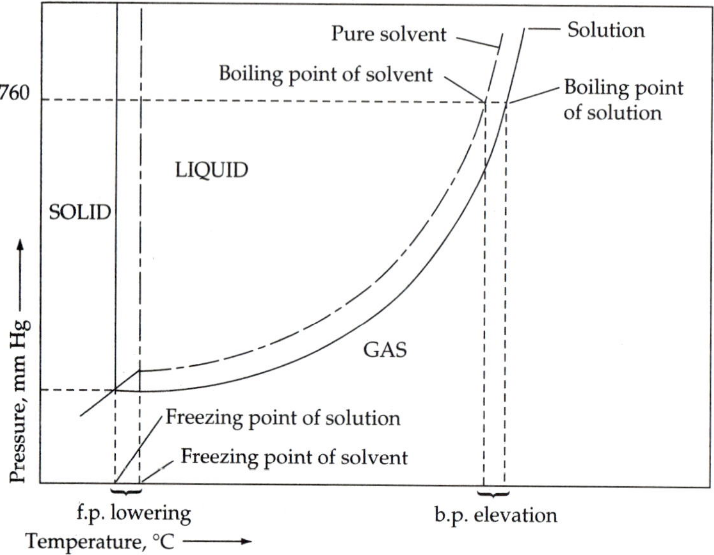

▲ **FIGURE 19.1** Phase diagram for a solvent and a solution.

of the pure solvent by amounts that are directly proportional to the molal concentration of the solute. This relationship is expressed by Equation [1] for the freezing-point lowering and boiling-point elevation

$$\Delta T = Km \qquad [1]$$

where ΔT is the freezing-point lowering or boiling-point elevation, K is a constant that is specific for each solvent, and m is the molality of the solution (number of moles solute per kg solvent). Some representative constants, boiling points, and freezing points are given in Table 19.1. For naphthalene, the solvent used in this experiment, the molal freezing-point depression constant (K_{fp}) has a value of 6.9°C/m.

Example 19.1

What would be the freezing point of a solution containing 19.5 g of biphenyl ($C_{12}H_{10}$) dissolved in 100 g of naphthalene if the normal freezing point of naphthalene is 80.6°C?

TABLE 19.1 Molal Freezing-Point and Boiling-Point Constants

Solvent	Freezing point (°C)	K_{fp}(°C/m)	Boiling point(°C)	K_{bp}(°C/m)
CH_3COOH (acetic acid)	16.6	3.90	118.1	2.93
C_6H_6 (benzene)	5.5	5.12	80.1	2.53
$CHCl_3$ (chloroform)	−63.5	4.68	61.2	3.63
C_2H_5OH (ethyl alcohol)	−114.6	1.99	78.4	1.22
H_2O (water)	0.0	1.86	100.0	0.51
$C_{10}H_8$ (naphthalene)	80.6	6.9	218	—
C_6H_{12} (cyclohexane)	6.6	20.4	80.7	2.79

SOLUTION:

$$\text{moles } C_{12}H_{10} = \frac{19.5 \text{ g}}{154 \text{ g/mol}} = 0.127 \text{ mol}$$

$$\frac{\text{moles } C_{12}H_{10}}{1 \text{ kg naphthalene}} = \left(\frac{0.127 \text{ mol}}{100 \text{ g}}\right)\left(\frac{1000 \text{ g}}{1 \text{ kg}}\right)$$

$$= 1.27\ m$$

$$\Delta T = (6.9°\text{C}/m)(1.27\ m)$$

$$= 8.8°\text{C}$$

Since the freezing point is lowered, the observed freezing point of this solution will be

$$80.6°\text{C} - 8.8°\text{C} = 71.8°\text{C}$$

Since the molal freezing-point-depression constant is known, it is possible to obtain the molar mass of a solute by measuring the freezing point of a solution and the mass of both the solute and solvent.

EXAMPLE 19.2

What is the molar mass of urea if the freezing point of a solution containing 15 g of urea in 100 g of naphthalene is 63.3°C?

SOLUTION: The freezing point of pure naphthalene is 80.6°C. Therefore, the freezing-point lowering (ΔT) is:

$$\Delta T = 80.6°\text{C} - 63.3°\text{C} = 17.3°\text{C}$$

From Equation [1] above,

$$17.3°\text{C} = K_{\text{fp}}m$$

$$m = \frac{17.3°\text{C}}{K_{\text{fp}}}$$

We know that K_{fp} for naphthalene is 6.9°C/m. Therefore, the molality of this solution is

$$m = \frac{17.3°\text{C}}{6.9°\text{C}/m} = 2.5\ m$$

Remember that molality is the number of moles of solute per kg of solvent. In our solution there are 15 g urea in 100 g of naphthalene, or 150 g of urea in 1000 g of naphthalene. Thus

$$150 \text{ g} = 2.5 \text{ mol}$$

$$1 \text{ mol} = 60 \text{ g}$$

The molar mass of urea is, therefore, 60 g/mol.

In this experiment you will determine the molar mass of either sulfur or an unknown. You will do this by determining the freezing-point depression of a naphthalene solution having a known concentration of either sulfur or your unknown. The freezing temperature is difficult to ascertain by direct visual observation because of a phenomenon called supercooling and also because solidification of solutions usually occurs over a broad temperature range. Temperature-time graphs, called *cooling curves*, reveal freezing temperatures rather clearly. Therefore, you will study the rate at which liquid

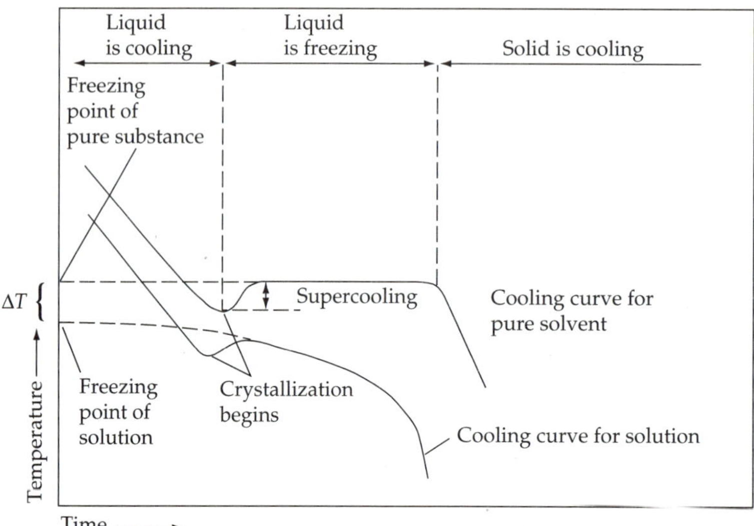

▲ **FIGURE 19.2** Cooling curves for a solvent and a solution.

naphthalene and its solutions cool and will construct a cooling curve similar to the one shown in Figure 19.2.

You will construct cooling curves for both the pure solvent and the solution. Figure 19.2 shows how the freezing point of a solution must be determined by extrapolation of the cooling curve. Extrapolation is necessary because as the solution freezes, the solid that is formed is essentially pure solvent and the remaining solution becomes more and more concentrated. Thus its freezing point lowers continuously. Clearly, supercooling produces an ambiguity in the freezing point and should be minimized. Stirring the solution helps to minimize supercooling.

PROCEDURE

Instructor: Cyclohexane can be substituted for naphthalene, but it must be cooled with an ice bath and KEPT AWAY FROM FLAMES.

A. Cooling Curve for Pure Naphthalene

Weigh a large test tube to the nearest 0.01 g. Add about 15 to 20 g of naphthalene and weigh again. The difference in mass is the mass of naphthalene. Assemble the apparatus as shown in Figure 19.3; be certain to use a split two-hole rubber stopper. Carefully insert the thermometer into the hole that has been slit. Bend the stirrer so that the loop encircles the thermometer.

Fill your 600-mL beaker nearly full of water and heat it to about 85°C. Clamp the test tube in the water bath as shown in Figure 19.3. When most of the naphthalene has melted, insert the stopper containing the thermometer and stirrer into the test tube. Make certain that the thermometer is not resting on the bottom of or touching the sides of the test tube. When all of the naphthalene has melted, stop heating, remove the beaker of water, and dry the outside of the test tube with a cloth towel. Place the test tube in a wide-mouthed bottle that contains a piece of crumpled paper in the bottom to lessen the chance that impact of the test tube with the bottle will cause the bottle to break. The purpose of the wide-mouth bottle is to minimize drafts. Record temperature readings every 30 s while you are stirring. When the freezing point is reached, crystals will start to form and the temperature

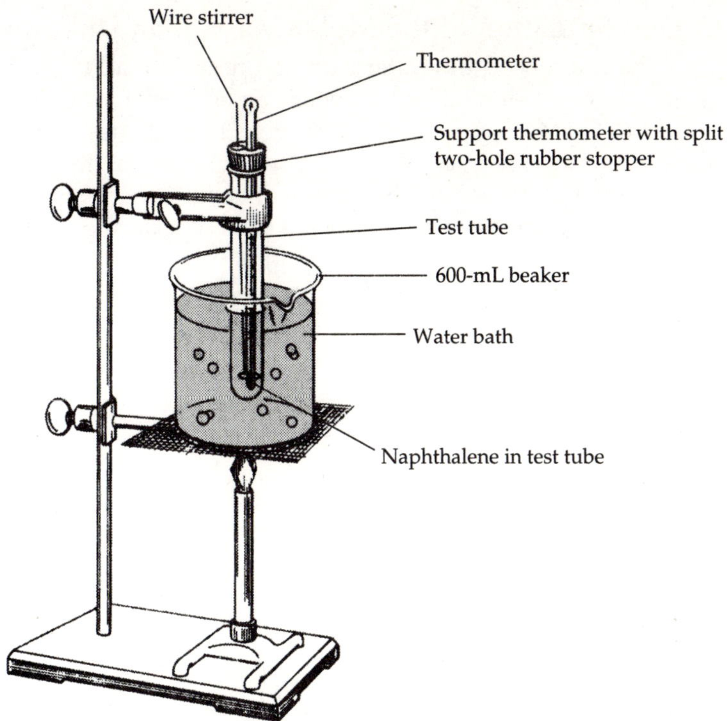

▲ **FIGURE 19.3** Apparatus for determination of cooling curve.

will remain constant. Shortly after this, the naphthalene will solidify to the point where you can no longer stir it.

Your lab instructor will direct you to perform either procedure B or procedure C.

Instructor: Instruct the students whether to perform procedure B or C.

B. Determination of the Molar Mass of Sulfur

Using weighing paper, weigh to the nearest 0.01 g about 1.2 to 1.5 g of sulfur. CLEAN UP ANY SULFUR SPILLS NEAR THE BALANCE. Replace the test tube in the water bath and heat until all the naphthalene has melted. Gently remove the stopper, making sure that no naphthalene is lost, and add the sulfur to the test tube. Replace the stopper and stir gently until all the sulfur has dissolved. Remove the water bath, dry the test tube with a towel, and insert the test tube in a wide-mouth glass bottle containing a crumpled piece of paper. Record the temperature every 30 s until all the naphthalene has solidified.

Cleanup To clean out the test tube at the *end* of the experiment, cautiously heat the test tube in a water bath until the naphthalene just melts. **(CAUTION:** ***Do not heat the thermometer beyond its temperature range. Be careful, because naphthalene is flammable.*****)** Remove the stopper and pour the molten naphthalene onto a crumpled wad of paper. When the naphthalene has solidified, throw both the paper and solid naphthalene into a designated waste receptacle. DO NOT POUR LIQUID NAPHTHALENE INTO THE SINK!

C. Determination of the Molar Mass of an Unknown

Place the test tube in the water bath and heat until all the naphthalene has melted. Using weighing paper, weigh about 2 g of your unknown to the nearest 0.01 g. Gently remove the stopper from the test tube, making sure that no naphthalene is lost, and add your unknown to the test tube. Replace the stopper and stir gently until all the unknown has dissolved. Remove the water bath, dry the test tube with a towel, and insert the test tube in a wide-mouth glass bottle containing a crumpled piece of paper. Record the temperature every 30 s until all the naphthalene has solidified. Clean up as described in Part B.

REVIEW QUESTIONS

Before beginning this experiment in the laboratory, you should be able to answer the following questions:

1. Distinguish between *solute* and *solvent.*
2. List three colligative properties and suggest a rationale for the choice of the word *colligative* to describe these properties.
3. Distinguish between volatile and nonvolatile substances.
4. What effect does the presence of a nonvolatile solute have upon (a) the vapor pressure of a solution, (b) the freezing point, and (c) the boiling point?
5. What is the molality of a solution that contains 3.0 g urea (molecular weight = 60 amu) in 100 g of benzene?
6. What is supercooling? How can it be minimized?
7. Calculate the freezing point of a solution containing 2.50 g of benzene in 120 g of chloroform.
8. A solution containing 2.00 g of an unknown substance in 25.0 g of naphthalene was found to freeze at 75.4°C. What is the molar mass of the unknown substance?
9. How many grams of $NaNO_3$ would you add to 250 g of H_2O in order to prepare a solution that is 0.500 molal in $NaNO_3$?
10. Define the terms *molality* and *molarity.*

Name ______________________ Desk ______________________

Date ______________ Laboratory Instructor ______________________

REPORT SHEET | EXPERIMENT 19

Colligative Properties: Freezing-Point Depression and Molar Mass

1. Mass of test tube + naphthalene 51.36 g
2. Mass of test tube 28.45 g
3. Mass of naphthalene 22.91 g

4. Mass of paper + sulfur or unknown 2.82 g
5. Mass of paper 0.75 g
6. Mass of sulfur or unknown 2.07 g

Cooling-curve data

Pure naphthalene		*Naphthalene 1 sulfur or unknown*	
Temp.	*Time*	*Temp.*	*Time*
83.5	30 s	82.0	30 s
82.0	1 min	80.0	1 min
81.0	1.5 min	79.0	1.5 min
80.75	2 min	78.5	2 min
80.75	2.5 min	78.25	2.5 min
no change	3 min	78.25	3 min
no change	3.5 min*	78.25	3.5 min
no change	4 min	78.25	4 min
no change	4.5 min	78.25	4.5 min
no change	5 min	78.25	5 min

(*Pure naphthalene: Crystals started to form at 3.5 min.)

7. Freezing point of pure naphthalene, from cooling curve 80.75 °C
8. Freezing point of solution of sulfur or unknown in naphthalene 78.25; ΔT = 2.50°C
9. Molality of sulfur or unknown (show calculations) 0.36 m

$$m = \frac{2.50°\text{C}}{6.9°\text{C}/m} = 0.36\ m$$

10. Molar mass of sulfur or unknown (show calculations) 2.5×10^2

$$\text{g/mol} = \left(\frac{2.07 \text{ g S}}{22.91 \text{ g naphthalene}}\right)\left(\frac{1000 \text{ g naph}}{1 \text{ kg naph}}\right)\left(\frac{1 \text{ kg naph}}{0.36 \text{ mol S}}\right) = \frac{2.5 \times 10^2 \text{ g S}}{\text{mol S}} = \mathcal{M}$$

Hence, the MW of sulfur is 2.5×10^2 amu and because the AW of sulfur is 32 amu, 250 amu/32 amu = 7.8, or S_8

HAND IN YOUR COOLING CURVES WITH YOUR REPORT SHEET.

QUESTIONS

1. What are the major sources of error in this experiment?

   Measurement of the freezing points. A thermometer with small subdivisions would allow for more accurate readings.

2. Suppose your thermometer consistently read a temperature 1.2° lower than the correct temperature throughout the experiment. How would this have affected the molar mass you found?

   It would have no effect because you are measuring a temperature difference, not an absolute temperature.

3. If the freezing point of the solution had been incorrectly read 0.3° lower than the true freezing point, would the calculated molar mass of the solute be too high or too low? Explain your answer.

   If the measured freezing point was lower than the real freezing point, the measured ΔT would be too large. Since $\Delta T = Km$, this would give a molality that is too large. Since m is proportional to mass/molar mass, this would give a molar mass that is too small.

4. Arrange the following aqueous solutions in order of increasing freezing points (lowest to highest temperature): 0.10 m glucose, 0.10 m $BaCl_2$, 0.20 m NaCl, and 0.20 m Na_2SO_4.

   0.10 m glucose $<$ 0.10 m $BaCl_2$ $<$ 0.20 m NaCl $<$ 0.20 m Na_2SO_4

   Total solute molality 0.1 $m < 0.3\ m < 0.4\ m < 0.6\ m$

   Na_2SO_4 would freeze at the lowest temperature. The order is: Na_2SO_4 < NaCl < $BaCl_2$ < glucose.

5. What mass of NaCl is dissolved in 150 g of water in a 0.050 m solution?

$$\left(\frac{0.050 \text{ mol}}{1000 \text{ g}}\right)(150 \text{ g}) = 0.0075 \text{ mol NaCl}$$

   (0.0075 mol NaCl)(58.45 g/mol) = 0.44 g

6. Calculate the molalities of some commercial reagents from the following data:

	HCl	$\mathbf{HC_2H_3O_2}$	$\mathbf{NH_3}$***(aq)***
Formula weight (amu)	36.465	60.05	17.03
Density of solution (g/mL)	1.19	1.05	0.90
Weight %	37.2	99.8	28.0
Molarity	12.1	17.4	14.8

HCl: Consider 1 L of acid:

1 L of acid weighs 1190 g and contains $0.372 \times 1190 = 443$ g HCl, and it contains 1190 g − 443 = 747 g H_2O. The number of grams of HCl per 1000 g H_2O:

$$\left(\frac{443 \text{ g HCl}}{747 \text{ g H}_2\text{O}}\right)(1000 \text{ g H}_2\text{O}) = 593 \text{ g HCl.} \quad (593 \text{ g HCl})\left(\frac{1 \text{ mol HCl}}{36.465 \text{ g}}\right) = 16.3 \text{ mol or solution is 16.3 molal.}$$

Similar reasoning shows $HC_2H_3O_2$ is 8,310 molal and NH_3 to be 22.8 molal.

7. A solution of 2.00 g of para-dichlorobenzene (a clothes moth repellant) in 50.0 g of cyclohexane freezes at 1.05°C. What is the molar mass of this substance?

$\Delta T_f = 6.60°\text{C} - 1.05°\text{C} = 5.55°\text{C} \qquad \Delta T_f = K_f m$

$$m = \frac{5.55°\text{C}}{20.4°\text{C}/m} = 0.272 \text{ molal}$$

$$\frac{\text{grams solute}}{\text{kg cyclohexane}} = \left(\frac{2.00 \text{ g solute}}{50.0 \text{ g cyclohexane}}\right)\left(\frac{10^3 \text{ g cyclohexane}}{1 \text{ kg cyclohexane}}\right)$$

$$= \frac{40.0 \text{ g solute}}{1 \text{ kg cyclohexane}}$$

$$0.272 \text{ mol} = 40.0 \text{ g}$$

$$1 \text{ mol} = \frac{40.0 \text{ g}}{0.272 \text{ mol}} = 147 \text{ g/mol}$$

COOLING CURVE FOR PURE NAPHTHALENE

COOLING CURVE FOR SOLUTION OF SULFUR OR UNKNOWN IN NAPHTHALENE

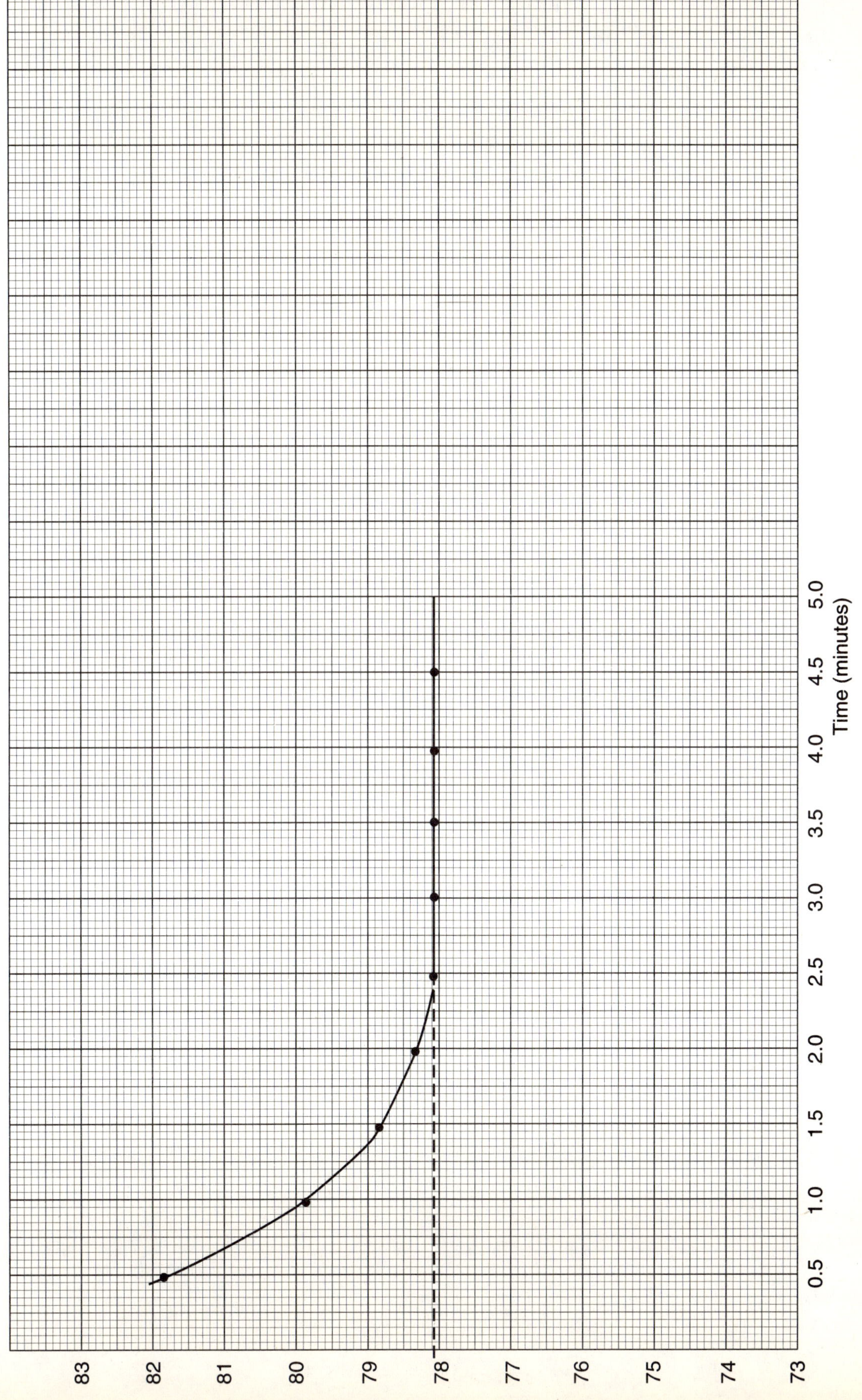

Titration of Acids and Bases

OBJECTIVE

To become familiar with the techniques of titration, a volumetric method of analysis; to determine the amount of acid in an unknown.

APPARATUS AND CHEMICALS

Apparatus

50-mL buret	balance
600-mL beaker	Bunsen burner and hose
500-mL Erlenmeyer flask	1-pint bottle with rubber stopper
250-mL Erlenmeyer flasks (3)	wash bottle
weighing bottle	buret clamp and ring stand
ring stand and ring	wire gauze

Chemicals

10 *M* NaOH	phenophthalein solution
potassium hydrogen phthalate (KHP, primary standard)	unknown acid

DISCUSSION

One of the most common and familiar reactions in chemistry is the reaction of an acid with a base. This reaction is termed *neutralization,* and the essential feature of this process in aqueous solution is the combination of hydronium ions with hydroxide ions to form water:

$$H_3O^+(aq) + OH^-(aq) \longrightarrow 2H_2O(l)$$

In this experiment, you will use this reaction to determine accurately the concentration of a sodium hydroxide solution that you have prepared. The process of determining the concentration of a solution is called *standardization.* Next you will measure the amount of acid in an unknown. To do this, you will accurately measure, with a buret, the volume of your standard base that is required to exactly neutralize the acid present in the unknown. The technique of accurately measuring the volume of a solution required to react with another reagent is termed *titration.*

An indicator solution is used to determine when an acid has exactly neutralized a base, or vice versa. A suitable indicator changes colors when equivalent amounts of acid and base are present. The color change is termed the *end point* of the titration. Indicators change colors at different pH values. Phenolphthalein, for example, changes from colorless to pink at a pH of about 9. In slightly more acidic solutions it is colorless, whereas, in more alkaline solutions it is pink.

In this experiment your solution of NaOH will be standardized by titrating it against a very pure sample of potassium hydrogen phthalate, $KHC_8H_4O_4$, of known weight. Potassium hydrogen phthalate (often abbreviated as KHP) has only one acidic hydrogen. Its structure is shown here. It is

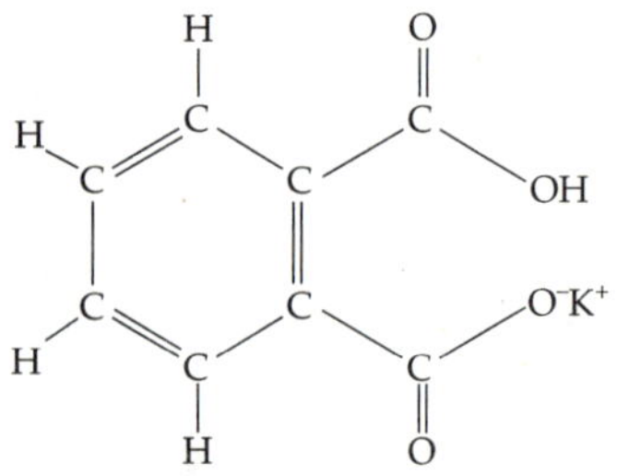

KHP

a monoprotic acid with the acidic hydrogen bonded to oxygen and has a molar mass of 204.2 g.

The balanced equation for the neutralization of KHP is given in Equation [1]:

$$KHC_8H_4O_4(aq) + NaOH(aq) \longrightarrow H_2O(l) + KNaC_8H_4O_4(aq) \qquad [1]$$

In the titration of the base NaOH against KHP, an equal number of moles of KHP and NaOH are present at the equivalence point. In other words, at the equivalence point

$$\text{moles NaOH} = \text{moles KHP} \qquad [2]$$

The point at which stoichiometrically equivalent quantities are brought together is known as the *equivalence point* of the titration.

It should be noted that the equivalence point in a titration is a theoretical point. It can be estimated by observing some physical change associated with the condition of equivalence, such as the change in color of an indicator, which is termed the end point.

The most common way of quantifying concentrations is molarity (symbol M), which is defined as the number of moles of solute per liter of solution, or the number of millimoles of solute per milliliter of solution:

$$M = \frac{\text{moles solute}}{\text{volume of solution in liters}} = \frac{10^{-3}\text{ mole}}{10^{-3}\text{ liter}} = \frac{\text{mmol}}{\text{mL}} \qquad [3]$$

From Equation [3] the moles of solute (or mmol solute) are related to the molarity and the volume of the solution as follows:

$$M \times \text{liters} = \text{moles solute and } M \times \text{mL} = \text{mmol solute} \qquad [4]$$

Thus, if one measures the volume of base, NaOH, required to neutralize a known weight of KHP, it is possible to calculate the molarity of the NaOH solution.

EXAMPLE 20.1

Calculate the molarity of a solution that is made by dissolving 16.7 g of sodium sulfate, Na_2SO_4, in enough water to form 125 mL of solution.

SOLUTION:

$$\text{molarity} = \frac{\text{moles } Na_2SO_4}{\text{liters soln}}$$

Using the formula weight of Na_2SO_4, we calculate the number of moles of Na_2SO_4:

$$\text{moles } Na_2SO_4 = (16.7 \text{ g } Na_2SO_4)\left(\frac{1 \text{ mol } Na_2SO_4}{142 \text{ g } Na_2SO_4}\right) = 0.118 \text{ mol } Na_2SO_4$$

Changing the volume of the solution to liters:

$$125 \text{ mL} \times (1\text{L}/1000 \text{ mL}) = 0.125 \text{ L}$$

Thus the molarity is:

$$\text{molarity} = \frac{0.118 \text{ mol } Na_2SO_4}{0.125 \text{ L}} = 0.941\ M\ Na_2SO_4$$

EXAMPLE 20.2

What is the molarity of a NaOH solution if 35.75 mL of it is required to neutralize 1.070 g of KHP?

SOLUTION: Recall from Equation [2] that at the equivalence point, the number of moles of NaOH equals the number of moles of KHP.

$$\text{moles KHP} = (1.070 \text{ g KHP})\left(\frac{1 \text{ mol KHP}}{204.2 \text{ g KHP}}\right) = 5.240 \times 10^{-3} \text{ mol KHP}$$

Because this is exactly the number of moles of NaOH that is contained in 35.75 mL of solution, its molarity is:

$$\text{molarity} = \frac{5.240 \times 10^{-3} \text{ mol NaOH}}{0.03575 \text{ L}} = 0.1466\ M \text{ NaOH}$$

Once the molarity of the NaOH solution is accurately known, the base can be used to determine the amount of KHP or any other acid present in a known weight of an impure sample. The percentage of KHP in an impure sample is

$$\% \text{ KHP} = \frac{\text{g KHP}}{\text{mass of sample}} \times 100$$

In this experiment an acid-base indicator, phenolphthalein, is used to signal the end point in the titration. This indicator was chosen because its color change coincides so closely with the equivalence point.

EXAMPLE 20.3

What is the percentage of KHP in an impure sample of KHP that weighs 2.537 g and requires 32.77 mL of 0.1466 M NaOH to neutralize it?

SOLUTION: The number of grams of KHP in the sample must first be determined. Remember that at the equivalence point, the number of millimoles of NaOH equals the number of millimoles of KHP.

$$\text{mmol NaOH} = (32.77 \text{ mL NaOH})(0.1466 \text{ mmol NaOH/mL NaOH})$$
$$= 4.804 \text{ mmol NaOH}$$

Thus, there are 4.804 mmol of KHP in the sample, which corresponds to the following number of grams of KHP in the sample:

$$\text{grams KHP} = 4.804 \text{ mmol KHP}\left(\frac{1 \text{ mol KHP}}{1000 \text{ mmol KHP}}\right)\left(\frac{204.2 \text{ g KHP}}{1 \text{ mol KHP}}\right)$$
$$= 0.9810 \text{ g KHP}$$

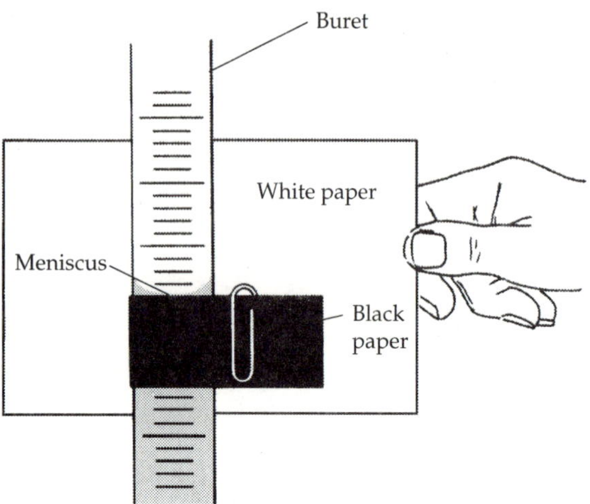

▲ **FIGURE 20.1** Reading a buret.

Therefore,

$$\% \text{ KHP} = \frac{0.9810 \text{ g}}{2.537 \text{ g}} \times 100 = 38.67\%$$

PROCEDURE

Instructor: Time may be saved in this experiment if the students are provided with an approximately 0.1 *M* NaOH solution.

Preparation of Approximately 0.100 *M* Sodium Hydroxide (NaOH) Heat 500 mL of distilled water to boiling in a 600-mL flask,* and *after cooling under the water tap*, transfer to a 1-pint bottle fitted with a rubber stopper.† Add 3 mL of stock solution of carbonate-free NaOH (approximately 10 *M*) and shake vigorously for at least 1 min.

Preparation of a Buret for Use Clean a 50-mL buret with soap solution and a buret brush and thoroughly rinse with tap water. Then rinse with at least five 10-mL portions of distilled water. The water must run freely from the buret without leaving any drops adhering to the sides. Make sure that the buret does not leak and that the stopcock turns freely.

Reading a Buret All liquids, when placed in a buret, form a curved meniscus at their upper surfaces. In the case of water or water solutions, this meniscus is concave (Figure 20.1), and the most accurate buret readings are obtained by observing the position of the lowest point on the meniscus on the graduated scales.

To avoid parallax errors when taking readings, the eye must be on a level with the meniscus. Wrap a strip of paper around the buret and hold the top edges of the strip evenly together. Adjust the strip so that the front and back edges are in line with the lowest part of the meniscus and take the reading by estimating to the nearest tenth of a marked division (0.01 mL). A simple way of doing this for repeated readings on a buret is illustrated in Figure 20.1.

*The water is boiled to remove carbon dioxide (CO_2), which would react with the NaOH and change its molarity.

†A rubber stopper should be used for a bottle containing NaOH solution. A strongly alkaline solution tends to cement a glass stopper so firmly that it is difficult to remove.

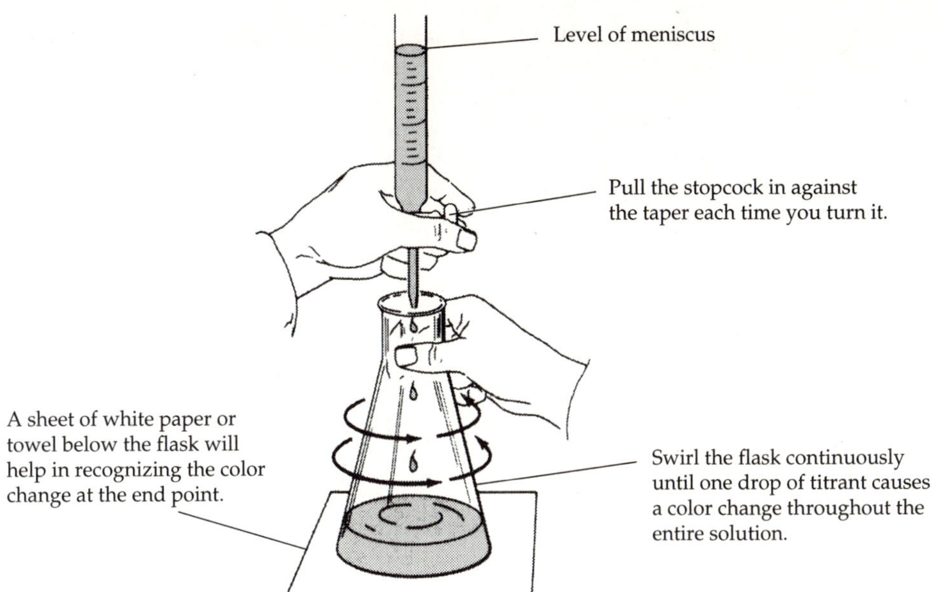

▲ **FIGURE 20.2** Titration procedure.

A. Standardization of Sodium Hydroxide (NaOH) Solution

Prepare about 400 to 450 mL of CO_2-free water by boiling for about 5 min. To save time, make an additional 400 mL of CO_2- free water for Part B by boiling it now. Weigh from a weighing bottle (your lab instructor will show you how to use a weighing bottle if you don't already know) triplicate samples of between 0.4 and 0.6 g each of pure potassium hydrogen phthalate (KHP) into three separate 250-mL Erlenmeyer flasks; accurately weigh to four significant figures.* Do not weigh the flasks. Record the masses and label the three flasks in order to distinguish among them. Add to each sample about 100 mL of distilled water that has been freed from CO_2 by boiling, and warm gently with swirling until the salt is completely dissolved. Add to each flask two drops of phenolphthalein indicator solution.

Instructor: Demonstrate how to use a weighing bottle.

Rinse the previously cleaned buret with at least four 5-mL portions of the approximately 0.100 *M* NaOH solution that you have prepared. Discard each portion into the designated receptacle. *Do not return any of the washings to the bottle.* Completely fill the buret with the solution and remove the air from the tip by running out some of the liquid into an empty beaker. Make sure that the lower part of the meniscus is at the zero mark or slightly lower. Allow the buret to stand for at least 30 sec before reading the exact position of the meniscus. Remove any hanging drop from the buret tip by touching it to the side of the beaker used for the washings. Record the initial buret reading on your report sheet.

Slowly add the NaOH solution to one of your flasks of KHP solution while gently swirling the contents of the flask, as illustrated in Figure 20.2. As the NaOH solution is added, a pink color appears where the drops of the

*In cases where the mass of a sample is larger than 1 g, it is necessary to weigh only to the nearest milligram to obtain four significant figures. Buret readings can be read only to the nearest 0.02 mL, and for readings greater than 10 mL, this represents four significant figures.

base come in contact with the solution. This coloration disappears with swirling. As the end point is approached, the color disappears more slowly, at which time the NaOH should be added drop by drop. It is most important that the flask be swirled constantly throughout the entire titration. The end point is reached when one drop of the NaOH solution turns the entire solution in the flask from colorless to pink. The solution should remain pink when it is swirled. Allow the titrated solution to stand for at least 1 min so the buret will drain properly. Remove any hanging drop from the buret tip by touching it to the side of the flask and wash down the sides of the flask with a stream of water from the wash bottle. Record the buret reading on your report sheet. Repeat this procedure with the other two samples. Dispose of the neutralized solutions as instructed.

From the data you obtain in the three titrations, calculate the molarity of the NaOH solution to four significant figures as in Example 20.2.

The three determinations should agree within 1.0%. If they do not, the standardization should be repeated until agreement is reached. The average of the three acceptable determinations is taken as the molarity of the NaOH. Calculate the standard deviation of your results. SAVE your standardized solution for the unknown determination.

B. Analysis of an Unknown Acid

Calculate the approximate mass of unknown that should be taken to require about 20 mL of your standardized NaOH, assuming that your unknown sample is 75% KHP.

From a weighing bottle, weigh by difference triplicate portions of the sample to four significant figures and place them in three separate 250-mL flasks. The sample size should be about the amount determined by the above computation. Dissolve the sample in 100 mL of CO_2-free distilled water (prepared by boiling) and add two drops of phenolphthalein indicator solution. Titrate with your standard NaOH solution to the faintest visible shade of pink (not red) as described above in the standardization procedure. Calculate the percentage of KHP in the samples as in Example 20.3. For good results, the three determinations should agree within 1.0%. Your answers should have four significant figures. Compute the standard deviation of your results.

Test your results by computing the average deviation from the mean. If one result is noticeably different from the others, perform an additional titration. If any result is more than two standard deviations away from the mean, discard it and titrate another sample.

REVIEW QUESTIONS

Before beginning this experiment in the laboratory, you should be able to answer the following questions:

1. Define *standardization* and state how you would go about doing it.
2. Define the term *titration.*
3. Define the term *molarity.*
4. Why do you weigh by difference?
5. What are equivalence points and end points, and how do they differ?

6. What is parallax, and why should you avoid it?
7. Why is it necessary to rid the distilled water of CO_2?
8. What is the molarity of a solution that contains 3.78 g of $H_2C_2O_4 \cdot 2H_2O$ in 200 mL of solution?
9. If 50.0 mL of NaOH solution is required to react completely with 0.62 g KHP, what is the molarity of the NaOH solution?
10. In the titration of an impure sample of KHP, it was found that 30.6 mL of 0.100 *M* NaOH was required to react completely with 0.745 g of sample. What is the percentage of KHP in this sample?

NOTES AND CALCULATIONS

Name ______________________ Desk ______________________

Date ______________ Laboratory Instructor ______________________

REPORT SHEET | EXPERIMENT

Titration of Acids and Bases | 20

A. Standardization of Sodium Hydroxide (NaOH) Solution

	Trial 1	*Trial 2*	*Trial 3*
Mass of bottle + KHP	1.3760 g	1.4045 g	1.3967 g
Mass of bottle	1.0369 g	1.0369 g	1.0369 g
Mass of KHP used	0.3391 g	0.3676 g	0.3598 g
Final buret reading	19.30 mL	35.60 mL	45.10 mL
Initial buret reading	4.90 mL	20.10 mL	30.10 mL
mL of NaOH used	14.40 mL	15.50 mL	15.00 mL
Molarity of NaOH	0.1153 *M*	0.1161 *M*	0.1175 *M*

Average molarity 0.1163 Standard deviation ± 0.0011

Show your calculations for molarity and standard deviation:

$$\text{moles KHP} = (0.3391 \text{ g KHP})\left(\frac{1 \text{ mol KHP}}{204.2 \text{ g KHP}}\right) = 1.661 \times 10^{-3} \text{ mol KHP} = \text{moles NaOH}$$

$$M_1 = \frac{1.661 \times 10^{-3} \text{ mol NaOH}}{0.01440 \text{ L}} = 0.1153\ M \text{ etc.} \qquad M_{ave} = \frac{0.1153 + 0.1161 + 0.1175}{3} = 0.1163$$

$$\text{SD} = \sqrt{\frac{\Sigma|X_i - \mu|^2}{m-1}} = \sqrt{\frac{(0.001)^2 + (0.0002)^2 + (0.0012)^2}{2}} = 1.1 \times 10^{-3}$$

B. Analysis of an Unknown Acid

	Trial 1	*Trial 2*	*Trial 3*
Mass of bottle + unknown	1.9762 g	1.9557 g	1.9813 g
Mass of bottle	1.0317 g	1.0317 g	1.0317 g
Mass of unknown used	0.9445 g	0.9240 g	0.9496 g

	Trial 1	*Trial 2*	*Trial 3*
Final buret reading	21.40 mL	39.90 mL	41.05 mL
Initial buret reading	1.05 mL	20.05 mL	20.65 mL
mL of NaOH used	20.35 mL	19.85 mL	20.40 mL
Mass of KHP in unknown	0.4833 g	0.4714 g	0.4845 g
Percent of KHP in unknown	51.12%	51.02%	51.02%

Average percent of KHP 51.15% Standard deviation ± 0.06%

Calculations of percent KHP and standard deviation (show using equations with units):

mmol NaOH = (20.35 mL NaOH)(0.1163 mmol NaOH/mL NaOH) = 2.367 mmol NaOH

$$\text{grams KHP} = (2.367\text{ mmol KHP})\left(\frac{1\text{ mol KHP}}{10^3\text{ mmol KHP}}\right)\left(\frac{204.2\text{ g KHP}}{1\text{ mol KHP}}\right) = 0.4833\text{ g}$$

$$\%\text{ KHP} = \left(\frac{0.4833\text{ g}}{0.9455\text{ g}}\right)(100) = 51.12\%\text{ etc.}\quad \text{Avg \% KHP} = \frac{51.12 + 51.02 + 51.02}{3} = 51.05$$

$$\text{SD} = \sqrt{\frac{(0.07)^2 + (0.03)^2 + (0.03)^2}{2}} = 0.06$$

QUESTIONS

1. Write the balanced chemical equation for the reaction of KHP with NaOH.

   $KHP + NaOH \longrightarrow NaKP + H_2O$

2. Suppose your laboratory instructor inadvertently gave you a sample of KHP contaminated with NaCl to standardize your NaOH. How would this affect the molarity you calculated for your NaOH solution? Justify your answer.

   The molarity would be too high because it would require too small a volume of NaOH for the given weight of KHP for neutralization.

3. A solution of malonic acid, $H_2C_3H_2O_4$, was standardized by titration with 0.100 *M* NaOH solution. If 20.76 mL of the NaOH solution were required to neutralize completely 13.15 mL of the malonic acid solution, what is the molarity of the malonic acid solution?

$$H_2C_3H_2O_4 + 2NaOH \longrightarrow Na_2C_3H_2O_4 + 2H_2O$$

mmol malonic acid = $^1/_2$ mmol NaOH

$$mL_A \times M_A = {}^1/_2\,(mL_B \times M_B)$$

$$13.15\text{ mL} \times M_A = {}^1/_2\,(20.76\text{ mL} \times 0.100\ M_B)$$

$$M_A = 0.7893\ M$$

4. Sodium carbonate is a reagent that may be used to standardize acids in the same way that you have used KHP in this experiment. In such a standardization it was found that a 0.498-g sample of sodium carbonate required 23.5 mL of a sulfuric acid solution to reach the end point for the reaction.

$$Na_2CO_3(aq) + H_2SO_4(aq) \longrightarrow H_2O(l) + CO_2(g) + Na_2SO_4(aq)$$

What is the molarity of the H_2SO_4?

$$\text{mol } H_2SO_4 = \text{mol } Na_2CO_3 = (0.498\text{ g})\left(\frac{1\text{ mol}}{106\text{ g}}\right) = 4.70 \times 10^{-3}\text{ mol}$$

$$\text{molarity} = \frac{4.70 \times 10^{-3}\text{ mol}}{0.0235\text{ L}} = 0.200\ M\ H_2SO_4$$

5. A solution contains 0.063 g of oxalic acid, $H_2C_2O_4 \cdot 2H_2O$, in 250 mL. What is the molarity of this solution?

$$\text{moles } H_2C_2O_4 \cdot 2H_2O = \left(\frac{0.063\text{ g } H_2C_2O_4 \cdot 2H_2O}{126\text{ g } H_2C_2O_4 \cdot 2H_2O\text{ g/mol}}\right) = 5.00 \times 10^{-4}\text{ mol}$$

$$\text{molarity} = \frac{5.00 \times 10^{-4}\text{ mol}}{0.250\text{ L}} = 2.00 \times 10^{-3}\ M$$

NOTES AND CALCULATIONS

Reactions in Aqueous Solutions: Metathesis Reactions and Net Ionic Equations

Experiment

OBJECTIVE

To become familiar with writing equations for metathesis reactions, including net ionic equations.

APPARATUS AND CHEMICALS

Apparatus

small test tubes (12)
evaporating dish
thermometer
filter paper
Bunsen burner and hose
magnifying glass
100-mL beaker (2)
600-mL beaker
funnels (2)
short-stem funnel
funnel support
ring stand and ring

Chemicals

sodium nitrate
potassium chloride
0.1 *M* potassium chloride
0.1 *M* barium chloride
0.1 *M* sodium phosphate
0.1 *M* silver nitrate
0.1 *M* nickel chloride
0.1 *M* sodium sulfide
0.1 *M* cadmium chloride
ice
0.1 *M* sodium acetate
0.1 *M* lead nitrate
0.1 *M* copper(II) sulfate
0.1 *M* sodium nitrate
1.0 *M* sulfuric acid
1.0 *M* ammonium chloride
1.0 *M* sodium hydroxide
1.0 *M* hydrochloric acid
1.0 *M* sodium carbonate

ALL SOLUTIONS SHOULD BE PROVIDED IN DROPPER BOTTLES.

DISCUSSION

In molecular equations for many aqueous reactions, cations and anions appear to exchange partners. These reactions conform to the following general equation:

$$AX + BY \longrightarrow AY + BX \quad [1]$$

Such reactions are known as *metathesis reactions*. For a metathesis reaction to lead to a net change in solution, ions must be removed from the solution. In general, three chemical processes can lead to the removal of ions from solution, thus serving as a *driving force* for metathesis to occur:

1. The formation of a precipitate
2. The formation of a weak electrolyte or nonelectrolyte
3. The formation of a gas that escapes from solution

The reaction of barium chloride with silver nitrate is a typical example:

$$BaCl_2(aq) + 2AgNO_3(aq) \longrightarrow Ba(NO_3)_2(aq) + 2AgCl(s) \quad [2]$$

This form of the equation for this reaction is referred to as the *molecular equation*. Because we know that the salts $BaCl_2$, $AgNO_3$, and $Ba(NO_3)_2$ are strong electrolytes and are completely dissociated in solution, we can more realistically write the equation as follows:

$$Ba^{2+}(aq) + 2Cl^-(aq) + 2Ag^+(aq) + 2NO_3^-(aq) \longrightarrow Ba^{2+}(aq) + 2NO_3^-(aq) + 2AgCl(s) \quad [3]$$

This form, in which all ions are shown, is known as the *complete ionic equation*. Reaction [2] occurs because the insoluble substance AgCl precipitates out of solution. The other product, barium nitrate, is soluble in water and remains in solution. We see that Ba^{2+} and NO_3^- ions appear on both sides of the equation and thus do not enter into the reaction. Such ions are called *spectator ions*. If we eliminate or omit them from both sides, we obtain the *net ionic equation*:

$$Ag^+(aq) + Cl^-(aq) \longrightarrow AgCl(s) \quad [4]$$

This equation focuses our attention on the salient feature of the reaction: the formation of the precipitate AgCl. It tells us that solutions of any soluble Ag^+ salt and any soluble Cl^- salt, when mixed, will form insoluble AgCl. When writing net ionic equations, remember that only *strong electrolytes* are written in the ionic form. Solids, gases, nonelectrolytes, and weak electrolytes are written in the molecular form. Frequently the symbol (*aq*) is omitted from ionic equations. The symbols (*g*) for gas and (*s*) for solid should not be omitted. Thus, Equation [4] can be written as

$$Ag^+ + Cl^- \longrightarrow AgCl(s) \quad [5]$$

Consider mixing solutions of KCl and $NaNO_3$. The ionic equation for the reaction is

$$K^+(aq) + Cl^-(aq) + Na^+(aq) + NO_3^-(aq) \longrightarrow K^+(aq) + NO_3^-(aq) + Na^+(aq) + Cl^-(aq) \quad [6]$$

Because all the compounds are water-soluble and are strong electrolytes, they have been written in the ionic form. They completely dissolve in water. If we eliminate spectator ions from the equation, nothing remains. Hence, there is no reaction:

$$K^+(aq) + Cl^-(aq) + Na^+(aq) + NO_3^-(aq) \longrightarrow \text{no reaction} \quad [7]$$

Metathesis reactions occur when a precipitate, a gas, a weak electrolyte, or a nonelectrolyte is formed. The following equations are further illustrations of such processes.

FORMATION OF A GAS

Molecular equation:

$$2HCl(aq) + Na_2S(aq) \longrightarrow 2NaCl(aq) + H_2S(g)$$

Complete ionic equation:

$$2H^+(aq) + 2Cl^-(aq) + 2Na^+(aq) + S^{2-}(aq) \longrightarrow 2Na^+(aq) + 2Cl^-(aq) + H_2S(g)$$

Net ionic equation:

$$2H^+(aq) + S^{2-}(aq) \longrightarrow H_2S(g)$$

or

$$2H^+ + S^{2-} \longrightarrow H_2S(g)$$

FORMATION OF A WEAK ELECTROLYTE

Molecular equation:

$$HNO_3(aq) + NaOH(aq) \longrightarrow H_2O(l) + NaNO_3(aq)$$

Complete ionic equation:

$$H^+(aq) + NO_3^-(aq) + Na^+(aq) + OH^-(aq) \longrightarrow H_2O(l) + Na^+(aq) + NO_3^-(aq)$$

Net ionic equation:

$$H^+(aq) + OH^-(aq) \longrightarrow H_2O(l)$$

In order to decide if a reaction occurs, we need to be able to determine whether or not a precipitate, a gas, a nonelectrolyte, or a weak electrolyte will be formed. The following brief discussion is intended to aid you in this regard. Table 21.1 summarizes solubility rules and should be consulted while performing this experiment.

The common gases are CO_2, SO_2, H_2S, and NH_3. Carbon dioxide and sulfur dioxide may be regarded as resulting from the decomposition of their corresponding weak acids, which are initially formed when carbonate and sulfite salts are treated with acid:

$$H_2CO_3(aq) \longrightarrow H_2O(l) + CO_2(g)$$

and

$$H_2SO_3(aq) \longrightarrow H_2O(l) + SO_2(g)$$

Ammonium salts form NH_3 when they are treated with strong bases:

$$NH_4^+(aq) + OH^- \longrightarrow NH_3(g) + H_2O(l)$$

TABLE 21.1 Solubility Rules

Water-soluble salts	
Na^+, K^+, NH_4^+	All sodium, potassium, and ammonium salts are soluble.
NO_3^-, ClO_3^-, $C_2H_3O_2^-$	All nitrates, chlorates, and acetates are soluble.
Cl^-	All chlorides are soluble except AgCl, Hg_2Cl_2, and $PbCl_2$*.
Br^-	All bromides are soluble except AgBr, Hg_2Br_2, $PbBr_2$,* and $HgBr_2$*.
I^-	All iodides are soluble except AgI, Hg_2I_2, PbI_2, and HgI_2.
SO_4^{2-}	All sulfates are soluble except $CaSO_4$,* $SrSO_4$, $BaSO_4$, Hg_2SO_4, $PbSO_4$, and Ag_2SO_4.
Water-insoluble salts	
CO_3^{2-}, SO_3^{2-}, PO_4^{3-}, CrO_4^{2-}	All carbonates, sulfites, phosphates, and chromates are insoluble except those of alkali metals and NH_4^+.
OH^-	All hydroxides are insoluble except those of alkali metals and $Ca(OH)_2$,* $Sr(OH)_2$,* and $Ba(OH)_2$.
S^{2-}	All sulfides are insoluble except those of the alkali metals, alkaline earths, and NH_4^+.

*Slightly soluble.

TABLE 21.2 Strong Electrolytes

Salts	All common soluble salts
Acids	$HClO_4$, HCl, HBr, HI, HNO_3, and H_2SO_4 are strong electrolytes; all others are weak.
Bases	Alkali metal hydroxides, $Ca(OH)_2$, $Sr(OH)_2$, and $Ba(OH)_2$ are strong electrolytes; all others are weak.

Which are the weak electrolytes? The easiest way of answering this question is to identify all of the strong electrolytes, and if the substance does not fall in that category it is a weak electrolyte. Note, water is a nonelectrolyte. Strong electrolytes are summarized in Table 21.2.

In the first part of this experiment, you will study some metathesis reactions. In some instances it will be very evident that a reaction has occurred, whereas in others it will not be so apparent. In the doubtful case, use the guidelines above to decide whether or not a reaction has taken place. You will be given the names of the compounds to use but not their formulas. This is being done deliberately to give practice in writing formulas from names.

In the second part of this experiment, you will study the effect of temperature on solubility. The effect that temperature has on solubility varies from salt to salt. We conclude that mixing solutions of KCl and $NaNO_3$ resulted in no reaction (see Equations [6] and [7]). What would happen if we cooled such a mixture? The solution would eventually become saturated with respect to one of the salts, and crystals of that salt would begin to appear as its solubility was exceeded. Examination of Equation [6] reveals that crystals of any of the following salts could appear initially: KNO_3, KCl, $NaNO_3$, or NaCl.

Consequently, if a solution containing Na^+, K^+, Cl^-, and NO_3^- ions is evaporated at a given temperature, the solution becomes more and more concentrated and will eventually become saturated with respect to one of the four compounds. If evaporation is continued, that compound will crystallize out, removing its ions from solution. The other ions will remain in solution and increase in concentration. Before beginning this laboratory exercise you are to plot the solubilities of the four salts given in Table 21.3 on the graph on your report sheet.

PROCEDURE

A. Metathesis Reactions

The report sheet lists 16 pairs of chemicals that are to be mixed. Use about 1 mL of the reagents to be combined as indicated on the report sheet. Mix the solutions in small test tubes and record your observations on the report

TABLE 21.3 Molar Solubilities of NaCl, $NaNO_3$, KCl, and KNO_3 (mol/L)

Compound	0°C	20°C	40°C	60°C	80°C	100°C
NaCl	5.4	5.4	5.5	5.5	5.5	5.6
$NaNO_3$	6.7	7.6	8.5	9.4	10.4	11.3
KCl	3.4	4.0	4.6	5.1	5.5	5.8
KNO_3	1.3	3.2	5.2	7.0	9.0	11.0

sheet. If there is no reaction, write N.R. (The reactions need not be carried out in the order listed. Congestion at the reagent shelf can be avoided if everyone does not start with reagents for reaction 1.) Dispose of the contents of your test tubes in the designated receptacles.

B. Solubility, Temperature, and Crystallization

Place 8.5 g of sodium nitrate and 7.5 g of potassium chloride in a 100-mL beaker and add 25 mL of water. Warm the mixture, stirring, until the solids completely dissolve. Assuming a volume of 25 mL for the solution, calculate the molarity of the solution with respect to $NaNO_3$, KCl, NaCl, and KNO_3, and record these molarities (1).

Cool the solution to about 10°C by dipping the beaker in ice water in a 600-mL beaker and stir the solution carefully with a thermometer, being careful not to break it. (SHOULD THE THERMOMETER BREAK, IMMEDIATELY CONSULT YOUR INSTRUCTOR.) When no more crystals form, at approximately 10°C, filter the cold solution quickly and allow the filtrate to drain thoroughly into an evaporating dish. Dry the crystals between two dry pieces of filter paper or paper towels. Examine the crystals with a magnifying glass (or fill a Florence flask with water and look at the crystals through it). Describe the shape of the crystals—that is, needles, cubes, plates, rhombs, and so forth (2). Based upon your solubility graph, which compound crystallized out of solution (3)?

Instructor: It is often easier to observe the crystal shape in the flask or on the filter paper than after they have been dried.

Evaporate the filtrate to about half of its volume using a Bunsen burner and ring stand. A second crop of crystals should form. Record the temperature (4) and rapidly filter the hot solution, collecting the filtrate in a clean 100-mL beaker. Dry the second batch of crystals between two pieces of filter paper and examine their shape. Compare their shape with the first batch of crystals (5). Based upon your solubility graph, what is this substance (6)?

Finally, cool the filtrate to 10°C while stirring carefully with a thermometer to obtain a third crop of crystals. Carefully observe their shapes and compare them with those of the first and second batches (7). What compound is the third batch of crystals (8)? Dispose of the chemicals in the designated receptacles.

REVIEW QUESTIONS

Before beginning this experiment in the laboratory, you should be able to answer the following questions:

1. Write molecular, complete ionic, and net ionic equations for the reactions that occur, if any, when solutions of the following substances are mixed:
   (a) nitric acid and barium carbonate
   (b) zinc chloride and lead nitrate
   (c) acetic acid and sodium hydroxide
   (d) calcium nitrate and sodium carbonate
   (e) ammonium chloride and potassium hydroxide
2. Which of the following are not water-soluble:
   $Ba(NO_3)_2$, $FeCl_3$, $CuCO_3$, $CuSO_4$, ZnS, $ZnSO_4$?
3. Write equations for the decomposition of H_2CO_3 and H_2SO_3.
4. At what temperature (from your graph) do KNO_3 and NaCl have the same molar solubility?

5. Which of the following are strong electrolytes: $BaCl_2$, $AgNO_3$, HCl, HNO_3, $HC_2H_3O_2$?

6. Which of the following are weak electrolytes: HNO_3, HF, HCl, $NH_3(aq)$, NaOH?

7. For each of the following water-soluble compounds, indicate the ions present in an aqueous solution: NaI, K_2SO_4, NaCN, $Ba(OH)_2$, $(NH_4)_2SO_4$.

8. Write a balanced chemical equation showing how you could prepare each of the following salts from an acid-base reaction: $NaNO_3$, KCl, $BaSO_4$.

Name ______________________ Desk ______________

Date ______________ Laboratory Instructor ______________________

REPORT SHEET | EXPERIMENT

Reactions in Aqueous Solutions: Metathesis Reactions and Net Ionic Equations

21

A. Metathesis Reactions

1. Copper(II) sulfate + sodium carbonate
   - Observations: A pale blue (turquoise) precipitate forms.
   - Molecular equation: $CuSO_4 + Na_2CO_3 \longrightarrow CuCO_3(s) + Na_2SO_4$
   - Complete ionic equation: $Cu^{2+} + SO_4^{2-} + 2Na^+ + CO_3^{2-} \longrightarrow CuCO_3(s) + 2Na^+ + SO_4^{2-}$
   - Net ionic equation: $Cu^{2+} + CO_3^{2-} \longrightarrow CuCO_3(s)$
2. Copper(II) sulfate + barium chloride
   - Observations: A white precipitate forms.
   - Molecular equation: $CuSO_4 + BaCl_2 \longrightarrow CuCl_2 + BaSO_4(s)$
   - Complete ionic equation: $Cu^{2+} + SO_4^{2-} + Ba^{2+} + 2Cl^- \longrightarrow Cu^{2+} + 2Cl^- + BaSO_4(s)$
   - Net ionic equation: $Ba^{2+} + SO_4^{2-} \longrightarrow BaSO_4(s)$
3. Copper(II) sulfate + sodium phosphate
   - Observations: A light blue precipitate forms.
   - Molecular equation: $3CuSO_4 + 2Na_3PO_4 \longrightarrow Cu_3(PO_4)_2(s) + 3Na_2SO_4$
   - Complete ionic equation: $3Cu^{2+} + 3SO_4^{2-} + 6Na^+ + 2PO_4^{3-} \longrightarrow Cu_3(PO_4)_2(s) + 6Na^+ + 3SO_4^{2-}$
   - Net ionic equation: $3Cu^{2+} + 2PO_4^{3-} \longrightarrow Cu_3(PO_4)_2(s)$
4. Sodium carbonate + sulfuric acid
   - Observations: A gas (bubble) forms along with a colorless solution.
   - Molecular equation: $Na_2CO_3 + H_2SO_4 \longrightarrow Na_2SO_4 + H_2O + CO_2(g)$
   - Complete ionic equation: $2Na^+ + CO_3^{2-} + 2H^+ + SO_4^{2-} \longrightarrow 2Na^+ + SO_4^{2-} + H_2O + CO_2(g)$
   - Net ionic equation: $2H^+ + CO_3^{2-} \longrightarrow H_2O + CO_2(g)$
5. Sodium carbonate + hydrochloric acid
   - Observations: A gas (bubble) forms along with a colorless solution.
   - Molecular equation: $Na_2CO_3 + 2HCl \longrightarrow 2NaCl + H_2O + CO_2(g)$
   - Complete ionic equation: $2Na^+ + CO_3^{2-} + 2H^+ + 2Cl^- \longrightarrow 2Na^+ + 2Cl^- + H_2O + CO_2(g)$
   - Net ionic equation: $2H^+ + CO_3^{2-} \longrightarrow H_2O + CO_2(g)$
6. Cadmium chloride + sodium sulfide
   - Observations: A yellow precipitate forms.
   - Molecular equation: $CdCl_2 + Na_2S \longrightarrow CdS(s) + 2NaCl$
   - Complete ionic equation: $Cd^{2+} + 2Cl^- + 2Na^+ + S^{2-} \longrightarrow CdS(s) + 2Na^+ + 2Cl^-$
   - Net ionic equation: $Cd^{2+} + S^{2-} \longrightarrow CdS(s)$

7. Cadmium chloride + sodium hydroxide
   - Observations: A yellow precipitate forms.
   - Molecular equation: $CdCl_2 + 2NaOH \longrightarrow Cd(OH)_2(s) + 2NaCl$
   - Complete ionic equation: $Cd^{2+} + 2Cl^- + 2Na^+ + 2OH^- \longrightarrow Cd(OH)_2(s) + 2Na^+ + 2Cl^-$
   - Net ionic equation: $Cd^{2+} + 2OH^- \longrightarrow Cd(OH)_2(s)$
8. Nickel chloride + silver nitrate
   - Observations: A white precipitate forms in a green solution.
   - Molecular equation: $NiCl_2 + 2AgNO_3 \longrightarrow 2AgCl(s) + Ni(NO_3)_2$
   - Complete ionic equation: $Ni^{2+} + 2Cl^- + 2Ag^+ + 2NO_3^- \longrightarrow 2AgCl(s) + Ni^{2+} + 2NO_3^-$
   - Net ionic equation: $Ag^+ + Cl^- \longrightarrow AgCl(s)$
9. Nickel chloride + sodium carbonate
   - Observations: A pale green precipitate forms.
   - Molecular equation: $NiCl_2 + Na_2CO_3 \longrightarrow NiCO_3(s) + 2NaCl$
   - Complete ionic equation: $Ni^{2+} + 2Cl^- + 2Na^+ + CO_3^{2-} \longrightarrow NiCO_3(s) + 2Na^+ + 2Cl^-$
   - Net ionic equation: $Ni^{2+} + CO_3^{2-} \longrightarrow NiCO_3(s)$
10. Hydrochloric acid + sodium hydroxide
    - Observations: A colorless solution was formed; gets warm.
    - Molecular equation: $HCl + NaOH \longrightarrow NaCl + H_2O$
    - Complete ionic equation: $H^+ + Cl^- + Na^+ + OH^- \longrightarrow Na^+ + Cl^- + H_2O$
    - Net ionic equation: $H^+ + OH^- \longrightarrow H_2O$
11. Ammonium chloride + sodium hydroxide
    - Observations: The odor of NH_3 was noted.
    - Molecular equation: $NH_4Cl + NaOH \longrightarrow NH_3(g) + H_2O + NaCl$
    - Complete ionic equation: $NH_4^+ + Cl^- + Na^+ + OH^- \longrightarrow NH_3(g) + H_2O + Na^+ + Cl^-$
    - Net ionic equation: $NH_4^+ + OH^- \longrightarrow NH_3(g) + H_2O$
12. Sodium acetate + hydrochloric acid
    - Observations: The odor of vinegar was noted.
    - Molecular equation: $NaC_2H_3O_2 + HCl \longrightarrow HC_2H_3O_2 + NaCl$
    - Complete ionic equation: $Na^+ + C_2H_3O_2^- + H^+ + Cl^- \longrightarrow HC_2H_3O_2 + Na^+ + Cl^-$
    - Net ionic equation: $H^+ + C_2H_3O_2^- \longrightarrow HC_2H_3O_2$
13. Sodium sulfide + hydrochloric acid
    - Observations: A gas with a rotten-egg odor formed.
    - Molecular equation: $Na_2S + 2HCl \longrightarrow 2NaCl + H_2S(g)$
    - Complete ionic equation: $2Na^+ + S^{2-} + 2H^+ + 2Cl^- \longrightarrow 2Na^+ + 2Cl^- + H_2S(g)$
    - Net ionic equation: $2H^+ + S^{2-} \longrightarrow H_2S(g)$
14. Lead nitrate + sodium sulfide
    - Observations: A dark black-brown precipitate formed.
    - Molecular equation: $Pb(NO_3)_2 + Na_2S \longrightarrow PbS(s) + 2NaNO_3$
    - Complete ionic equation: $Pb^{2+} + 2NO_3^- + 2Na^+ + S^{2-} \longrightarrow PbS(s) + 2Na^+ + 2NO_3^-$
    - Net ionic equation: $Pb^{2+} + S^{2-} \longrightarrow PbS(s)$

15. Lead nitrate + sulfuric acid
    Observations A white precipitate formed.
    Molecular equation $Pb(NO_3)_2 + H_2SO_4 \longrightarrow PbSO_4(s) + 2HNO_3$
    Complete ionic equation $Pb^{2+} + 2NO_3^- + 2H^+ + SO_4^{2-} \longrightarrow PbSO_4(s) + 2H^+ + 2NO_3^-$
    Net ionic equation $Pb^{2+} + SO_4^{2-} \longrightarrow PbSO_4(s)$
16. Potassium chloride + sodium nitrate
    Observations There was no apparent reaction.
    Molecular equation $KCl + NaNO_3 \longrightarrow NaCl + KNO_3$
    Complete ionic equation $K^+ + Cl^- + Na^+ + NO_3^- \longrightarrow K^+ + Cl^- + Na^+ + NO_3^-$
    Net ionic equation none

B. Solubility, Temperature, and Crystallization

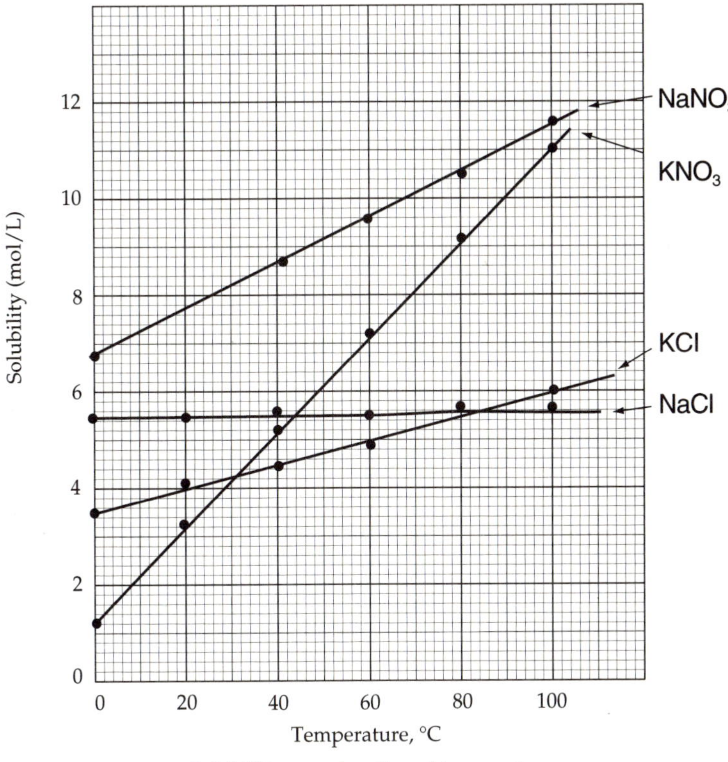

Solubilities as a function of temperature

1. Molarities
   4.0 *M* $NaNO_3$, 4.0 *M* KCl, 4.0 *M* NaCl, 4.0 *M* KNO_3
2. Crystal shape fine needles
3. Identity of crystals KNO_3
4. Temperature of filtrate 95°–100°C
5. Crystal shape of second batch small cubes

6. Identity of second batch of crystals NaCl
7. Crystal shape of third batch needles
8. Identity of third batch of crystals KNO_3

QUESTIONS

1. Which of the following reactions are metathesis reactions?
   (a) $2KClO_3 \longrightarrow 2KCl + 3O_2$ No (redox)
   (b) $Cu(NO_3)_2 + Zn \longrightarrow Cu + Zn(NO_3)_2$ No (redox)
   (c) $BaCO_3 + 2HCl \longrightarrow BaCl_2 + H_2O + CO_2$ Yes
   (d) $Na_2CO_3 + CuSO_4 \longrightarrow Na_2SO_4 + CuCO_3$ Yes

2. How many grams of each of the following substances will dissolve in 100 mL of cold water? Consult a handbook.

   $Ce(IO_3)_4$, $RaSO_4$, $Pb(NO_3)_2$, $(NH_4)_2SeO_4$

   0.015 g; 2×10^{-6} g; 37.65 g; 117 g

3. Suppose you have a solution that might contain any or all of the following cations: Cu^{2+}, Ag^+, Ba^{2+}, and Mn^{2+}. Addition of HCl causes a precipitate to form. After the precipitate is filtered off, H_2SO_4 is added to the supernate and another precipitate forms. This precipitate is filtered off, and a solution of NaOH is added to the supernatant liquid until it is strongly alkaline. No precipitate is formed. Which ions are present in each of the precipitates? Which cations are not present in the original solution?

   $Ag^+ + Cl^- \longrightarrow AgCl$ first precipitate

   $Ba^{2+} + SO_4^{2-} \longrightarrow BaSO_4$ second precipitate

   Since both $Cu(OH)_2$ and $Mn(OH)_2$ are sparingly soluble, the fact that the NaOH addition did not cause precipitation suggests their absence. The first two reactions show that Ag^+ and Ba^{2+} are initially present.

4. Write balanced net ionic equations for the reactions, if any, that occur between (a) FeS(*s*) and HBr(*aq*); (b) K_2CO_3(*aq*) and $CuCl_2$(*aq*); (c) $Fe(NO_3)_2$(*aq*) and HCl(*aq*); (d) $Bi(OH)_3$(*s*) and HNO_3(*aq*).

   a) $FeS(s) + 2H^+(aq) \longrightarrow Fe^{2+}(aq) + H_2S(g)$

   b) $Cu^{2+}(aq) + CO_3^{2-}(aq) \longrightarrow CuCO_3(s)$

   c) No reaction

   d) $3H^+(aq) + Bi(OH)_3(s) \longrightarrow 3H_2O(l) + Bi^{3+}(aq)$

Experiment

Determination of the Dissociation Constant of a Weak Acid

OBJECTIVE

To become familiar with the operation of a pH meter and quantitative equilibrium constants.

APPARATUS AND CHEMICALS

Apparatus

pH meter with electrodes
balance
150-mL beakers (4)
buret
1-pint bottle and stopper
Bunsen burner and hose
wire gauze
600-mL Erlenmeyer flask
250-mL Erlenmeyer flacks (3)
25-mL pipet and pipet bulb
buret clamp and ring stand
weighing bottle
ring stand and ring

Chemicals

potassium hydrogen phthalate (KHP)
standard buffer solution
unknown solution of a weak acid (~0.1 *M*)
0.1 *M* NaOH or 10 *M* NaOH
phenolphthalein indicator solution

DISCUSSION

Acid-Base Equilibria

According to the Brønsted-Lowry acid-base theory, the strength of an acid is related to its ability to donate protons. All acid-base reactions are then competitions between bases of various strengths for these protons. For example, the strong acid HCl reacts with water according to Equation [1]:

$$HCl(aq) + H_2O(l) \longrightarrow H_3O^+(aq) + Cl^-(aq) \qquad [1]$$

This acid is a strong acid and is completely dissociated—in other words, 100% dissociated—in dilute aqueous solution. Consequently, the $[H_3O^+]$ concentration of 0.1 *M* HCl is 0.1 *M*.

By contrast, acetic acid, $HC_2H_3O_2$ (abbreviated HOAc), is a weak acid and is only slightly dissociated, as shown in Equation [2]:

$$H_2O(l) + HOAc(aq) \rightleftharpoons H_3O^+(aq) + OAc^-(aq) \qquad [2]$$

Its acid dissociation constant, as shown by Equation [3], is therefore small:

$$K_a = \frac{[H_3O^+][OAc^-]}{[HOAc]} = 1.8 \times 10^{-5} \qquad [3]$$

Acetic acid only partially dissociates in aqueous solution, and an appreciable quantity of undissociated acetic acid remains in solution.

For the general weak acid HA, the dissociation reaction and dissociation constant expression are

$$HA(aq) + H_2O(l) \rightleftarrows H_3O^+(aq) + A^-(aq) \qquad [4]$$

$$K_a = \frac{[H_3O^+][A^-]}{[HA]} \qquad [5]$$

Recall that pH is defined as

$$-\log [H_3O^+] = pH \qquad [6]$$

Solving Equation [5] for $[H_3O^+]$ and substituting this quantity into Equation [6] yields

$$[H_3O^+] = K_a\frac{[HA]}{[A^-]} \qquad [7]$$

$$-\log [H_3O^+] = -\log K_a - \log\frac{[HA]}{[A^-]} \qquad [8]$$

$$pH = pK_a - \log\frac{[HA]}{[A^-]} \qquad [9]$$

$$\text{where } pK_a = -\log K_a$$

If we titrate the weak acid HA with a base, there will be a point in the titration at which the number of moles of base added is half the number of moles of acid initially present. This is the point at which 50% of the acid has been titrated to produce A^- and 50% remains as HA. At this point $[HA] = [A^-]$, the ratio $[HA]/[A^-] = 1$, and $\log [HA]/[A^-] = 0$. Hence, at this point in a titration, that is, at half the equivalence point, Equation [9] becomes

$$pH = pK_a \qquad [10]$$

By titrating a weak acid with a strong base and recording the pH versus the volume of base added, we can determine the ionization constant of the weak acid. From the resultant titration curve we obtain the ionization constant, as explained in the following paragraph.

From the titration curve (Figure 22.1), we see that at the point denoted as half the equivalence point, where $[HA] = [A^-]$, the pH is 4.3. Thus, from Equation [10], at this point $pH = pK_a$, or

$$pK_a = 4.3$$

$$-\log K_a = 4.3$$

$$\log K_a = -4.3$$

$$K_a = 5 \times 10^{-5}$$

Instructor: Demonstration

(Your instructor will show you a graphical method for locating the equivalence point on your titration curves.)

Operation of the pH Meter

To measure the pH during the course of the titration, we shall use an electronic instrument called a pH meter. This device consists of a meter and two electrodes, as illustrated in Figure 22.2.

The main variations among different pH meters involve the positions of the control knobs and the types of electrodes and electrode-mounting devices. The measurement of pH requires two electrodes: a sensing electrode

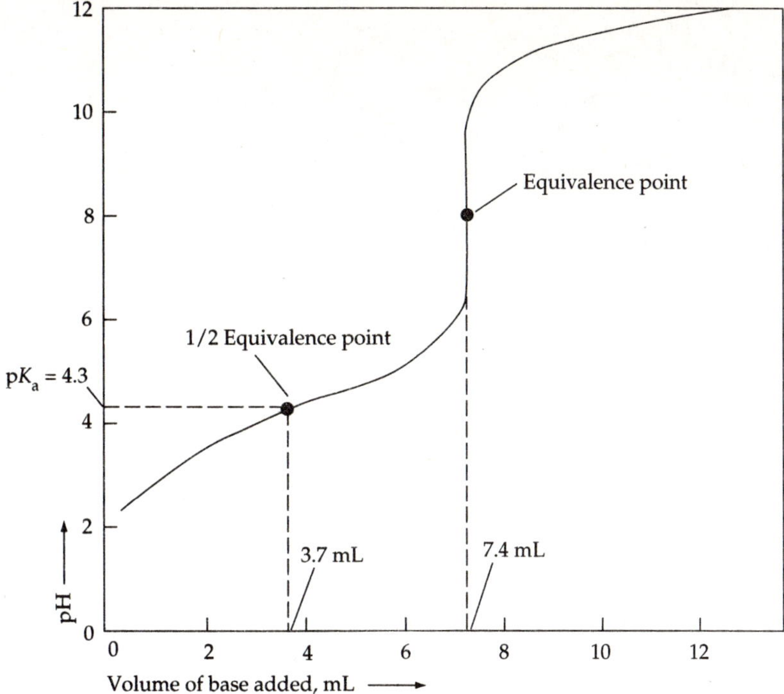

▲ **FIGURE 22.1** Exemplary titration curve for the titration of a weak acid HA with a strong base.

that is sensitive to H_3O^+ concentrations and a reference electrode. This is because the pH meter is really just a voltmeter that measures the electrical potential of a solution. Typical sensing and reference electrodes are illustrated in Figure 22.3.

The reference electrode is an electrode that develops a known potential that is essentially independent of the contents of the solution into which it is

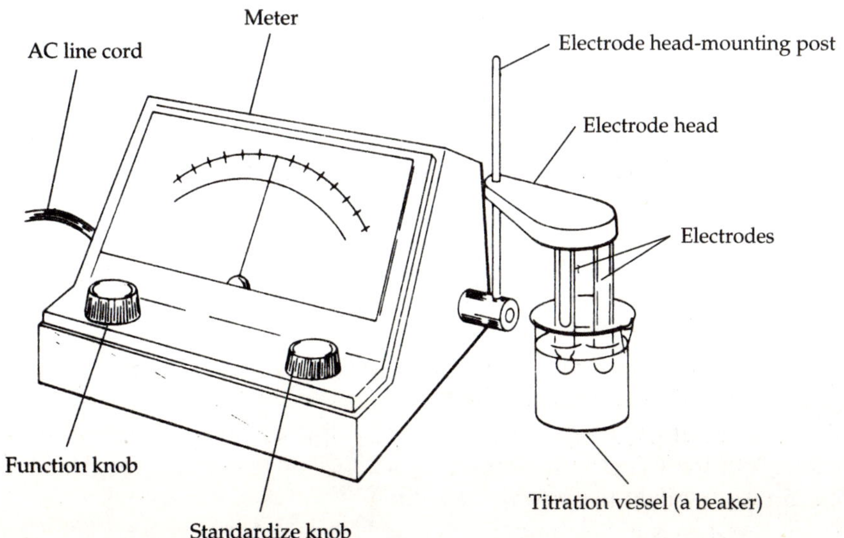

▲ **FIGURE 22.2** An analog pH meter.

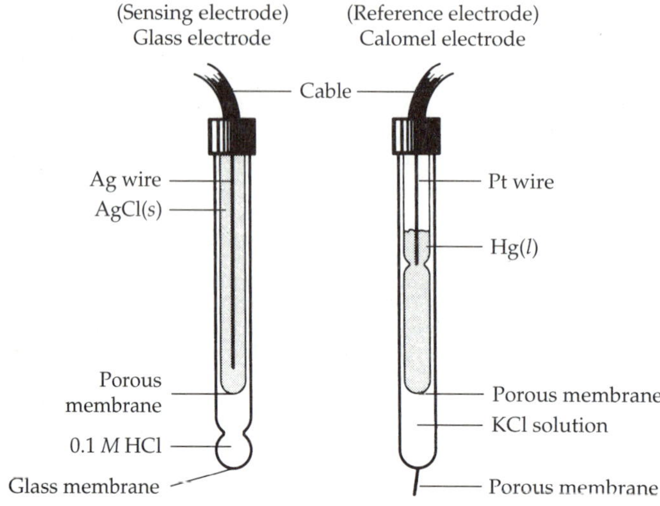

▲ **FIGURE 22.3** A typical sensing and reference electrode.

placed. The glass electrode is sensitive to the H_3O^+ concentration of the solution into which it is placed; its potential is a function of $[H_3O^+]$. It operates by transport of H_3O^+ ions through the glass membrane. This can be more precisely described, but for our purposes here it is sufficient for you to understand that two electrodes are required. These two electrodes are sometimes combined into an electrode called a combination electrode, which appears to be a single electrode. The combination electrode, however, does contain both a reference and a sensing electrode.

Preliminary Operations with the pH Meter

1. Obtain a buffer solution of known pH.
2. Plug in the pH meter to line current and allow at least 10 min for warm-up. It should be left plugged in until you are completely finished with it. *This does not apply to battery-operated meters.*
3. Turn the function knob on the pH meter to the standby position.
4. *Prepare the electrodes.* Make certain that the solution in the reference electrode extends well above the internal electrode. If it does not, ask your instructor to fill it with saturated KCl solution. Remove the rubber tip and slide down the rubber collar on the reference electrode. Rinse the outside of the electrodes well with distilled water.

Instructor: Caution students about the proper use of the electrodes.

5. *Standardize the pH meter.* Carefully immerse the electrodes in the buffer solution contained in a small beaker. *Remember that the glass electrode is very fragile; it breaks easily!* DO NOT touch the bottom of the beaker with the electrodes!! Turn the function knob to "read" or "pH." Turn the standardize knob until the pH meter indicates the exact pH of the buffer solution. Wait 5 s to be certain that the reading remains constant. *Once you have standardized the pH meter, don't readjust the standardize knob.* Turn the function knob to standby. Carefully lift the electrodes from the buffer and rinse them with distilled water. The pH meter is now ready to use to measure pH.

RECORD ALL DATA DIRECTLY ONTO THE REPORT SHEETS.

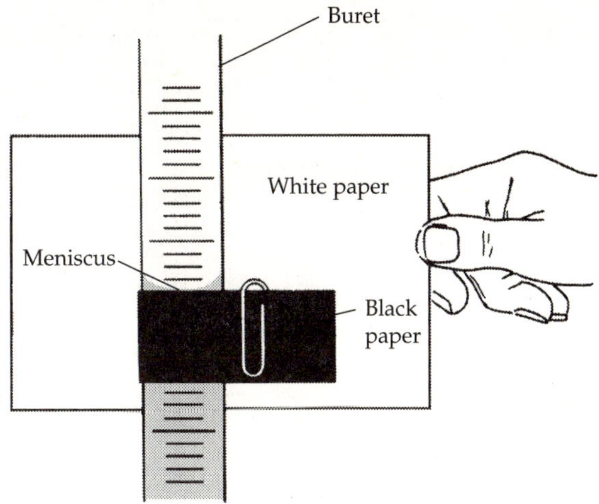

▲ **FIGURE 22.4** Reading a buret.

PROCEDURE

Instructor: Time may be saved in this experiment if the students are provided with an approximately 0.1 *M* of NaOH solution.

A. Preparation of Approximately 0.100 *M* Sodium Hydroxide (NaOH)

Heat 500 mL of distilled water to boiling in a 600-mL flask* and, *after cooling under the water tap*, transfer to a 1-pint bottle fitted with a rubber stopper.† Add 3 mL of stock solution of carbonate-free NaOH (approximately 10 *M*) and shake vigorously for at least 1 min.

Preparation of a Buret for Use Clean a 50-mL buret with soap solution and a buret brush and thoroughly rinse with tap water. Then rinse with at least five 10-mL portions of distilled water. The water must run freely from the buret without leaving any drops adhering to the sides. Make sure that the buret does not leak and that the stopcock turns freely.

Reading a Buret All liquids, when placed in a buret, form a curved meniscus at their upper surfaces. In the case of water or water solutions, this meniscus is concave (Figure 22.4), and the most accurate buret readings are obtained by observing the position of the lowest point on the meniscus on the graduated scales.

To avoid parallax errors when taking readings, the eye must be on a level with the meniscus. Wrap a strip of paper around the buret and hold the top edges of the strip evenly together. Adjust the strip so that the front and back edges are in line with the lowest part of the meniscus and take the reading by estimating to the nearest tenth of a marked division (0.01 mL). A simple way of doing this for repeated readings on a buret is illustrated in Figure 22.4.

*The water is boiled to remove carbon dioxide (CO_2), which would react with the NaOH and change its molarity.
†A rubber stopper should be used for a bottle containing NaOH solution. A strongly alkaline solution tends to cement a glass stopper so firmly that it is difficult to remove.

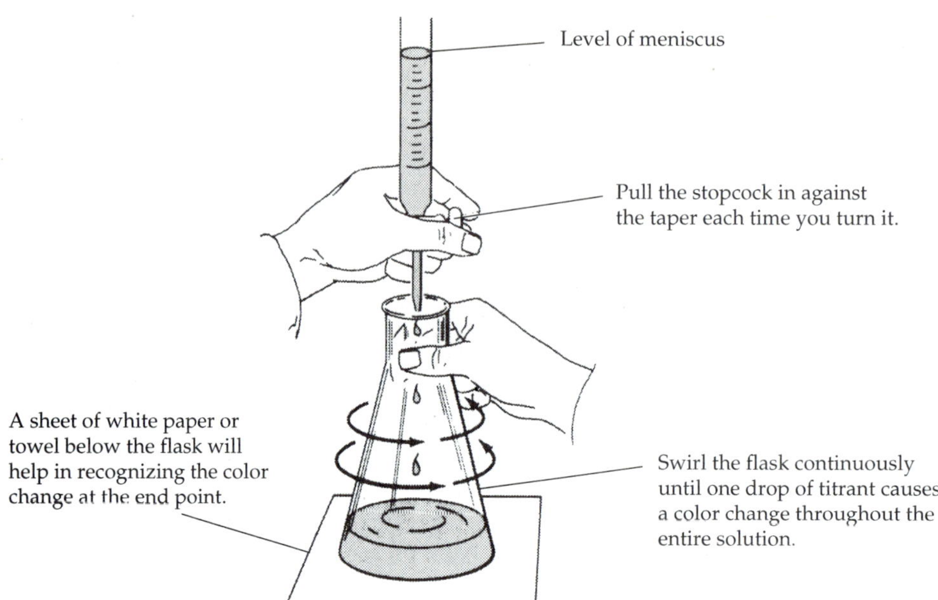

▲ FIGURE 22.5 Titration procedure.

B. Standardization of Sodium Hydroxide (NaOH) Solution

Instructor: Demonstrate how to use a weighing bottle.

Prepare about 400 to 450 mL of CO_2-free water by boiling for about 5 min. Weigh from a weighing bottle (your lab instructor will show you how to use a weighing bottle if you don't already know) triplicate samples of between 0.4 and 0.6 g each of pure potassium hydrogen phthalate (KHP) into three separate 250-mL Erlenmeyer flasks; accurately weigh to four significant figures.* Do not weigh the flasks. Record the masses and label the three flasks in order to distinguish among them. Add to each sample about 100 mL of distilled water that has been freed from CO_2 by boiling, and warm gently with swirling until the salt is completely dissolved. Add to each flask two drops of phenolphthalein indicator solution.

Rinse the previously cleaned buret with at least four 5-mL portions of the approximately 0.100 *M* NaOH solution that you have prepared. Discard each portion into the designated receptacle. *Do not return any of the washings to the bottle.* Completely fill the buret with the solution and remove the air from the tip by running out some of the liquid into an empty beaker. Make sure that the lower part of the meniscus is at the zero mark or slightly lower. Allow the buret to stand for at least 30 s before reading the exact position of the meniscus. Remove any hanging drop from the buret tip by touching it to the side of the beaker used for the washings. Record the initial buret reading on your report sheet.

Slowly add the NaOH solution to one of your flasks of KHP solution while gently swirling the contents of the flask, as illustrated in Figure 22.5. As the NaOH solution is added, a pink color appears where the drops of the

*In cases where the mass of a sample is larger than 1 g, it is necessary to weigh only to the nearest milligram to obtain four significant figures. Buret readings can be read only to the nearest 0.02 mL, and for readings greater than 10 mL this represents four significant figures.

base come in contact with the solution. This coloration disappears with swirling. As the end point is approached, the color disappears more slowly, at which time the NaOH should be added drop by drop. It is most important that the flask be swirled constantly throughout the entire titration. The end point is reached when one drop of the NaOH solution turns the entire solution in the flask from colorless to pink. The solution should remain pink when it is swirled. Allow the titrated solution to stand for at least 1 min so the buret will drain properly. Remove any hanging drop from the buret tip by touching it to the side of the flask and wash down the sides of the flask with a stream of water from the wash bottle. Record the buret reading on your report sheet. Repeat this procedure with the other two samples.

From the data you obtain in the three titrations, calculate the molarity of the NaOH solution to four significant figures.

The three determinations should agree within 1.0%. If they do not, the standardization should be repeated until agreement is reached. The average of the three acceptable determinations is taken as the molarity of the NaOH. Calculate the standard deviation of your results. SAVE your standardized solution for the determination of the pK_a of the unknown acid.

C. Determination of pK_a of Unknown Acid

With the aid of a pipet bulb, pipet a 25-mL aliquot of your unknown acid solution into a 250-mL beaker and carefully immerse the previously rinsed electrodes in this solution. Measure the pH of this solution by turning the function knob to "read" or "pH." Record the pH on the report sheet. Begin your titration by adding 1 mL of your standardized base from a buret and record the volume of titrant and pH. Repeat with successive additions of 1 mL of base until you approach the end point; then add 0.1-mL increments of base and record the pH and milliliters of NaOH added. When the pH no longer changes upon addition of NaOH, your titration is completed. From these data, plot a titration curve of pH versus mL titrant added. Repeat the titration with two more 25-mL aliquots of your unknown acid and plot the titration curves. From these curves calculate the ionization constant. Time may be saved if the first titration is run with larger-volume increments of the titrant to locate an approximate equivalence point; then the second and third titrations may be run with the small increments indicated above. Turn the function knob to standby, rinse the electrodes with distilled water, and wipe them with a clean, dry tissue.

D. Concentration of Unknown Acid

Using the volume of base at the equivalence point, its molarity, and the fact that you used 25.0 mL of acid, calculate the concentration of the unknown acid and record this on the report sheet.

REVIEW QUESTIONS

Before beginning this experiment in the laboratory, you should be able to answer the following questions:

1. Define Brønsted-Lowry acids and bases.
2. Differentiate between the dissociation constant and equilibrium constant for the dissociation of a weak acid, HA, in aqueous solution.
3. Why isn't the pH at the equivalence point always equal to 7 in a neutralization titration? When would it be 7?

4. What is the pK_a of an acid whose K_a is 7.5×10^{-5}?
5. Why must two electrodes be used to make an electrical measurement such as pH?
6. What is a buffer solution?
7. The pH at one half the equivalence point in an acid-base titration was found to be 6.57. What is the value of K_a for this unknown acid?
8. If 30.15 mL of 0.0995 *M* NaOH is required to neutralize 0.216 g of an unknown acid, HA, what is the molecular weight of the unknown acid?
9. If K_a is 1.85×10^{-5} for acetic acid, calculate the pH at one half the equivalence point and at the equivalence point for a titration of 50 mL of 0.100 *M* acetic acid with 0.100 *M* NaOH.

Name ______________________ Desk ______________

Date ____________ Laboratory Instructor ______________

Unknown no. or letter ______________

REPORT SHEET | EXPERIMENT

Determination of the Dissociation Constant of a Weak Acid

22

Typical student data; results will vary from sample to sample.

B. Standardization of Sodium Hydroxide (NaOH) Solution

	Trial 1	*Trial 2*	*Trial 3*
Mass of bottle + KHP	28.4809 g	27.6812 g	26.7978 g
Mass of bottle	27.6812 g	26.7978 g	26.0447 g
Mass of KHP used	0.7997 g	0.8834 g	0.7531 g
Final buret reading	39.85 mL	44.33 mL	39.65 mL
Initial buret reading	0.00 mL	0.19 mL	2.07 mL
mL of NaOH used	39.85 mL	44.14 mL	37.58 mL
Molarity of NaOH	0.09827 *M*	0.09801 *M*	0.09814 *M*

Average molarity (show calculations and standard deviation) 0.09814 *M*

Standard deviation (see Experiment 8) 1.3×10^{-4}

$$M = \frac{0.7997 \times 10^3\text{mg}}{(39.85\text{ mL})(204.2\text{ mg/mmol})} = 0.09827\text{ mmol/mL}$$

$$M = \frac{0.8834 \times 10^3\text{ mg}}{(44.14\text{ mL})(204.2\text{ mg/mmol})} = 0.09801\text{ mmol/mL}$$

$$M = \frac{0.7531 \times 10^3\text{ mg}}{(37.58\text{ mL})(204.2\text{ mg/mmol})} = 0.09814\text{ mmol/mL}$$

$$\text{ave} = \frac{0.09827 + 0.09801 + 0.09814}{3} = 0.09814$$

$$SD = \sqrt{\frac{(1.3 \times 10^{-4})^2 + (1.3 \times 10^{-4})^2 + (0)^2}{2}}$$

$$= 1.3 \times 10^{-4}$$

C. Determination of pK_a of Unknown Acid

First determination		*Second determination*		*Third determination*	
mL NaOH	*pH*	*mL NaOH*	*pH*	*mL NaOH*	*pH*
0	3.08	0		0	
0.50	3.35				
2.45	3.80				
3.70	4.00				
4.42	4.05				
5.88	4.20				
7.31	4.28				
8.91	4.39				
10.75	4.49				
12.88	4.62				
14.61	4.72				
15.83	4.78				
17.45	4.88				
19.95	5.05				
20.98	5.15		SIMILAR DATA		
22.43	5.31				
23.78	5.50				
24.45	5.65				
24.77	5.75				
25.13	6.00				
25.82	6.30				
26.03	6.75				
26.20	7.45				
26.32	8.85				
27.01	10.40				
27.38	10.55				
28.25	10.85				
28.98	10.95				
30.18	11.05				
31.50	11.15				
32.78	11.24				

Volume at equivalence point 26.30 mL 26.40 mL 26.40 mL

Volume at one half equivalence point 13.15 mL 13.20 mL 13.20 mL

pK_a 4.65 pK_a 4.65 pK_a 4.60

K_a 2.23×10^{-5} K_a 2.23×10^{-5} K_a 2.51×10^{-5}

Average K_a (show calculations) 2.32×10^{-5} Standard deviation of K_a 0.16×10^{-5}

$$\frac{(2.23 + 2.23 \times 2.51)}{3} \times 10^{-5} = 2.32 \times 10^{-5}$$

D. Concentration of Unknown Acid

	Trial 1	*Trial 2*	*Trial 3*
Volume of unknown acid	25.0 mL	25.0 mL	25.0 mL
Average molarity of NaOH from above	0.09814	0.09814	0.09814
mL of NaOH at equivalence point	26.3 mL	26.4 mL	26.4 mL
Molarity of unknown acid	0.103 M	0.104 M	0.104 M

Average molarity (show calculations) 0.104 *M* Standard deviation 0.7×10^{-3}

$$M = \frac{(26.3 \text{ mL})(0.09814\ M)}{(25.0 \text{ mL})} = 0.103$$

$$M = \frac{(26.4 \text{ mL})(0.09814\ M)}{(25.0 \text{ mL})} = 0.104$$

$$M_{\text{ave}} = \frac{0.103 + 0.104 + 0.104}{3} = 0.104$$

$$SD = \sqrt{\frac{(0.001)^2 + (0)^2 + (0)^2}{2}}$$

$$= 0.7 \times 10^{-3}$$

QUESTIONS

1. What are the largest sources of error in this experiment?

   Reading the buret and calibrating the pH meter

2. What is the pH of the solution obtained by mixing 60.00 mL of 0.250 *M* HCl and 60.00 mL of 0.125 *M* NaOH?

$$NaOH + HCl \longrightarrow + NaCl + H_2O$$

mmol HCl = 60.00 mL × 0.250 mmol/mL = 15.0 mmol

mmol NaOH = 60.00 mL × 0.125 mmol/mL = 7.50 mmol; HCl is in excess

M_{HCl} = (15.0 mmol − 7.50 mmol)/120.00 mL = 0.0625 *M* HCl

$\therefore [H^+] = 0.0625\ M$ and pH = 1.20

3. What is the pH of a solution that is 0.75 *M* in sodium acetate and 0.50 *M* in acetic acid? (K_a for acetic acid is 1.85×10^{-5}.)

$$HOAc \rightleftharpoons H^+ + OAc \qquad K_a = \frac{[H^+][OAc^-]}{[HOAc]} \qquad [H^+] = \frac{[HOAc]K_a}{[OAc^-]}$$

$$[H^+] = \frac{(0.50\ M)(1.85 \times 10^{-5})}{(0.75\ M)} = 1.23 \times 10^{-5}\ M$$

pH = 4.91

4. Calculate the pH of a solution prepared by mixing 15.0 mL of 0.50 *M* NaOH and 30.0 mL of 0.50 *M* benzoic acid solution. (Benzoic acid is monoprotic; its dissociation constant is 6.5×10^{-5}.)
   NaOH + HBz = NaBz + H_2O; 15.0 mmol HBz and 7.5 mmol NaOH yield 45.0 mL of solution that contains 7.5 mmol NaBz. The acid is half neutralized; hence pH = pK_a = $-\log(6.5 \times 10^{-5})$
   pH = 4.19

5. K_a for hypochlorous acid, HClO, is 3.0×10^{-8}. Calculate the pH after 10.0, 20.0, 30.0, and 40.0 mL of 0.100 *M* NaOH have been added to 40.0 mL of 0.100 *M* HClO.
   NaOH + HClO $\longrightarrow$ NaClO + H_2O
   After 10.0 mL:
   mmol HClO = (40.0 mL)(0.100 mmol/mL) = 4.00 mmol
   mmol NaOH = (10.0 mL)(0.100 mmol/mL) = 1.00 mmol
   This solution will contain 3.00 mmol HClO in 50.0 mL and 1.00 mL NaClO in 50.0 mL solution.

$$[H^+] = \frac{[HClO]}{[NaClO]} 3.0 \times 10^{-8} = \frac{[3.00 \text{ mmol}/50 \text{ mL}]}{[1.00 \text{ mmol}/50 \text{ mL}]} 3.0 \times 10^{-8} = 9.0 \times 10^{-8}\,M \quad \text{pH} = 7.05$$

After 20 mL, pH = pK_a = 7.52; After 30.0 mL, pH = 8.0; After 40.0 mL, solution contains only NaClO. Then the hydrolysis of ClO^- needs to be considered as follows:

[NaClO] = 4.00 mmol/80.0 mL = 0.0500 *M*

$$\begin{array}{lcl} OCl^- + H_2O & \rightleftarrows & HOCl + OH^- \\ 0.0500 - X & & X \qquad X \end{array}$$

$$K_b = \frac{K_w}{K_a} = \frac{10^{-14}}{3.0 \times 10^{-8}} = 3.3 \times 10^{-7} = \frac{[HOCl][OH^-]}{[OCl^-]}$$

$$3.3 \times 10^{-7} = \frac{x^2}{[0.0500 - x]} \simeq \frac{x^2}{0.0500}$$ because x is small compared to 0.05

$$\therefore x^2 = 1.6 \times 10^{-8}$$
$$x = 1.26 \times 10^{-4} = [OH^-]$$
$$\text{pOH} = 3.90$$
$$\text{pH} = 14.0 - 3.90 = 10.1$$

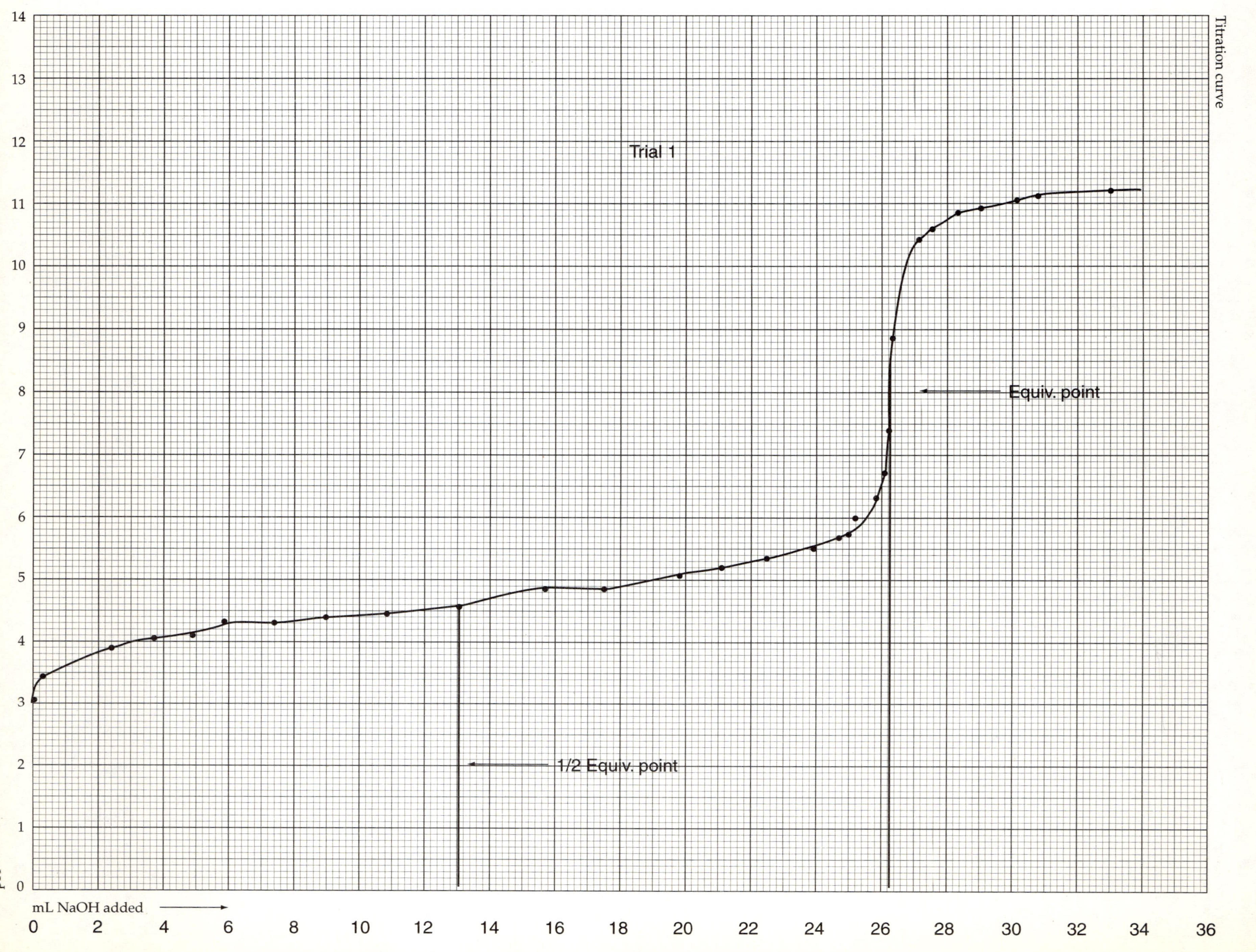

Titration curve
Trial 1
Equiv. point
1/2 Equiv. point
mL NaOH added
pH

Titration curve

pH →

14 13 12 11 10 9 8 7 6 5 4 3 2 1 0

mL NaOH added →

Titration curve
14
13
12
11
10
9
8
7
6
5
4
3
2
1
0
pH
mL NaOH added

NOTES AND CALCULATIONS

Experiment

Titration Curves of Polyprotic Acids

OBJECTIVE

To become familiar with consecutive equilibria, acid dissociation constants, and molarity.

APPARATUS AND CHEMICALS

Apparatus

pH meter with electrodes
100-mL beakers (3)
600-mL beaker
25-mL pipet
iron ring and wire gauze
Bunsen burner
balance
150-mL beaker
1-pint bottle with rubber stopper
buret, buret clamp, and ring stand
weighing bottle
wash bottle

Chemicals

sodium hydroxide solution (0.1 *M* or 10 *M* stock solution)
unknown solution of a polyprotic acid (approximately 0.1 *M* solution)
potassium hydrogen phthalate
phenolphthalein indicator solution
standard buffer solution

DISCUSSION

In this experiment, an acid-base titration will be performed in order to determine the dissociation constants of an unknown polyprotic acid, H_nA. Consider the triprotic acid H_3PO_4. It undergoes the following dissociations in aqueous solution:

$$H_3PO_4 + H_2O \rightleftharpoons H_2PO_4^- + H_3O^+ \qquad K_{a1} = \frac{[H_2PO_4^-][H_3O^+]}{[H_3PO_4]} \qquad [1]$$

$$H_2PO_4^- + H_2O \rightleftharpoons HPO_4^{2-} + H_3O^+ \qquad K_{a2} = \frac{[HPO_4^{2-}][H_3O^+]}{[H_2PO_4^-]} \qquad [2]$$

$$HPO_4^{2-} + H_2O \rightleftharpoons PO_4^{3-} + H_3O^+ \qquad K_{a3} = \frac{[PO_4^{3-}][H_3O^+]}{[HPO_4^{2-}]} \qquad [3]$$

The acid H_3PO_4 possesses three dissociable protons, and for this reason it is termed a triprotic acid. If you were to perform a titration of H_3PO_4 with NaOH, the following reactions would occur in turn:

$$H_3PO_4 + NaOH \rightleftharpoons NaH_2PO_4 + H_2O \qquad [4]$$

$$NaH_2PO_4 + NaOH \rightleftharpoons Na_2HPO_4 + H_2O \qquad [5]$$

$$Na_2HPO_4 + NaOH \rightleftharpoons Na_3PO_4 + H_2O \qquad [6]$$

The resultant titration curve, when plotted as pH versus milliliters of NaOH added, would be similar to that shown in Figure 23.1. At the point at

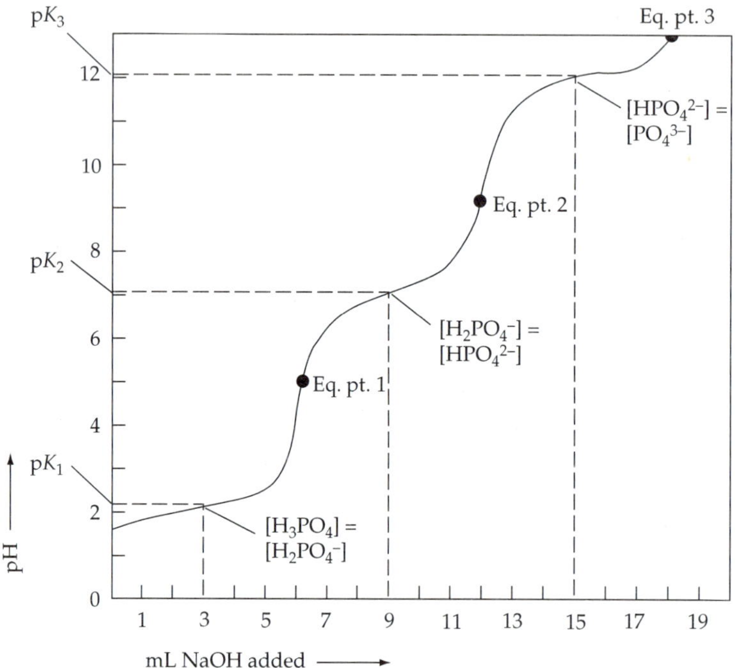

▲ **FIGURE 23.1** Titration curve for titration of H_3PO_4 with NaOH.

which half of the protons in the first dissociation step of H_3PO_4 have been titrated with NaOH (i.e., halfway to the first equivalence point), the H_3PO_4 concentration is equal to the $H_2PO_4^-$ concentration. Substituting $[H_3PO_4] = [H_2PO_4^-]$ into the expression for K_{a1} (Equation [1]) yields $K_{a1} = [H_3O^+]$, or $pH = pK_{a1}$ at this point.

Similarly, at one half the second equivalence point, one half of the $H_2PO_4^-$ has been neutralized and $[H_2PO_4^-] = [HPO_4^{2-}]$. Substituting this into the expression for K_{a2} (Equation [2]) yields $K_{a2} = [H_3O^+]$, or $pK_{a2} = pH$ at this point.

In the same manner, at one-half the third equivalence point, $[HPO_4^{2-}] = [PO_4^{3-}]$. Substituting this into the expression for K_{a3} (Equation [3]), we obtain the expression $K_{a3} = [H_3O^+]$, or $pK_{a3} = pH$ (see Figure 23.1).

The same type of result is obtained for any polyprotic acid. If a titration of the acid is performed with a pH meter, the dissociation constants may be obtained from titration curves as long as the dissociation constants exceed the ion product of water, which you should recall is 10^{-14} for the reaction

$$2H_2O \rightleftharpoons H_3O^+ + OH^-$$

In practice, if the acidity of the acid being studied approaches that of water, as in the case for the titration of the third proton of H_3PO_4 for which K_{a3} is 4.2×10^{-13}, it is difficult to determine the dissociation constant in this manner. Thus for H_3PO_4, both K_{a1} and K_{a2} are readily obtained in this way, but K_{a3} is not.

In this experiment you will determine the dissociation constants K_{a1} and K_{a2} of a diprotic acid and its molarity. Recall that the definition for molarity is

$$\text{molarity } (M) = \frac{\text{number of moles solute}}{\text{liter solution}} = \frac{\text{mmoles solute}}{\text{mL solution}}$$

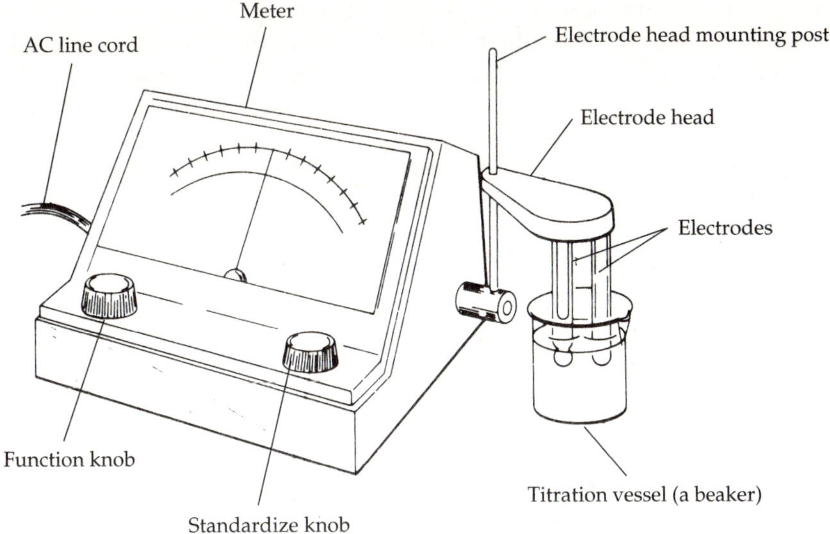

▲ **FIGURE 23.2** An analog pH meter.

Operation of the pH Meter

To measure the pH during the course of the titration, we shall use an electronic instrument called a pH meter. This device consists of a meter and two electrodes, as illustrated in Figure 23.2.

The main variations among different pH meters involve the positions of the control knobs and the types of electrodes and electrode-mounting devices. The measurement of pH requires two electrodes: a sensing electrode that is sensitive to H_3O^+ concentrations, and a reference electrode. This is because the pH meter is really just a voltmeter that measures the electrical potential of a solution. Typical sensing and reference electrodes are illustrated in Figure 23.3.

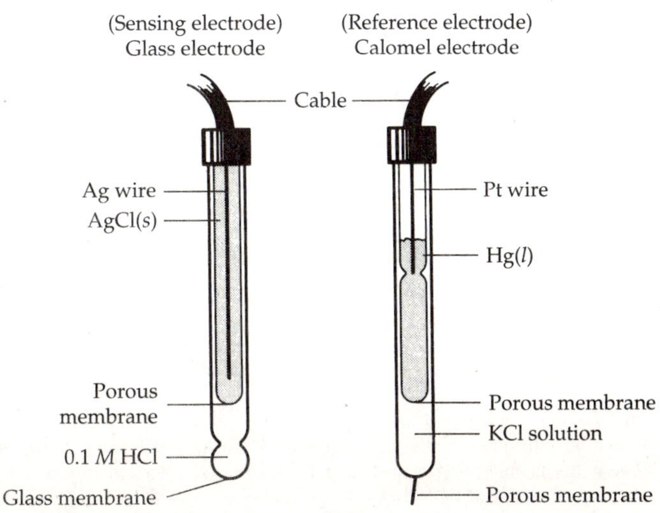

▲ **FIGURE 23.3** A typical sensing and reference electrode.

The reference electrode is an electrode that develops a known potential that is essentially independent of the contents of the solution into which it is placed. The glass electrode is sensitive to the H_3O^+ concentration of the solution into which it is placed; its potential is a function of $[H_3O^+]$. It operates by transport of H_3O^+ ions through the glass membrane. This can be more precisely described, but for our purposes here it is sufficient for you to understand that two electrodes are required. These two electrodes are sometimes combined into an electrode called a combination electrode, which appears to be a single electrode. The combination electrode, however, does contain both a reference and a sensing electrode.

Preliminary Operations with the pH Meter

1. Obtain a buffer solution of known pH.
2. Plug in the pH meter to line current and allow at least 10 min for warm-up. It should be left plugged in until you are completely finished with it. *This does not apply to battery-operated meters.*
3. Turn the function knob on the pH meter to the standby position.
4. *Prepare the electrodes.* Make certain that the solution in the reference electrode extends well above the internal electrode. If it does not, ask your instructor to fill it with saturated KCl solution. Remove the rubber tip and slide down the rubber collar on the reference electrode. Rinse the outside of the electrodes well with distilled water.
5. *Standardize the pH meter.* Carefully immerse the electrodes in the buffer solution contained in a small beaker. *Remember that the glass electrode is very fragile; it breaks easily!* DO NOT touch the bottom of the beaker with the electrodes! Turn the function knob to "read" or "pH." Turn the standardize knob until the pH meter indicates the exact pH of the buffer solution. Wait 5 s to be certain that the reading remains constant. *Once you have standardized the* pH *meter, don't readjust the standardize knob.* Turn the function knob to standby. Carefully lift the electrodes from the buffer and rinse them with distilled water. The pH meter is now ready to use to measure pH.

Instructor: Explain the proper use of electrodes

RECORD ALL DATA DIRECTLY ONTO THE REPORT SHEETS.

PROCEDURE

A. Preparation of Approximately 0.100 *M* Sodium Hydroxide (NaOH)*

Heat 500 mL of distilled water to boiling in a 600-mL flask† and, *after cooling under the water tap,* transfer to a 1-pint bottle fitted with a rubber stopper.‡ Add 3 mL of stock solution of carbonate-free NaOH (approximately 10 *M*) and shake vigorously for at least 1 min.

Instructor: If you do not provide your students with standardized 0.1 *M* NaOH, this experiment requires two lab periods.

Preparation of a Buret for Use Clean a 50-mL buret with soap solution and a buret brush and thoroughly rinse with tap water. Then rinse with at least

*Time may be saved in this experiment if the students are provided with an approximately 0.1 *M* NaOH solution, or if they use the NaOH solution that they standardized in Experiment 19.

†The water is boiled to remove carbon dioxide, which would react with the NaOH and change its molarity. (Recall that $H_2O + CO_2 \longrightarrow ?$)

‡A rubber stopper should be used for a bottle containing NaOH solution. A strongly alkaline solution tends to cement a glass stopper so firmly that it is difficult to remove.

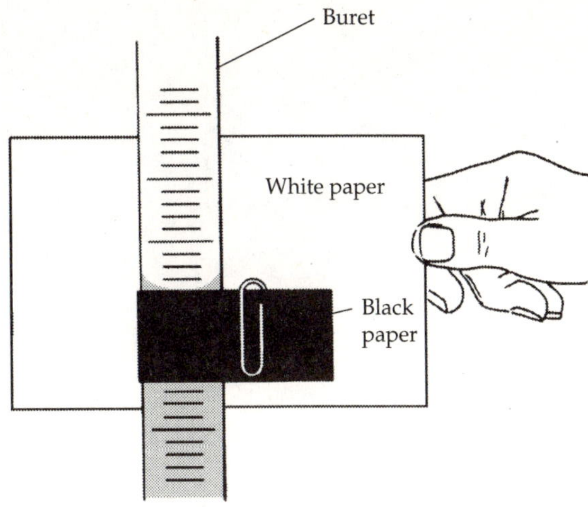

▲ **FIGURE 23.4** Reading a buret.

five 10-mL portions of distilled water. The water must run freely from the buret without leaving any drops adhering to the sides. Make sure that the buret does not leak and that the stopcock turns freely.

Reading a Buret All liquids, when placed in a buret, form a curved meniscus at their upper surfaces. In the case of water or water solutions, this meniscus is concave (Figure 23.4), and the most accurate buret readings are obtained by observing the position of the lowest point on the meniscus on the graduated scales.

To avoid parallax errors when taking readings, the eye must be on a level with the meniscus. Wrap a strip of paper around the buret and hold the top edges of the strip evenly together. Adjust the strip so that the front and back edges are in line with the lowest part of the meniscus and take the reading by estimating to the nearest tenth of a marked division (0.01 mL). A simple way of doing this for repeated readings on a buret is illustrated in Figure 23.4.

B. Standardization of Sodium Hydroxide (NaOH) Solution

Prepare about 400 to 450 mL of CO_2-free water by boiling for about 5 min. Weigh from a weighing bottle (your lab instructor will show you how to use a weighing bottle if you don't already know) triplicate samples of between 0.4 and 0.6 g each of pure potassium hydrogen phthalate (KHP) into three separate 250-mL Erlenmeyer flasks; accurately weigh to four significant figures. Do not weigh the flasks. Record the masses and label the three flasks in order to distinguish among them. Add to each sample about 100 mL of distilled water that has been freed from CO_2 by boiling and warm gently with swirling until the salt is completely dissolved. Add to each flask two drops of phenolphthalein indicator solution.

Instructor: Demonstrate use of weighing bottle.

Rinse the previously cleaned buret with at least four 5-mL portions of the approximately 0.100 *M* NaOH solution that you have prepared. Discard each portion. *Do not return any of the washings to the bottle.* Completely fill the buret with the solution and remove the air from the tip by running out some of the liquid into an empty beaker. Make sure that the lower part of the

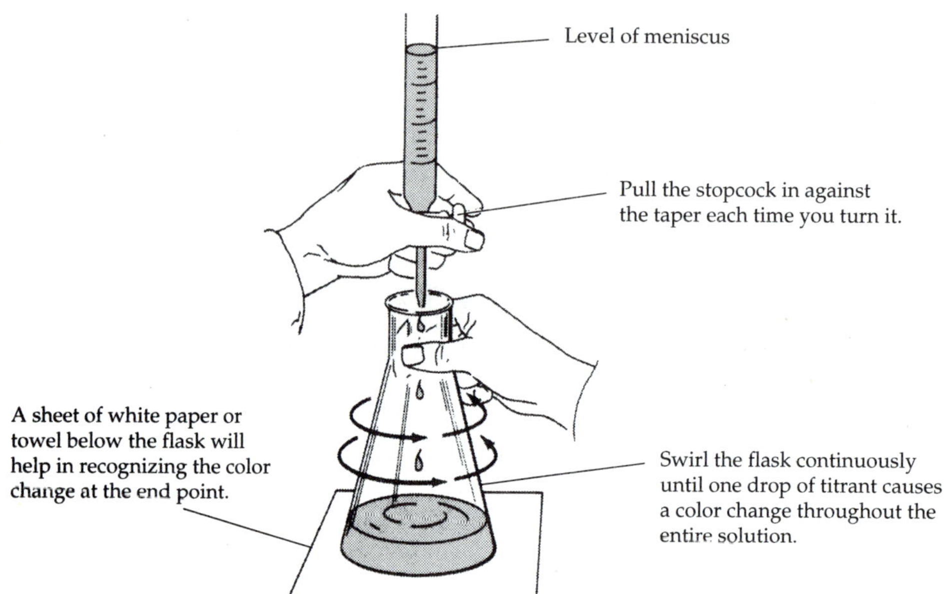

▲ **FIGURE 23.5** Titration procedure.

meniscus is at the zero mark or slightly lower. Allow the buret to stand for at least 30 s before reading the exact position of the meniscus. Remove any hanging drop from the buret tip by touching it to the side of the beaker used for the washings. Record the initial buret reading.

Slowly add the NaOH solution to one of your flasks of KHP solution while gently swirling the contents of the flask, as illustrated in Figure 23.5. As the NaOH solution is added, a pink color appears where the drops of the base come in contact with the solution. This coloration disappears with swirling. As the end point is approached, the color disappears more slowly, at which time the NaOH should be added drop by drop. It is most important that the flask be swirled constantly throughout the entire titration. The end point is reached when one drop of the NaOH solution turns the entire solution in the flask from colorless to pink. The solution should remain pink when it is swirled. Allow the titrated solution to stand for at least 1 min so the buret will drain properly. Remove any hanging drop from the buret tip by touching it to the side of the flask and wash down the sides of the flask with a stream of water from the wash bottle. Record the buret reading. Repeat this procedure with the other two samples.

From the data you obtain in the three titrations, calculate the molarity of the NaOH solution to four significant figures.

The three determinations should agree within 1.0%. If they do not, the standardization should be repeated until agreement is reached. The average of the three acceptable determinations is taken as the molarity of the NaOH. Calculate the standard deviation of your results. SAVE your standardized solution for the determination of the acid dissociation constants.

C. Determination of the Acid Dissociation Constants and the Molarity of the Unknown Acid

Titrate three separate 25-mL aliquots of the unknown acid in three separate 100-mL beakers using a pH meter. Plot the titration curves as shown in Figure 23.1. From the volumes at the two equivalence points, calculate the volumes halfway to each equivalence point and determine the pH's at these two points. These two pH's are equal to the two pK_a's for the unknown acid. Record these pK_a's and calculate the K_a's. Calculate the molarity of your unknown acid from the first equivalence point volume. For this calculation, recall that at the first equivalence point, moles acid equal moles base so that (mL acid)(M acid) = (mL base)(M base) at this point.

(HINT: You can save time if you do your first titration rapidly so that you know the approximate volumes of the equivalence points; then you can do the next two titrations with large-volume increments away from the equivalence points and small-volume increments near the equivalence points.)

REVIEW QUESTIONS

Before beginning this experiment in the laboratory, you should be able to answer the following questions:

1. What is a polyprotic acid?
2. If 16.5 mL of 0.120 M NaOH is required to reach the first equivalence point of a solution of citric acid ($H_3C_6H_5O_7$), how many mL of NaOH are required to completely neutralize this solution?
3. How many mmol of NaOH will react completely with 50 mL of 3.6 M $H_2C_2O_4$?
4. How many moles of H_3O^+ are present in 250 mL of a 0.15 M solution of H_2SO_4?
5. Why is it necessary to standardize a pH meter?
6. If the pH at one half the first and second equivalence points of a dibasic acid is 4.20 and 7.34, respectively, what are the values for pK_{a1} and pK_{a2}? From pK_{a1} and pK_{a2} calculate the K_{a1} and K_{a2}.
7. Derive the relationship between pH and pK_a at one half the equivalence point for the titration of a weak acid with a strong base.
8. Could K_b for a weak base be determined in the same way that K_a for a weak acid is determined in this experiment?
9. What is the relationship of the successive equivalence-point volumes in the titration of a polyprotic acid?
10. If the pK_{a1} of a diprotic acid is 2.90, what is the pH of a 0.10 M solution of this acid that has been one quarter neutralized?

NOTES AND CALCULATIONS

Name ______________________ Desk ______________________

Date ______________________ Laboratory Instructor ______________________

REPORT SHEET | EXPERIMENT

Titration Curves of Polyprotic Acids | 23

B. Standardization of Sodium Hydroxide (NaOH) Solution

	Trial 1	*Trial 2*	*Trial 3*
Mass of bottle + KHP	7.3033 g	10.3538 g	10.3029 g
Mass of bottle	7.0002 g	10.0004 g	10.0000 g
Mass of KHP used	0.3031 g	0.3534 g	0.3029 g
Final buret reading	16.50 mL	45.00 mL	25.70 mL
Initial buret reading	0.00 mL	27.70 mL	9.20 mL
mL of NaOH used	16.50 mL	17.30 mL	16.50 mL
Molarity of NaOH	0.08996 *M*	0.08967 *M*	0.08990 *M*

Average molarity (show calculations) 0.08984

$$M_{ave} = \frac{0.08996 + 0.08967 + 0.08990}{3} = 0.08984$$

(N.B. different base used below)

Standard deviation (show calculations) 1.53×10^{-4}

$$SD = \sqrt{\frac{(0.00012)^2 + (0.00017)^2 + (0.0006)^2}{2}} = 1.53 \times 10^{-4}$$

C. Determination of the Acid Dissociation Constants and the Molarity of the Unknown Acid

	Trial 1	*Trial 2*	*Trial 3*
Volume of unknown acid	25 mL	25 mL	25 mL
Molarity of NaOH from above	0.1065 *M*	0.1092 *M*	0.1097 *M*
mL of NaOH at eq. point 1	22.7 mL	22.8 mL	22.8 mL
mL of NaOH at eq. point 2	45.4 mL	45.6 mL	45.3 mL
Molarity of unknown acid	0.0967 *M*	0.0996 *M*	0.1000 *M*

Average molarity <u>0.0988 *M*</u>

at the second equiv. point:

mmol NaOH = 2 (mmol acid)

45.4 mL × 0.1065 mmol/mL = 2(25.0 mL × M_{acid})

M_{acid} = 2(45.4 mL × 0.1065 mmol/mL)/25.0 mL = 0.0967 M_1 etc.

Standard deviation (show calculations) <u>1.8×10^{-3}</u>

$$M_{ave} = \frac{0.0967 + 0.0996 + 0.1000}{3} = 0.0988$$

$$SD = \sqrt{\frac{(0.0021)^2 + (0.0008)^2 + (0.0012)^2}{2}} = 1.8 \times 10^{-3}$$

Determination of pK_a Values of Unknown Acid

First determination		*Second determination*		*Third determination*	
mL NaOH	*pH*	*mL NaOH*	*pH*	*mL NaOH*	*pH*
0.0	1.55	0		0	
0.60	1.60				
1.00	1.60				
1.50	1.60				
2.00	1.60				
2.50	1.60				
3.00	1.65				
3.50	1.68				
4.00	1.70				
4.50	1.70				
5.00	1.75				
6.00	1.85				
7.00	1.88				
8.00	1.92				
8.90	2.00	SIMILAR DATA		SIMILAR DATA	
9.90	2.00				
10.90	2.10				
11.90	2.15				
12.90	2.20				
13.90	2.30				
15.00	2.38				
16.00	2.48				
17.00	2.55				
18.00	2.65				
19.00	2.80				
20.00	3.20				
22.00	3.70				
22.50	4.15				

First determination		*Second determination*		*Third determination*	
mL NaOH	*pH*	*mL NaOH*	*pH*	*mL NaOH*	*pH*
23.0	5.20				
24.0	5.90				
25.0	6.10				
26.0	6.30				
27.2	6.50				
28.3	6.60				
29.0	6.70				
30.0	6.80				
31.0	6.90				
32.0	7.00				
33.0	7.05				
34.0	7.15				
35.0	7.20				
36.0	7.30				
37.0	7.40				
38.0	7.48				
39.0	7.55				
40.0	7.65				
41.0	7.80				
42.0	7.90				
43.0	8.10				
44.0	8.40				
45.0	9.00				
46.0	10.60				
47.0	11.15				
48.0	11.40				
49.0	11.70				
50.0	11.70				

	First	Second	Third
pK_{a1}	2.11	2.12	2.12
K_{a1}	7.76×10^{-3}	7.59×10^{-3}	7.59×10^{-3}

Average K_{a1} 7.65×10^{-3} Standard deviation 9.8×10^{-5}

	First	Second	Third
pK_{a2}	7.18	7.20	7.14
K_{a2}	6.61×10^{-8}	6.31×10^{-8}	7.24×10^{-8}

Average K_{a2} 6.72×10^{-8} Standard deviation 4.75×10^{-9}

pH of Unknown Acid

Using the measured K_1 and the concentration, calculate the pH of your unknown acid solution and compare it with that measured.

Calculated pH 1.60 Calculated pH 1.60 Calculated pH ______

Measured pH 1.55 Measured pH 1.55 Measured pH ______

Average: Calculated pH 1.60 Measured pH 1.55

Identity of Unknown Acid

Consult a reference book such as the *Handbook of Chemistry and Physics,* and using tables of acid dissociation constants contained therein, identify your unknown acid. phosphoric acid

Handbook lists: $pK_{a1} = 2.12$
$pK_{a2} = 7.21$
pK_{a3} is difficult to determine this way.

Also see Appendix E.

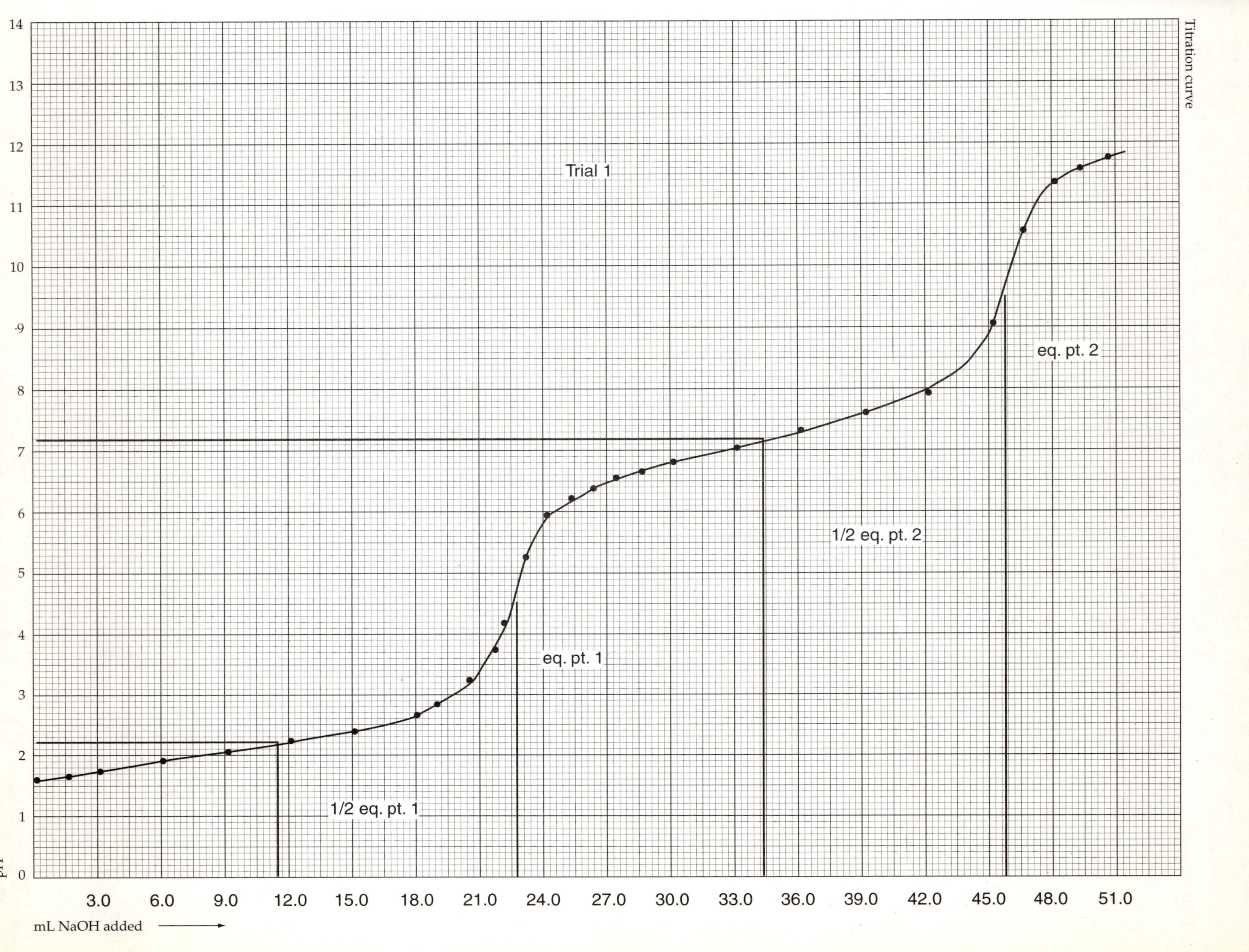
Titration curve
Trial 1
eq. pt. 2
1/2 eq. pt. 2
eq. pt. 1
1/2 eq. pt. 1
14
13
12
11
10
9
8
7
6
5
4
3
2
1
0
pH
3.0
6.0
9.0
12.0
15.0
18.0
21.0
24.0
27.0
30.0
33.0
36.0
39.0
42.0
45.0
48.0
51.0
mL NaOH added

Titration curve

Titration curve

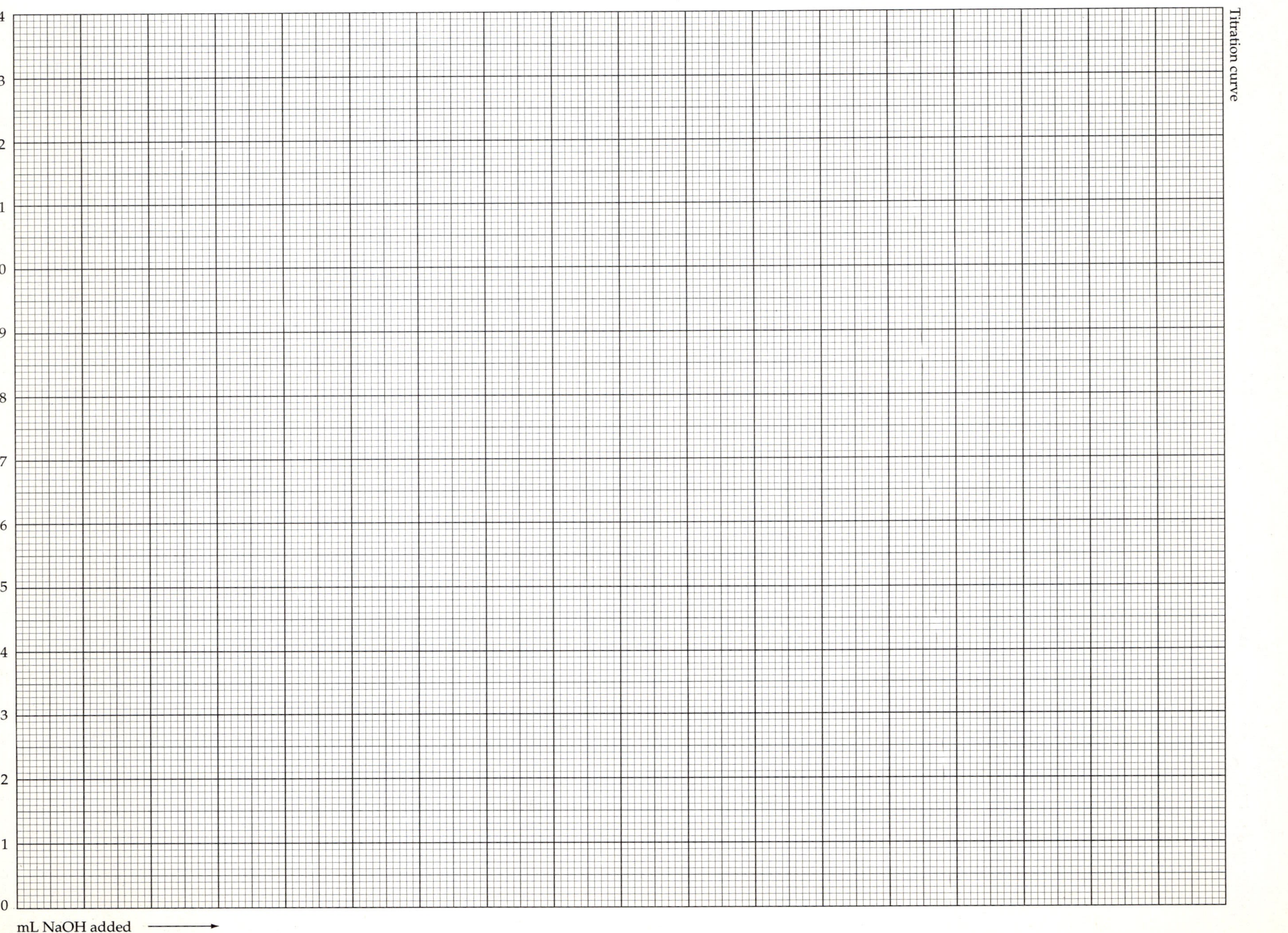

NOTES AND CALCULATIONS

Experiment

Hydrolysis of Salts and pH of Buffer Solutions

OBJECTIVE

To learn about the concept of hydrolysis and to gain familiarity with acid-base indicators and the behavior of buffer solutions.

APPARATUS AND CHEMICALS

Apparatus

500-mL Erlenmeyer flask
150-mL beakers (2)
10- and 100-mL graduated cylinders
test tubes (6)
test-tube rack
wire gauze
stirring rods (2)
pH meter
balance
1-mL pipet
Bunsen burner and hose
ring stand and iron ring
plastic wash bottle

Chemicals

$NaC_2H_3O_2 \cdot 3H_2O$
0.1 *M* $ZnCl_2$
0.1 *M* NH_4Cl
0.1 *M* $KAl(SO_4)_2$
0.1 *M* Na_2CO_3
0.1 *M* NaCl
0.1 *M* $NaC_2H_3O_2$
6.0 *M* HCl
3.0 *M* $HC_2H_3O_2$
6.0 *M* NaOH
dropper bottles of:
methyl orange
methyl red
bromothymol blue
phenolphthalein
alizarin yellow-R
standard buffer solution (pH 4.5)

DISCUSSION

We expect solutions of substances such as HCl and HNO_2 to be acidic and solutions of NaOH and NH_3 to be basic. However, we may be somewhat surprised at first to discover that aqueous solutions of some salts such as sodium nitrite, $NaNO_2$, and potassium acetate, $KC_2H_3O_2$, are basic, whereas others such as NH_4Cl and $FeCl_3$ are acidic. Recall that salts are the products formed in neutralization reactions of acids and bases. For example, when NaOH and HNO_2 (nitrous acid) react, the salt $NaNO_2$ is formed:

$$NaOH(aq) + HNO_2(aq) \longrightarrow NaNO_2(aq) + H_2O(l)$$

Nearly all salts are strong electrolytes and exist as ions in aqueous solutions. Many ions react with water to produce acidic or basic solutions. The reactions of ions with water are frequently called *hydrolysis reactions*. We will see that anions such as CN^- and $C_2H_3O_2^-$ that are the conjugate bases of the weak acids HCN and $HC_2H_3O_2$, respectively, react with water to form OH^- ions. Cations such as NH_4^+ and Fe^{3+} come from weak bases and react with water to form H^+ ions.

Hydrolysis of Anions: Basic Salts

Let us consider the behavior of anions first. Anions of weak acids react with proton sources. When placed in water these anions react to some extent with water to accept protons and generate OH^- ions and thus cause the solution pH to be greater than 7. Recall that proton acceptors are Brønsted-Lowry bases. Thus, the anions of weak acids are basic in two senses: They are proton acceptors, and their aqueous solutions have pH's above 7. The nitrite ion, for example, reacts with water to increase the concentration of OH^- ions:

$$NO_2^-(aq) + H_2O(l) \rightleftharpoons HNO_2(aq) + OH^-(aq)$$

This reaction of the nitrite ion is similar to that of weak bases such as NH_3 with water:

$$NH_3(aq) + H_2O(l) \rightleftharpoons NH_4^+(aq) + OH^-(aq)$$

Thus, both NH_3 and NO_2^- are bases and as such have a basicity or base-dissociation constant, K_b, associated with their corresponding equilibria.

According to the Brønsted-Lowry theory, the nitrite ion is the conjugate base of nitrous acid. Let's consider the conjugate acid-base pair HNO_2 and NO_2^- and their behavior in water:

$$HNO_2 \rightleftharpoons H^+ + NO_2^- \qquad K_a = \frac{[H^+][NO_2^-]}{[HNO_2]}$$

$$NO_2^- + H_2O \rightleftharpoons HNO_2 + OH^- \qquad K_b = \frac{[HNO_2][OH^-]}{[NO_2^-]}$$

Multiplication of these dissociation constants yields:

$$K_a \times K_b = \left(\frac{[H^+]\cancel{[NO_2^-]}}{\cancel{[HNO_2]}}\right)\left(\frac{\cancel{[HNO_2]}[OH^-]}{\cancel{[NO_2^-]}}\right) = [H^+][OH^-] = K_w$$

where K_w is the ion-product constant of water.

Thus, the product of the acid-dissociation constant for an acid and the base-dissociation constant for its conjugate base is the ion-product constant for water:

$$K_a \times K_b = K_w = 1.0 \times 10^{-14} \qquad [1]$$

Knowing the K_a for a weak acid, we can easily find the K_b for the anion of the acid:

$$K_b = \frac{K_w}{K_a} \qquad [2]$$

By consulting a table of acid-dissociation constants, such as Appendix E, we can find that K_a for nitrous acid is 4.5×10^{-4}. Using this value, we can readily determine K_b for NO_2^-:

$$K_b = \frac{1.0 \times 10^{-14}}{4.5 \times 10^{-4}} = 2.2 \times 10^{-11}$$

We note that the stronger the acid is, the larger its K_a, and the weaker its conjugate base, the smaller its K_b. Likewise, the weaker the acid (the smaller the K_a), the stronger the conjugate base (the larger the K_b).

Anions derived from *strong acids,* such as Cl^- from HCl, do not react with water to affect the pH. Nor do Br^-, I^-, NO_3^-, SO_4^{2-}, and ClO_4^- affect the pH, for the same reason. They are spectator ions in the acid-base sense and can be described as neutral ions. Similarly, cations from strong bases, such as Na^+ from NaOH or K^+ from KOH, do not react with water to affect the pH. Hydrolysis of an ion occurs only when it can form a molecule or ion that is a weak electrolyte in the reaction with water. Strong acids and bases do not exist as molecules in dilute water solutions.

Example 24.1

What is the pH of a 0.10 *M* NaClO solution if K_a for HClO is 3.0×10^{-8}?

SOLUTION: The salt NaClO exists as Na^+ and ClO^-. The Na^+ ions are spectator ions, but ClO^- ions undergo hydrolysis to form the weak acid HClO. Let x equal the equilibrium concentration of HClO (and OH^-):

$$ClO^-(aq) + H_2O(l) \rightleftharpoons HClO(aq) + OH^-(aq)$$
$$(0.10 - x)M \qquad\qquad xM \qquad xM$$

The value of K_b for the reaction is $(1.0 \times 10^{-14})/(3.0 \times 10^{-8}) = 3.3 \times 10^{-7}$. Because K_b is so small, we can neglect x in comparison with 0.10 and thus $0.10 - x \simeq 0.10$.

$$\frac{[HClO][OH^-]}{[ClO^-]} = K_b$$

$$\frac{x^2}{0.10} = 3.3 \times 10^{-7}$$

$$x^2 = 3.3 \times 10^{-8}$$

$$x = 1.8 \times 10^{-4}\,M$$

$$pOH = 3.74$$

$$\text{and } pH = 14 - 3.74 = 10.26$$

Anions with ionizable protons such as HCO_3^-, $H_2PO_4^-$, and HPO_4^{2-} may be either acidic or basic, depending on the relative values of K_a and K_b for the ion. We will not consider such ions in this experiment.

Hydrolysis of Cations: Acidic Salts

Cations that are derived from weak bases react with water to increase the hydrogen-ion concentration; they form acidic solutions. The ammonium ion is derived from the weak base NH_3 and reacts with water as follows:

$$NH_4^+(aq) + H_2O(l) \rightleftharpoons H_3O^+(aq) + NH_3(aq)$$

This reaction is completely analogous to the dissociation of any other weak acid, such as acetic acid or nitrous acid. We can represent this acid-dissociation of NH_4^+ more simply:

$$NH_4^+(aq) \rightleftharpoons NH_3(aq) + H^+(aq)$$

Here too the acid-dissociation constant is related to the K_b of NH_3, which is the conjugate base of NH_4^+:

$$NH_3(aq) + H_2O(l) \rightleftharpoons NH_4^+(aq) + OH^-(aq)$$

Knowing the value of K_b for NH_3, we can readily calculate the acid dissociation constant from Equation [3]:

$$K_a = \frac{K_w}{K_b} \qquad [3]$$

Cations of the alkali metals (Group 1A) and the larger alkaline earth ions, Ca^{2+}, Sr^{2+}, and Ba^{2+}, do not react with water, because they come from strong bases. Thus, these ions have no influence on the pH of aqueous solutions. They are merely spectator ions in acid-base reactions. Consequently, they are described as being neutral in the acid-base sense. The cations of most other metals do hydrolyze to produce acidic solutions. Metal cations are coordinated with water molecules, and it is the hydrated ion that serves as the proton donor. The following equations illustrate this behavior for the hexaaqua iron (III) ion:

$$Fe(H_2O)_6^{3+}(aq) + H_2O(l) \rightleftharpoons Fe(OH)(H_2O)_5^{2+}(aq) + H_3O^+(aq) \qquad [4]$$

We frequently omit the coordinated water molecules from such equations. For example, Equation [4] may be written as

$$Fe^{3+}(aq) + H_2O(l) \rightleftharpoons Fe(OH)^{2+}(aq) + H^+(aq) \qquad [5]$$

Additional hydrolysis reactions can occur to form $Fe(OH)_2^+$ and even lead to the precipitation of $Fe(OH)_3$. The equilibria for such cations are often complex, and not all species have been identified. However, equations such as [4] and [5] serve to illustrate the acidic character of dipositive and tripositive ions and account for most of the H^+ in these solutions.

Summary of Hydrolysis Behavior

Whether a solution of a salt will be acidic, neutral, or basic can be predicted on the basis of the strengths of the acid and base from which the salt was formed.

1. *Salt of a strong acid and a strong base:* Examples: NaCl, KBr, and $Ba(NO_3)_2$. Neither the cation nor anion hydrolyzes, and the solution has a pH of 7.
2. *Salt of a strong acid and a weak base:* Examples: NH_4Br, $ZnCl_2$, and $Al(NO_3)_3$. The cation hydrolyzes, forming H^+ ions, and the solution has a pH less than 7.
3. *Salt of a weak acid and a strong base:* Examples: $NaNO_2$, $KC_2H_3O_2$, and $Ca(OCl)_2$. The anion hydrolyzes, forming OH^- ions, and the solution has a pH greater than 7.
4. *Salt of a weak acid and a weak base:* Examples: NH_4F, $NH_4C_2H_3O_2$, and $Zn(NO_2)_2$. Both ions hydrolyze. The pH of the solution is determined by the relative extent to which each ion hydrolyzes.

In this experiment, we will test the pH of water and of several aqueous salt solutions to determine whether these solutions are acidic, basic, or neutral. In each case, the salt solution will be 0.1 *M*. Knowing the concentration of the salt solution and the measured pH of each solution allows us to calculate K_a or K_b for the ion that hydrolyzes. Example 24.2 illustrates such calculations.

EXAMPLE 24.2

Calculate K_b for OBr^- if a 0.10 *M* solution of NaOBr has a pH of 10.85.

SOLUTION: The spectator ion is Na^+. Alkali metal ions do not react with water and have no influence on pH. The ion OBr^- is the anion of a weak acid and thus reacts with water to produce OH^- ions:

$$OBr^- + H_2O \rightleftharpoons HOBr + OH^-$$

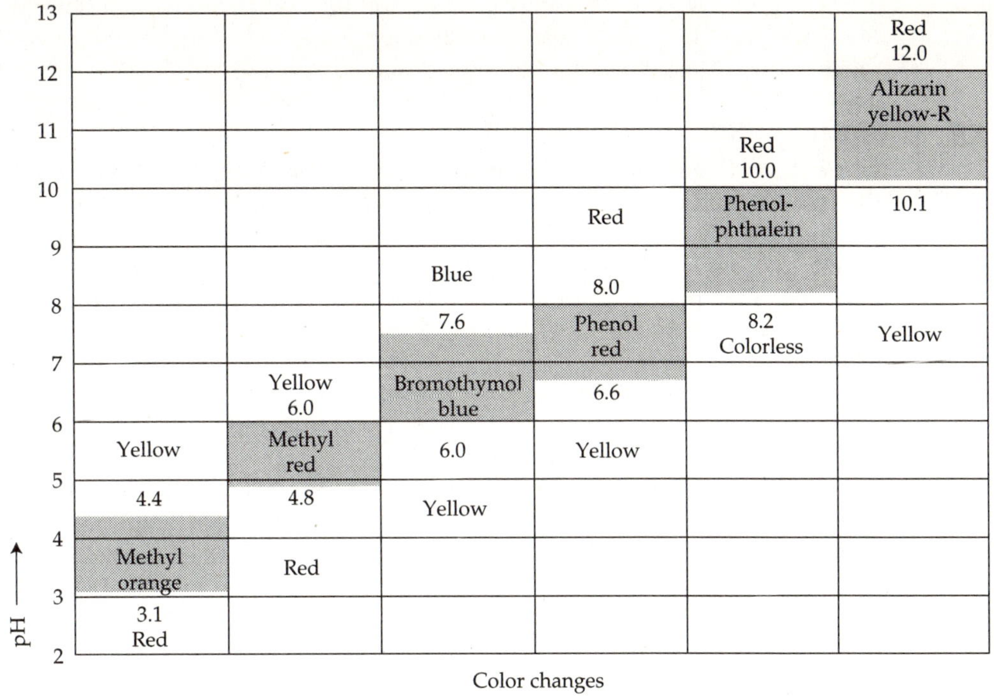

pH → 2 3 4 5 6 7 8 9 10 11 12 13

Methyl orange	Methyl red	Bromothymol blue	Phenol red	Phenolphthalein	Alizarin yellow-R
Red	Red	Yellow	Yellow	Colorless	Yellow
3.1	4.8	6.0	6.6	8.2	10.1
Methyl orange	Methyl red	Bromothymol blue	Phenol red	Phenol-phthalein	Alizarin yellow-R
4.4	6.0	7.6	8.0	10.0	12.0
Yellow	Yellow	Blue	Red	Red	Red

Color changes

▲ **FIGURE 24.1** The color behavior of indicators.

and the corresponding expression for the base dissociation constant is

$$K_b = \frac{[HOBr][OH^-]}{[OBr^-]} \quad [6]$$

If the pH is 10.85, then

$$pOH = 14.00 - 10.85 = 3.15$$

and

$$[OH^-] = \text{antilog}\,(-3.15) = 7.1 \times 10^{-4}\, M$$

The concentration of HOBr that is formed along with OH^- must also be $7.1 \times 10^{-4}\, M$. The concentration of OBr^- that has not hydrolyzed is

$$[OBr^-] = 0.10\, M - 0.00071\, M \simeq 0.10\, M$$

Substituting these values into Equation [6] for K_b yields

$$K_b = \frac{[7.1 \times 10^{-4}][7.1 \times 10^{-4}]}{[0.10]}$$

$$= 5.0 \times 10^{-6}$$

We will use a set of indicators to determine the pH of various salt solutions. The dark areas in Figure 24.1 denote the transition ranges for the indicators you will use.

We will generally find that the solutions that we test will be more acidic than we would predict them to be. A major reason for this increased acidity is the occurrence of CO_2 dissolved in the solutions. CO_2 reacts with water to generate H^+:

$$CO_2(g) + H_2O(l) \rightleftharpoons H_2CO_3(aq) \rightleftharpoons H^+(aq) + HCO_3^-(aq)$$

The solubility of CO_2 is greatest in basic solutions, intermediate in neutral ones, and least in acidic ones. Even distilled water will therefore be somewhat acidic, unless it is boiled to remove the dissolved CO_2.

Buffer Solutions

Chemists, biologists, and environmental scientists frequently need to control the pH of aqueous solutions. The effects on pH caused by the addition of a small amount of a strong acid or base to water are dramatic. The addition of a mere 0.001 mole of HCl to 1 L of water causes the pH to drop instantly from 7.0 to 3.0 as the hydronium-ion concentration increases from 1×10^{-7} to 1×10^{-3} mol/L. And, on the other hand, the addition of 0.001 mole of NaOH to 1 L of water will cause the pH to increase from 7.0 to 11.0. That life could not exist without some mechanism for controlling or absorbing excess acid or base is indicated by the narrow normal range of blood pH, 7.35 to 7.45.

The control of pH is often accomplished by use of *buffer solutions* (often simply called *buffers*. A buffer solution has the important property of resisting large changes in pH upon the addition of small amounts of strong acids or bases. A buffer solution must have two components—one that will react with H^+, and the other that will react with OH^-. The two components of a buffer solution are usually a weak acid and its conjugate base, such as $HC_2H_3O_2$-$C_2H_3O_2^-$ or NH_4^+-NH_3. Thus, buffers are often prepared by mixing a weak acid or a weak base with a salt of that acid or base. For example, the $HC_2H_3O_2$-$C_2H_3O_2^-$ buffer can be prepared by adding $NaC_2H_3O_2$ to a solution of $HC_2H_3O_2$; the NH_4^+-NH_3 buffer can be prepared by adding NH_4Cl to a solution of NH_3. By the appropriate choice of components and their concentrations, buffer solutions of virtually any pH can be made.

Let's examine how a buffer works. Consider a buffer composed of a hypothetical weak acid HX and one of its salts MX, where M^+ could be Na^+, K^+, or other cations. The acid-dissociation equilibrium in this buffer involves both the acid, HX, and its conjugate base X^-:

$$HX(aq) \rightleftharpoons H^+(aq) + X^-(aq) \qquad [1]$$

The corresponding acid-dissociation-constant expression is

$$K_a = \frac{[H^+][X^-]}{[HX]} \qquad [2]$$

Solving this expression for $[H^+]$, we have

$$[H^+] = K_a \frac{[HX]}{[X^-]} \qquad [2]$$

We see from this expression that the hydrogen-ion concentration and therefore the pH is determined by two factors: the value of K_a for the weak acid component of the buffer, and the ratios of the concentrations of the weak acid and its conjugate base, $[HX]/[X^-]$.

If OH^- ions are added to the buffered solution, they react with the acid component of the buffer:

$$OH^-(aq) + HX(aq) \longrightarrow H_2O(l) + X^-(aq) \qquad [4]$$

This reaction results in a slight decrease in the [HX] and a slight increase in the $[X^-]$, as long as the amounts of HX and X^- in the buffer are large compared to the amount of the added OH^-. In that case, the ratio $[HX]/[X^-]$ doesn't change much, and thus the change in the pH is small.

If H^+ ions are added to the buffered solution, they react with the base component of the buffer:

$$H^+(aq) + X^-(aq) \longrightarrow HX(aq) \qquad [5]$$

This reaction causes a slight decrease in the $[X^-]$ and a slight increase in the $[HX]$. Once again, as long as the change in the ratio $[HX]/[X^-]$ is small, the change in the pH will be small.

Buffers resist changes in pH most effectively when the concentrations of the conjugate acid-base pair, HX and X^-, are about the same. We see from examining Equation [3] that under these conditions their ratio is close to one, and thus the $[H^+]$ is approximately equal to K_a. For this reason we try to select a buffer whose acid form has a pK_a close to the desired pH.

Because we are interested in pH, let's take the negative logarithm of both sides of Equation [3] and obtain

$$-\log[H^+] = -\log K_a - \log\frac{[HX]}{[X^-]}$$

Because $-\log[H^+] = \text{pH}$ and $-\log[K_a] = pK_a$, we have

$$\text{pH} = pK_a - \log\frac{[HX]}{[X^-]}$$

and making use of the properties of logarithms (see Appendix A) we have

$$\text{pH} = pK_a + \log\frac{[X^-]}{[HX]} \qquad [6]$$

and in general

$$\text{pH} = pK_a + \log\frac{[\text{conjugate base}]}{[\text{weak acid}]} \qquad [7]$$

This relationship is known as the *Henderson-Hasselbalch equation.* Biochemists, biologists, and others who frequently work with buffers often use this equation to calculate the pH of buffers. What makes this equation particularly convenient is that we can normally neglect the amounts of the acid and base of the buffer that ionize. Therefore, we can use the *initial concentrations* of the acid and conjugate base components of the buffer directly in Equation [7].

EXAMPLE 24.3

What is the pH of a buffer that is 0.120 *M* in lactic acid, $HC_3H_5O_3$, and 0.100 *M* in sodium lactate, $NaC_3H_5O_3$? For lactic acid, $K_a = 1.4 \times 10^{-4}$.

SOLUTION: Because lactic acid is a weak acid, we will assume that its initial concentration is 0.120 *M* and that none of it has dissociated. We will also assume that the lactate ion concentration is that of the salt, sodium lactate, 0.100 *M*. Let x represent the concentration in mol/L of the lactic acid that dissociate. The initial and equilibrium concentrations involved in this equilibrium are

$$HC_3H_5O_3(aq) \rightleftharpoons H^+(aq) + C_3H_5O_3^-(aq)$$

	$HC_3H_5O_3(aq)$	$H^+(aq)$	$C_3H_5O_3^-(aq)$
Initial	0.120 *M*	0	0.100 *M*
Change	$-x$ *M*	$+x$ *M*	$+x$ *M*
Equilibrium	$(0.120 - x)$ *M*	x *M*	$(0.100 + x)$ *M*

The equilibrium concentrations are governed by the equilibrium expression

$$K_a = 1.4 \times 10^{-4} = \frac{[H^+][C_3H_5O_3^-]}{[HC_3H_5O_3]} = \frac{x\,(0.100 + x)}{0.120 - x}$$

Because K_a is small and the presence of a common ion, we expect x to be small relative to 0.12 and 0.10 M. Thus, our equation can be simplified to give

$$1.4 \times 10^{-4} = \frac{x\,(0.100)}{0.120}$$

Solving for x gives a value that justifies our neglecting it:

$$x = [H^+] = \left(\frac{0.120}{0.100}\right)(1.4 \times 10^{-4}) = 1.7 \times 10^{-4}\,M$$

$$pH = -\log(1.7 \times 10^{-4}) = 3.77$$

Alternatively, we could have used the Henderson-Hasselbalch equation to calculate the pH directly:

$$pH = pK_a + \log\left(\frac{[\text{conjugate base}]}{[\text{weak acid}]}\right) = 3.85 + \log\left(\frac{0.100}{0.120}\right)$$

$$= 3.85 + (-0.08) = 3.77$$

Addition of Strong Acids or Bases to Buffers

Let's consider in a quantitative way the manner in which a buffer solution responds to the addition of a strong acid or base. Consider a buffer that consists of the weak acid HX and its conjugate base X^- (from the salt NaX). When a strong acid is added to this buffer, the H^+ is consumed by the X^- to produce HX; thus [HX] increases and $[X^-]$ decreases. Whereas, when a strong base is added to the buffer, the OH^- is consumed by HX to produce X^-; in this case, [HX] decreases and X^- increases. There are basically two steps involved in calculating how the pH of the buffer responds to the addition of a strong acid or base. First consider the acid-base neutralization reaction and determine its effect on [HX] and $[X^-]$. And second, after performing this stoichiometric calculation, use K_a and the new concentrations of [HX] and $[X^-]$ to calculate the $[H^+]$. This second step in the calculation is a standard equilibrium calculation. This procedure is illustrated in Example 24.4.

EXAMPLE 24.4

A buffer is made by adding 0.120 mol $HC_3H_5O_3$ and 0.100 mol $NaC_3H_5O_3$ to enough water to make 1.00 liter of solution The pH of the buffer is 3.77 (see Example 24.3). Calculate the pH of the solution after 0.001 mol NaOH is added (neglect any volume changes).

SOLUTION: Solving this problem involves two steps.

Stoichiometric calculation: The OH^- provided by the NaOH reacts with the $HC_3H_5O_3$, the weak-acid component of the buffer. The following table summarizes the concentrations before and after the neutralization reaction:

$$HC_3H_5O_3(aq) + OH^-(aq) \longrightarrow H_2O(l) + C_3H_5O_3^-(aq)$$

Before reaction	0.120 *M*	0.001 *M*	—	0.100 *M*
Change	−0.001 *M*	−0.001 *M*	—	+0.001 *M*
After reaction	0.119 *M*	0.0 *M*	—	0.101 *M*

Equilibrium calculation: After neutralization the solution contains different concentrations for the weak acid–conjugate base pair. We next consider the proton-transfer equilibrium in order to determine the pH of the solution:

$$HC_3H_5O_3(aq) + H_2O(aq) \rightleftharpoons H_3O^+(aq) + C_3H_5O_3^-(aq)$$

Before reaction	0.119 *M*	—	0	0.101 *M*
Change	$-x$ *M*	—	$+x$ *M*	$+x$ *M*
After reaction	$(0.119 - x)$ *M*	—	x *M*	$(0.101 + x)$ *M*

$$K_a = \frac{[H_3O^+][C_3H_5O_3^-]}{[HC_3H_5O_3]} = \frac{(x)(0.101 + x)}{0.119 - x} \approx \frac{(x)(0.101)}{0.119} = 1.4 \times 10^{-4}$$

$$x = [H_3O^+] = \frac{(0.119)(1.4 \times 10^{-4})}{(0.101)} = 1.65 \times 10^{-4} M$$

$$\text{pH} = -\log(1.65 \times 10^{-4}) = 3.78$$

Note how the buffer resists a change in its pH. The addition of 0.001 mol of NaOH to a liter of this buffer results in a change in the pH of only 0.01 units, whereas the addition of the same amount of NaOH to a liter of water results in a change of pH from 7.0 to 11.0, a change of 4.0 pH units!

A. Hydrolysis of Salts

PROCEDURE

Boil approximately 450 mL of distilled water for about 10 min to expel dissolved carbon dioxide. Allow the water to cool to room temperature. While the water is boiling and subsequently cooling, add about 5 mL of unboiled distilled water to each of six test tubes. Add three drops of a different indicator to each of these six test tubes (one indicator per tube) and record the colors on the report sheet. From these colors and the data given in Figure 24.1, determine the pH of the unboiled water to the nearest pH unit. (Remember that we would expect its pH to be below 7 because of dissolved CO_2.) Empty the contents of the test tubes and rinse the test tubes three times with about 3 mL of boiled distilled water. Then pour about 5 mL of the boiled distilled water into each of the six test tubes and add three drops of each of the indicators (one indicator per tube) to each tube. Record the colors and determine the pH. Empty the contents of the test tubes and rinse each tube three times with about 3 mL of boiled distilled water.

Repeat the same procedure to determine the pH of each of the following solutions that are 0.1 *M*: NaCl, $NaC_2H_3O_2$, NH_4Cl, $ZnCl_2$, $KAl(SO_4)_2$, and Na_2CO_3. Use 5 mL of each of these solutions per test tube. Do not forget to rinse the test tubes with boiled distilled water when you go from one solution to the next.

From the pH values that you determined, calculate the hydrogen- and hydroxide-ion concentrations for each solution. Complete the tables on the report sheets and calculate the K_a or K_b as appropriate.

Dispose of chemicals in designated receptacles.

B. pH of Buffer Solutions

1. Preparation of Acetic Acid–Sodium Acetate Buffer

Weigh about 3.5 g of $NaC_2H_3O_2 \cdot 3H_2O$ to the nearest 0.01 g, record its mass, and add it to a 150-mL beaker. Using a 10-mL graduated cylinder, measure 8.8 mL of 3.0 *M* acetic acid and add it to the beaker containing the sodium acetate. Using a graduated cylinder, measure 55.6 mL of distilled water and add it to the solution of acetic acid and sodium acetate. Stir the solution until all of the sodium acetate is dissolved. The pH of this solution will be measured using a pH meter. The operation and calibration of the pH meter are described in the next section. Calibrate the pH meter using a standard buffer with a pH of 4.5. After you have calibrated the pH meter, measure the pH of the buffer solution you have prepared and record the value. Save this buffer solution for part 2 below.

Operation and Calibration of the pH Meter

Instructor: Inform students about proper use and care of electrodes.

1. Obtain a buffer solution of known pH.
2. Plug in the pH meter to the line current and allow at least 10 min for warming up. It should be left plugged in until you are completely finished with it. *This does not apply to battery-operated meters.*
3. Turn the function knob on the pH meter to the standby position.
4. *Prepare the electrodes.* Make certain that the solution in the reference electrode extends well above the internal electrode. If it does not, ask your instructor to fill it with saturated KCl solution. Remove the rubber tip and slide down the rubber collar on the reference electrode. Rinse the outside of the electrodes well with distilled water.
5. *Standardize the pH meter.* Carefully immerse the electrodes in the buffer solution contained in a small beaker. *Remember that the glass electrode is very fragile; it breaks easily!* Do not touch the bottom of the beaker with the electrodes!! Turn the function knob to "read" or "pH." Turn the standardize knob until the pH meter indicates the exact pH of the buffer solution. Wait 5 s to be certain that the reading remains constant. *Once you have standardized the* pH *meter, do not readjust the standardize knob.* Turn the function knob to standby. Carefully lift the electrodes from the buffer and rinse them with distilled water. The pH meter is now ready to use to measure pH.

2. Effect of Acid and Base on the Buffer pH

Pour half (32 mL) of the buffer solution you prepared above into another 150-mL beaker. Label the two beakers 1 and 2. Pipet 1.0 mL of 6.0 *M* HCl into beaker 1, mix, and then measure the pH of the resultant solution and record the pH. Remember to rinse the electrodes between pH measurements. Similarly, pipet 1.0 mL of 6.0 *M* NaOH into beaker 2, mix, and then measure and record the pH of the resultant solution. Calculate the pH values of the original buffer solution, the values after the additions of the HCl and NaOH.

How do the measured and calculated values compare? Dispose of the chemicals in the designated receptacles.

REVIEW QUESTIONS

Before beginning this experiment in the laboratory, you should be able to answer the following questions:

1. Define Brønsted-Lowry acids and bases.
2. Which of the following ions will react with water in a hydrolysis reaction: Na^+, Ca^{2+}, Cu^{2+}, Zn^{2+}, F^-, SO_3^{2-}, Br^-?
3. For those ions in question 2 that undergo hydrolysis, write net ionic equations for the hydrolysis reaction.
4. The K_a for HCN is 4.9×10^{-10}. What is the value of K_b for CN^-?
5. What are the conjugate base and conjugate acid of $H_2PO_4^-$?
6. From what acid and what base were the following salts made: $CaSO_4$, NH_4Br, and $BaCl_2$?
7. Define the term *salt*.
8. Tell whether 0.1 *M* solutions of the following salts would be acidic, neutral, or basic: $BaCl_2$, $CuSO_4$, $(NH_4)_2SO_4$, $ZnCl_2$, NaCN.
9. If the pH of a solution is 8, what are the hydrogen- and hydroxide-ion concentrations?
10. The pH of a 0.1 *M* MCl (M^+ is an unknown cation) was found to be 5.3. Write a net ionic equation for the hydrolysis of M^+ and its corresponding equilibrium expression K_b. Calculate the value of K_b.
11. What is the pH of a solution that is 0.20 *M* $HC_2H_3O_2$ and 0.40 *M* $NaC_2H_3O_2$? K_a for acetic acid 1.8×10^{-5}.

NOTES AND CALCULATIONS

Name ______________________ Desk ______________________

Date ______________ Laboratory Instructor ______________________

REPORT SHEET | EXPERIMENT

Hydrolysis of Salts and pH of Buffer Solutions | 24

A. Hydrolysis of Salts

Solution	Ion expected to hydrolyze (if any)	Spectator ion(s) (if any)
0.1 *M* NaCl	—	Na^+, Cl^-
0.1 *M* Na_2CO_3	CO_3^{2-}	Na^+
0.1 *M* $NaC_2H_3O_2$	$C_2H_3O_2^-$	Na^+
0.1 *M* NH_4Cl	NH_4^+	Cl^-
0.1 *M* $ZnCl_2$	Zn^{2+}	Cl^-
0.1 *M* $KAl(SO_4)_2$	Al^{3+}	K^+, SO_4^{2-}

Solution	Indicator Color*								
	Methyl orange	Methyl red	Bromo-thymol blue	Phenol red	Phenol-phtha-lein	Alizarin yellow-R	pH	$[H^+]$	$[OH^+]$
H_2O (unboiled)	org	red	yell	yell	—	yell	4.0	10^{-4}	10^{-10}
H_2O (boiled)	yell	yell	yell	yell	—	yell	6.0	10^{-6}	10^{-8}
NaCl	yell	yell	yell	yell	—	yell	6.0	10^{-6}	10^{-8}
$NaC_2H_3O_2$	yell	yell	blue	red	pink	yell	9.1	7.9×10^{-10}	1.3×10^{-5}
NH_4Cl	yell	red	yell	yell	—	yell	4.5	3.2×10^{-5}	3.1×10^{-10}
$ZnCl_2$	yell	red	yell	yell	—	yell	4.5	3.2×10^{-5}	3.1×10^{-10}
$KAl(SO_4)_2$	org	red	yell	yell	—	yell	3.9	1.3×10^{-4}	7.7×10^{-11}
Na_2CO_3	yell	yell	blue	red	red	org	11	10^{-11}	10^{-3}

* color key: org = orange; ppl = purple; — = colorless; yell = yellow.

CALCULATIONS

Solution	Net-ionic equation for hydrolysis	Expression for equilibrium constant (K_a or K_b)	Value of K_a or K_b
$NaC_2H_3O_2$	$C_2H_3O_2^- + H_2O = HC_2H_3O_2 + OH^-$	$K_b = [HOAc][OH^-]/[OAc^-]$	1.7×10^{-9}
Na_2CO_3	$CO_3^{2-} + H_2O = HCO_3^- + OH^-$	$K_b = [HCO_3^-][OH^-]/[CO_3^{2-}]$	1×10^{-5}
NH_4Cl	$NH_4^+ = NH_3 + H^+$	$K_a = [H^+][NH_3]/[NH_4^+]$	1.02×10^{-8}
$ZnCl_2$	$Zn^{2+} + H_2O = Zn(OH)^+ + H^+$	$K_a = [Zn(OH)^+][H^+]/[Zn^{2+}]$	1.02×10^{-8}
$KAl(SO_4)_2$	$Al^{3+} + H_2O = Al(OH)^{2+} + H^+$	$K_a = [Al(OH)^{2+}][H^+]/[Al^{3+}]$	1.7×10^{-7}

QUESTIONS

1. Using the K_a's for $HC_2H_3O_2$ and HCO_3^- (from Appendix E), calculate the K_b's for the $C_2H_3O_2^-$ and CO_3^{2-} ions. Compare these values with those calculated from your measured pH's.

   $C_2H_3O_2^-$:
   $K_b = 1 \times 10^{-14}/1.8 \times 10^{-5} = 5.6 \times 10^{-10}$
   Found: 1.7×10^{-9}

   CO_3^{2-}:
   $K_b = 1 \times 10^{-14}/5.6 \times 10^{-1} = 1.8 \times 10^{-4}$
   Found: 1×10^{-5}

2. Using K_b for NH_3 (from Appendix F), calculate K_a for the NH_4^+ ion. Compare this value with that calculated from your measured pH's.

   $$K_a \times \frac{1 \times 10^{-14}}{1.8 \times 10^{-5}} = 5.6 \times 10^{-10}$$

   Found: 1.02×10^{-8}.

3. How should the pH of a 0.1 *M* solution of $NaC_2H_3O_2$ compare with that of a 0.1 *M* solution of $KC_2H_3O_2$? Explain briefly.

   The pH should be the same because neither Na^+ or K^+ hydrolyze. The acetate ion that is common to both salts hydrolyzes.

4. What is the greatest source of error in this experiment? How could you minimize this source of error?

   Judgment of colors and probably the pH of water. Use of a pH meter to determine the pH of solutions would help. (Absorption of CO_2 as solutions stand also causes errors.)

B. pH of Buffer Solutions

Mass of $NaC_2H_3O_2 \cdot 3H_2O$ (FW = 136 g/mol) 3.50 g

pH of Original buffer solution 4.74

pH of Buffer + HCl 4.38

pH Buffer + NaOH 5.11

Calculate pH of original buffer (Show calculations below.) 4.74

Total volume of solution = 8.8 mL + 55.6 mL = 64.4 mL

$$[HC_2H_3O_2] = \frac{8.8 \text{ mL}}{64.4 \text{ mL}} \times 3.0\ M = 0.41\ M$$

$$[NaC_2H_3O_2 \cdot 3H_2O] = [C_2H_3O_2^-] = \frac{3.50 \text{ g}}{0.0644 \text{ L}} \times \frac{1 \text{ mol}}{136 \text{ g}} = 0.40\ M$$

$$K_a = 1.8 \times 10^{-5} = \frac{[H^+][C_2H_3O_2^-]}{[HC_2H_3O_2]} = \frac{x(0.40)}{(0.41)} \quad \text{where } x = [H^+] = 1.8 \times 10^{-5}\ M$$

$$pH = -\log(1.8 \times 10^{-5}) = 4.74$$

Calculate pH of buffer + HCl (Show calculations below.) 4.30

Dilution:

Total volume = 1.0 mL + 32.2 mL = 33.2 mL; $[HCl]_{before\ rxn} = \frac{1.0\ mL}{33.2\ mL} \times 6.0\ M = 0.18\ M$

$$[C_2H_3O_2^-]_{before\ rxn} = \frac{32.2\ mL}{33.2\ mL} \times 0.40\ M = 0.39\ M$$

$$[HC_2H_3O_2]_{before\ rxn} = \frac{32.2\ mL}{33.2\ mL} \times 0.41\ M = 0.40\ M$$

Acid-base reaction:

Reaction of HCl with base component of buffer ($C_2H_3O_2^-$) gives

$[HC_2H_3O_2] = 0.40\ M + 0.18\ M = 0.58\ M$

$[C_2H_3O_2^-] = 0.39\ M - 0.18\ M = 0.21\ M$

Equilibrium:

$$1.8 \times 10^{-5} = \frac{[H^+][C_2H_3O_2^-]}{[HC_2H_3O_2]} = \frac{x(0.21)}{(0.58)} \quad \text{where } x = [H^+] = 5.0 \times 10^{-5}\ M$$

$pH = -\log(5.0 \times 10^{-5}) = 4.30$

Calculate pH of buffer + NaOH (Show calculations below.) 5.16

Dilution:

Total volume = 1.0 mL + 32.2 mL = 33.2 mL; $[NaOH]_{before\ rxn} = \frac{1.0\ mL}{33.2\ mL} \times 6.0\ M = 0.18\ M$

$$[C_2H_3O_2^-]_{before\ rxn} = \frac{32.2\ mL}{33.2\ mL} \times 0.40\ M = 0.39\ M$$

$$[HC_2H_3O_2]_{before\ rxn} = \frac{32.2\ mL}{33.2\ mL} \times 0.41\ M = 0.40\ M$$

Acid-base reaction:

Reaction of NaOH with acid component of buffer ($HC_2H_3O_2$) gives

$[HC_2H_3O_2] = 0.40\ M - 0.18\ M = 0.22\ M$

$[C_2H_3O_2^-] = 0.39\ M + 0.18\ M = 0.57\ M$

Equilibrium:

$$1.8 \times 10^{-5} = \frac{[H^+][C_2H_3O_2^-]}{[HC_2H_3O_2]} = \frac{x(0.57)}{(0.22)} \quad \text{where } x = [H^+] = 6.9 \times 10^{-6}\ M$$

$pH = -\log(6.9 \times 10^{-6}) = 5.16$

NOTES AND CALCULATIONS

Experiment

Determination of the Solubility-Product Constant for a Sparingly Soluble Salt

OBJECTIVE

To become familiar with equilibria involving sparingly soluble substances by determining the value of the solubility-product constant for a sparingly soluble salt.

APPARATUS AND CHEMICALS

Apparatus

spectrophotometer and cuvettes	100-mL volumetric flasks (4)
5-mL pipets (2)	buret
75-mm test tubes (3)	centrifuge
150-mm test tubes (3)	ring stand and buret clamp
no. 1 corks (3)	

Chemicals

0.0024 M K_2CrO_4	0.004 M $AgNO_3$
0.25 M $NaNO_3$	

DISCUSSION

Inorganic substances may be broadly classified into three different categories: acids, bases, and salts. According to the Brønsted-Lowry theory, acids are proton donors, and bases are proton acceptors. When an acid reacts with a base in aqueous solution, the products are a salt and water, as illustrated by the reaction of H_2SO_4 and $Ba(OH)_2$:

$$H_2SO_4(aq) + Ba(OH)_2(aq) \longrightarrow BaSO_4(s) + 2H_2O(l) \quad [1]$$

With but a few exceptions, nearly all common salts are strong electrolytes. The solubilities of salts span a broad spectrum, ranging from slightly or sparingly soluble to very soluble. This experiment is concerned with heterogeneous equilibria of slightly soluble salts. For a true equilibrium to exist between a solid and solution, the solution must be saturated. Barium sulfate is a slightly soluble salt, and in a saturated solution this equilibrium may be represented as follows:

$$BaSO_4(s) \rightleftharpoons Ba^{2+}(aq) + SO_4^{2-}(aq) \quad [2]$$

The equilibrium constant for Equation [2] is

$$K_c = \frac{[Ba^{2+}][SO_4^{2-}]}{[BaSO_4]} \quad [3]$$

The terms in the numerator refer to the molar concentration of ions in solution. The term in the denominator refers to the "concentration" of solid $BaSO_4$. Because the concentration of a pure solid is a constant, $[BaSO_4]$ can be combined with K_c to give a new equilibrium constant, K_{sp}, which is called the solubility-product constant.

$$K_{sp} = K_c[BaSO_4] = [Ba^{2+}][SO_4^{2-}]$$

At a given temperature the value of K_{sp} is a constant. The solubility product for a sparingly soluble salt can be easily calculated by determining the solubility of the substance in water. Suppose, for example, we determined that 2.42×10^{-4} g of $BaSO_4$ dissolves in 100 mL of water. The molar solubility of this solution (that is, the molarity of the solution) is

$$\left(\frac{2.42 \times 10^{-4}\text{ g } BaSO_4}{100\text{ mL}}\right)\left(\frac{1000\text{ mL}}{\text{liter}}\right)\left(\frac{1\text{ mol } BaSO_4}{233.4\text{ g } BaSO_4}\right) = 1.04 \times 10^{-5}\,M$$

We see from Equation [2] that for each mole of $BaSO_4$ that dissolves, one mole of Ba^{2+} and one mole of SO_4^{2-} are formed. It follows, therefore, that

$$\begin{aligned} \text{solubility of } BaSO_4 \text{ in moles/liter} &= [Ba^{2+}] \\ &= [SO_4^{2-}] \\ &= 1.04 \times 10^{-5}\,M \end{aligned}$$

and

$$\begin{aligned} K_{sp} &= [Ba^{2+}][SO_4^{2-}] \\ &= [1.04 \times 10^{-5}][1.04 \times 10^{-5}] \\ &= 1.08 \times 10^{-10} \end{aligned}$$

In a saturated solution the product of the molar concentrations of Ba^{2+} and SO_4^{2-} cannot exceed 1.08×10^{-10}. If the ion product $[Ba^{2+}][SO_4^{2-}]$ exceeds 1.08×10^{-10}, precipitation of $BaSO_4$ would occur until this product is reduced to the value of K_{sp}. Or if a solution of Na_2SO_4 is added to a solution of $Ba(NO_3)_2$, $BaSO_4$ would precipitate if the ion product $[Ba^{2+}][SO_4^{2-}]$ is greater than K_{sp}.

Similarly, if we determine that the solubility of Ag_2CO_3 is 3.49×10^{-3} g/100 mL, we could calculate the solubility-product constant for Ag_2CO_3 as follows. The solubility equilibrium involved is

$$Ag_2CO_3(s) \rightleftharpoons 2\,Ag^+(aq) + CO_3^{2-}(aq) \qquad [4]$$

and the corresponding solubility-product expression is

$$K_{sp} = [Ag^+]^2[CO_3^{2-}]$$

The rule for writing the solubility-product expression states that K_{sp} is equal to the product of the concentration of the ions involved in the equilibrium, each raised to the power of its coefficient in the equilibrium equation.

The solubility of Ag_2CO_3 in moles per liter is

$$\left(\frac{3.49 \times 10^{-3}\text{ g } Ag_2CO_3}{100\text{ mL}}\right)\left(\frac{1000\text{ mL}}{\text{liter}}\right)\left(\frac{1\text{ mol } Ag_2CO_3}{278.5\text{ g } Ag_2CO_3}\right) = 1.27 \times 10^{-4}\,M$$

so that

$$[CO_3^{2-}] = 1.27 \times 10^{-4}\,M \qquad \text{(from Equation [4])}$$

and

$$\begin{aligned} [Ag^+] &= 2(1.27 \times 10^{-4}) \\ &= 2.54 \times 10^{-4}\,M \qquad \text{(from Equation [4])} \\ K_{sp} &= [Ag^+]^2[CO_3^{2-}] \\ &= [2.54 \times 10^{-4}]^2[1.27 \times 10^{-4}] \\ &= 8.19 \times 10^{-12} \end{aligned}$$

To determine the solubility of Ag_2CrO_4, you will first prepare it by the reaction of $AgNO_3$ with K_2CrO_4:

$$2AgNO_3(aq) + K_2CrO_4(aq) \rightleftharpoons Ag_2CrO_4(s) + 2KNO_3(aq)$$

If a solution of $AgNO_3$ is added to a solution of K_2CrO_4, precipitation will occur if the ion product $[Ag^+]^2[CrO_4^{2-}]$ numerically exceeds the value of K_{sp}; if not, no precipitation will occur.

Example 25.1

If the K_{sp} for PbI_2 is 7.1×10^{-9}, will precipitation of PbI_2 occur when 10 mL of 1.0×10^{-4} *M* $Pb(NO_3)_2$ is mixed with 10 mL of 1.0×10^{-3} *M* KI?

SOLUTION:

$$PbI_2(s) \rightleftharpoons Pb^{2+}(aq) + 2I^-(aq)$$

$$K_{sp} = [Pb^{2+}][I^-]^2 = 7.1 \times 10^{-9}$$

Precipitation will occur if $[Pb^{2+}][I^-]^2 > 7.1 \times 10^{-9}$

$$[Pb^{2+}] = \left(\frac{10\text{ mL}}{20\text{ mL}}\right)(1.0 \times 10^{-4}\ M)$$

$$= 5.0 \times 10^{-5}\ M$$

$$[I^-] = \left(\frac{10\text{ mL}}{20\text{ mL}}\right)(1.0 \times 10^{-3}\ M)$$

$$= 5.0 \times 10^{-4}\ M$$

$$[Pb^{2+}][I^-]^2 = [5.0 \times 10^{-5}][5.0 \times 10^{-4}]^2$$

$$= 125 \times 10^{-13}$$

$$= 1.3 \times 10^{-11}$$

Since $1.3 \times 10^{-11} < 7.1 \times 10^{-9}$, no precipitation will occur. However, if 10 mL of 1.0×10^{-2} *M* $Pb(NO_3)_2$ is added to 10 mL of 2.0×10^{-2} *M* KI, then

$$[Pb^{2+}] = \left(\frac{10\text{ mL}}{20\text{ mL}}\right)(1.0 \times 10^{-3}\ M)$$

$$= 5.0 \times 10^{-3}$$

$$[I^-] = \left(\frac{10\text{ mL}}{20\text{ mL}}\right)(2.0 \times 10^{-3})$$

$$= 1.0 \times 10^{-2}\ M$$

and

$$[Pb^{2+}][I^-]^2 = [5.0 \times 10^{-3}][1.0 \times 10^{-2}]^2 = 5.0 \times 10^{-7}$$

Because $5.0 \times 10^{-7} > 7.1 \times 10^{-9}$, precipitation of PbI_2 will occur in this solution.

To determine the solubility-product constant for a sparingly soluble substance, we need only determine the concentration of one of the ions, because the concentration of the other ion is related to the first ion's concentration by a simple stoichiometric relationship. Any method that we could use to accurately determine the concentration would be suitable. In this experiment, you will determine the solubility-product constant for Ag_2CrO_4. This substance contains the yellow chromate ion, CrO_4^{2-}. You will determine the concentration of the chromate ion spectrophotometrically at 375 nm.

Although the eye can discern differences in color intensity with reasonable accuracy, an instrument known as a *spectrophotometer,* which eliminates

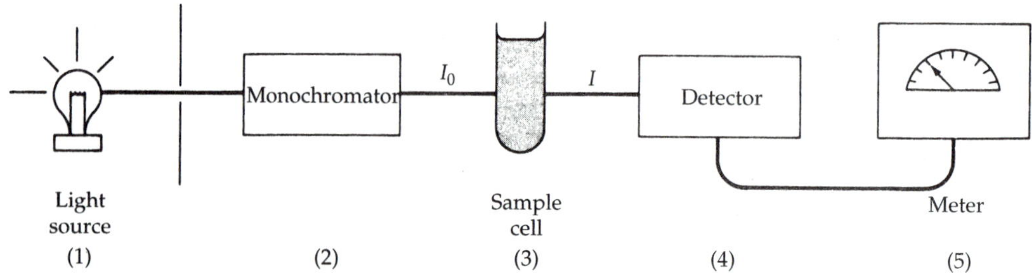

▲ **FIGURE 25.1** Schematic representation of a spectrophotometer.

the "human" error, is commonly used for this purpose. Basically, it is an instrument that measures the fraction I/I_0 of an incident beam of light of a particular wavelength and of intensity I_0 that is transmitted by a sample. (Here, I is the intensity of the light transmitted by the sample.) A schematic representation of a spectrophotometer is shown in Figure 25.1. The instrument has these five fundamental components:

1. A light source that produces light with a wavelength range from about 375 to 650 nm.
2. A monochromator, which *selects* a particular wavelength of light and sends it to the sample cell with an intensity of I_0.
3. The sample cell, which contains the solution being analyzed.
4. A detector that measures the intensity, I, of the light transmitted from the sample cell. If the intensity of the incident light is I_0 and the solution absorbs light, the intensity of the transmitted light, I, is less than I_0.
5. A meter that indicates the intensity of the transmitted light.

For a given substance, the amount of light absorbed depends on the

1. concentration
2. cell or path length
3. wavelength of light
4. solvent

Plots of the amount of light absorbed versus wavelength are called *absorption spectra.* There are two common ways of expressing the amount of light absorbed. One is in terms of *percent transmittance,* $\%T$, which is defined as

$$\%T = \frac{I}{I_0} \times 100 \qquad [5]$$

As the term implies, percent transmittance corresponds to the percentage of light transmitted. When the sample in the cell is a solution, I is the intensity of light transmitted by the solution, and I_0 is intensity of light transmitted when the cell only contains solvent. Another method of expressing the amount of light absorbed is in terms of *absorbance, A,* which is defined by

Instructor: Explain the relation between A and $\%T$:
$A = -\log(\%T/100)$
$= 2 - \log \%T$

$$A = \log \frac{I_0}{I} \qquad [6]$$

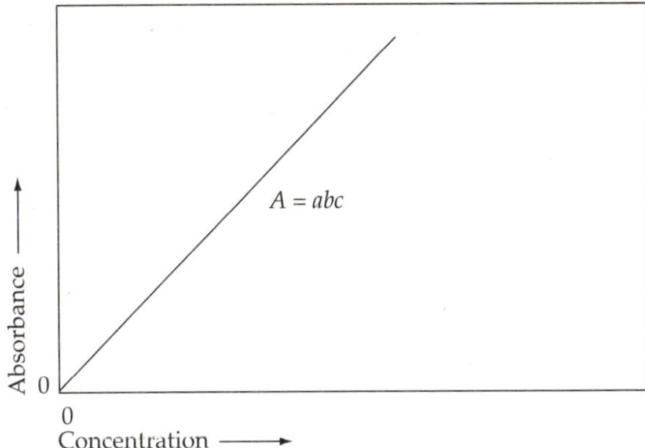

▲ **FIGURE 25.2** Relationship between absorbance and concentration according to the Beer-Lambert law.

The term *optical density*, OD, is synonymous with absorbance. If there is no absorption of light by a sample at a given wavelength, the percent transmittance is 100, and the absorbance is 0. On the other hand, if the sample absorbs all of the light, $\%T = 0$ and $A = \infty$.

Absorbance is related to concentration by the Beer-Lambert law:

$$A = abc$$

where A is absorbance, b is solution path length, c is concentration in moles per liter, and a is molar absorptivity or molar extinction coefficient. There is a linear relationship between absorbance and concentration when the Beer-Lambert law is obeyed, as illustrated in Figure 25.2. However, because deviations from this law occasionally occur, it is wise to construct a calibration curve of absorbance versus concentration.

PROCEDURE

A. Preparation of a Calibration Curve

WORK IN GROUPS OF FOUR TO OBTAIN YOUR CALIBRATION CURVE, BUT EVALUATE YOUR DATA INDIVIDUALLY. Using a buret, add 1, 5, 10, and 15 mL of standardized 0.0024 *M* K_2CrO_4 to each of four clean, dry 100-mL volumetric flasks and dilute to the 100 mL mark with 0.25 *M* $NaNO_3$. Calculate the CrO_4^{2-} concentration in each of these solutions. Measure the absorbance of these solutions at 375 nm and plot the absorbance versus concentration to construct your calibration curve as shown in Figure 25.2.

Operating Instructions for Spectronic 20

1. Turn the wavelength-control knob (Figure 25.3) to the desired wavelength.
2. Turn on the instrument by rotating the power control clockwise, and allow the instrument to warm up about 5 min. With no sample in the holder but with the cover closed, turn the zero adjust to bring the meter needle to zero on the "percent transmittance" scale.
3. Fill the cuvette about halfway with distilled water (or solvent blank) and insert it in the sample holder, aligning the line on the cuvette with

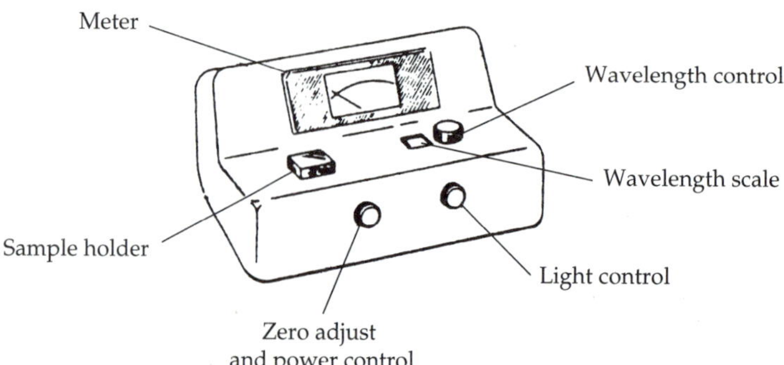

▲ **FIGURE 25.3** Spectrophotometer controls.

that of the sample holder. Close the cover and rotate the light-control knob until the meter reads 100% transmittance.

4. Remove the blank from the sample holder and replace it with the cuvette containing the sample whose absorbance is to be measured. Align the lines on the cuvette with the holder and close the cover. Read the percent transmittance or optical density from the meter.

B. Determination of the Solubility-Product Constant

Accurately prepare three separate solutions in separate 150-mm test tubes by adding 5 mL of 0.004 *M* $AgNO_3$ to 5 mL of 0.0024 *M* K_2CrO_4.

Stopper each test tube. Shake the solutions thoroughly at periodic intervals for about 15 min to establish equilibrium between the solid phase and the ions in solution. Transfer approximately 3 mL of each solution along with most of the insoluble Ag_2CrO_4 to 75-mm test tubes and centrifuge. Discard the supernatant liquid and retain the precipitate. To each of the test tubes add 2 mL of 0.25 *M* $NaNO_3$. Shake each test tube thoroughly and centrifuge again. Discard the supernatant liquid, then add 2 mL of 0.25 *M* $NaNO_3$ to each of the test tubes and shake the test tubes vigorously and periodically for about 15 min to establish an equilibrium between the solid and the solution. There must be some solid Ag_2CrO_4 remaining in these test tubes. If there is not, start over again. After shaking the test tubes for about 15 min, centrifuge the mixtures. Transfer the clear, pale yellow supernatant liquid from each of the three test tubes to a clean, dry cuvette. Measure and record the absorbance of the three solutions. Using your calibration curve, calculate the molar concentration of CrO_4^{2-} in each solution.

Note on Calculations

You are determining the K_{sp} of Ag_2CrO_4 in this experiment. The equilibrium reaction for the dissolution of Ag_2CrO_4 is

$$Ag_2CrO_4(s) \rightleftharpoons 2Ag^+(aq) + CrO_4^{2-}(aq)$$

for which $K_{sp} = [Ag^+]^2[CrO_4^{2-}]$.

You should note that at equilibrium $[Ag^+] = 2[CrO_4^{2-}]$; hence, having determined the concentration of chromate ions, you know the silver-ion concentration.

Waste Disposal Instructions Because chromates are hazardous, all chromate solutions should be treated with care. Avoid spilling or touching these solutions. All excess K_2CrO_4 solution from part A should be returned to a specially marked waste container, not to the original stock solution. All chromate solutions from part B of the experiment should be placed in the same container. Likewise, all the Ag_2CrO_4 samples should be disposed of in the second specially marked waste container. Silver nitrate ($AgNO_3$) solution is also hazardous. Any $AgNO_3$ solution that is spilled on the skin will cause discoloration after a few minutes. All excess $AgNO_3$ solution should be returned to a third specially marked container. The other solution used in this experiment, $NaNO_3$, is harmless; any excess can be disposed of in the sinks.

REVIEW QUESTIONS

Before beginning this experiment in the laboratory, you should be able to answer the following questions:

1. Write the solubility equilibrium and the solubility-product constant expression for the slightly soluble salt, CaF_2.
2. Calculate the number of moles of Ag^+ in 5 mL of 0.004 *M* $AgNO_3$ and the number of moles of CrO_4^{2-} in 5 mL of 0.0024 *M* K_2CrO_4.
3. If 10 mL of 0.004 *M* $AgNO_3$ is added to 10 mL of 0.0024 *M* K_2CrO_4, is either Ag^+ or CrO_4^{2-} in stoichiometric excess? If so, which is in excess?
4. The K_{sp} for $BaCrO_4$ is 1.2×10^{-10}. Will $BaCrO_4$ precipitate upon mixing 20 mL of 1×10^{-4} *M* $Ba(NO_3)_2$ with 20 mL of 1×10^{-4} *M* K_2CrO_4?
5. The K_{sp} for $BaCO_3$ is 5.1×10^{-9}. How many grams of $BaCO_3$ will dissolve in 1000 mL of water?
6. Distinguish between the equilibrium-constant expression and K_{sp} for the dissolution of a sparingly soluble salt.
7. List as many experimental techniques as you can that may be used to determine K_{sp} for a sparingly soluble salt.
8. Why must some solid remain in contact with a solution of a sparingly soluble salt in order to ensure equilibrium?
9. In general, when will a sparingly soluble salt precipitate from solution?
10. Sparingly soluble bases and salts, such as $Fe(OH)_2$ and $FeCO_3$, are more soluble in acidic than in neutral solutions. Why?

NOTES AND CALCULATIONS

Name ______________________ Desk ______________________

Date ______________ Laboratory Instructor ______________________

REPORT SHEET | EXPERIMENT

Determination of the Solubility-Product Constant for a Sparingly Soluble Salt

25

A. Preparation of a Calibration Curve

Initial $[CrO_4^{2-}]$ 0.0024 *M*

Volume of 0.0024 M K_2CrO_4	*Total volume*	*$[CrO_4^{2-}]$*	*Absorbance*
1. 1 mL	100 mL	2.4×10^{-5}	0.041
2. 5 mL	100 mL	1.2×10^{-4}	0.37
3. 10 mL	100 mL	2.4×10^{-4}	0.85
4. 15 mL	100 mL	3.6×10^{-4}	1.13

Molar absorption coefficient for $[CrO_4^{2-}]$

1. 1708 2. 3083 3. 3542 4. 3139

Average molar absorption coefficient 2868

$$a_{ave} = \frac{1708 + 3083 + 3542 + 3139}{4} = 2868$$

Standard deviation (show calculations) 800

$$SD = \sqrt{\frac{(1160)^2 + (215)^2 + (674)^2 + (271)^2}{3}} = 800$$

B. Determination of the Solubility-Product Constant

Absorbance	*$[CrO_4^{2-}]$*	*$[Ag^+]$*	*K_{sp} of Ag_2CrO_4*
1. 0.92; %T = 11.9	3.2×10^{-4} *M*	6.4×10^{-4} *M*	1.3×10^{-10}
2. 0.94; %T = 11.5	3.3×10^{-4} *M*	6.6×10^{-4} *M*	1.4×10^{-10}
3. 1.10; %T = 7.9	3.8×10^{-4} *M*	7.6×10^{-4} *M*	2.2×10^{-10}

Average K_{sp} (show calculations) 1.6×10^{-10}

Standard deviation 5×10^{-11}

(Show calculations)

$$K_{sp_{ave}} = \frac{(1.3 + 1.4 + 2.2)\ 10^{-10}}{3} = 1.6 \times 10^{-10}$$

$$SD = \sqrt{\frac{(0.3 \times 10^{-10})^2 + (0.2 \times 10^{-10})^2 + (0.6 \times 10^{-10})^2}{2}}$$

$$= 5 \times 10^{-11}$$

QUESTIONS

1. If the standard solutions had unknowingly been made up to be 0.0024 *M* $AgNO_3$ and 0.0040 *M* K_2CrO_4, would this have affected your results? How?

   The extinction coefficient for CrO_4^{2-} would be too large. This would result in a value of K_{sp} that is smaller than it should be. The concentrations that are determined using this extinction coefficient will be too small.

2. If your cuvette had been dirty, how would this have affected the value of K_{sp}?

   Dirty cuvettes would result in high absorbances and hence higher concentrations. Thus, K_{sp} would be larger than it should be.

3. Using your determined value of K_{sp}, calculate how many milligrams of Ag_2CrO_4 will dissolve in 10.0 mL of H_2O.

$$Ag_2CrO_4 \rightleftharpoons 2Ag^+ + CrO_4^{2-}$$

$$K_{sp} = [Ag^+]^2[CrO_4^{2-}] \qquad \text{let } [CrO_4^{2-}] = x \qquad \text{then } K_{sp} = [2x]^2[x] = 4x^3$$

$$4x^3 = 1.6 \times 10^{-10};\ x = 3.4 \times 10^{-4} = [CrO_4^{2-}] = \text{moles } Ag_2CrO_4 \text{ that dissolved}$$

$$\text{mg } Ag_2CrO_4 = (10.0\ \text{mL})\left(\frac{1\ \text{L}}{1000\ \text{mL}}\right)\left(\frac{3.4 \times 10^{-4}\ \text{mol}}{\text{L}}\right)\left(\frac{331.7\ \text{g}}{\text{mol}}\right)\left(\frac{10^3\ \text{mg}}{1\ \text{g}}\right) = 1.1\ \text{mg}$$

4. The experimental procedure for this experiment has you add 5 mL of 0.004 *M* $AgNO_3$ to 5 mL of 0.0024 *M* K_2CrO_4. Is either of these reagents in excess, and if so, which one?

   5 mL × 0.004 mol/1000 mL = 0.02 mmol $AgNO_3$
   5 mL × 0.0024 mol/1000 mL = 0.012 mmol K_2CrO_4
   There should be twice as many moles of $AgNO_3$ as K_2CrO_4, but there isn't. Thus, K_2CrO_4 is in excess.

5. Use your experimentally determined value of K_{sp} and show, by calculation, that Ag_2CrO_4 should precipitate when 5 mL of 0.004 *M* $AgNO_3$ are added to 5 mL of 0.0024 *M* K_2CrO_4.

   Because $Q > K_{sp}$ $(5 \times 10^{-9} > 1.6 \times 10^{-10})$, precipitation occurs.

6. Look up the accepted value of K_{sp} for Ag_2CrO_4 in the back of your textbook. Calculate the percentage error in your experimentally determined value for K_{sp}.

   Accepted value is 1.1×10^{-12}.

   $$\frac{1.6 \times 10^{-10} - 1.1 \times 10^{-12}}{1.1 \times 10^{-12}} \times 100 = 1.5 \times 10^{4} \text{ percent}$$

7. Although Ag_2CrO_4 is insoluble in water, it is soluble in dilute HNO_3. Explain, using chemical equations.

   I. $Ag_2CrO_4(s) \rightleftharpoons 2Ag^{+} + CrO_4^{-2}$ II. $2H^{+} + 2CrO_4^{2-} \rightleftharpoons Cr_2O_7^{2-} + H_2O$

   Chromate ions are removed from Equation I by forming $Cr_2O_7^{2-}$. This causes the equilibrium in Equation I to shift to the right, and thus the precipitate dissolves.

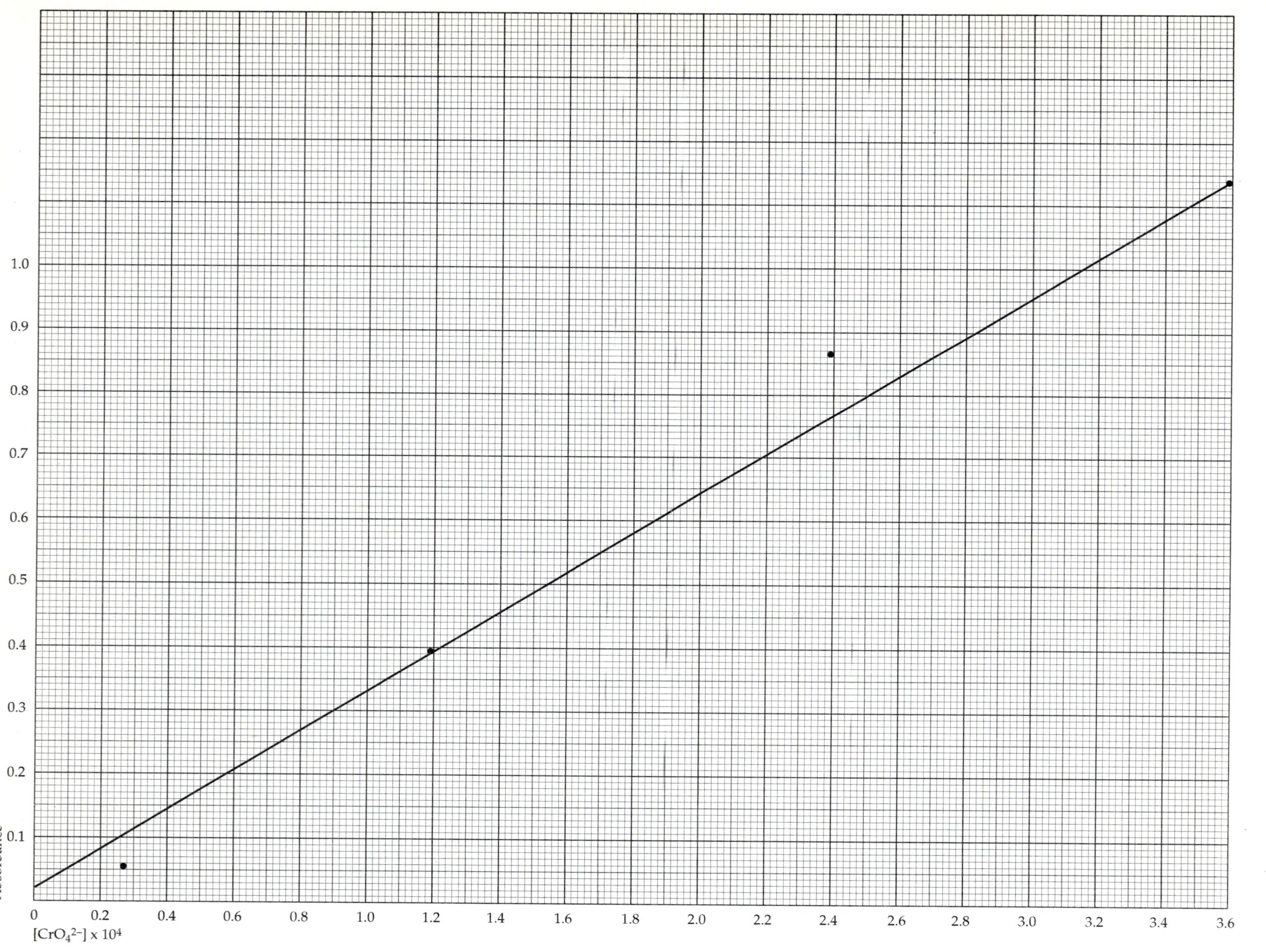

Absorbance
0.1
0.2
0.3
0.4
0.5
0.6
0.7
0.8
0.9
1.0
0
0.2
0.4
0.6
0.8
1.0
1.2
1.4
1.6
1.8
2.0
2.2
2.4
2.6
2.8
3.0
3.2
3.4
3.6
$[CrO_4^{2-}] \times 10^4$

Heat of Neutralization

OBJECTIVE

To measure, using a calorimeter, the energy changes accompanying neutralization reactions.

APPARATUS AND CHEMICALS

Apparatus

Bunsen burner
Styrofoam cups (2)
cardboard square with hole in center
400-mL beaker
wire gauze
thermometers (2)
50-mL graduated cylinder
split one-hole rubber stopper
250-mL beaker
ring stand and ring

Chemicals

1 *M* HCl
1 *M* acetic acid ($HC_2H_3O_2$)
1 *M* NaOH

WORK IN PAIRS, BUT EVALUATE YOUR DATA INDIVIDUALLY.

DISCUSSION

Every chemical change is accompanied by a change in energy, usually in the form of heat. The energy change of a reaction that occurs at constant pressure is termed the *heat of reaction* or the *enthalpy change.* The symbol ΔH (the symbol Δ means "change in") is used to denote the enthalpy change. If heat is evolved, the reaction is *exothermic* ($\Delta H < 0$); and if heat is absorbed, the reaction is *endothermic* ($\Delta H > 0$). In this experiment, you will measure the heat of neutralization (or the enthalpy of neutralization) when an acid and a base react to form water.

This quantity of heat is measured experimentally by allowing the reaction to take place in a thermally insulated vessel called a *calorimeter.* The heat liberated in the neutralization will cause an increase in the temperature of the solution and of the calorimeter. If the calorimeter were perfect, no heat would be radiated to the laboratory. The calorimeter you will use in this experiment is shown in Figure 26.1.

Because we are concerned with the heat of the reaction and because some heat is absorbed by the calorimeter itself, we must know the amount of heat absorbed by the calorimeter. This requires that we determine the heat capacity of the calorimeter. By "heat capacity of the calorimeter" we mean the amount of heat (that is, the number of joules) required to raise its temperature 1 kelvin, which is the same as 1°C. In this experiment, the temperature of the calorimeter and its contents is measured before and after the reaction. The change in the enthalpy, ΔH, is equal to the negative product of the temperature change, ΔT, times the heat capacity of the calorimeter and its contents:

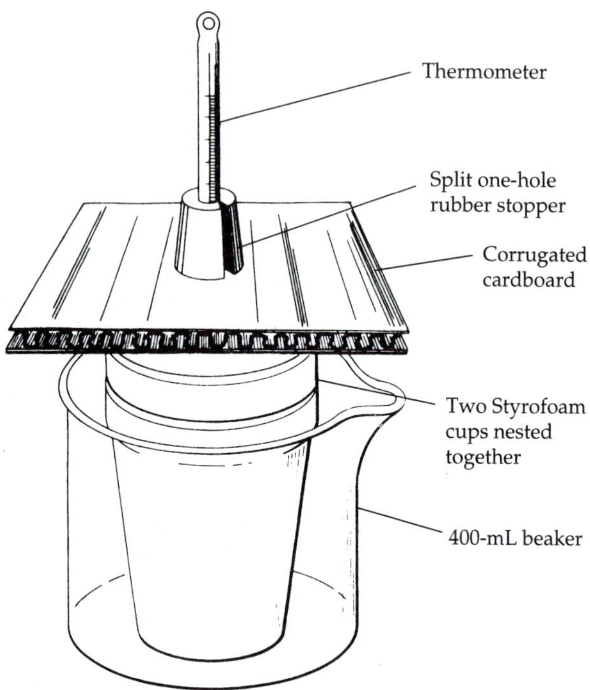

▲ **FIGURE 26.1** A simple calorimeter.

$$\Delta H = -\ \Delta T \text{ (heat capacity of calorimeter + heat capacity of contents)} \quad [1]$$

Note that the *numerical difference* on the Celsius scale is the same as the *numerical difference* on the kelvin scale where ΔT is the difference between the final and initial temperatures: $\Delta T = T_f - T_i$. Because ΔH is negative for an exothermic reaction whereas ΔT is positive, a negative sign is required in Equation [1].

The heat capacity of the calorimeter is determined by measuring the temperature change that occurs when a known amount of hot water is added to a known amount of cold water in the calorimeter. The heat lost by the warm water is equal to the heat gained by the cold water and the calorimeter. (We assume no heat is lost to the laboratory.) For example, if T_1 equals the temperature of a calorimeter and 50 mL of cooler water, if T_2 equals the temperature of 50 mL of warmer water added to it, and if T_f equals the temperature after mixing, then the heat lost by the warmer water is

$$\text{heat lost by warmer water} = (T_2 - T_f) \times 50 \text{ g} \times 4.18 \text{ J/K-g} \quad [2]$$

The specific heat of water is 4.184 J/K-g, and the density of water is 1.00 g/mL. The heat gained by the cooler water is

$$\text{heat gained by cooler water} = (T_f - T_1) \times 50 \text{ g} \times 4.18 \text{ J/K-g} \quad [3]$$

The heat lost to the calorimeter is the difference between heat lost by the warmer water and that gained by the cooler water:

(heat lost by warmer water) − (heat gained by cooler water)
= heat gained by the calorimeter

Substituting Equations [2] and [3] we have

$$[(T_2 - T_f) \times 50 \text{ g} \times 4.18 \text{ J/K-g}] - [(T_f - T_1) \times 50 \text{ g} \times 4.18 \text{ J/K-g}]$$
$$= (T_f - T_1) \times \text{heat capacity of calorimeter} \quad [4]$$

Note that the heat lost to the colorimeter equals its temperature change times its heat capacity. Thus by measuring T_1, T_2, and T_f, the heat capacity of the calorimeter can be calculated from Equation [4]. This is illustrated in Example 26.1.

EXAMPLE 26.1

Given the following data, calculate the heat lost by the warmer water, the heat lost to the cooler water, the heat lost to the calorimeter, and the heat capacity of the calorimeter:

Temperature of 50.0 mL warmer water: 37.92°C = T_2

Temperature of 50.0 mL cooler water: 20.91°C = T_1

Temperature after mixing: 29.11°C = T_f

SOLUTION: The heat lost by the warmer water, where ΔT = 37.92°C − 29.11°C, is

$$8.81 \text{ K} \times 50 \text{ g} \times 4.18 \text{ J/K-g} = 1840 \text{ J}$$

The heat gained by the cooler water, where ΔT = 29.11°C − 20.91°C, is

$$8.20 \text{ K} \times 50 \text{ g} \times 4.18 \text{ J/K-g} = 1710 \text{ J}$$

The heat lost to the calorimeter is

$$1840 \text{ J} - 1710 \text{ J} = 130 \text{ J}$$

The heat capacity of the calorimeter is, therefore,

$$130 \text{ J}/8.20 \text{ K} = 16.0 \text{ J/K}$$

Once the heat capacity of the calorimeter is determined, Equation [1] can be used to determine the ΔH for the neutralization reaction. Example 26.2 illustrates such a calculation.

EXAMPLE 26.2

Given the following data, calculate the heat gained by the solution, the heat gained by the calorimeter, and the heat of reaction:

Temperature of 50.0 mL of acid before mixing: 21.02°C

Temperature of 50.0 mL of base before mixing: 21.02°C

Temperature of 100.0 mL of solution after mixing: 27.53°C

Assume that the density of these solutions is 1.00 g/mL.

SOLUTION: The heat gained by the solution, where ΔT = 27.53°C − 21.02°C, is

$$6.51 \text{ K} \times 100 \text{ g} \times 4.18 \text{ J/K-g} = 2720 \text{ J}$$

The heat gained by the calorimeter, where ΔT = 27.53°C − 21.02°C, is

$$6.51 \text{ K} \times 16.0 \text{ J/K} = 104 \text{ J}$$

The heat of reaction is therefore

$$2720 \text{ J} + 104 \text{ J} = 2824 \text{ J}$$

or

$$2.82 \text{ kJ}$$

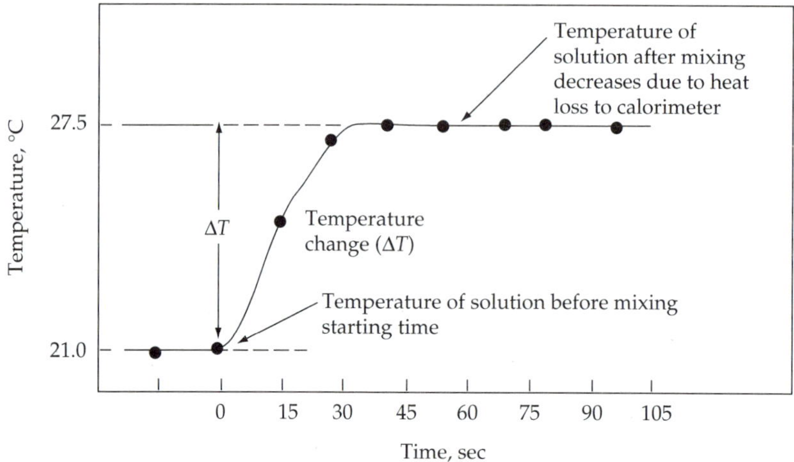

▲ **FIGURE 26.2** Temperature as a function of time.

PROCEDURE

Instructor: Students often have difficulty here with the thermometers. Do not use the one in the hot water to measure the temperature in the calorimeter.

A. Heat Capacity of Calorimeter

Construct a calorimeter similar to the one shown in Figure 26.1 by nesting two Styrofoam cups together. Use a cork borer to make a hole in the lid just big enough to admit the thermometer and slip the thermometer into a split one-hole rubber stopper to prevent the thermometer from entering too deeply into the calorimeter. The thermometer should not touch the bottom of the cup. Rest the entire apparatus in a 400-mL beaker to provide stability.

Place exactly 50.0 mL of tap water in the calorimeter cup and replace the cover and thermometer. Allow 5 to 10 min for the system to reach thermal equilibrium; then record the temperature to the nearest 0.1°C.

Place exactly 50.0 mL of water in a 250-mL beaker and heat the water with a low flame until the temperature is approximately 15° to 20°C above room temperature. Do not heat to boiling, or appreciable water will be lost, leading to an erroneous result. Allow the hot water to stand for a minute or two; quickly record its temperature to the nearest 0.1°C and pour it as completely as possible into the calorimeter. Replace the lid with the thermometer and carefully stir the water with the thermometer. Observe the temperature for the next 3 min and record the temperature every 15 s. Plot the temperature as a function of time, as shown in Figure 26.2. Determine ΔT from your curve and then do the calculations indicated on the report sheet.

B. Heat of Neutralization of HCl—NaOH

Dry the calorimeter and the thermometer with a towel. Carefully measure 50.0 mL of 1.0 *M* NaOH and add it to the calorimeter. Place the lid on the calorimeter but leave the thermometer out. Measure out exactly 50.0 mL of 1.0 *M* HCl into a dry beaker. Allow it to stand near the calorimeter for 3 to 4 min. Measure the temperature of the acid, rinse the thermometer with tap water, and wipe dry. Insert the thermometer into the calorimeter and measure the temperature of the NaOH solution.

The temperatures of the NaOH and the HCl should not differ by more than 0.5°C. If the difference is greater than 0.5°C, adjust the temperature of the HCl by *either* warming it by holding the beaker in your hands or cooling the outside of the beaker with tap water until the temperature of the HCl is within 0.5°C of that of the NaOH.

Instructor: Students may have difficulty with the thermometers. The one in the calorimeter should not have HCl on it.

Record the temperature of the NaOH solution. Lift the lid and carefully add the 1.0 *M* HCl all at once. Be careful not to splash any on the upper sides of the cup. Stir the solution gently with the thermometer and record the temperature as a function of time every 15 s for the next 3 min. Construct a temperature-versus-time curve and determine ΔT. Calculate the heat of neutralization per mole of water formed. You may assume that the NaCl solution has the same density and specific heat as water.

C. Heat of Neutralization of $HC_2H_3O_2$—NaOH

Follow the same procedure as in Part B, but substitute 1.0 *M* $HC_2H_3O_2$ for 1.0 *M* HCl. Calculate the heat of neutralization per mole of water formed.

Waste Disposal Instructions Handle the stock solutions carefully. You may use a wet sponge or paper towel to clean up any spills. The reaction mixtures produced in the Styrofoam cups contain harmless salts. They should be disposed of in the designated receptacles.

REVIEW QUESTIONS

Before beginning this experiment in the laboratory, you should be able to answer the following questions:

1. Define endothermic and exothermic reactions in terms of the sign of ΔH.
2. A 625-mL sample of water was cooled from 50.0°C to 10.0°C. How much heat was lost?
3. Define the term *heat capacity*.
4. How many joules are required to change the temperature of 60.0 g of water from 23.3°C to 38.8°C?
5. Define the term *specific heat*.
6. Calculate the final temperature when 50 mL of water at 60°C are added to 25 mL of water at 20°C.
7. Describe how you could determine the specific heat of a metal by using the apparatus and techniques in this experiment.
8. A piece of metal weighing 5.10 g at a temperature of 48.6°C was placed in a calorimeter into 20.00 mL of water at 22.1°C, and the final equilibrium temperature was found to be 28.2°C. What is the specific heat of the metal?
9. If the specific heat of methanol is 2.51 J/K-g, how many joules are necessary to raise the temperature of 250 g of methanol from 18°C to 33°C?
10. When a 3.25-g sample of solid sodium hydroxide was dissolved in a calorimeter in 100.0 g of water, the temperature rose from 23.9°C to 32.0°C. Calculate ΔH (in kJ/mol NaOH) for the solution process:

$$NaOH(s) \longrightarrow Na^+(aq) + OH(aq)$$

Assume it's a perfect calorimeter and that the specific heat of the solution is the same as that of pure water.

NOTES AND CALCULATIONS

Name ____________________ Desk ____________________

Date ____________________ Laboratory Instructor ____________________

REPORT SHEET | EXPERIMENT

Heat of Neutralization | 26

A. Heat Capacity of Calorimeter

1. Temp. of calorimeter and water before mixing ___22.9___ °C
2. Temp. of warm water ___43.0___ °C
3. Maximum temp. determined from your curve ___32.7___ °C
4. Heat lost by warm water (temp decrease × 50.0 g × 4.18 J/K-g)= ___2150___ J

   10.3 K × 50.0 g × 4.18 J/K-g = 2150 J

5. Heat gained by cooler water (temp. increase × 50.0 g × 4.18 J/K-g)= ___2050___ J

   9.8 K × 50.0 g × 4.18 J/K-g = 2050 J

6. Heat gained by the calorimeter [(4)−(5)]= ___100___ J
7. Heat capacity of calorimeter:

   $$\frac{\text{Heat gained by the calorimeter}}{\text{Temperature increase}} = \frac{100\text{ J}}{9.8\text{ K}}$$ ___10.2___ J/K

B. Heat of Neutralization of HCl—NaOH

1. Temp. of calorimeter and NaOH ___24.1___ °C
2. ΔT determined from your curve after adding HCl to the NaOH ___6.8___ °C
3. Heat gained by solution (temperature increase × 100 g × 4.18 J/K-g) ___2800___ J

   6.8 K × 100 g × 4.18 J/K-g = 2800 J

4. Heat gained by calorimeter (temperature increase × heat capacity of calorimeter) = ___69___ J

   6.8 K × 10.2 J/K = 69 J

5. Total joules released by reaction [(3) + (4)] = ___2869___ J
6. Complete: HCl + NaOH ⟶ ___NaCl + H_2O___

7. The number of moles of HCl in 50 mL of 1.0 *M* HCl (show calculations): 0.050 L × 1.0 mol/L ____0.050____ mol

8. The number of moles of H_2O produced in reaction of 50 mL 1.0 *M* HCl and 50 mL 1.0 *M* NaOH (show calculations): ____0.050____ mol

9. Joules released per mole of water formed:

$$\frac{\text{Total joules released (5)}}{\text{Number of moles water produced (8)}} =$$ ____57____ kJ/mol

$$\frac{2869\text{ J}}{0.050\text{ mol}} = 57\text{ kJ/mol}$$

C. Heat of Neutralization of $HC_2H_3O_2$—NaOH

1. Temperature of calorimeter and NaOH ____23.9____ °C
2. ΔT determined from cooling curve after adding $HC_2H_3O_2$ to NaOH ____6.5____ °C
3. Heat gained by solution (temp. increase × 100 g × 4.18 J/K-g) = ____2720____ J

   6.5 K × 100 g × 4.18 J/K-g = 2720 J

4. Heat gained by calorimeter (temp. increase × heat capacity of calorimeter) = ____70____ J

   6.5 K × 10.7 J/K = 70 J

5. Total joules released by reaction [(3)+(4)] = ____2790____ J

6. Complete: $HC_2H_3O_2$ + NaOH ⟶ $NaC_2H_3O_2 + H_2O$
7. The number of moles of H_2O produced in reaction of 50 mL 1.0 *M* $HC_2H_3O_2$ and 50 mL 1.0 *M* NaOH (show calculations): ____0.050____ mol

8. Joules released per mole of water formed:

$$\frac{\text{Total joules released (5)}}{\text{Number of moles water produced (7)}} =$$ ____56____ kJ/mol

$$\frac{2790\text{ J}}{0.050\text{ mol}} = 56\text{ kJ/mol*}$$

*P. J. Cerutti et al, Can. J. Chem., (1978) 56, 3084 report 13,340 cal/mol = 55.81 kJ/mol

QUESTIONS

1. What is the largest source of error in the experiment?
   Inaccuracy of thermometer—a thermometer with finer subdivisions would help. Inefficient calorimeter.

2. How should the two heats of reaction for the neutralization of NaOH and the two acids compare? Why?
   One might expect them to be similar. The net reaction is: $H^+ (aq) + OH^- (aq) \longrightarrow H_2O(l)$. The heat would be the same except that acetic acid is a weak acid and thus they will differ by the heat of ionization for acetic acid.

3. The experimental procedure has you wash your thermometer and dry it after you measure the temperature of NaOH solution and before you measure the temperature of the HCl solution. Why?
   Any NaOH adhering to the thermometer would neutralize the HCl, changing its concentration and thus resulting in an error.

4. A 50.0-mL sample of a 1.00 *M* solution of $CuSO_4$ is mixed with 50.0 mL of 2.00 *M* KOH in a calorimeter. The temperature of both solutions was 20.2°C before mixing and 26.3°C after mixing. The heat capacity of the calorimeter is 12.1 J/K. From these data calculate ΔH for the process

$$CuSO_4(1\ M) + 2KOH(2\ M) \longrightarrow Cu(OH)_2(s) + K_2SO_4(0.5\ M)$$

Assume the specific heat and density of the solution after mixing are the same as those of pure water.

Answer:

$(299.3\ K - 293.2\ K)[(4.18\ J/K\text{-}g)(100\ g) + 12.1\ J/K] = 2.6\ kJ$

$\Delta H = -2.6$ kJ for the formation of 0.050 mol $Cu(OH)_2$; the reaction is exothermic.

$$\text{Thus, } \Delta H = \frac{-2.6\ kJ}{0.050\ mol} = -52\ kJ/mol.$$

Temperature, °C

0 15 30 45 60 75 90 105 120 135 150 165 180

Time, s

Temperature, °C

0 15 30 45 60 75 90 105 120 135 150 165 180

Time, s

Temperature, °C

0 15 30 45 60 75 90 105 120 135 150 165 180

Time, s

Rates of Chemical Reactions I: A Clock Reaction

OBJECTIVE

To measure the effect of concentration upon the rate of the reaction of peroxydisulfate ion with iodide ion; to determine the order of the reaction with respect to the reactant concentrations; and to obtain the rate law for the chemical reaction.

APPARATUS AND CHEMICALS

Apparatus

burets (2)
1-mL pipets (2)
clock or watch with second hand
pipet bulb
buret clamp
ring stand
25-mL pipet
50-mL pipet
test tubes (8)
250-mL Erlenmeyer flasks (4)
100-mL beakers (4)

Chemicals

0.2 *M* KI
1% starch solution, boiled
0.2 *M* $(NH_4)_2S_2O_8$ (freshly prepared)
0.4 *M* $Na_2S_2O_3$ (freshly prepared)
0.1 *M* solution of Na_2H_2EDTA
0.2 *M* KNO_3

WORK IN PAIRS, BUT EVALUATE YOUR DATA INDIVIDUALLY.

DISCUSSION

Factors Affecting Rates of Reactions

On the basis of the experiments you've performed, you probably have already noticed that reactions occur at varying speeds. There is an entire spectrum of speeds of reactions, ranging from very slow to extremely fast. For example, the rusting of iron is reasonably slow, whereas the decomposition of TNT is extremely fast. The branch of chemistry that is concerned with the rates of reactions is called *chemical kinetics.* Experiments show that rates of homogeneous reactions in solution depend upon:

1. The nature of the reactants
2. The concentration of the reactants
3. The temperature
4. Catalysis

Before a reaction can occur, the reactants must come into direct contact via collisions of the reacting particles. However, even then, the reacting particles (ions or molecules) must collide with sufficient energy to result in a reaction. If they do not, their collisions are ineffective and analogous to collisions of billiard balls. With these considerations in mind, we can qualitatively explain how the various factors influence the rates of reactions.

Concentration Changing the concentration of a solution alters the number of particles per unit volume. The more particles present in a given volume, the greater the probability of their colliding. Hence, increasing the concentration of a solution increases the number of collisions per unit time and therefore the rate of reaction.

Temperature Since temperature is a measure of the average kinetic energy, an increase in temperature increases the kinetic energy of the particles. An increase in kinetic energy increases the velocity of the particles and therefore the number of collisions between them in a *given period of time.* Thus, the rate of reaction increases. Also, an increase in kinetic energy results in a greater proportion of the collisions having the required energy for reaction. As a rule of thumb, for each 10°C increase in temperature, the rate of reaction doubles.

Catalyst Catalysts, in some cases, are believed to increase reaction rates by bringing particles into close juxtaposition in the correct geometrical arrangement for reaction to occur. In other instances, catalysts offer an alternative route to the reaction, one that requires less energetic collisions between reactant particles. If less energy is required for a successful collision, a larger percentage of the collisions will have the requisite energy, and the reaction will occur faster. Actually, the catalyst may take an active part in the reaction, but at the end of the reaction, the catalyst can be recovered chemically unchanged.

Order of Reaction Defined

Let's examine precisely what is meant by the expression *rate of reaction.* Consider the hypothetical reaction

$$A + B \longrightarrow C + D \qquad [1]$$

The rate of this reaction may be measured by observing the rate of disappearance of either of the reactants A and B or the rate of appearance of either of the products C and D. In practice, then, one measures the change of concentration with time of either A, B, C, or D. Which species you choose to observe is a matter of convenience. For example, if A, B, and D are colorless and C is colored, you could conveniently measure the rate of appearance of C by observing an increase in the intensity of the color of the solution as a function of time. Mathematically, the rate of reaction may be expressed as follows:

$$\text{rate of disappearance of A} = \frac{\text{change in concentration of A}}{\text{time required for change}} = \frac{-\Delta[A]}{\Delta t}$$

$$\text{rate of appearance of C} = \frac{\text{change in concentration of C}}{\text{time required for change}} = \frac{\Delta[C]}{\Delta t}$$

In general, the rate of the reaction will depend on the concentration of the reactants. Thus, the rate of our hypothetical reaction may be expressed as

$$\text{rate} = k[A]^x[B]^y \qquad [2]$$

where [A] and [B] are the molar concentrations of A and B, x and y are the powers to which the respective concentrations must be raised to describe the rate, and k is the *specific rate constant.* One of the objectives of chemical kinetics is to determine the rate law. Stated slightly differently, one goal of measuring the rate of the reaction is to determine the numerical values of x and y. Suppose that we found $x = 2$ and $y = 1$ for this reaction. Then,

$$\text{rate} = k[\text{A}]^2[\text{B}] \qquad [3]$$

would be the rate law. It should be evident from Equation [3] that doubling the concentration of B (keeping [A] the same) would cause the reaction rate to double. On the other hand, doubling the concentration of A (keeping [B] the same) would cause the rate to increase by a factor of 4, because the rate of the reaction is proportional to the *square* of the concentration of A. The powers to which the concentrations in the rate law are raised are termed the *order of the reaction.* In this case, the reaction is said to be second order in A and first order in B. The *overall order* of the reaction is the sum of the exponents, 2 + 1 = 3, or a third-order reaction. It is possible to determine the order of the reaction by noting the effects of changing reagent concentrations on the rate of the reaction. Note that the order of a reaction may be (and frequently is) different from the stoichiometry of the reaction.

It should be emphasized that k, the specific rate constant, has a definite value that is independent of the concentration. It is characteristic of a given reaction and depends only on temperature. Once the rate is known, the value of k can be calculated.

Reaction of Peroxydisulfate Ion with Iodide Ion

In this experiment you will measure the rate of the reaction

$$S_2O_8^{2-} + 2I^- \longrightarrow I_2 + 2SO_4^{2-} \qquad [4]$$

and you will determine the rate law by measuring the amount of peroxydisulfate, $S_2O_8^{2-}$, that reacts as a function of time. The rate law to be determined is of the form

$$\text{Rate of disappearance of } S_2O_8^{2-} = k[S_2O_8^{2-}]^x[I^-]^y \qquad [5]$$

or

$$\frac{\Delta[S_2O_8^{2-}]}{\Delta t} = k[S_2O_8^{2-}]^x\,[I^-]^y$$

Your goal will be to determine the values of x and y as well as the specific rate constant, k.

You will add to the solution a small amount of another reagent (sodium thiosulfate, $Na_2S_2O_3$), which will cause a change in the color of the solution. The amount is such that the color change will occur when 2×10^{-4} mol of $S_2O_8^{2-}$ has reacted. For reasons to be explained shortly, the solution will turn blue-black when 2×10^{-4} mol of $S_2O_8^{2-}$ has reacted. You will quickly add another portion of $Na_2S_2O_3$ after the appearance of the color, and the blue-black color will disappear. When the blue-black color reappears the second time, *another* 2×10^{-4} mol of $S_2O_8^{2-}$ has reacted, making a total of $2(2 \times 10^{-4})$ mol of $S_2O_8^{2-}$ that has reacted. You will repeat this procedure several times, keeping *careful* note of the time for the appearance of the blue-black colors. By graphing the amount of $S_2O_8^{2-}$ consumed versus time, you will be able to determine the rate of the reaction. By changing the initial concentrations of $S_2O_8^{2-}$ and I^- and observing the effects upon the rate of the reaction, you will determine the order of the reaction with respect to $S_2O_8^{2-}$ and I^-.

The blue-black color that will appear in the reaction is due to the presence of a starch-iodine complex that is formed from iodine, I_2, and starch in the solution. The color therefore will not appear until a detectable amount of I_2 is formed according to Equation [4]. The thiosulfate that is added to the solution reacts *extremely rapidly* with the iodine, as follows:

$$I_2 + 2S_2O_3^{2-} \longrightarrow 2I^- + S_4O_6^{2-} \quad [6]$$

Consequently, until the same amount of $S_2O_3^{2-}$ that is added is all consumed, there will not be a sufficient amount of I_2 in the solution to yield the blue-black color. You will add 4×10^{-4} mol of $S_2O_3^{2-}$ each time (these equal portions are termed *aliquots*). From the stoichiometry of Equations [4] and [6] you can verify that when this quantity of $S_2O_3^{2-}$ has reacted, 2×10^{-4} mol of $S_2O_8^{2-}$ has reacted. Note also that although iodide, I^-, is consumed according to Equation [4], it is rapidly regenerated according to Equation [6] and therefore its concentration does not change during a given experiment.

Graphical Determination of Rate

The more rapidly the 2×10^{-4} mol of $S_2O_8^{2-}$ is consumed, the faster is the reaction. To determine the rate of the reaction, a plot of moles of $S_2O_8^{2-}$ that have reacted versus the time required for the reaction is made, as shown in Figure 27.1. The best straight line passing through the origin is drawn, and the slope is determined. The slope, $\Delta S_2O_8^{2-}/\Delta t$, corresponds to the moles of $S_2O_8^{2-}$ that have been consumed per second and is proportional to the rate. Since the rate corresponds to the change in the concentration of $S_2O_8^{2-}$ per second, dividing the slope by the volume of the solution yields the rate of disappearance of $S_2O_8^{2-}$, that is, $\Delta[S_2O_8^{2-}]\Delta t$. If the total volume of the solution in this example were 75 mL, the rate would be as follows:

$$\frac{4.5 \times 10^{-5}\ \text{mol/s}}{0.075\ \text{L}} = 6.0 \times 10^{-4}\ \text{mol/L–s}$$

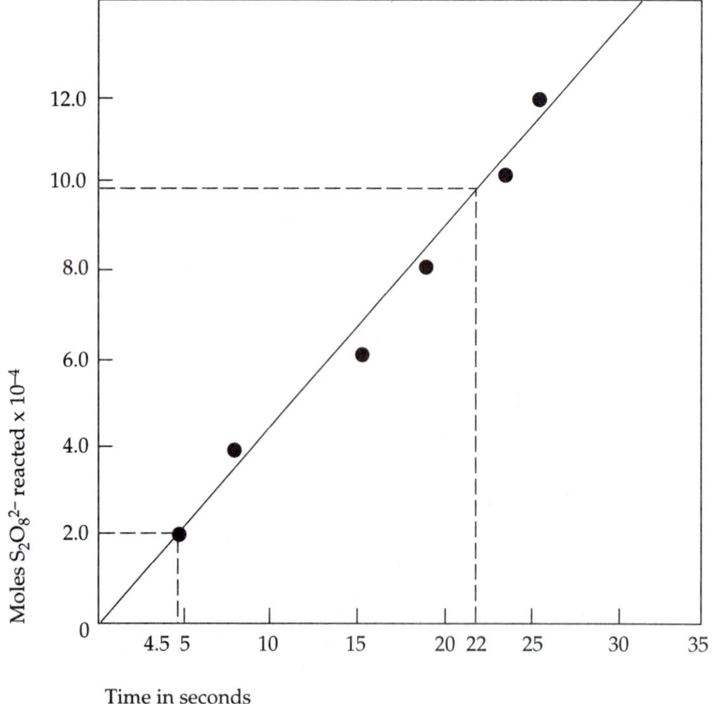

$$\text{Slope} = \frac{\Delta S_2O_8^{2-}}{\Delta t} = \frac{(9.8 - 2.0) \times 10^{-4}\ \text{mol}}{(22.0 - 4.5)\ \text{s}} = \frac{7.8 \times 10^{-4}\ \text{mol}}{17.5\ \text{s}} = 4.5 \times 10^{-5}\ \text{mol/s}$$

▲ **FIGURE 27.1** Graphical determination of rate.

If we obtain a rate of 6.0×10^{-4} mol/L-s when $[S_2O_8^{2-}] = 2.0\ M$ and $[I^-] = 2.0\ M$, and a rate of 3.0×10^{-4} mol/L-s when $[S_2O_8^{2-}] = 1.0\ M$ and $[I^-] = 2.0\ M$, we then know that doubling the concentration of $S_2O_8^{2-}$ doubles the rate of the reaction and the reaction is first order in $S_2O_8^{2-}$. By varying the initial concentrations of $S_2O_8^{2-}$ and I^-, you can, via the above analysis, determine the order of the reaction with respect to both of these species.

Helpful Comments

1. According to the procedure of this experiment, the solution will turn blue-black when exactly 2×10^{-4} mol of $S_2O_8^{-2}$ has reacted.
2. The purpose of the KNO_3 solution in this reaction is to keep the *reaction medium* the same in each run in terms of the concentration of ions; it does not enter into the reaction in any way.
3. The reaction studied in this experiment is catalyzed by metal ions. The purpose of the drop of the EDTA solution is to minimize the effects of trace quantities of metal ion impurities that would cause spurious effects on the reaction.
4. You will perform a few preliminary experiments to become acquainted with the observations in this experiment so that you will know what to expect in the reactions.
5. The initial concentrations of the reactants have been provided for you on the report sheet.

PROCEDURE

A. Preliminary Experiments

1. Dilute 5.0 mL of 0.2 *M* KI solution with 10.0 mL of distilled water in a test tube, add three drops of starch solution and mix thoroughly, and then add 5.0 mL of 0.2 *M* $(NH_4)_2S_2O_8$ solution. Mix. Wait a while and observe color changes.
2. Repeat the procedure in (1), but when the solution changes color add four drops of 0.4 *M* $Na_2S_2O_3$, mix the solution, and note the effect that the addition of $Na_2S_2O_3$ has on the color.

B. Kinetics Experiment

Solution Preparation Prepare four reaction solutions as follows (one at a time):

Solution 1: 25.0 mL KI solution
1.0 mL starch solution
1.0 mL $Na_2S_2O_3$ solution
48.0 mL KNO_3 solution
1 drop EDTA solution
Total volume = 75.0 mL

Solution 2: 25.0 mL KI solution
1.0 mL starch solution
1.0 mL $Na_2S_2O_3$ solution
23.0 mL KNO_3 solution
1 drop EDTA solution
Total volume = 50.0 mL

Solution 3: 50.0 mL KI solution
1.0 mL starch solution
1.0 mL $Na_2S_2O_3$ solution
23.0 mL KNO_3 solution
1 drop EDTA solution
Total volume = 75.0 mL

Solution 4: 12.5 mL KI solution
1.0 mL starch solution
1.0 mL $Na_2S_2O_3$ solution
35.5 mL KNO_3 solution
1 drop EDTA solution
Total volume = 50.0 mL

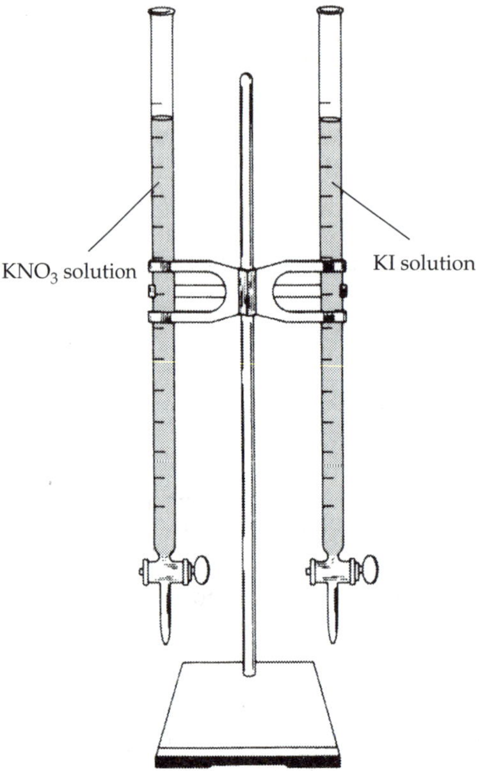

▲ FIGURE 27.2

Instructor: The $(NH_4)_2S_2O_8$ solution can also be dispensed from a buret.

Equipment Setup Set up two burets held by a clamp on a ring stand as shown in Figure 27.2. Use these burets to accurately measure the volumes of the KI and KNO_3 solutions. Use two separate 1-mL pipets for measuring the volumes of the $Na_2S_2O_3$ and starch solutions, and use 25-mL and 50-mL pipets to measure the volumes of the $(NH_4)_2S_2O_8$ solutions. Each solution must be freshly prepared to begin the rate study—that is, *prepare solutions 1, 2, 3, and 4 one at a time as you make your measurements.*

Rate Measurements Prepare solution 1 in a 250-mL Erlenmeyer flask that has been scrupulously cleaned and dried. Pipet 25.0 mL of $(NH_4)_2S_2O_8$ solution into a clean, dry 100-mL beaker. *Be ready* to begin timing the reaction when the solutions are mixed (READ AHEAD). The reaction starts the moment the solutions are mixed! BE PREPARED! ZERO TIME! Quickly pour the 25.0 mL of $(NH_4)_2S_2O_8$ solution into solution 1 and swirl vigorously; note the time you begin mixing to the nearest second. At the instant when the blue-black color appears, 2×10^{-4} mol of $S_2O_8^{2-}$ has reacted. IMMEDIATELY (be prepared!) add a 1-mL aliquot of $Na_2S_2O_3$ solution from the pipet and swirl the solution; the color will disappear. If you fill each of seven clean, dry test tubes with 1 mL of $Na_2S_2O_3$ solution, you then can add these aliquots to your reactions at the appearance of the blue color without loss of time.

Record the time for the reappearance of the blue-black color. Add another 1-mL aliquot of $Na_2S_2O_3$ solution and note the time for the reappearance of the color. The time interval being measured is that between the appearance of the blue-black color. For good results, these aliquots of $Na_2S_2O_3$ must be measured as quickly, accurately, and reproducibly as possible. Continue this procedure until you have added seven (7) aliquots to solution 1.

You are finished with solution 1 when you have recorded all your times on the report sheet. (The time intervals are cumulative.)

Solutions 2, 3, and 4 should be treated in exactly the same manner except that 50.0-mL portions of $(NH_4)_2S_2O_8$ solutions should be added to solutions 2 and 4 and 25 mL of $(NH_4)_2S_2O_8$ solution should be added to solution 3. (CAUTION: *Be on guard—solution 3 will react much more rapidly than solution 1.)* In each of these reactions the final total solution volume is 100 mL.

Instructor: Time may be saved in this experiment by setting up several burets filled with $(NH_4)_2S_2O_8$, KNO_3, KI, and $Na_2S_2O_3$ solutions on side tables in the laboratory. The students may obtain the solutions from these burets and refill the burets from bottles of stock solutions also kept there.

Calculations

Tabulate on the data sheet for each aliquot of $Na_2S_2O_3$ added to each of the four solutions:

1. The time interval from the start of the reaction (addition of $S_2O_8^{2-}$) to the appearance of color for the first aliquot of $S_2O_3^{2-}$ and the time interval from the preceding color appearance for each succeeding aliquot (column 2).
2. The cumulative time from the start of the reaction to each appearance of color (column 3).
3. The corresponding numbers of moles $S_2O_8^{2-}$ consumed (column 4).

For each solution, plot on the graph paper provided the moles of $S_2O_8^{2-}$ consumed (as the ordinate, vertical axis) versus time in seconds (as the abscissa, horizontal axis), using the data in columns 3 and 4. Calculate the slope of each plot, and from these calculations answer the questions on your report sheet.

Waste-Disposal Instructions No wastes from this experiment should be flushed down the sink. Locate the special containers placed in the laboratory for the disposal of excess iodide and peroxydisulfate solutions as well as for the reaction mixtures from the test tubes or flasks. All wastes should be disposed of in these containers.

REVIEW QUESTIONS

Before beginning this experiment in the laboratory, you should be able to answer the following questions:

1. What factors influence the rate of a chemical reaction?
2. What is the general form of a rate law?
3. What is the order of reaction with respect to A and B for a reaction that obeys the rate law rate $= k[A]^2[B]^3$?
4. Write the chemical equations involved in this experiment and show that the rate of disappearance of $[S_2O_8^{2-}]$ is proportional to the rate of appearance of the blue-black color of the starch-iodine complex.
5. It is found for the reaction $A + B \longrightarrow C$ that doubling the concentration of either A or B quadruples the rate of the reaction. Write the rate law for this reaction.
6. If 2×10^{-4} moles of $S_2O_8^{2-}$ in 50 mL of solution is consumed in 188 seconds, what is the rate of consumption of $S_2O_8^{2-}$?

7. Why are chemists concerned with the rates of chemical reactions? What possible practical value does this type of information have?

8. Suppose you were dissolving a metal such as zinc with hydrochloric acid. How would the particle size of the zinc affect the rate of its dissolution?

9. Assuming that a chemical reaction doubles in rate for each 10° temperature increase, by what factor would the rate increase if the temperature were increased 40°C?

10. A reaction between the substances A and B has been found to give the following data:

$$3A + 2B \longrightarrow 2C + D$$

[A] (mol/L)	*[B] (mol/L)*	*Rate of appearance of C (mol/L-hr)*
1.0×10^{-2}	1.0	0.3×10^{-6}
1.0×10^{-2}	3.0	8.1×10^{-6}
2.0×10^{-2}	3.0	3.24×10^{-5}
2.0×10^{-2}	1.0	1.20×10^{-6}
3.0×10^{-2}	3.0	7.30×10^{-5}

Using the above data, determine the order of the reaction with respect to A and B and the rate law and calculate the specific rate constant.

Name ______________________ Desk ______________________

Date ______________ Laboratory Instructor ______________________

REPORT SHEET | EXPERIMENT 27

Rates of Chemical Reactions I: A Clock Reaction

A. Preliminary Experiments

1. What are the colors of the following ions: K^+ colorless ; I^- colorless
2. The color of the starch · I_2 complex is blue-black

B. Kinetics Experiment

Solution 1. Initial $[S_2O_8^{2-}] = 0.05\ M$; initial $[I^-] = 0.05\ M$. Time experiment started ______________

Aliquot no.	*Time (s) between appearances of color*	*Cumulative times(s)*	*Total moles of $S_2O_8^{2-}$ consumed*
1	188	188	2.0×10^{-4}
2	190	378	4.0×10^{-4}
3	200	578	6.0×10^{-4}
4	215	793	8.0×10^{-4}
5	235	1028	10.0×10^{-4}
6	255	1283	12.0×10^{-4}
7	274	1557	14.0×10^{-4}

Solution 2. Initial $[S_2O_8^{2-}] = 0.10\ M$; initial $[I^-] = 0.05\ M$. Time experiment started ______________

Aliquot no.	*Time (s) between appearances of color*	*Cumulative time (s)*	*Total moles of $S_2O_8^{2-}$ consumed*
1	91	91	2.0×10^{-4}
2	95	186	4.0×10^{-4}
3	97	283	6.0×10^{-4}
4	99	382	8.0×10^{-4}
5	100	482	10.0×10^{-4}
6	100	582	12.0×10^{-4}
7	102	684	14.0×10^{-4}

Solution 3. Initial $[S_2O_8^{2-}] = 0.05\ M$; initial $[I^-] = 0.10\ M$. Time experiment started ____________

Aliquot no.	*Time (s) between appearances of color*	*Cumulative time (s)*	*Total moles of $S_2O_8^{2-}$ consumed*
1	122	122	2.0×10^{-4}
2	126	248	4.0×10^{-4}
3	136	384	6.0×10^{-4}
4	143	527	8.0×10^{-4}
5	153	680	10.0×10^{-4}
6	160	840	12.0×10^{-4}
7	170	1010	14.0×10^{-4}

Solution 4. Initial $[S_2O_8^{2-}] = 0.10\ M$; initial $[I^-] = 0.025\ M$. Time experiment started ____________

Aliquot no.	*Time (s) between appearances of color*	*Cumulative time (s)*	*Total moles of $S_2O_8^{2-}$ consumed*
1	192	192	2.0×10^{-4}
2	198	390	4.0×10^{-4}
3	203	593	6.0×10^{-4}
4	209	802	8.0×10^{-4}
5	209	1011	10.0×10^{-4}
6	218	1229	12.0×10^{-4}
7	221	1450	14.0×10^{-4}

Calculations

1. Rate of reaction, $\Delta[S_2O_8^{2-}]/\Delta t$, as calculated from graphs (that is, from slopes of lines):

   Solution 1 0.092×10^{-4} Solution 3 0.15×10^{-4} (All have the unit: mol/L-s)

   Solution 2 0.20×10^{-4} Solution 4 0.11×10^{-4}

2. What effect does doubling the concentration of I^- have on the rate of this reaction?
   From 1 and 3, rate increases by factor of 1.6 (should double).

3. What effect does changing the $[S_2O_8^{2-}]$ have on the reaction?
   From 1 and 2, rate increases by a factor of 2; doubling concentration doubles rate.

4. Write the rate law for this reaction that is consistent with your data.
   rate $= k[I^-][S_2O_8^{2-}]$

5. From your knowledge of x and y in the equation (as well as the rate in a given experiment from your graph), calculate k from your data. Rate $= k[S_2O_8^{2-}]^x[I^-]^y$
   See graphs. average $k = 3.7 \times 10^{-3}$ L/mol–s
   $k_1 = \text{Rate}_1/[S_2O_8^{2-}][I^-] = 3.7 \times 10^{-3}$ L/mol–s
   $k_2 = 4.0 \times 10^{-3}$ L/mol–s
   $k_3 = 3.0 \times 10^{-3}$ L/mol–s
   $k_4 = 4.0 \times 10^{-3}$ L/mol–s

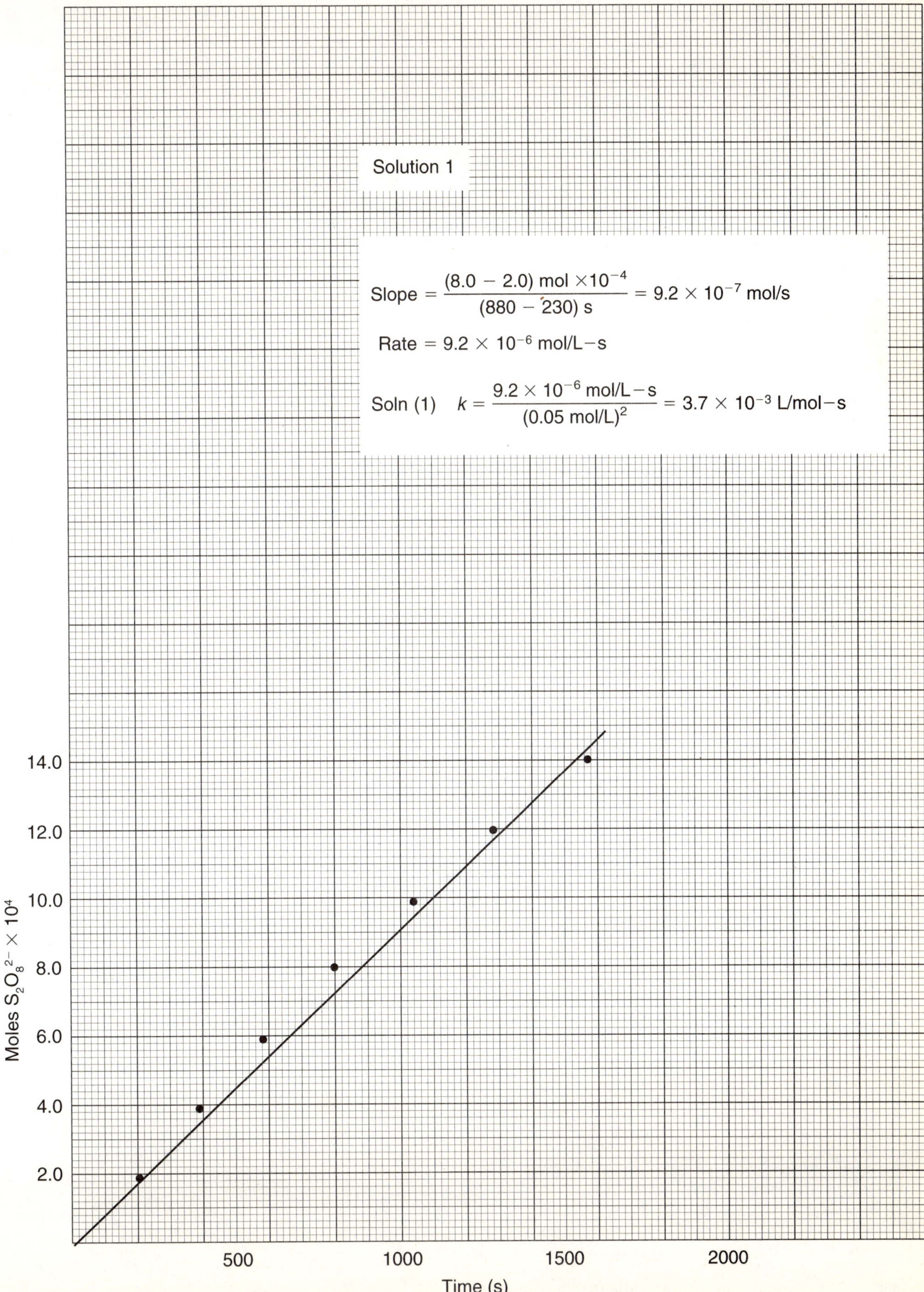
Solution 1
Slope = (8.0 − 2.0) mol ×10^−4 / (880 − 230) s = 9.2 × 10^−7 mol/s
Rate = 9.2 × 10^−6 mol/L−s
Soln (1) k = 9.2 × 10^−6 mol/L−s / (0.05 mol/L)^2 = 3.7 × 10^−3 L/mol−s
14.0
12.0
10.0
8.0
6.0
4.0
2.0
Moles S2O8 2− × 10^4
500
1000
1500
2000
Time (s)

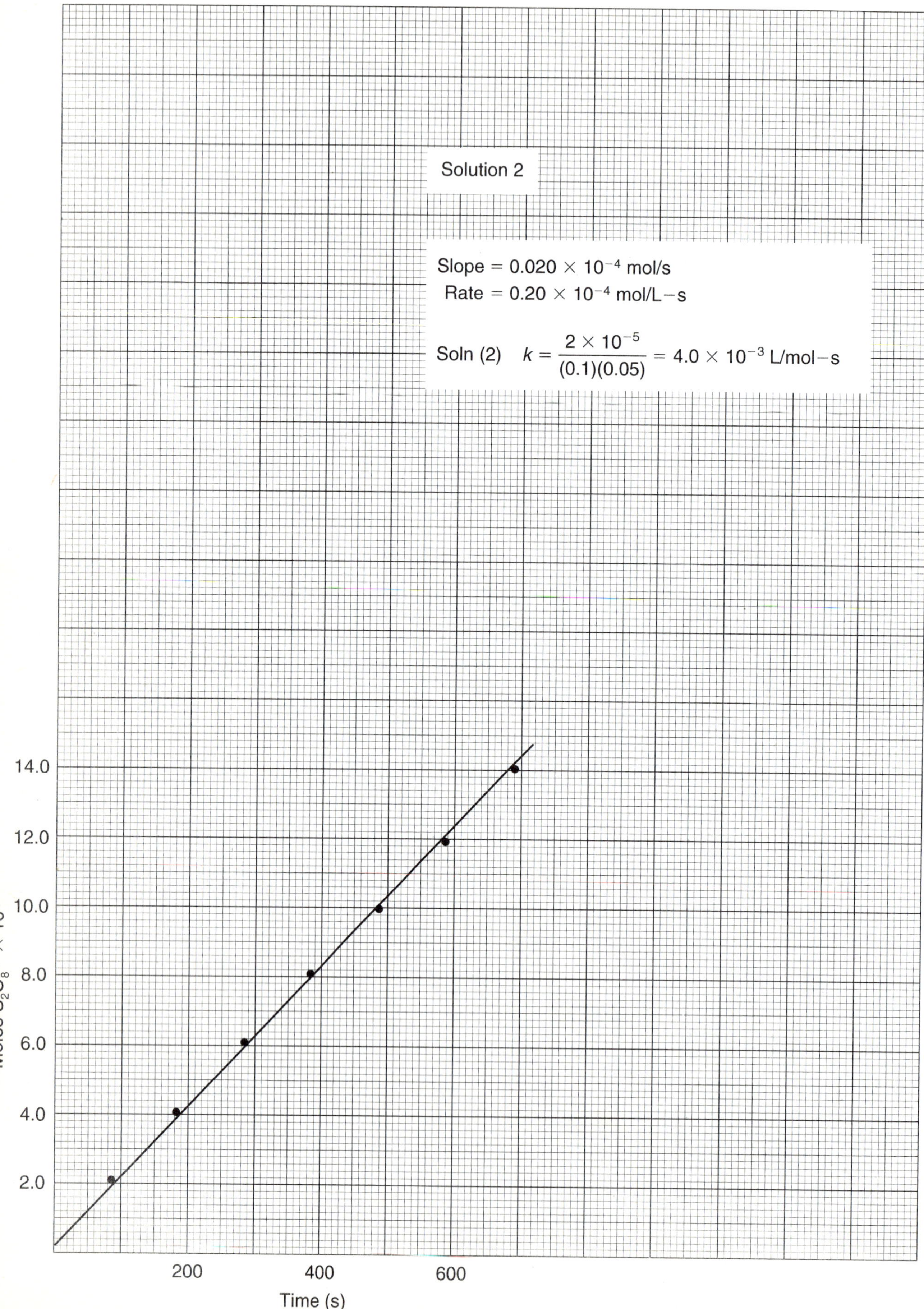
Solution 2
Slope = 0.020 × 10^−4 mol/s
Rate = 0.20 × 10^−4 mol/L−s
Soln (2) k = 2 × 10^−5 / (0.1)(0.05) = 4.0 × 10^−3 L/mol−s
14.0
12.0
10.0
8.0
6.0
4.0
2.0
Moles S2O8 2− × 10^4
200
400
600
Time (s)

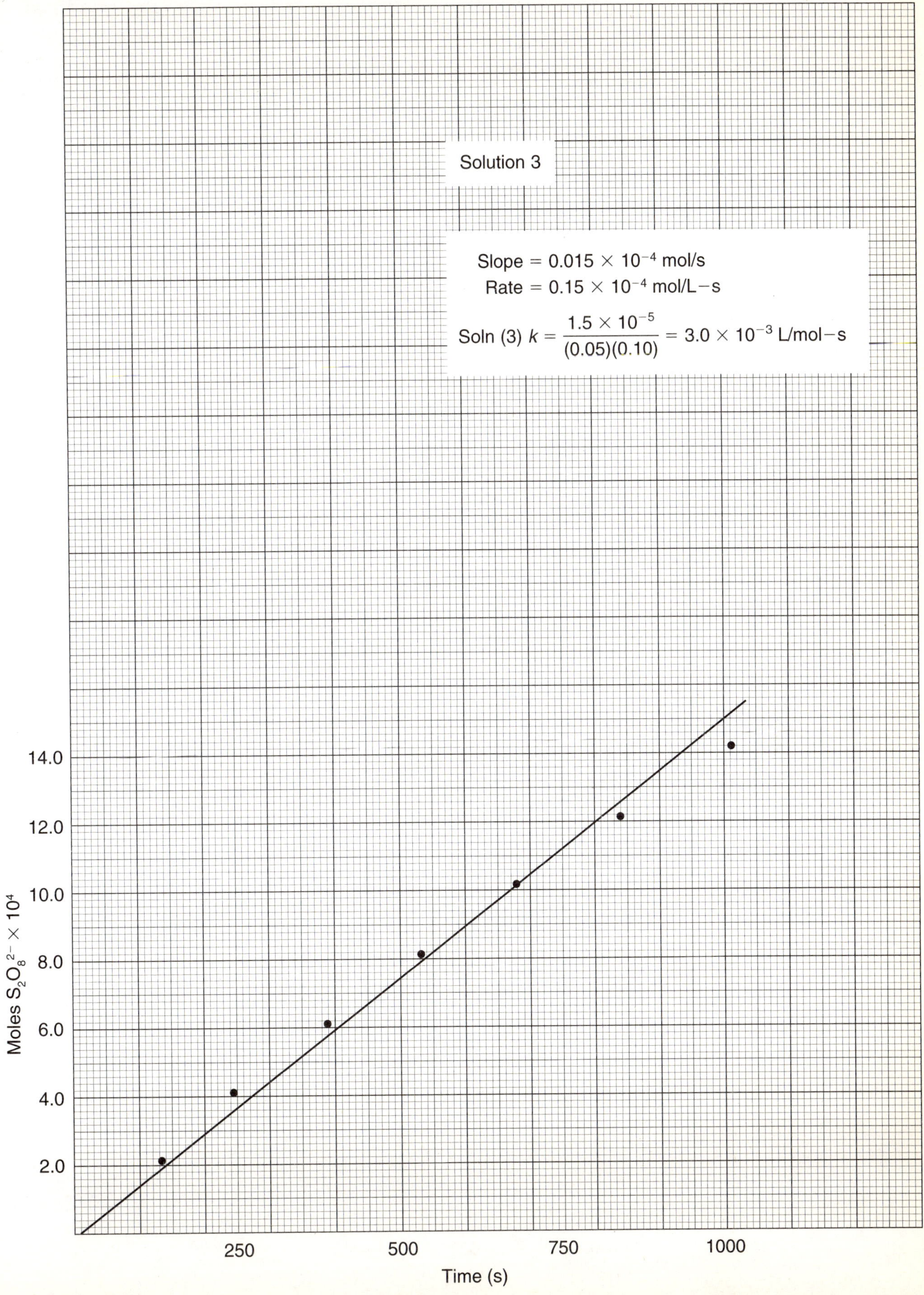

Solution 3
Slope = 0.015 × 10⁻⁴ mol/s
Rate = 0.15 × 10⁻⁴ mol/L−s
Soln (3) $k = \frac{1.5 \times 10^{-5}}{(0.05)(0.10)} = 3.0 \times 10^{-3}$ L/mol−s
14.0
12.0
10.0
8.0
6.0
4.0
2.0
Moles $S_2O_8^{2-} \times 10^4$
250
500
750
1000
Time (s)

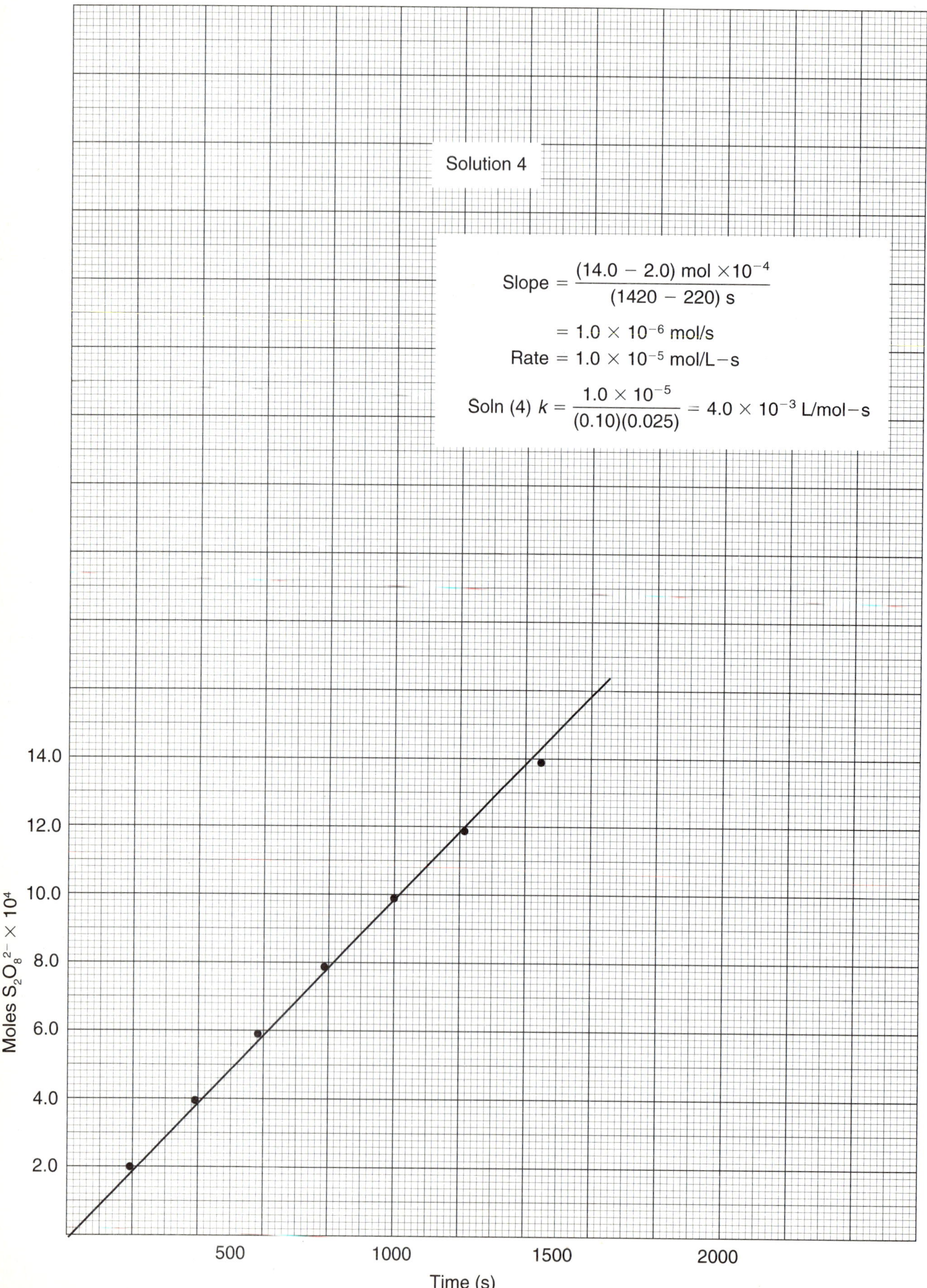

Solution 4
Slope = (14.0 − 2.0) mol ×10^−4 / (1420 − 220) s
= 1.0 × 10^−6 mol/s
Rate = 1.0 × 10^−5 mol/L−s
Soln (4) k = 1.0 × 10^−5 / (0.10)(0.025) = 4.0 × 10^−3 L/mol−s
14.0
12.0
10.0
8.0
6.0
4.0
2.0
Moles S2O8 2− × 10^4
500
1000
1500
2000
Time (s)

Experiment

Rates of Chemical Reactions II: Rate and Order of H_2O_2 Decomposition

OBJECTIVE

To determine the rate and order of reaction for the decomposition of hydrogen peroxide.

APPARATUS AND CHEMICALS

Apparatus

Mohr buret or standard buret
leveling bulb or funnel
125-mL Erlenmeyer flask
buret clamp
graduated cylinder
rubber tubing
test tube
medicine droppers (2)
barometer
ring stand and iron ring
wristwatch or timer
thermometer
pneumatic trough
No. 00 and No. 6 one-hole stoppers

Chemicals

3% H_2O_2
0.10 *M* $Hg(NO_3)_2$
0.10 *M* KI

WORK IN PAIRS, BUT EVALUATE YOUR DATA INDEPENDENTLY.

DISCUSSION

We have learned that much information may be obtained from a chemical equation–for example, the identity of the products formed from the specified reactants and the quantitative mass relationships between the substances involved in the reaction. The equation, however, does not tell us anything about the conditions required for the reaction to occur, nor does it give any information about the speed or rate at which the reaction takes place. The area of chemistry concerned with rates of chemical reactions is called *chemical kinetics*.

Experiments show that rates of homogeneous reactions in solution depend on:

1. The nature of the reactants
2. The concentration of the reactants
3. The temperature
4. Catalysis

Before a reaction can occur, the reactants must come into direct contact via collisions of the reacting particles. However, even then, the reacting particles (ions or molecules) must collide with sufficient energy to result in a reaction. If they do not, their collisions are ineffective and analogous to collisions of billiard balls. With these considerations in mind, we can qualitatively explain how the various factors influence the rates of reactions.

Concentration Changing the concentration of a solution alters the number of particles per unit volume. The more particles present in a given volume, the greater the probability of their colliding. Hence, increasing the concentration of a solution increases the number of collisions per unit time and therefore the rate of reaction.

Temperature Since temperature is a measure of the average kinetic energy, an increase in temperature increases the kinetic energy of the particles. An increase in kinetic energy increases the velocity of the particles and therefore the number of collisions between them in a *given period of time.* Thus, the rate of reaction increases. Also, an increase in kinetic energy results in a greater proportion of the collisions having the required energy for reaction. As a rule of thumb, for each 10°C increase in temperature, the rate of reaction doubles.

Catalyst Catalysts, in some cases, are believed to increase reaction rates by bringing particles into close juxtaposition in the correct geometrical arrangement for reaction to occur. In other instances, catalysts offer an alternative route to the reaction, one that requires less energetic collisions between reactant particles. If less energy is required for a successful collision, a larger percentage of the collisions will have the requisite energy, and the reaction will occur faster. Actually, the catalyst may take an active part in the reaction, but at the end of the reaction, the catalyst can be recovered chemically unchanged.

Order of Reaction Defined

Let's examine precisely what is meant by the expression *rate of reaction.* Consider the hypothetical reaction

$$A + B \longrightarrow C + D \qquad [1]$$

The rate of this reaction may be measured by observing the rate of disappearance of either of the reactants A and B, or the rate of appearance of either of the products C and D. In practice, then, one measures the change of concentration with time of either A, B, C, or D. Which species you choose to observe is a matter of convenience. For example, if A, B, and D are colorless and C is colored, you could conveniently measure the rate of appearance of C by observing an increase in the intensity of the color of the solution as a function of time. Mathematically, the rate of reaction may be expressed as follows:

$$\text{rate of disappearance of A} = \frac{\text{change in concentration of A}}{\text{time required for change}} = \frac{-\Delta[\text{A}]}{\Delta t}$$

$$\text{rate of appearance of C} = \frac{\text{change in concentration of C}}{\text{time required for change}} = \frac{\Delta[\text{C}]}{\Delta t}$$

In general, the rate of the reaction will depend on the concentration of the reactants. Thus, the rate of our hypothetical reaction may be expressed as

$$\text{rate} = k[\text{A}]^x[\text{B}]^y \qquad [2]$$

where [A] and [B] are the molar concentrations of A and B, x and y are the powers to which the respective concentrations must be raised to describe the rate, and k is the *specific rate constant*. One of the objectives of chemical kinetics is to determine the rate law. Stated slightly differently, one goal of measuring the rate of the reaction is to determine the numerical values of x and y. Suppose that we found $x = 2$ and $y = 1$ for this reaction. Then,

$$\text{rate} = k[A]^2[B] \qquad [3]$$

would be the rate law. It should be evident from Equation [3] that doubling the concentration of B (keeping [A] the same) would cause the reaction rate to double. On the other hand, doubling the concentration of A (keeping [B] the same) would cause the rate to increase by a factor of 4, because the rate of the reaction is proportional to the *square* of the concentration of A. The powers to which the concentrations in the rate law are raised are termed the *order of the reaction*. In this case, the reaction is said to be second order in A and first order in B. The *overall order* of the reaction is the sum of the exponents, $2 + 1 = 3$, or a third-order reaction. It is possible to determine the order of the reaction by noting the effects of changing reagent concentrations on the rate of the reaction. Note that the order of a reaction may be (and frequently is) different from the stoichiometry of the reaction.

It should be emphasized that k, the specific rate constant, has a definite value that is independent of the concentration. It is characteristic of a given reaction and depends only on temperature. Once the rate is known, the value of k can be calculated.

The term *rate of reaction* refers to the speed at which reactants are being consumed or products are being formed. This must be determined experimentally by measuring the time rate of change in the concentration of one of the reactants or one of the products or some physical property (such as the volume of a gas or color intensity of a solution) that is directly proportional to one of these concentrations. The rate may be expressed, for example, as moles per liter of product being formed per minute, milliliters of gas being produced per minute, or moles per liter of reactant being consumed per second. In this experiment, you will study the rate of the iodide-catalyzed decomposition of hydrogen peroxide to form water and oxygen:

$$2H_2O_2(aq) \xrightarrow{I^-} O_2(g) + 2H_2O(l) \qquad [4]$$

The standard enthalpy change, ΔH°_{298}, and standard free-energy change, ΔG°_{298}, for the reaction in Equation [4] are −196.1 kJ and −233.5 kJ, respectively. Even though ΔG° is negative and we can conclude that the reaction will proceed spontaneously at room temperature, we cannot predict anything about the *rate* at which H_2O_2 decomposes. Reactions generally involve an energy barrier (Figure 28.1) that must be overcome in going from reactants to products. This energy barrier is called the *activation energy* and is commonly designated by the term E_a. It corresponds to the minimum energy that molecules must have for the reaction to occur. The magnitude of E_a determines the rate of the reaction, whereas the magnitude of the free-energy change determines the extent of the reaction. By observing the volume of oxygen formed as a function of time, you will determine how the rate of reaction [4] is affected by different initial concentrations of hydrogen peroxide and iodide ion (which is introduced into the reaction in the form of the strong electrolyte KI). Although iodide does not appear in the chemical equation, it does have a pronounced effect on the rate of the reaction, for it serves as a catalyst.

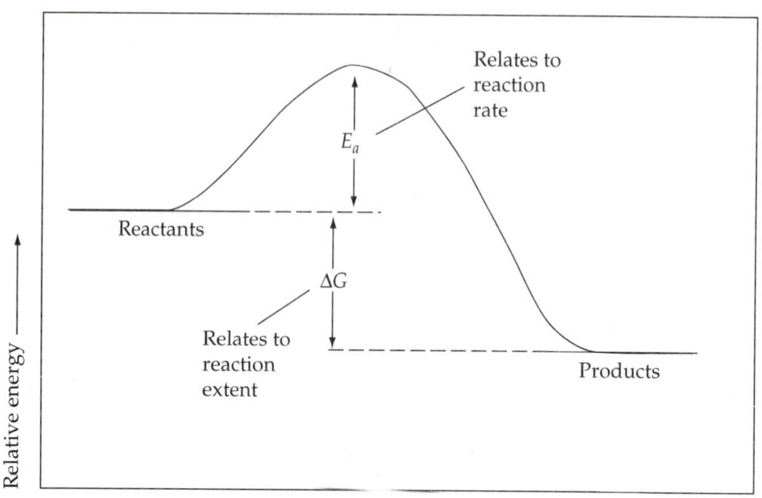

▲ **FIGURE 28.1** Reaction profile for an exothermic reaction illustrating ΔG and E_a.

Your ultimate goal in this experiment is to deduce a rate law for the reaction, showing the dependence of the rate on the concentrations of H_2O_2 and I^-. Your rate law will be of the form.

$$\text{rate of oxygen production} = k[H_2O_2]^x[I^-]^y$$

where k is the specific rate constant and depends only on temperature. Thus your objective is to determine the numerical values of the exponents x and y.

You will also make a qualitative study of the effect that temperature has on the rate of decomposition of H_2O_2. At higher temperatures, the average kinetic energy of molecules is greater. Hence, the fraction of molecules with kinetic energy sufficient for reaction is larger, and the rate of reaction is correspondingly larger. Arrhenius noted a nonlinear relation between rate and temperature and found that most reaction-rate data obeyed the equation

$$\log k = \log A - \frac{E_a}{2.30\,RT}$$

where k is the specific rate constant, E_a the activation energy, R the gas constant, and T the absolute temperature. A is a constant, or nearly so, as temperature is varied. It is called the frequency factor and is related to the frequency of collisions and the probability that the molecules are suitably oriented for reaction.

PROCEDURE

A. Order of Reaction

Assemble an apparatus for the collection of oxygen similar to that shown in Figure 28.2. (The rubber tube connections must be snug!) Fill the trough with water at room temperature and record the temperature. It may be necessary to add some hot water to achieve room temperature. Add room-temperature water to the assembly until the height of water in the buret is about 10 mL from the top when the water in the leveling bulb is at the same level (see Figure 28.2). Check for leaks in the apparatus by lowering

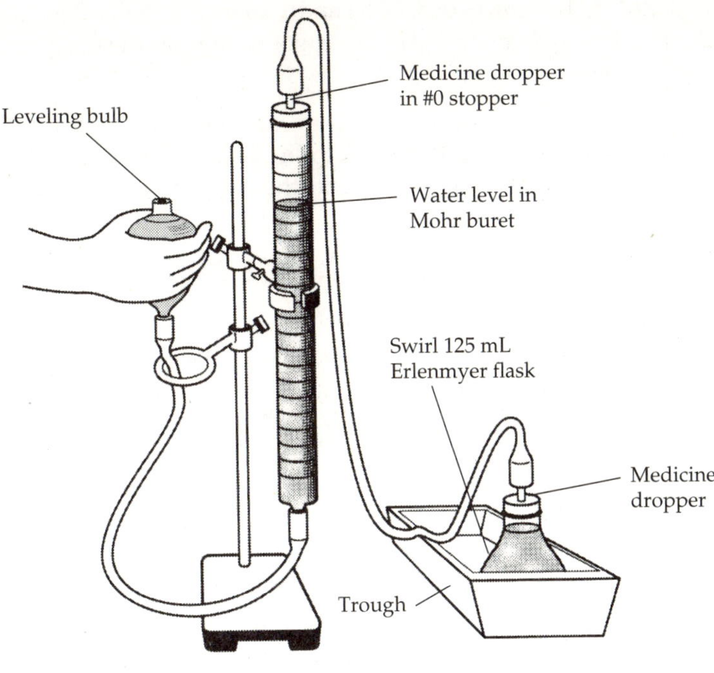

▲ FIGURE 28.2

the bulb with the system closed. In the absence of leaks, only small changes in the water level in the buret due to pressure changes will occur as the leveling bulb is lowered.

Solution 1 Add 10.0 mL of 0.10 *M* KI and 15.0 mL of distilled water to a clean Erlenmeyer flask. Carefully swirl the flask for a few minutes so that the solution attains the bath temperature in the trough. Add 5.0 mL of 3% H_2O_2 and quickly stopper the flask. One student should keep swirling the flask in the bath as vigorously as possible throughout the remainder of the experiment; the other student should observe the volume of oxygen produced during the reaction at various times. The recording of volume and time should commence after approximately 2 mL of gas has been evolved. Because it is important to measure the volume of oxygen evolved at constant pressure, it is necessary to measure the volume with the level of water in the leveling bulb the same as that in the buret. One student should match the water levels in the buret and leveling bulb (by lowering the leveling bulb) and also read the volume. The other student should continue to swirl the flask and record on the report sheet the total elapsed time at the instant the gas volume is read. Time and volume readings should be taken at approximately 2-mL intervals until a total of about 14 mL of oxygen has been evolved. Although it is not necessary, you will find it convenient to use the stopwatch function of a digital wristwatch for noting the time intervals.

Instructor: Remind students that they should keep track of cumulative, not incremental, times here.

Solution 2 Rinse the Erlenmeyer flask and graduated cylinder thoroughly with distilled water and allow them to drain completely. Be certain that the bath temperature is the same as that used for solution 1. It may be necessary to adjust it by adding a little warm water. Repeat the experiment, this time first adding 10.0 mL of 0.10 *M* KI and 10.0 mL of distilled water, swirling, and then adding 10.0 mL of 3% H_2O_2. Quickly stopper the flask and take volume-time readings as before.

Solution 3 Rinse the flask and graduated cylinder. Check bath temperature and adjust if necessary. Repeat as follows: First add 20.0 mL of 0.10 *M* KI and 5.0 mL of distilled water, swirl, and then add 5.0 mL of H_2O_2. Take volume-time readings as before.

B. Effect of Temperature

Adjust the bath temperature so that it is approximately 10°C to 12°C higher than what it was for Part A above. Then repeat the experiment, following the directions given in Part A for solution 1. After you have collected 14 mL of O_2, do not take any more volume-time readings, but allow the reaction to continue to completion, that is, until no more oxygen is evolved. Save the solution for Part C. Record the total volume of O_2 collected and the barometric pressure on your report sheet.

C. Identification of the Catalyst

You can show that iodide has been a catalyst in the decomposition of H_2O_2 by noting the fact that it has not been consumed during the reaction even though it appears in the rate law. You can identify the iodide ion by precipitating it as HgI_2. Pour about 1 mL of the reaction mixture that you saved from Part B into a small test tube. Add 10 drops of 0.1 *M* $Hg(NO_3)_2$ to the test tube, mix, and record your observations. **(CAUTION:** ***Mercuric nitrate is toxic, and you should not get any on yourself. Should you come in contact with the $Hg(NO_3)_2$ solution, immediately wash it off with copious amounts of water.)*** Dilute about 0.2 mL of 0.10 *M* KI with about 5 mL of distilled water. Add two drops of this solution to 10 drops of 0.1 *M* $Hg(NO_3)_2$ and record your observations.

Waste Disposal Instructions Any of the mercury-containing solutions should be disposed of in a designated waste container. The other solutions are harmless and should be disposed of according to the instructor's directions.

Calculations

After you have completed Parts A, B, and C, graph your data, showing for each run the volume of oxygen on the ordinate (vertical axis) versus the elapsed time along the abscissa (horizontal axis). Zero time for each run corresponds to the time when the first 2 mL of O_2 were evolved. Draw the best straight line through your points for each run. Your line need not pass through all of the points. Label each line. Calculate the slopes for each of your four lines. The slopes of the lines correspond to the initial rates of the reaction.

Decide how the initial rate of oxygen formation is affected by doubling the concentrations of H_2O_2 and I^-. How do the concentrations of H_2O_2 and I^- change from solution 1 to solution 2 and then solution 3? Write a rate law for the reaction using the values of x and y you have determined by rounding them off to whole numbers.

REVIEW QUESTIONS

Before beginning this experiment in the laboratory, you should be able to answer the following questions:

1. What four factors influence the rate of a reaction?
2. If the rate law for a reaction is rate = $k[A]^2[B]$,
   (a) What is the overall order of the reaction?
   (b) If the concentration of both A and B are doubled, how will this affect the rate of the reaction?
   (c) What name is given to k?
   (d) How will doubling the concentration of A, while the concentration of B is kept constant, affect the value of k (assume temperature does not change)? How is the rate affected?
3. Define the term *catalyst*.
4. Write the chemical equation for the iodide-catalyzed decomposition of H_2O_2.
5. It is found for the reaction $2A + B \longrightarrow C$ that doubling the concentration of either A or B quadruples the reaction rate. Write a rate law for the reaction.
6. When the concentration of a substance is doubled, what effect does it have on the rate if the order with respect to that reactant is (a) 0; (b) 1; (c) 2; (d) 3; (e) $^1/_2$?
7. The following data were collected for the volume of O_2 produced in the decomposition of H_2O_2.

Time (s)	*mL O_2*
0.0	0.0
45.0	2.0
88.0	3.9
131.0	5.8

   Calculate the average rate of reaction for each of the time intervals between measurements.

NOTES AND CALCULATIONS

Name ______________________ Desk ______________________

Date ______________________ Laboratory Instructor ______________________

REPORT SHEET | EXPERIMENT

Rates of Chemical Reactions II: Rate and Order of H_2O_2 Decomposition | 28

A. Order of Reaction

BATH TEMPERATURE

21.1 °C			21.5 °C			21.0 °C		
Solution 1			***Solution 2***			***Solution 3***		
Buret reading (mL)	*Vol. of O_2 (mL)*	*Time*	*Buret reading (mL)*	*Vol. of O_2 (mL)*	*Time*	*Buret reading (mL)*	*Vol. of O_2 (mL)*	*Time*
12.0	0.0	0:00	24.4	0.0	0:00	24.5	0.0	0:00
14.0	2.0	0:59	26.4	2.0	1:01	26.5	2.0	0:54
16.0	4.0	1:50	28.4	4.0	1:30	28.5	4.0	1:27
18.0	6.0	2:42	30.4	6.0	1:57	30.5	6.0	1:55
20.0	8.0	3:39	32.4	8.0	2:21	32.5	8.0	2:22
22.0	10.0	4:31	34.4	10.0	2:46	34.5	10.0	2:48
24.0	12.0	5:26	36.4	12.0	3:10	36.5	12.0	3:15
26.0	14.0	6:25	38.4	14.0	3:36	38.5	14.0	3:42

Slope $= \dfrac{14.0 \text{ ml}}{385\ s}$

$= 3.64 \times 10^{-2}$ mL/s

Slope $= \dfrac{13.0 - 3.0 \text{ mL}}{205 - 77\ s}$

$= \dfrac{10.0 \text{ mL}}{128\ s} = 7.81 \times 10^{-2}$ mL/s

Slope $= \dfrac{14.0 - 1.0 \text{ mL}}{222 - 40\ s}$

$= \dfrac{13.0 \text{ mL}}{182\ s}$

$= 7.14 \times 10^{-2}$ mL/s

Rate Data

Solution	mL H_2O_2	mL KI	Rate (mL O_2/s)
1	5.0	10.0	3.64×10^{-2}
2	10.0	10.0	7.81×10^{-2}
3	5.0	20.0	7.14×10^{-2}

Rate law:

Rate of formation of $O_2 = k[H_2O_2][I^-]$

B. Effect of Temperature

BATH TEMPERATURE 33.1°C

Buret reading (mL)	*Vol. of O_2 (mL)*	*Time*
24.2	0.0	0:00
26.2	2.0	0:29
28.2	4.0	0:50
30.2	6.0	1:10
32.2	8.0	1:27
34.2	10.0	1:45
36.2	12.0	2:04
38.2	14.0	2:22
40.6	16.4	4:20

Total volume of O_2 collected 16.4 mL
Barometric pressure 752 mm Hg
Vapor pressure of water at bath temperature (see Appendix K) 37.9 mm Hg

$$\text{Slope} = \frac{14.0 - 2.0 \text{ ml}}{142 - 29 \; s} = \frac{12.0 \text{ mL}}{113 \; s} = 10.6 \times 10^{-2} \text{ mL}/s$$

Compared with the rate found for solution 1, the rate is 2.91 times faster. $\frac{10.6}{3.64} = 2.91$
Using the ideal-gas law, calculate the moles of O_2 collected 6.13×10^{-4} mol O_2
(show calculations)

$$n = \frac{PV}{RT} = \frac{0.939 \text{ atm} \times 0.0164 \text{ L}}{0.0821 \text{ L-atm/mol-}K \times 306 \; K} = 6.13 \times 10^{-4} \text{ mol}$$

$273 + 33 = 306 \; K$; $752 - 38 = 714$ mm

$$\frac{714 \text{ mm}}{760 \text{ mm/atm}} = 0.939 \text{ atm}$$

Based on the moles of O_2 evolved, calculate the molar concentration of the *original* 3% H_2O_2 solution (show calculations).

Original volume of H_2O_2 = 5.0 mL = 0.0050 L

$$\frac{2(0.613 \text{ mmol})}{5.0 \text{ mL}} = 0.25 \; M$$

Note: $2H_2O_2 \longrightarrow O_2 + 2H_2O$ or $\frac{6.13 \times 10^{-4} \text{ mol } O_2}{0.0050 \text{ L}} \times \frac{2 \text{ mol } H_2O_2}{1 \text{ mol } O_2} = 2.5 \times 10^{-1} \; M$

mol $H_2O_2 = 2 \times$ mol O_2

C. Identification of the Catalyst

Observation when $Hg(NO_3)_2$ is added to
(a) Solution in which reaction has gone to completion:

Precipitate of HgI_2: red-orange; test works better if $Hg(NO_3)_2$ is in excess—i.e., add KI to $Hg(NO_3)_2$.

(b) Original KI solution:

Precipitate; red-orange

Conclusion regarding nature of KI:

KI is a catalyst; it is not consumed in the reaction.

QUESTIONS

1. How does the rate of formation of O_2 compare with (a) the rate of formation of H_2O and (b) the rate of disappearance of H_2O_2 for reaction [4]?

   $2H_2O_2(l) \longrightarrow O_2(g) + 2H_2O(l)$

   For each mole of O_2 formed, two moles of H_2O_2 are consumed, and two moles of H_2O are formed. The rate of formation of O_2 is half the rate of formation of H_2O and consumption of H_2O_2.

2. Why should the levels of water in the leveling bulb and buret be kept the same?

   The levels should be the same to ensure that the pressure inside the buret is constant (and equal to atmospheric) in order to obtain volume measurements that are proportional to the extent of the reaction.

3. Why were you instructed to keep swirling the Erlenmeyer flask?

   Oxygen is slightly soluble in water, and shaking helps reduce the solubility, avoiding formation of a super-saturated solution.

4. If you use 0.20 *M* KI instead of 0.10 *M* KI, how would this affect (a) the slopes of your curves, (b) the rate of the reactions, and (c) the numerical value of *k* for the reaction?

   (a) Value of slope would double.
   (b) Rate of reaction would double.
   (c) *k* is a constant and would not change; *k* only depends on temperature.

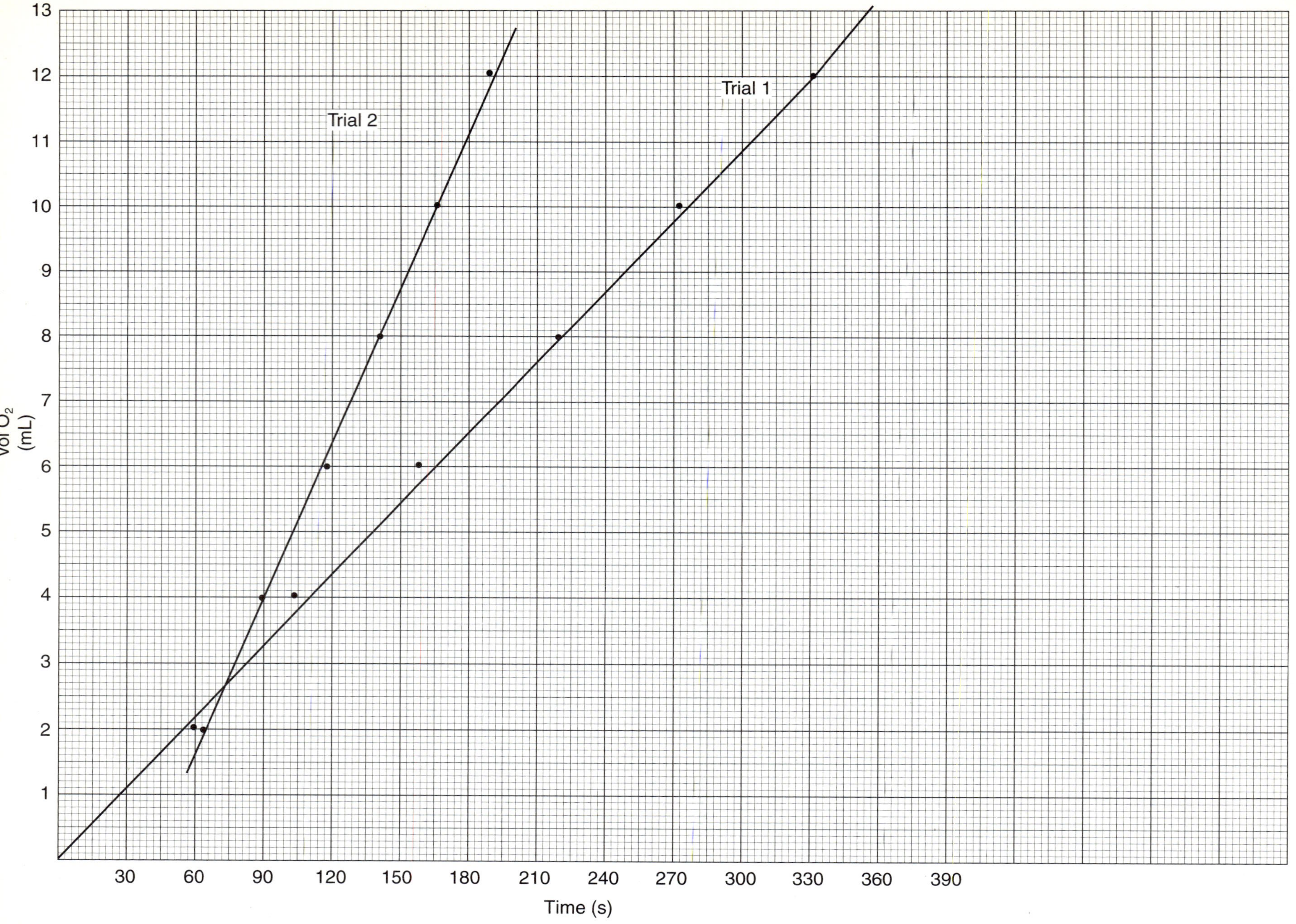

Vol O_2
(mL)
13
12
11
10
9
8
7
6
5
4
3
2
1
30
60
90
120
150
180
210
240
270
300
330
360
390
Time (s)
Trial 1
Trial 2

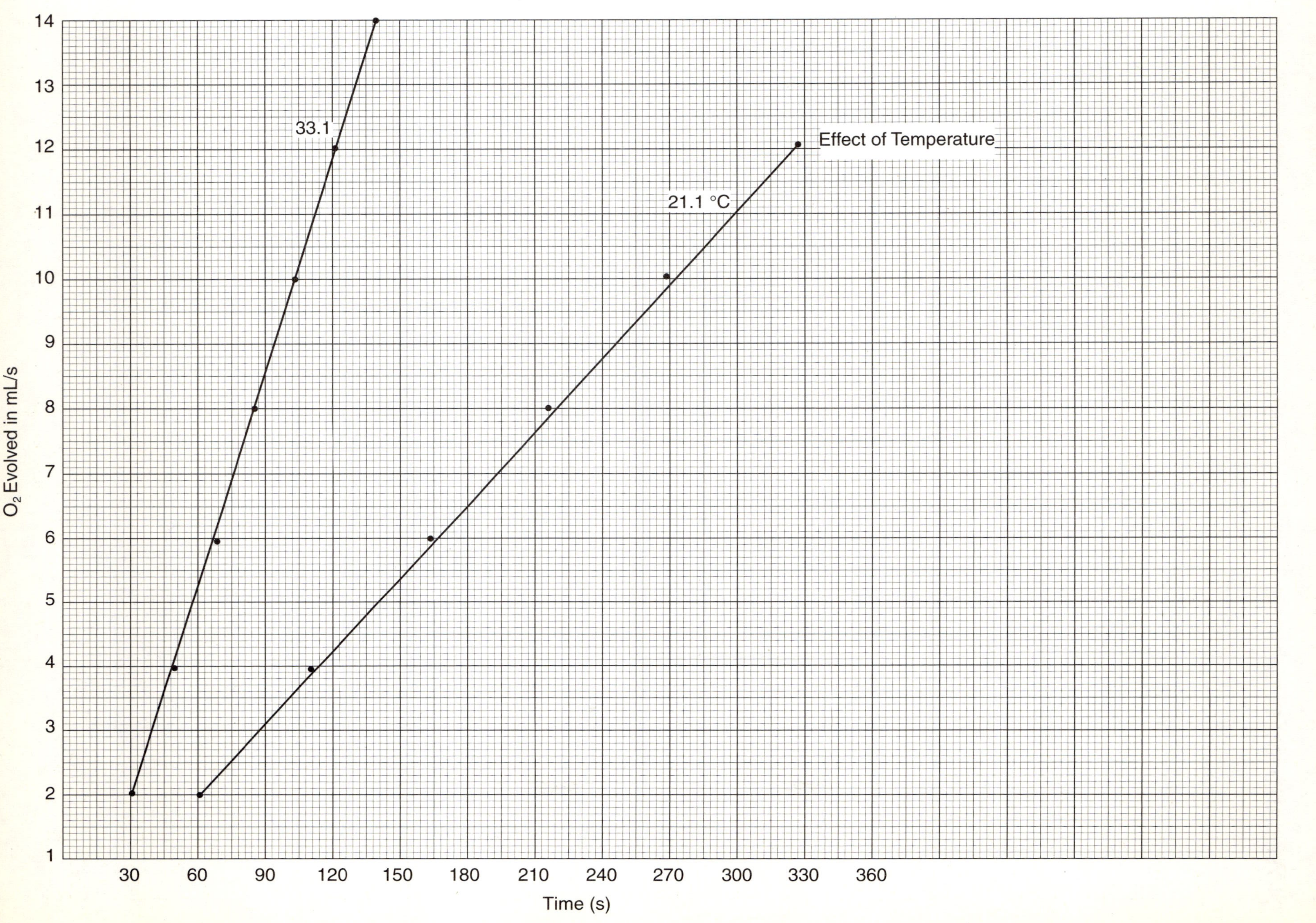
Effect of Temperature
33.1
21.1 °C
O2 Evolved in mL/s
14
13
12
11
10
9
8
7
6
5
4
3
2
1
30
60
90
120
150
180
210
240
270
300
330
360
Time (s)

NOTES AND CALCULATIONS

Experiment

29

Introduction to Qualitative Analysis

CATIONS: Na^+, NH_4^+, Ag^+, Fe^{3+}, Al^{3+}, Cr^{3+}, Ca^{2+}, Mg^{2+}, Ni^{2+}, Zn^{2+}
ANIONS: SO_4^{2-}, NO_3^-, CO_3^{2-}, Cl^-, Br^-, I^-

OBJECTIVE

To become acquainted with the chemistry of several elements and the principles of qualitative analysis.

APPARATUS AND CHEMICALS

Apparatus

small test tubes (12)
centrifuge
evaporating dish or crucible
litmus
droppers (6)
Bunsen burner and hose
10-cm Nichrome wire with loop at end

Chemicals

6 *M* H_2SO_4
18 *M* H_2SO_4
3 *M* NaOH
6 *M* HCl
6 *M* NH_3
15 *M* NH_3
6 *M* HNO_3
15 *M* HNO_3
5 *M* NH_4Cl
3% H_2O_2
0.2 *M* $K_4Fe(CN)_6$
0.1% Aluminon reagent*
1% dimethylglyoxime†
0.2 *M* $BaCl_2$
0.3 *M* $(NH_4)_2C_2O_4$
Magnesium reagent‡
0.2 *M* $FeSO_4$ (stabilized with iron wire and 0.01 *M* H_2SO_4)
0.2 *M* $AgNO_3$
Cl_2 water
mineral oil
0.1 *M* $Ba(OH)_2$
Solids: $ZnSO_4$, $NaNO_3$, Na_2CO_3, NaCl, NaBr, NaI
0.1 *M* NH_4NO_3
0.1 *M* $AgNO_3$
0.1 *M* $Fe(NO_3)_3$
0.1 *M* $Cr(NO_3)_3$
0.1 *M* $Al(NO_3)_3$
0.1 *M* $Ca(NO_3)_2$
0.1 *M* $NaNO_3$
0.1 *M* $Zn(NO_3)_2$
0.1 *M* $Ni(NO_3)_2$
0.1 *M* $Mg(NO_3)_2$
unknown cation solution
unknown anion salt, solid

ALL SOLUTIONS SHOULD BE PROVIDED IN DROPPER BOTTLES.

DISCUSSION

Qualitative analysis is concerned with the identification of the constituents contained in a sample of unknown composition. Inorganic qualitative analysis deals with the detection and identification of the elements that are pres-

*One gram of ammonium aurintricarboxylic acid in 1 L H_2O.
†In 95% ethyl alcohol solution.
‡A solution of 0.1% (0.1 g/L) *p*-nitrobenzene-azoresorcinol in 0.025 M NaOH.

ent in a sample of material. Frequently this is accomplished by making an aqueous solution of the sample and then determining which cations and anions are present on the basis of chemical and physical properties. In this experiment you will become familiar with some of the chemistry of 10 cations (Ag^+, Fe^{3+}, Cr^{3+}, Al^{3+}, Ca^{2+}, Mg^{2+}, Ni^{2+}, Zn^{2+}, Na^+, NH_4^+) and six anions (SO_4^{2-}, NO_3^-, Cl^-, Br^-, I^-, CO_3^{2-}), and you will learn how to test for their presence or absence. Because there are many other elements and ions than those which we shall consider, we call this experiment an "abbreviated" qualitative-analysis scheme.

PART I: CATIONS

If a substance contains only a single cation (or anion), its identification is a fairly simple and straightforward process, as you may have witnessed in Experiments 7 and 10. However, even in this instance additional confirmatory tests are sometimes required to distinguish between two cations (or anions) that have similar chemical properties. The detection of a particular ion in a sample that contains several ions is somewhat more difficult because the presence of the other ions may interfere with the test. For example, if you are testing for Ba^{2+} with K_2CrO_4 and obtain a yellow precipitate, you may draw the erroneous conclusion that because Pb^{2+} is present, it also will form a yellow precipitate. Thus, the presence of lead ions interferes with this test for barium ions. This problem can be circumvented by first precipitating the lead as PbS with H_2S, thereby removing the lead ions from solution prior to testing for Ba^{2+}.

The successful analysis of a mixture containing 10 or more cations centers about the systematic separation of the ions into groups containing only a few ions. It is a much simpler task to work with two or three ions than with 10 or more. Ultimately, the separation of cations depends upon the differences in their tendencies to form precipitates, to form complex ions, or to exhibit amphoterism.

The chart in Figure 29.1 illustrates how the 10 cations you will study are separated into groups. Three of the 10 cations are colored: Fe^{3+} (rust to yellow), Cr^{3+} (blue-green), and Ni^{2+} (green). Therefore, a preliminary examination of an unknown that can contain any of the 10 cations under consideration yields valuable information. If the solution is colorless, you know immediately that iron, chromium, and nickel are absent. You are to take advantage of all clues that will aid you in identifying ions. However, in your role as a detective in identifying ions, be aware that clues can sometimes be misleading. For example, if Fe^{3+} and Cr^{3+} are present together, what color would you expect this mixture to display? Would the color depend upon the proportions of Fe^{3+} and Cr^{3+} present? Could you assign a *definite* color to such a mixture?

The group-separation chart (Figure 29.1) shows that silver can be separated from all the other cations (what are they?) as an insoluble chloride by the addition of hydrochloric acid. Iron, chromium, and aluminum are separated from the decantate as their corresponding hydroxides by precipitating them from a buffered ammonium hydroxide solution. Next, calcium can be isolated by precipitating it as insoluble calcium oxalate. Finally, magnesium and nickel can be separated from the remaining ions (Zn^{2+} and Na^+) as their insoluble hydroxides. Examination of the chart shows that in achieving these separations, reagents containing sodium and ammonium ions are used; therefore, tests for these ions must be made prior to their introduction into the solution.

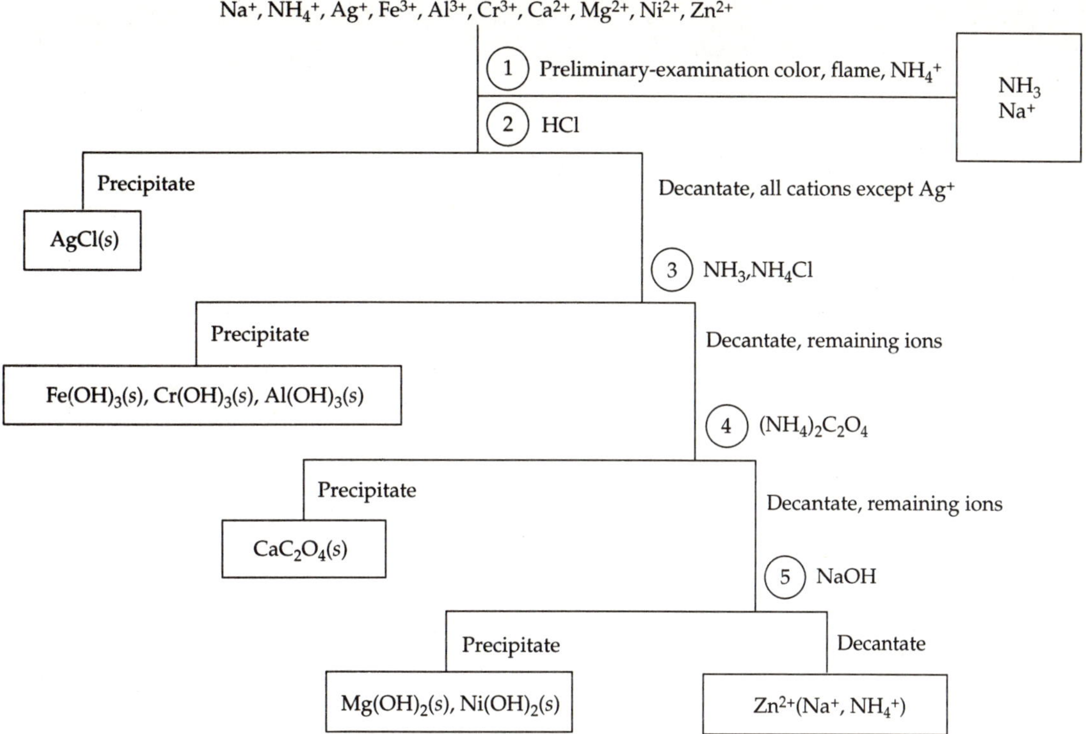

▲ **FIGURE 29.1** Flow chart for group separation.

To derive the maximum benefit from this exercise, you should be thoroughly familiar with the group-separation chart in Figure 29.1. You should know not only which 10 cations (by formula and charge) you are studying but also how they are separated into groups. More details regarding the identification of these cations are provided in the flow chart for cations (Figure 29.2) and in the following discussion about the chemistry of the analytical scheme. Make frequent referrals to the flow charts while learning about the chemistry of the qualitative analysis scheme.

CHEMISTRY OF THE QUALITATIVE ANALYSIS SCHEME

1 Detection of Sodium and Ammonium

Sodium salts and ammonium salts are added as reagents in the analysis of your general unknown. Hence, tests for Na^+ and NH_4^+ must be made on the original sample before performing tests for the other cations. Remember, your unknown may contain up to 10 cations, and you do not want inadvertently to introduce any of them into your unknown.

Sodium Most sodium salts are water-soluble. The simplest test for the sodium ion is a flame test. Sodium salts impart a characteristic yellow color to a flame. The test is *very* sensitive, and because of the prevalence of sodium ions, much care must be exercised to keep equipment clean and free from contamination by these ions.

Ammonium The ammonium ion, NH_4^+, is the conjugate acid of the base ammonia, NH_3. The test for NH_4^+ takes advantage of the following equilibrium:

$$NH_4^+(aq) + OH^-(aq) \rightleftharpoons NH_3(g) + H_2O(l)$$

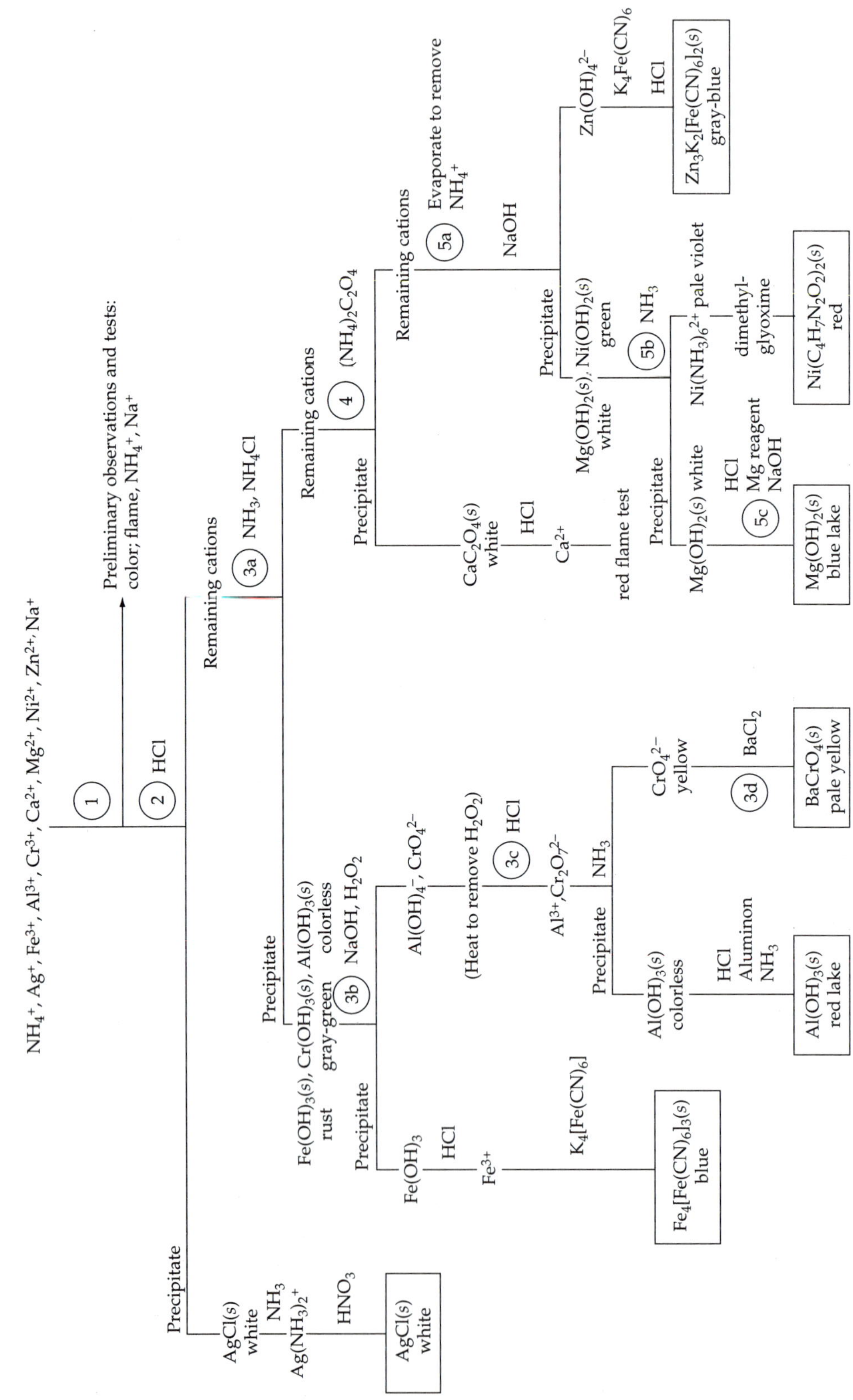

▲ **FIGURE 29.2** Cation flow chart.

Thus, when a strong base such as sodium hydroxide is added to a solution of an ammonium salt, and this solution is heated, NH_3 gas is evolved. The NH_3 gas can easily be detected by its effect upon moist red litmus paper.

2 Separation and Detection of Silver

All chloride salts are soluble in water except those of Pb^{2+}, Hg_2^{2+}, and Ag^+. Silver can be precipitated and separated from the other nine cations that we are considering by the addition of HCl to the original unknown:

$$Ag^+(aq) + Cl^-(aq) \longrightarrow \underset{\text{white}}{\mathbf{AgCl}(s)}$$

A slight excess of HCl is used to ensure the complete precipitation of silver ions and to reduce their solubility by the common-ion effect; an excess of chloride ions drives the above equilibrium to the right. However, a large excess of chloride ions must be avoided because AgCl tends to dissolve by forming a *soluble-complex ion:*

$$\mathbf{AgCl}(s) + Cl^-(aq) \longrightarrow AgCl_2^-(aq)$$

To be absolutely certain that the white precipitate is AgCl ($PbCl_2$ and Hg_2Cl_2 are also insoluble, and they are also white), NH_3 is added to the precipitate. If the precipitate is indeed AgCl, it will dissolve and then re-precipitate when the ammonical solution is made acidic:

$$\mathbf{AgCl}(s) + 2NH_3(aq) \longrightarrow Ag(NH_3)_2^+(aq) + Cl^-(aq)$$

$$Ag(NH_3)_2^+(aq) + 2H^+(aq) + Cl^-(aq) \longrightarrow \mathbf{AgCl}(s) + 2NH_4^+(aq)$$

The other two insoluble chlorides do not behave this way. Thus we can be assured that the white chloride precipitate is silver chloride.

3a Separation and Detection of Iron, Aluminum, and Chromium

Iron, aluminum, and chromium can be separated from the other ions (Ca^{2+}, Mg^{2+}, Ni^{2+}, and Zn^{2+}) by making the solution alkaline and precipitating these cations as their corresponding hydroxides:

$$Fe^{3+}(aq) + 3NH_3(aq) + 3H_2O(l) \longrightarrow \underset{\text{rust}}{\mathbf{Fe(OH)_3}(s)} + 3NH_4^+(aq)$$

$$Cr^{3+}(aq) + 3NH_3(aq) + 3H_2O(l) \longrightarrow \underset{\text{gray-green}}{\mathbf{Cr(OH)_3}(s)} + 3NH_4^+(aq)$$

$$Al^{3+}(aq) + 3NH_3(aq) + 3H_2O(l) \longrightarrow \underset{\text{colorless}}{\mathbf{Al(OH)_3}(s)} + 3NH_4^+(aq)$$

The hydroxide-ion concentration required to precipitate these three ions must be carefully controlled, because if it is too high, $Mg(OH)_2$ will also precipitate. An alkaline buffer of NH_3 and NH_4Cl provides a hydroxide-ion concentration that is high enough to precipitate Fe^{3+}, Cr^{3+}, and Al^{3+} and yet is low enough to prevent precipitation of $Mg(OH)_2$. Aqueous ammonia is a weak base:

$$NH_3(aq) + H_2O(l) \rightleftharpoons NH_4^+(aq) + OH^-(aq)$$

By itself, it would provide too high of a hydroxide-ion concentration, and $Mg(OH)_2$ would precipitate along with the other cations. However, the NH_4^+ ions derived from the NH_4Cl causes this equilibrium to shift to the left; this reduces the hydroxide-ion concentration sufficiently to prevent Mg^{2+} from precipitating.

3b Separation and Detection of Iron

Iron hydroxide can be separated from the other hydroxides by treating the precipitate with the strong base NaOH and hydrogen peroxide, H_2O_2. These reagents do not react with the insoluble $Fe(OH)_3$; however, $Al(OH)_3$ is amphoteric and dissolves, forming the soluble complex ion $Al(OH)_4^-$. The $Cr(OH)_3$ also dissolves, being oxidized by H_2O_2 to form CrO_4^{2-}:

$$\mathbf{Al(OH)_3}(s) + OH^-(aq) \longrightarrow Al(OH)_4^-(aq)$$

$$\mathbf{2Cr(OH)_3}(s) + 3H_2O_2(aq) + 4OH\ (aq) \longrightarrow \underset{\text{yellow}}{2CrO_4^{2-}(aq)} + 8H_2O(l)$$

The rust-colored $Fe(OH)_3$ remains undissolved. That the rust-colored precipitate is in fact iron hydroxide can be confirmed by dissolving it in acid and then adding potassium hexacyanoferrate(II), $K_4[Fe(CN)_6]$, and noting the formation of a dark blue precipitate (Prussian blue):

$$\mathbf{Fe(OH)_3}(s) + 3H^+(aq) \longrightarrow Fe^{3+}(aq) + 3H_2O(l)$$

$$4Fe^{3+}(aq) + 3Fe(CN)_6^-(aq) \longrightarrow \underset{\text{blue}}{\mathbf{Fe_4[Fe(CN)_6]_3}(s)}$$

3c and 3d Separation and Detection of Chromium and Aluminum

While iron was being precipitated as the hydroxide, chromium was oxidized to the yellow chromate ion, CrO_4^{2-}, and aluminum was converted to the soluble complex aluminate ion, $Al(OH)_4^-$, which is colorless. These two ions are in the supernatant liquid, which upon acidification converts the chromate ion to the orange dichromate ion and the aluminate to the colorless solvated aluminum ion:

$$2CrO_4^{2-}(aq) + 2H^+(aq) \rightleftharpoons H_2O(l) + Cr_2O_7^{2-}(aq)$$

$$4H^+(aq) + Al(OH)_4^-(aq) \longrightarrow 4H_2O(l) + Al^{3+}(aq)$$

When this solution is treated with aqueous ammonia, aluminum precipitates as $Al(OH)_3$, which can be separated from the chromate ion in the supernatant liquid. (Note: the CrO_4^{2-} ion is stable in neutral or alkaline solution but is reversibly converted to the dichromate ion, $Cr_2O_7^{2-}$, in acidic solution.) The formation of a yellow precipitate, $BaCrO_4$, upon the addition of barium chloride, confirms the presence of chromium:

$$Ba^{2+}(aq) + CrO_4^{2-}(aq) \longrightarrow \underset{\text{yellow}}{\mathbf{BaCrO_4}(s)}$$

Aluminum hydroxide is a clear, colorless substance and is difficult to see in this analysis. A confirmatory test for aluminum involves dissolving the aluminum hydroxide in acid and then re-precipitating it again with ammonia in the presence of Aluminon reagent. As the aluminum hydroxide precipitates, it absorbs the Aluminon reagent and assumes a red coloration known as a "lake."

$$Al^{3+}(aq) + 3NH_3(aq) + 3H_2O(l) + \underset{\text{reagent}}{\text{Aluminon}} \longrightarrow \underset{\text{red lake}}{\mathbf{Al(OH)_3}(s)} + 3NH_4^+(aq)$$

4 Separation and Detection of Calcium

Calcium Calcium can be separated from the remaining cations (Mg^{2+}, Ni^{2+}, and Zn^{2+}) by precipitating it as an insoluble oxalate salt:

$$Ca^{2+}(aq) + C_2O_4^{2-}(aq) \longrightarrow \mathbf{CaC_2O_4}(s)$$

If magnesium is present, it is possible that it can precipitate and be mistaken for calcium. To confirm that the precipitate is that of calcium and not magnesium, it is dissolved in acid and a flame test is performed on the solution.

$$\mathbf{CaC_2O_4}(s) + 2H^+(aq) \longrightarrow Ca^{2+}(aq) + H_2C_2O_4(aq)$$

A transient brick-red flame verifies the presence of calcium ions. Magnesium ions do not impart any color to a flame.

5a Separation and Detection of Zinc

Separation of zinc from magnesium and nickel can be accomplished by precipitating $Mg(OH)_2$ and $Ni(OH)_2$ from a strongly alkaline solution. Excess ammonium ions must be removed before the precipitation of these hydroxides because ammonium ions interfere with the confirmatory test for magnesium. Ammonium ions are easily removed by evaporating the solution, which has been acidified with nitric acid to dryness:

$$NH_4Cl(s) \xrightarrow{\Delta} NH_3(g) + HCl(g)$$

$$NH_4NO_3(s) \xrightarrow{\Delta} N_2O(g) + 2H_2O(g)$$

Although magnesium and nickel form insoluble hydroxides in the presence of a strong base, zinc is amphoteric and forms the soluble complex ion $Zn(OH)_4^{2-}$:

$$Ni^{2+}(aq) + 2OH^-(aq) \longrightarrow \mathbf{Ni(OH)_2}(s)$$

$$Mg^{2+}(aq) + 2OH^-(aq) \longrightarrow \mathbf{Mg(OH)_2}(s)$$

$$Zn^{2+}(aq) + 4OH^-(aq) \longrightarrow Zn(OH)_4^{2-}(aq)$$

The presence of zinc is confirmed by precipitating it from an acidic solution as a gray-blue salt, $Zn_3K_2[Fe(CN)_6]_2$:

$$3Zn(OH)_4^{2-}(aq) + 12H^+(aq) + 2K^+(aq) + 2Fe(CN)_6^{4-}(aq) \longrightarrow \underset{\text{gray-blue}}{\mathbf{Zn_3K_2[Fe(CN)_6]_2}(s)} + 12H_2O(l)$$

5b Separation and Detection of Nickel

Because nickel forms the soluble complex ion hexaamminenickel(II), $Ni(NH_3)_6^{2+}$, in the presence of aqueous ammonia, it can be separated from $Mg(OH)_2$:

$$\mathbf{Ni(OH)_2}(s) + 6NH_3(aq) \longrightarrow Ni(NH_3)_6^{2+}(aq) + 2OH^-(aq)$$

It is confirmed by forming a strawberry red precipitate, $Ni(C_4H_7N_2O_2)_2$, with an organic reagent, dimethylglyoxime.

$$Ni(NH_3)_6^{2+}(aq) + 2HC_4H_7N_2O_2(aq) \longrightarrow \underset{\text{strawberry red}}{\mathbf{Ni(C_4H_7N_2O_2)_2}(s)} + 4NH_3(aq) + 2NH_4^+(aq)$$

5c Detection of Magnesium

Magnesium is confirmed by dissolving the hydroxide with acid and then reprecipitating it in the presence of an organic compound called "Magnesium reagent." The presence of Mg^{2+} is indicated by formation of a blue lake.

$$\mathbf{Mg(OH)_2}(s) + 2H^+(aq) \longrightarrow Mg^{2+}(aq) + 2H_2O(l)$$

$$Mg^{2+}(aq) + 2OH^-(aq) + \text{Mg reagent} \longrightarrow \underset{\text{blue lake}}{\mathbf{Mg(OH)_2}(s)}$$

PROCEDURE

Instructor: Explain the philosophy of testing knowns before unknowns. Review the techniques involved in qualitative analysis.

First you will analyze a known that contains all ten cations. Record on your report sheet the reagents used in each step, your observations, and the equations for each precipitation reaction. After completing this practice analysis, obtain an unknown. Follow the same procedures as with the known, again recording reagents and observations. Also record conclusions regarding the presence or absence of all cations. *Before beginning this experiment, review the techniques used in qualitative analysis found in Appendix I: heating solutions, precipitation, centrifugation, washing precipitates, and testing acidity.*

Waste Disposal Instructions All waste from this experiment should be placed in appropriate containers in the laboratory.

1 Initial Observations and Tests for Sodium and Ammonium

Note the color of your sample and record any conclusions about what cations are present or absent on your report sheet.

The flame test for sodium is very sensitive, and traces of sodium ion will impart a characteristic yellow color to the flame. Just about every solution has a trace of sodium and thus will give a positive result. On the basis of the intensity and duration of the yellow color, you can decide whether Na^+ is merely a contaminant or present in substantial quantity. To perform the flame test, obtain a piece of platinum or Nichrome wire that has been sealed in a piece of glass tubing. Clean the wire by dipping it in 12 *M* HCl that is contained in a small test tube and heat the wire in the hottest part of your Bunsen burner flame. Repeat this operation until no color is seen when the wire is placed in the flame. Several cleanings will be required before this is achieved. Then place 10 drops of the solution to be analyzed in a clean test tube and perform a flame test on it. If the sample being tested is your unknown, run a flame test on distilled water and then another on a 0.2 *M* NaCl solution. Compare the tests; this should help you make a decision as to the presence of sodium in your unknown.

Place 10 drops of the original sample to be analyzed in an evaporating dish or a crucible. Moisten a strip of red or neutral litmus paper with distilled water and place the paper on the bottom of a small watch glass. Add 10 drops of 3 *M* NaOH to the unknown, swirl the evaporating dish or crucible, and immediately place the watch glass on it with the litmus paper down. Let stand for a few minutes. The presence of NH_4^+ ions is confirmed if the paper turns blue.

2 Separation and Detection of Silver

Place 10 drops of the original solution to be analyzed in a small test tube, add five drops of distilled water, and two drops of 6 *M* HCl. Stir well, centrifuge (consult Appendix I for techniques), and reserve the decantate for Procedure 3. Wash the precipitate with 10 drops of distilled water, centrifuge, and add the washings to the decantate. Dissolve the precipitate in four drops of 6 *M* NH_3*; then add 6 *M* HNO_3 to the ammoniacal solution until it is acidic to litmus. A curdy white precipitate of AgCl confirms the presence of Ag^+ ions.

3a Separation and Detection of Iron, Aluminum, and Chromium

To the decantate from Procedure 2 add two drops of NH_4Cl solution and 6 *M* NH_3 until the solution is basic to litmus. Centrifuge and reserve the decantate for Procedure 4. Wash the precipitate with 10 drops of distilled water, centrifuge, and add the washings to the decantate.

3b Separation and Detection of Iron

To the precipitate from Procedure 3a, add five drops of distilled water, 10 drops of 3 *M* NaOH, and five drops of 3% H_2O_2. Stir well, centrifuge, and reserve the decantate for Procedure 3c. Wash the precipitate with 10 drops of distilled water and add the washings to the decantate. Dissolve the reddish-brown precipitate in two drops of 6 *M* HCl and add 10 drops of $K_4[Fe(CN)_6]$ solution. A blue precipitate of $Fe_4[Fe(CN)_6]_3$ confirms the presence of Fe^{3+} ions.

3c Separation and Detection of Aluminum

If the decantate from Procedure 3b is yellow in color, place it in an evaporating dish or a crucible and evaporate almost to dryness to remove excess H_2O_2 (see Note 1 below). Add 10 drops of distilled water and 6 *M* HCl until acid to litmus. Then add 6 *M* NH_3 until the solution is basic to litmus. Centrifuge and reserve the decantate for Procedure 3d. Wash the precipitate with 10 drops of distilled water, centrifuge, and add the washings to the decantate. Dissolve the precipitate in two drops of 6 *M* HCl, add two drops of Aluminon reagent, and 6 *M* NH_3 until basic to litmus. Centrifuge the solution. A red "lake" confirms the presence of Al^{3+} ions.

Note 1 Hydrogen peroxide is a reducing agent in acid solutions, so CrO_4^{2-} ions could be reduced to Cr^{3+} when the solution is acidified. When NH_3 is added, $Cr(OH)_3$ will be precipitated and could be incorrectly reported as $Al(OH)_3$. Also, Cr^{3+} would not be confirmed in the proper procedure.

3d Separation and Detection of Chromium

Add two drops of $BaCl_2$ solution to the decantate from Procedure 3c. A yellow precipitate of $BaCrO_4$ confirms the presence of Cr^{3+} ions.

4 Separation and Detection of Calcium

Add three drops of $(NH_4)_2C_2O_4$ solution to the decantate from Procedure 3a and centrifuge. Reserve the decantate for Procedure 5a. Wash the precipitate with 10 drops of distilled water, centrifuge, and add the washings to the decantate. Dissolve the precipitate of CaC_2O_4 in two drops of 6 *M* HCl and

---

*Bottles may be labeled 6 *M* NH_4OH.

carry out a flame test with the solution. A brick-red flame confirms Ca^{2+} ions (see Note 2 below).

Note 2 If magnesium ions are present, they may be precipitated as MgC_2O_4. To determine if MgC_2O_4 is precipitated, test the acid solution for Mg^{2+} as described in Procedure 5c.

5a Separation and Detection of Zinc

Place the decantate from Procedure 4 in an evaporating dish and carefully evaporate to dryness using a burner flame. Add five to six drops of concentrated HNO_3 and heat again until no more fumes are observed (you are removing excess NH_4^+ ions that would interfere with the tests for Mg^{2+} ions). Dissolve the residue in five drops of 6 *M* HCl. Add 10 drops of distilled water to the HCl solution and transfer it to a small test tube. Wash out the evaporating dish with 10 drops of distilled water, and add the wash to the acid solution. Add 3 *M* NaOH until the solution is basic to litmus; centrifuge, and reserve the precipitate for Procedure 5b. Wash the precipitate with 10 drops of distilled water and add the washings to the decantate. Add three drops of $K_4[Fe(CN)_6]$ to the decantate and acidify with 6 *M* HCl; a gray-blue coloration or precipitate of $Zn_3K_2[Fe(CN)_6]_2$ confirms the presence of Zn^{2+} ions.

5b Separation and Detection of Nickel

Add five drops of distilled water and five drops of 15 *M* NH_3 to the precipitate from Procedure 5a; centrifuge, and reserve the precipitate for Procedure 5c. Wash the precipitate with 10 drops of distilled water, and add the wash to the decantate. Add one drop of dimethylglyoxime solution to the decantate; a strawberry-red precipitate of $Ni(C_4H_7N_2O_2)_2$ confirms the presence of Ni^{2+} ions (see Experiment 10).

5c Detection of Magnesium

If Ni^{2+} ions were confirmed in Procedure 5b, add 15 *M* NH_3 to the precipitate saved from Procedure 5b until a negative result for Ni^{2+} is obtained. Dissolve the hydroxide precipitate in two drops of 6 *M* HCl; add two drops of "Magnesium reagent" and 3 *M* NaOH until the solution is basic. A blue lake confirms the presence of Mg^{2+} ions.

REVIEW QUESTIONS

Before beginning Part I of this experiment in the laboratory, you should be able to answer the following questions:

1. What are the names and formulas of the 10 cations you will identify?
2. Why are confirmatory tests necessary in identifying ions?
3. Which of the 10 cations are colored, and what are their colors?
4. Which salt is insoluble: $FeCl_3$, $ZnCl_2$, NaCl, or AgCl?
5. How could you separate Fe^{3+} from Ag^+?
6. How could you separate Cr^{3+} from Mg^{2+}?
7. How could you separate Al^{3+} from Ag^+?
8. Complete and balance the following:

$$NH_4^+(aq) + OH^-(aq) \longrightarrow$$

$$AgCl(s) + NH_3(aq) \longrightarrow$$

PART II: ANIONS

DISCUSSION

A systematic scheme based on the kinds of principles involved in cation analysis can be designed for the analysis of anions. Because we shall limit our consideration to only six anions (SO_4^{2-}, NO_3^-, Cl^-, Br^-, I^-, and CO_3^{2-}) and will not consider mixtures of the ions, our method of analysis is quite simple and straightforward. It is based on specific tests for the individual ions and does not require special precaution to eliminate interferences that may arise in mixtures.

Initially you will make a general test on a solid salt with concentrated sulfuric acid (H_2SO_4). The results of this test should strongly suggest what the anion is. You will then confirm your suspicions by performing a specific test for the ion you believe to be present.

Table 29.1 summarizes the behavior of anions (as dry salts) with concentrated sulfuric acid.

PROCEDURE

Perform the general sulfuric acid test described below on the individual anions. Then perform the specific tests on each of the ions. Record your observations and equations for the reactions that occur. After completing these tests on the six anions, obtain a solid salt unknown and identify its anion. Record your observations and conclusion. Only one anion is present in the salt.

TABLE 29.1 **Behavior of Anions with Concentrated Sulfuric Acid, H_2SO_4**

A. Cold H_2SO_4

SO_4^{2-} No reaction.

NO_3^- No reaction.

CO_3^{2-} A colorless, odorless gas forms.

$$CO_3^{2-}(s) + 2H^+(aq) \longrightarrow H_2O(l) + CO_2(g)$$

Cl^- A colorless gas forms. It has a sharp-pungent odor, gives an acidic test result with litmus, and fumes in moist air.

$$Cl^-(s) + H^+(aq) \longrightarrow HCl(g)$$

Br^- A brownish-red gas forms. It has a sharp odor, gives an acidic test result with litmus, and fumes in moist air. The odor of SO_2 may be detected.

$$2Br^-(s) + 4H^+(aq) + SO_4^{2-}(aq) \longrightarrow Br_2(g) + SO_2(g) + 2H_2O(l)$$

(HBr is also liberated)

I^- Solid turns dark brown immediately with the slight formation of violet fumes. The gas has the odor of rotten eggs, gives an acidic test result with litmus, and fumes in moist air.

$$2I^-(s) + 4H^+(aq) + SO_4^{2-}(aq) \longrightarrow I_2(g) + SO_2(g) + 2H_2O(l)$$

(HI is also liberated)

B. Hot Concentrated H_2SO_4

There are no additional reactions with any of the anions except NO_3^-, which forms brown fumes of NO_2 gas.

$$4NO_3^-(s) + 4H^+(aq) \longrightarrow 4NO_2(g) + O_2(g) + 2H_2O(l)$$

Sulfuric Acid Test

Place a small amount of the solid (about the size of a pea) in a small test tube. Add one or two drops of 18 *M* H_2SO_4 and observe everything that occurs, especially the color and odor of gas formed. **(CAUTION:** ***Concentrated H_2SO_4 causes severe burns. Do not get it on your skin or clothing. If you come in contact with it, immediately wash the area with copious amounts of water.*)** DO NOT place your nose directly over the mouth of the test tube, but carefully fan gases toward your nose. Then *carefully* heat the test tube, but not so strongly as to boil the H_2SO_4. **(CAUTION:** ***If you heat the acid too strongly, it could come shooting out!*)** Note whether or not brown fumes of NO_2 are produced. **(CAUTION:** ***Do not look down into the test tube. Do not point the test tube at yourself or at your neighbors.*** **SAFETY GLASSES MUST BE WORN.)**

Specific Tests for Anions

When an anion is indicated by the preliminary test result with concentrated H_2SO_4, it is confirmed using the appropriate specific test. Make an aqueous solution of the solid unknown and perform the following tests on portions of this solution.

Sulfate Place 10 drops of a solution of the anion salt in a test tube, acidify with 6 *M* HCl, and add a drop of $BaCl_2$ solution. The formation of a white precipitate of $BaSO_4$ confirms SO_4^{2-} ions.

$$Ba^{2+}(aq) + SO_4^{2-}(aq) \longrightarrow \mathbf{BaSO_4}(s)$$

Nitrate Place 10 drops of a solution of the anion salt in a small test tube and add five drops of $FeSO_4$ solution; mix the solution. Carefully, without agitation, pour concentrated H_2SO_4 down the inside of the test tube so as to form two layers. The formation of a brown ring between the two layers confirms NO_3^- ions.

$$(1)\ 3Fe^{2+}(aq) + NO_3^-(aq) + 4H^+(aq) \longrightarrow 3Fe^{3+}(aq) + NO(g) + 2H_2O(l)$$

$$(2)\ NO(g) + Fe^{2+}(aq)\ \text{(excess)} \longrightarrow Fe(NO)^{2+}(aq)\ \text{(brown)}$$

Chloride Place 10 drops of a solution of the anion salt in a test tube and add a drop of $AgNO_3$ solution. A white, curdy precipitate confirms Cl^- ions.

$$Ag^+(aq) + Cl^-(aq) \longrightarrow \mathbf{AgCl}(s)$$

Bromide Place 10 drops of a solution of the anion salt in a test tube, add three drops of 6 *M* HCl, then add five drops of Cl_2 water and five drops of mineral oil. Shake well. Br^- ions are confirmed if the mineral-oil (top) layer is colored orange to brown. Wait 30 s for the layers to separate.

$$2Br^-(aq) + Cl_2(aq) \rightleftharpoons Br_2(aq) + 2Cl^-(aq)$$

Iodide Repeat the test as described for Br^- ions. If the mineral-oil layer is colored violet, I^- ions are confirmed.

$$2I^-(aq) + Cl_2(aq) \rightleftharpoons I_2(aq) + 2Cl^-(aq)$$

Carbonate Place a small amount of the solid anion salt in a small test tube and add a few drops of 6 *M* H_2SO_4. If a colorless, odorless gas evolves, hold a drop of $Ba(OH)_2$ solution over the mouth of the test tube using either an eyedropper or Nichrome wire loop; CO_3^{2-} ions are confirmed if the drop turns milky.

$$(1)\ 2H^+(aq) + CO_3^{2-}(s) \longrightarrow CO_2(g) + H_2O(l)$$

$$(2)\ CO_2(g) + Ba(OH)_2(aq) \longrightarrow \mathbf{BaCO_3}(s) + H_2O(l)$$

REVIEW QUESTIONS

Before beginning Part II of this experiment in the laboratory, you should be able to answer the following questions:

1. Give the names and formulas of the anions to be identified.
2. Describe the behavior of each solid containing the anions toward concentrated H_2SO_4.
3. If you had a mixture of NaCl and Na_2CO_3, would the action of concentrated H_2SO_4 allow you to decide that both Cl^- and CO_3^{2-} were present?

NOTES AND CALCULATIONS

Name ______________________ Desk ______________________

Date ______________ Laboratory Instructor ______________________

Unknown no. ______________________

REPORT SHEET | EXPERIMENT

Introduction to Qualitative Analysis

29

PART I: CATIONS

A. Known

Record the reagent used in each step, your observations, and the equations for each precipitation reaction.

	Reagent	*Observations*	*Equations*
1	NaOH	Litmus turns blue	$NH_4^+ (aq) + OH^- (aq) \rightleftarrows NH_3(g) + H_2O(l)$
2	HCl	White ppt. formed	$Ag^+(aq) + Cl^- (aq) \longrightarrow AgCl(s)$
3a	NH_4^+, NH_3	Dark ppt. formed	$Fe^{3+}(aq) + 3OH^-(aq) \longrightarrow Fe(OH)_3(s)$ $Cr^{3+}(aq) + 3OH^-(aq) \longrightarrow Cr(OH)_3(s)$ $Al^{3+}(aq) + 3OH^-(aq) \longrightarrow Al(OH)_3(s)$
3b	H_2O_2, NaOH	Supernatant liq. turned yellow, and ppt. has changed color to red-brown, due to $Fe(OH)_3$	$2Cr(OH)_3(s) + 3H_2O_2 + 4OH^-(aq) \longrightarrow 2CrO_4^{2-}(aq) + 8H_2O$
3c	$K_4[Fe(CN)_6]$ NH_3 Aluminon	Dark blue ppt. formed Red ppt. formed	$4FeCl_4^-(aq) + 3Fe(CN)_6^{4-} (aq) \longrightarrow Fe_4[Fe(CN)_6]_3(s) + 16Cl^-(aq)$ $Al^{3+}(aq) + 3OH^-(aq) \longrightarrow Al(OH)_3(s)$

	Reagent	*Observations*	*Equations*
3d	$BaCl_2$	A yellow ppt. formed	$Ba^{2+}(aq) + CrO_4^{2-}(aq) \longrightarrow BaCrO_4(s)$
4	$(NH_4)_2C_2O_4$	White ppt. formed Dark red flame	$Ca^{2+}(aq) + C_2O_4^{2-}(aq) \longrightarrow CaC_2O_4(s)$
5a	$K_4[Fe(CN)_6]$	Gray-blue ppt. formed	$Zn^{2+}(aq) + 4OH^- \longrightarrow Zn(OH)_4^{2-}(aq)$ $3Zn(OH)_4^{2-}(aq) + 12H^+(aq) + 2K^+(aq)$ $+ 2Fe(CN)_6^{4-}(aq) \longrightarrow 12H_2O(l)$ $+ Zn_3K_2[Fe(CN)_6]_2(s)$
5b	NaOH	Green ppt.	$Ni^{2+}(aq) + 2OH^-(aq) \longrightarrow Ni(OH)_2(s)$ $Mg^{2+}(aq) + 2OH^-(aq) \longrightarrow Mg(OH)_2(s)$
5c	DMG^-	Red ppt. formed	$Ni^{2+}(aq) + 2DMG^-(aq) \longrightarrow Ni(DMG)_2(s)$

B. Unknown (Cation unknown no. _______)

Record reagent used, your observations, and the equations for each precipitate formed.

Reagent	*Observations*	*Equations*
DMG^-	red ppt. formed	$Ni^{2+} + 2DMG \longrightarrow Ni(DMG)_2(s)$
NaOH magnesium reagent	blue ppt. formed	$Mg^{2+} + 2OH^- \longrightarrow Mg(OH)_2(s)$

Cations in unknown Ni^{2+} and Mg^{2+}

PART II: ANIONS

A. Known

Concentrated H_2SO_4 test.

Ion	*Observations and equations*
SO_4^{2-}	No change
NO_3^-	No change
CO_3^{2-}	Bubbles—gas formed, odorless and colorless $2H^+(aq) + CO_3^{2-}(aq) \longrightarrow H_2O(l) + CO_2(g)$
Cl^-	Bubbles—gas formed, colorless, pungent $H^+(aq) + Cl^-(aq) \longrightarrow HCl(g)$
Br^-	Solid darkens—red-brown gas formed $2Br^-(aq) + 4H^+(aq) + SO_4^{2-}(aq) \longrightarrow Br_2(g) + SO_2(g) + 2H_2O(l)$
I^-	Solid turned black-brown, and violet vapors formed $2I^-(aq) + 4H^+(aq) + SO_4^{2-}(aq) \longrightarrow I_2(g) + SO_2(g) + 2H_2O(l)$
NO_3^- (heated)	Brown gas produced $4H^+(aq) + 4NO_3^-(aq) \longrightarrow 4NO_2(g) + O_2(g) + 2H_2O(l)$

B. Known

Specific tests.

Ion	*Observations and equations*
SO_4^{2-}	White ppt. formed $Ba^{2+}(aq) + SO_4^{2-}(aq) \longrightarrow BaSO_4(s)$
NO_3^-	Dark brown "ring" formed between layers $3Fe^{2+}(aq) + NO_3^-(aq) + 4H^+(aq) \longrightarrow 3Fe^{3+}(aq) + NO(g) + 2H_2O(l)$ $Fe^{2+}(aq) + NO(g) \longrightarrow Fe(NO)^{2+}(aq)$
CO_3^{2-}	Gas (CO_2) formed that turned $Ba(OH)_2$ droplet white $2H^+(aq) + CO_3^{2-}(aq) \longrightarrow H_2O(l) + CO_2(g)$ $Ba^{2+}(aq) + 2OH^-(aq) + CO_2(g) \longrightarrow BaCO_3(s) + H_2O(l)$
Cl^-	White ppt. formed $Ag^+(aq) + Cl^-(aq) \longrightarrow AgCl(s)$
Br^-	Mineral oil (top layer) turned orange $Cl_2(aq) + 2Br^-(aq) \longrightarrow 2Cl^-(aq) + Br_2(aq)$
I^-	Mineral oil (top layer) turned violet $Cl_2(aq) + 2I^-(aq) \longrightarrow 2Cl^- + I_2(aq)$

C. Unknown (Anion unknown no. _______)

Observations and equations

1 H_2SO_4 test:

Colorless, odorless gas with cold acid

$2H^+(aq) + CO_3^{2-}(aq) \longrightarrow H_2O(l) + CO_2(g)$

No change on heating

2 Specific test(s):

$Ba(OH)_2$ turned milky

$2H^+(aq) + CO_3^{2-}(aq) \longrightarrow CO_2(g) + H_2O(l)$

$CO_2(g) + Ba(OH)_2(aq) \longrightarrow BaCO_3(s) + H_2O(l)$

Anions in unknown CO_3^{2-}

Experiment

Abbreviated Qualitative-Analysis Scheme

CATIONS: Ag^+, Pb^{2+}, Hg_2^{2+}, Cu^{2+}, Bi^{3+}, Sn^{4+}, Fe^{3+}, Mn^{2+}, Ni^{2+}, Al^{3+}, Ba^{2+}, NH_4^+, Na^+

ANIONS: SO_4^{2-}, NO_3^-, CO_3^{2-}, Cl^-, Br^-, I^-, CrO_4^{2-}, PO_4^{3-}, S^{2-}, SO_3^{2-}

OBJECTIVE

To become acquainted with the chemistry of several elements and the principles of qualitative analysis.

APPARATUS AND CHEMICALS

Apparatus

small test tubes (12)
centrifuge
evaporating dish
litmus
droppers (6)
10-cm Nichrome wire with loop at end
Bunsen burner and hose
aluminum wire, 26 gauge

Chemicals

6 *M* $HC_2H_3O_2$
2 *M* HCl
6 *M* HCl
12 *M* HCl
0.2 *M* $BaCl_2$
mineral oil
Cl_2 water
diethyl ether
1% dimethylglyoxime*
95% ethyl alcohol
0.2 *M* $Fe(NO_3)_2$
0.2 *M* $FeSO_4$
3% H_2O_2
3 *M* HNO_3
6 *M* HNO_3
6 *M* H_2SO_4
18 *M* H_2SO_4
6 *M* NH_3†
15 *M* NH_3†
Aluminon reagent‡
1 *M* $NH_4C_2H_3O_2$
2 *M* NH_4Cl
$(NH_4)_2MoO_4$§
$Ba(OH)_2$, saturated solution
6 *M* NaOH
0.2 *M* $SnCl_2$ (freshly prepared)
$(NH_4)_2S$‖
1 *M* K_2CrO_4
1 *M* $K_2C_2O_4$
0.2 *M* KNO_2
0.2 *M* KSCN
0.1 *M* $AgNO_3$
$NaBiO_3$, solid
Na_2CO_3, saturated solution

*In 95% ethyl alcohol solution.

†Reagent bottles may be labeled as 15 *M* or 6 *M* NH_4OH.

‡One gram of ammonium aurintricarboxylic acid in 1 L H_2O.

§Dissolve 20 g of MoO_3 in a mixture of 60 mL of distilled water and 30 mL of 15 *M* NH_3. Add this solution slowly, with constant stirring, to a mixture of 230 mL of water and 100 mL of 16 *M* HNO_3.

‖Add 1 volume of reagent-grade ammonium sulfide liquid to two volumes of water or saturate 5 *M* NH_3 with H_2S.

Na_2O_2, solid
starch solution
0.2 *M* $Pb(C_2H_3O_2)_2$
0.1 *M* $MnCl_2$
0.1 *M* $HgCl_2$
0.1 *M* $Hg(C_2H_3O_2)_2$
1 *M* thioacetamide
Zn, granulated
Groups 1, 2, 3, and 4 known solutions
solid sodium salt anion knowns
unknown cation solution
unknown anion salt, solid mixture

ALL SOLUTIONS SHOULD BE PROVIDED IN DROPPER BOTTLES.

DISCUSSION

Qualitative analysis is concerned with the identification of the constituents contained in a sample of unknown composition. Inorganic qualitative analysis deals with the detection and identification of the elements that are present in a sample of material. Frequently this is accomplished by making an aqueous solution of the sample and then determining which cations and anions are present on the basis of chemical and physical properties. In this experiment, you will become familiar with some of the chemistry of 14 cations (Ag^+, Pb^{2+}, Hg_2^{2+}, Cu^{2+}, Bi^{3+}, Sn^{4+}, Fe^{3+}, Mn^{2+}, Ni^{2+}, Al^{3+}, Ba^{2+}, Ca^{2+}, NH_4^+, Na^+) and 10 anions (CrO_4^{2-}, PO_4^{3-}, S^{2-}, SO_3^{2-}, SO_4^{2-}, NO_3^-, Cl^-, Br^-, I^-, CO_3^{2-}), and you will learn how to test for their presence or absence. Because there are many elements and ions other than those that we shall consider, we call this experiment an "abbreviated" qualitative-analysis scheme.

CATIONS

If a sample contains only a single cation (or anion), its identification is a fairly simple and straightforward process. However, even in this instance additional confirmatory tests are sometimes required to distinguish between two cations (or anions) that have similar chemical properties. The detection of a particular ion in a sample that contains several ions is somewhat more difficult because the presence of the other ions may interfere with the test. For example, if you are testing for Ba^{2+} with K_2CrO_4 and obtain a yellow precipitate, you may draw an erroneous conclusion because if Pb^{2+} is present, it also will form a yellow precipitate. Thus, lead ions interfere with this test for barium ions. This problem can be circumvented by first precipitating the lead as PbS with H_2S, thereby removing the lead ions from solution prior to testing for Ba^{2+}.

The successful analysis of a mixture containing 14 or more cations centers upon the systematic separation of the ions into groups containing only a few ions. It is a much simpler task to work with two or three ions than with 14 or more. Ultimately, the separation of cations depends upon the difference in their tendencies to form precipitates, to form complex ions, or to exhibit amphoterism.

The chart in Figure 30.1 illustrates how the 14 cations you will study are separated into groups. Four of the 14 cations are colored: Fe^{3+} (rust to yellow), Cu^{2+} (blue), Mn^{2+} (very faint pink), and Ni^{2+} (green). Therefore a preliminary examination of an unknown that can contain any of the 14 cations under consideration yields valuable information. If the solution is colorless, you know immediately that iron, copper, and nickel are absent. You are to take advantage of all clues that will aid you in identifying ions. However, in your role as a detective in identifying ions, be aware that clues can sometimes be misleading. For example, if Fe^{3+} and Ni^{2+} are present together, what color would you expect this mixture to display? Would the color depend upon the proportions of Fe^{3+} and Ni^{2+} present? Could you assign a *definite*

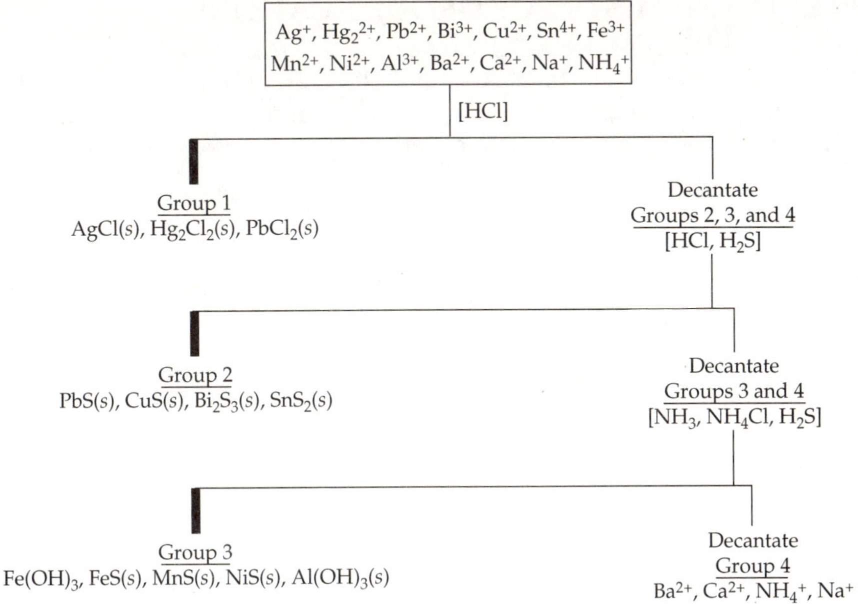

▲ **FIGURE 30.1** Flow chart for group separation.

color to such a mixture? The extremely faint pink color of Mn^{2+} may not be discerned unless you carefully examine the solution, and the presence of any of the other colored ions will definitely mask its color.

The group-separation chart (Figure 30.1) shows that silver, mercury(I), and lead can be separated from all the other cations as insoluble chlorides by the addition of hydrochloric acid. Addition of hydrogen sulfide and hydrochloric acid to the decantate results in the precipitation of the sulfides of lead, copper, bismuth, and tin(IV). The reason lead appears in both groups 1 and 2 is explained by the chemistry of these group cations.

To derive any benefit from this exercise, you must become thoroughly familiar with the group-separation chart in Figure 30.1. When you work with the four groups, you will see additional separation charts or flow schemes for the individual groups. Become familiar with these also. It is important to keep in mind what cations you are working with in each group. Know their formulas and charges and become familiar with how they are separated and identified.

You will analyze four known solutions, one for each group. Each of these solutions will contain all of the cations in the group you are studying. Then you will analyze a group unknown for each of the groups. Your instructor may tell you to analyze the known and unknown side by side. If you do, be careful not to mix up your test tubes! As you proceed with your analysis, record on your report sheets the reagents used in each step, your observations, and the equations for each precipitation reaction. After completing the four groups, you will analyze a general cation unknown that may contain any number of the 14 ions studied. You may then also analyze a simple salt and identify it.

Instructor: If time is limited, do not have students analyze a general cation unknown. Have them analyze a simple salt instead.

PART I: CHEMISTRY OF GROUP 1 CATIONS Pb^{2+}, Ag^+, Hg_2^{2+}

Because the chlorides of Pb^{2+}, Ag^+, and Hg_2^{2+} are insoluble, they may be precipitated and separated from the cations of groups 2, 3, and 4 by the addition of HCl. The following equations represent the reactions that occur:

$$Pb^{2+}(aq) + 2Cl^-(aq) \longrightarrow \mathbf{PbCl_2}(s) \quad \text{white} \quad [1]$$

$$Ag^+(aq) + Cl^-(aq) \longrightarrow \mathbf{AgCl}(s) \quad \text{white} \quad [2]$$

$$Hg_2^{2+}(aq) + 2Cl^-(aq) \longrightarrow \mathbf{Hg_2Cl_2}(s) \quad \text{white} \quad [3]$$

A slight excess of HCl is used to ensure complete precipitation of the cations and to reduce the solubility of the chlorides by the common-ion effect. However, a large excess of chloride must be avoided, because both AgCl and $PbCl_2$ tend to dissolve by forming soluble complex anions:

$$\mathbf{PbCl_2}(s) + 2Cl^-(aq) \longrightarrow PbCl_4^{2-}(aq) \quad [4]$$

$$\mathbf{AgCl}(s) + Cl^-(aq) \longrightarrow AgCl_2^-(aq) \quad [5]$$

$PbCl_2$ is appreciably more soluble than either AgCl or Hg_2Cl_2. Thus, even when $PbCl_2$ precipitates, a significant amount of Pb^{2+} remains in solution and is subsequently precipitated with the group 2 cations as the sulfide PbS. Because of its solubility, Pb^{2+} sometimes does not precipitate as the chloride, because either its concentration is too small or the solution is too warm.

Lead Lead chloride is much more soluble in hot water than in cold. It is separated from the other two insoluble chlorides by dissolving it in hot water. The presence of Pb^{2+} is confirmed by the formation of a yellow precipitate, $PbCrO_4$, upon the addition of K_2CrO_4:

$$Pb^{2+}(aq) + CrO_4^{2-}(aq) \longrightarrow \mathbf{PbCrO_4}(s) \quad \text{yellow} \quad [6]$$

Mercury(I) Silver chloride is separated from Hg_2Cl_2 by the addition of aqueous NH_3. Silver chloride dissolves because Ag^+ forms a soluble complex cation with NH_3:

$$\mathbf{AgCl}(s) + 2NH_3(aq) \longrightarrow Ag(NH_3)_2^+(aq) + Cl^-(aq) \quad [7]$$

Mercury(I) chloride reacts with aqueous ammonia in a disproportionation reaction to form a dark gray precipitate:

$$\mathbf{Hg_2Cl_2}(s) + \mathbf{2NH_3}(aq) \longrightarrow \mathbf{HgNH_2Cl}(s) + Hg(l) + NH_4^+(aq) + Cl^-(aq) \quad [8]$$

Although $HgNH_2Cl$ is white, the precipitate appears dark gray because of a colloidal dispersion of Hg(*l*).

Silver To verify the presence of Ag^+, the supernatant liquid from the last reaction is acidified and AgCl re-precipitates if Ag^+ is present. The acid decomposes $Ag(NH_3)_2^+$ by neutralizing NH_3 to form NH_4^+. It is necessary that the solution be acidic, or else the AgCl will not precipitate and Ag^+ can be missed.

$$Ag(NH_3)_2^+(aq) + 2H^+(aq) + Cl^-(aq) \longrightarrow \mathbf{AgCl}(s) + 2NH_4^+(aq) \quad [9]$$

The flow chart in Figure 30.2 shows how the group 1 cations are separated and identified. *You should become very familiar with it and consult it often as you perform your analysis.*

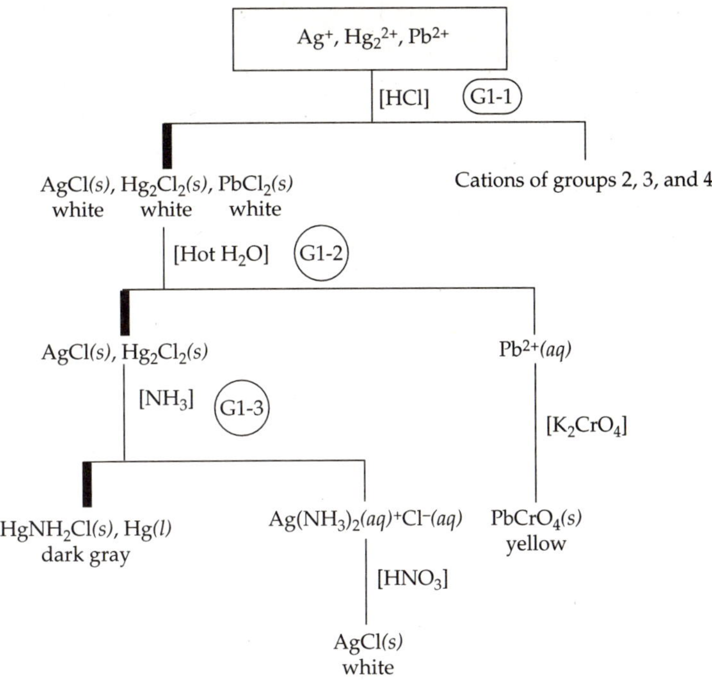

▲ **FIGURE 30.2** Group 1 flow chart. Procedure numbers are given in the circles.

PROCEDURE

First you will analyze a known that contains all three cations of group 1. Record on your report sheet the reagents used in each step, your observations, and the equation for each precipitation reaction. After completing the analysis of a known, obtain an unknown. Follow the same procedures as with the known, again recording reagents and observations. Also record conclusions regarding the presence or absence of all cations. *Before beginning this experiment, review the techniques used in qualitative analysis found in Appendix I: centrifugation, heating solutions, washing precipitates, and testing acidity.*

Waste Disposal Instructions All waste from this experiment should be placed in appropriate containers in the laboratory.

G1-1 Precipitation of Group 1 Cations

Measure out 10 drops (0.5 mL) of the test solution or the unknown into a small (10 mm × 75 mm) test tube. Add four drops of 6 *M* HCl, stir thoroughly, and then centrifuge. Test for completeness of precipitation by adding one drop of 6 *M* HCl to the clear supernate. If the supernate turns cloudy, this shows that not all of the group 1 cations have precipitated; add another two drops of 6 *M* HCl, stir, and centrifuge. Repeat this process until no more precipitate forms. All of the group 1 cations must be precipitated or else they will slip through and interfere with subsequent group analyses. If a general unknown is being analyzed, decant the supernate into a clean test tube and save it for analysis of group 2 cations; otherwise, discard it. Dispose of any wastes in the designated receptacles. Wash the precipitate by adding five drops of cold distilled water and stirring. Centrifuge and add the liquid to the supernate. Why is the precipitate washed? See Appendix I.

G1-2 Separation and Identification of Pb^{2+}

Add 15 drops of distilled water to the precipitate and place the test tube in a hot-water bath. Stir using a stirring rod and heat for 1 min or longer. Quickly centrifuge and decant the hot supernate into a clean test tube. Repeat this procedure two more times, combining the supernates, which should contain Pb^{2+} if it is present. Save the precipitate for Procedure G1-3. Add three drops of 1 *M* K_2CrO_4 to the supernate. The formation of a yellow precipitate, $PbCrO_4$, confirms the presence of Pb^{2+}.

G1-3 Separation and Identification of Ag^+ and Hg_2^{2+}

Add 10 drops of 6 *M* NH_3 to the precipitate from Procedure G1-2. The formation of a dark gray precipitate indicates the presence of mercury. Centrifuge and decant the clear supernate into a clean test tube. Add 20 drops of 6 *M* HNO_3 to the decantate. Stir the solution and test its acidity. Continue to add HNO_3 dropwise until the solution is acidic. A white cloudiness confirms the presence of Ag^+.

Dispose of all wastes as instructed by your instructor.

REVIEW QUESTIONS

Before beginning Part I of this experiment in the laboratory, you should be able to answer the following questions:

1. What are the symbols and charges of the group 1 cations?
2. Which chloride salt is insoluble in cold water but soluble in hot water?
3. Which chloride salt dissolves in aqueous NH_3?
4. How could you distinguish:
   (a) $BaCl_2$ from AgCl?
   (b) HNO_3 from HCl?
5. Complete and balance the following equations:
   (a) $AgCl(s) + NH_3(aq) \longrightarrow$
   (b) $Pb^{2+}(aq) + CrO_4^{2-}(aq) \longrightarrow$
   (c) $Hg_2Cl_2(s) + NH_3(aq) \longrightarrow$
   (d) $Ag(NH_3)_2^+(aq) + H^+(aq) + Cl^-(aq)$
6. What can you conclude if no precipitate forms when HCl is added to an unknown solution?
7. Why are precipitates washed?
8. How do you decant supernatant liquids from small test tubes?

PART II: CHEMISTRY OF GROUP 2 CATIONS Pb^{2+}, Cu^{2+}, Bi^{3+}, Sn^{4+}

Hydrogen sulfide is the precipitating agent for the group 2 cations, Pb^{2+}, Cu^{2+}, Bi^{3+}, and Sn^{4+}. We will generate H_2S from the hydrolysis of thioacetamide, CH_3CSNH_2:

$$CH_3CSNH_2(aq) + 2H_2O(l) \longrightarrow H_2S(aq) + CH_3CO_2^-(aq) + NH_4^+(aq) \qquad [1]$$

By controlling the hydrogen-ion concentration of the solution, we can control the sulfide-ion concentration. We see from Equation [2], which represents the overall ionization of H_2S, that the equilibrium will shift to the left if the hydrogen-ion concentration is increased by adding some strong acid to the solution.

$$H_2S(aq) \rightleftharpoons 2H^+(aq) + S^{2-}(aq) \qquad [2]$$

Under these conditions the sulfide-ion concentration is very small. Thus, by adjusting the pH of the solution to about 0.5, only the more insoluble sulfides will precipitate. These sulfides are those of group 2, PbS, CuS, Bi_2S_3, and SnS_2:

$$Pb^{2+}(aq) + S^{2-}(aq) \longrightarrow \mathbf{PbS}(s) \quad \text{black} \qquad [3]$$

$$Cu^{2+}(aq) + S^{2-}(aq) \longrightarrow \mathbf{CuS}(s) \quad \text{black} \qquad [4]$$

$$2Bi^{3+}(aq) + 3S^{2-}(aq) \longrightarrow \mathbf{Bi_2S_3}(s) \quad \text{brown} \qquad [5]$$

$$Sn^{4+}(aq) + 2S^{2-}(aq) \longrightarrow \mathbf{SnS_2}(s) \quad \text{yellow} \qquad [6]$$

The sulfides of group 3 are soluble in such acidic solutions and therefore do not precipitate because the sulfide-ion concentration is too small. Figure 30.3 summarizes how the cations of group 2 are separated and identified. Note the colors of the precipitates. *Consult this flow chart often to follow the remaining discussion.*

The cations are first treated with hydrogen peroxide (H_2O_2) and HCl to ensure that tin is in the +4 oxidation state:

$$Sn^{2+}(aq) + 2H^+(aq) + H_2O_2(aq) \longrightarrow Sn^{4+}(aq) + 2H_2O(l) \qquad [7]$$

Tin must be in the +4 oxidation state to form the soluble SnS_3^{2-} ion and, thereby, allow it to be separated from the insoluble sulfides PbS, CuS, and Bi_2S_3:

$$\mathbf{SnS_2}(s) + S^{2-}(aq) \longrightarrow SnS_3^{2-} \qquad [8]$$

These insoluble sulfides are brought into solution by treating them with hot nitric acid. Elemental sulfur is formed in the oxidation reactions:

$$\mathbf{3PbS}(s) + 8H^+(aq) + 2NO_3^-(aq) \longrightarrow 3Pb^{2+}(aq) + 3\mathbf{S}(s) + 2NO(g) + 4H_2O(l) \quad [9]$$

$$\mathbf{3CuS}(s) + 8H^+(aq) + 2NO_3^-(aq) \longrightarrow 3Cu^{2+}(aq) + 3\mathbf{S}(s) + 2NO(g) + 4H_2O(l) \quad [10]$$

$$\mathbf{Bi_2S_3} + 8H^+(aq) + 2NO_3^-(aq) \longrightarrow 2Bi^{3+}(aq) + 3\mathbf{S}(s) + 2NO(g) + 4H_2O(l) \quad [11]$$

Lead After the sulfides of lead, copper, and bismuth are brought into solution, lead is precipitated as the white sulfate by the addition of H_2SO_4:

$$Pb^{2+}(aq) + SO_4^{2-}(aq) \longrightarrow \mathbf{PbSO_4}(s) \quad \text{white} \qquad [12]$$

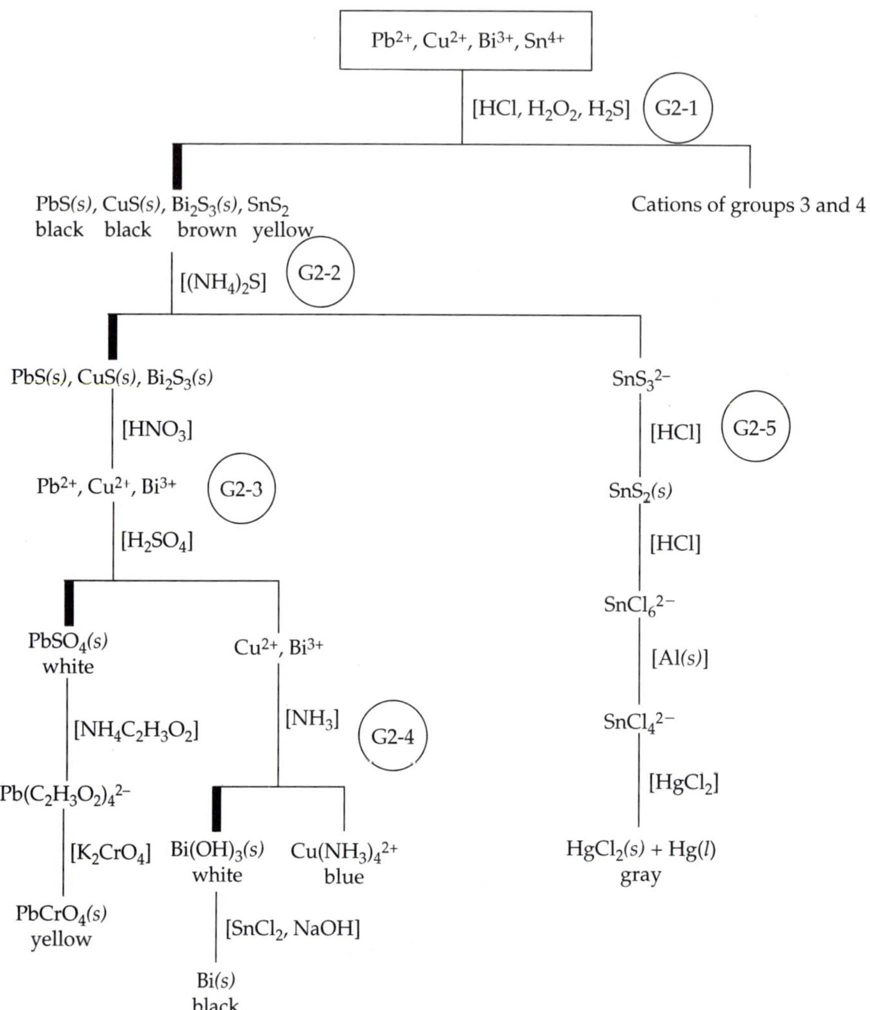

▲ **FIGURE 30.3** Group 2 flow chart.

The solution is heated strongly to drive off HNO_3, because $PbSO_4$ is soluble in the presence of nitric acid. To confirm the presence of Pb^{2+}, the $PbSO_4$ is dissolved with $NH_4C_2H_3O_2$ and precipitated as yellow $PbCrO_4$:

$$\mathbf{PbSO_4}(s) + 4C_2H_3O_2^-(aq) \longrightarrow Pb(C_2H_3O_2)_4^{2-}(aq) + SO_4^{2-}(aq) \quad [13]$$

$$Pb(C_2H_3O_2)_4^{2-} + CrO_4^{2-}(aq) \longrightarrow 4C_2H_3O_2^-(aq) + \mathbf{PbCrO_4}(s) \quad \text{yellow } [14]$$

Copper The flow chart of Figure 30.3 shows that addition of aqueous NH_3 to the solution containing Cu^{2+} and Bi^{3+} precipitates Bi^{3+} as white $Bi(OH)_3$ and forms a soluble deep-blue amine complex of copper, $Cu(NH_3)_4^{2+}$. The deep-blue color confirms the presence of copper and can be seen even when the concentration of copper is very low.

$$Bi^{3+}(aq) + 3NH_3(aq) + 3H_2O(l) \longrightarrow 3NH_4^+(aq) + \mathbf{Bi(OH)_3}(s) \text{ white} \quad [15]$$

$$Cu^{2+}(aq) + 4NH_3(aq) \longrightarrow Cu(NH_3)_4^{2+}(aq) \quad \text{deep blue } [16]$$

Bismuth The $Bi(OH)_3$ precipitate is often difficult to observe if the solution is blue. Bismuth is confirmed by separating the $Bi(OH)_3$ from the solution and reducing it with $SnCl_2$ in an alkaline solution. A black powder of finely divided bismuth is formed:

$$2Bi(OH)_3(s) + 3Sn(OH)_4^{2-}(aq) \longrightarrow 3Sn(OH)_6^{2-} + \mathbf{2Bi}(s) \quad \text{black} \qquad [17]$$

Tin Addition of concentrated HCl to a solution of SnS_3^{2-} first yields the precipitate SnS_2, which eventually dissolves in the acid solution:

$$SnS_3^{2-}(aq) + 2H^+(aq) \longrightarrow \mathbf{SnS_2}(s) + H_2S(g) \qquad [18]$$

$$\mathbf{SnS_2}(s) + 4H^+(aq) + 6Cl^-(aq) \longrightarrow SnCl_6^{2-}(aq) + 2H_2S(g) \qquad [19]$$

Aluminum metal reduces tin from the +4 to the +2 oxidation state. The Sn^{2+} (in the form of the complex $SnCl_4^{2-}$) in turn reduces $HgCl_2$ to insoluble Hg_2Cl_2, which is white, and Hg metal, which is black. Thus, the precipitate appears white to gray in color.

$$3SnCl_6^{2-} + 2Al \longrightarrow 3SnCl_4^{2-} + 2AlCl_3 \qquad [20]$$

$$2SnCl_4^{2-}(aq) + 3HgCl_2(aq) \longrightarrow 2SnCl_6^{2-}(aq) + \mathbf{Hg_2Cl_2}(s) + \mathbf{Hg}(l) \qquad [21]$$

PROCEDURE

G2-1 Oxidation of Sn^{2+} and Precipitation of Group 2

Place the decantate from Procedure G1-1 or seven drops of "known" or "unknown" solution in an evaporating dish. Add four drops of 3% H_2O_2 and four drops of 2 *M* HCl. Carefully boil the solution by passing it back and forth over the flame of your burner. Formation of brown areas on the bottom of the evaporating dish indicates overheating. If brown areas appear, swish the solution around until the brown areas disappear. When the volume of the solution is reduced to about five drops, stop heating and allow the heat from the evaporating dish to complete the evaporation. About three drops of solution should be present after cooling. Add 10 drops of 6 *M* HCl. IN THE HOOD evaporate the contents of the evaporating dish to a pasty mass, being careful again to avoid overheating. Let the evaporating dish cool and then add five drops of 2 *M* HCl and five drops of distilled water. Swish the contents to dissolve or suspend the residue and transfer it to a small test tube. The solution is evaporated to dryness or to a pasty mass in order to remove the unknown quantity of acid that is present. A known amount of HCl is then added, which is required for the group precipitation as sulfides.

Add 10 drops of 1 *M* thioacetamide to the solution in the test tube. Stir the mixture and heat the test tube in a boiling-water bath for 10 min. (See Appendix I for the use of a hot-water bath.) If excessive frothing occurs, temporarily remove the test tube from the bath. Occasionally stir the solution while it is being heated. After heating for 10 min, add 10 drops of *hot* water and 10 drops of 1 *M* thioacetamide and two drops of 1 *M* $NH_4C_2H_3O_2$ (ammonium acetate). Mix and heat in the boiling-water bath for 10 more minutes, stirring occasionally. Cool, centrifuge, and, using a pipet, decant (see Appendix I for the technique) into a test tube. The precipitate contains the group 2 insoluble sulfides, while the supernatant liquid may contain cations from groups 3 and 4.

Test the decantate for completeness of precipitation by adding three drops of H_2O, one drop of 1 *M* $NH_4C_2H_3O_2$, and two drops of 1 *M* thioacetamide. Mix and heat in the boiling-water bath for 1 min. If a colored precipitate forms, continue heating for three additional minutes. A faint cloudiness may develop because of the formation of colloidal sulfur. Repeat the precipitation procedure until precipitation is complete. If the decantate is to be analyzed for groups 3 and 4, transfer it to an evaporating dish and boil it to reduce the volume to about 0.5 mL. Otherwise discard it into a designated receptacle. Transfer the contents to a labeled test tube and save it for Procedure G3-1. Rinse the evaporating dish with six drops of water, and add the washing to the test tube and stopper it.

All precipitates should be combined into the test tube containing the original precipitate, using a few drops of water to aid in the transfer. Wash the precipitate three times, once with 10 drops of *hot* water and twice with 20-drop portions of a hot solution prepared using equal volumes of water and 1 *M* $NH_4C_2H_3O_2$. Be certain to stir the washing liquid and precipitate with a stirring rod before each centrifugation. Should the precipitate form a colloidal suspension, add 10 drops of 1 *M* $NH_4C_2H_3O_2$ and heat the suspension in the boiling water-bath. Discard the washings into a designated receptacle. Note: $NH_4C_2H_3O_2$ is often used in washing of precipitates. Its purpose is to help prevent the precipitate from becoming colloidal. Finely divided particles are difficult to settle.

G2-2 Separation of Sn^{4+} from PbS, CuS, and Bi_2S_3

To the precipitate from Procedure G2-1 add 10 drops of $(NH_4)_2S$ (ammonium sulfide) solution and stir well. Then heat for 3 to 4 min in the boiling-water bath. Remove the test tube from the water bath as necessary to avoid excessive frothing. Centrifuge, decant, and save the decantate. Repeat the treatment using seven drops of $(NH_4)_2S$. Centrifuge and combine the decantate with the first. Stopper the combined decantate, and save for analysis for tin (Procedure G2-5).

Wash the precipitate twice with 20-drop portions of a hot solution prepared by mixing equal volumes of water and 1 *M* $NH_4C_2H_3O_2$. The precipitate may contain PbS, CuS, and Bi_2S_3 and should be analyzed according to Procedure G2-3.

G2-3 Separation and Identification of Pb^{2+}

Add 1 mL (20 drops) of 3 *M* HNO_3 to the test tube containing the precipitate from Procedure G2-2. Mix thoroughly and transfer the contents to an evaporating dish. Boil the mixture gently for 1 min. Add more HNO_3, if necessary, to keep the amount of liquid constant. Cool, centrifuge, and discard any free sulfur that forms. Transfer the decantate to an evaporating dish and add six drops of 18 *M* H_2SO_4. (**CAUTION:** ***Concentrated*** H_2SO_4 ***causes severe burns. If you get any on yourself, immediately wash the area with copious amounts of water.***) IN A HOOD, evaporate the contents until the volume is about one drop and dense white fumes of SO_3 form. The fumes should be so dense that the bottom of the evaporating dish cannot be seen. The appearance of dense white fumes of SO_3 ensures that all HNO_3 has been removed. Cool, add 20 drops of water, and stir. Quickly transfer the contents to a test tube before the suspended material settles. Cool the test tube. A finely divided white precipitate, $PbSO_4$, indicates the presence of lead. Centrifuge and save the decantate for Procedure G2-4. Wash the precipitate twice with 10-drop portions of cold water and discard the washings. Add six drops of 1 *M* $NH_4C_2H_3O_2$ to the precipitate and stir for about 15 sec. Then add one drop of 1 *M* potassium chromate, K_2CrO_4. The formation of a yellow precipitate ($PbCrO_4$) confirms the presence of lead.

G2-4 Separation of Bi^{3+} and Identification of Bi^{3+} and Cu^{2+}

To the decantate from Procedure G2-3 carefully add 15 *M* aqueous NH_3 dropwise while constantly stirring until the solution is basic to litmus. **(CAUTION: *Avoid inhaling or skin contact with* NH_3.)** The appearance of a deep blue color of $Cu(NH_3)_4^{2+}$ confirms the presence of Cu^{2+}. Centrifuge and discard the supernatant liquid but save the white precipitate of $Bi(OH)_3$. Be careful in observing the white gelatinous $Bi(OH)_3$, for it may be somewhat difficult to see when the solution is colored.

Wash the precipitate once with 10 drops of hot water and discard the washings. Add six drops of 6 *M* NaOH and four drops of freshly prepared* 0.2 *M* $SnCl_2$ to the precipitate and stir. The formation of a jet-black precipitate confirms the presence of Bi^{3+}.

G2-5 Identification of Sn^{4+}

Transfer the decantate from Procedure G2-2 to an evaporating dish and boil for 1 min to expel H_2S; then add four drops of cold water. Add a 1-in. piece of 26-gauge aluminum wire and heat gently until the wire has dissolved. Continue to gently heat the solution for about two more minutes, replenishing the solution with 6 *M* HCl if necessary. There should be no dark residue at this stage; if there is, continue heating until it dissolves. (You cannot stop at this point.) Transfer the solution to a test tube and cool under running water. Immediately add three drops of 0.1 *M* mercuric chloride, $HgCl_2$, and mix. Allow the mixture to stand for 1 min. The formation of a white or gray precipitate confirms the presence of tin.

REVIEW QUESTIONS

Before beginning Part II of this experiment in the laboratory, you should be able to answer the following questions:

1. What are the symbols and charges of the group 2 cations?
2. How could CuS be separated from SnS_2?
3. How are Cu^{2+} and Bi^{3+} separated?
4. How can Ag^+ be separated from Cu^{2+}?
5. Complete and balance the following equations:
   (a) $Bi^{3+}(aq) + S^{2-}(aq) \longrightarrow$
   (b) $SnS_2(s) + S^{2-}(aq) \longrightarrow$
   (c) $PbS(s) + H^+(aq) + NO_3^-(aq) \longrightarrow$
   (d) $Bi^{3+}(aq) + NH_3(aq) + H_2O(l) \longrightarrow$
6. Why is H_2O_2 added in the initial step of the separation of group 2 cations?
7. What is the color of CuS? Of SnS_2? Of $PbSO_4$?

*Solid tin should be present in the bottle to ensure that tin is in the +2 oxidation state.

PART III: Chemistry of Group 3 Cations Fe^{3+}, Ni^{2+}, Mn^{2+}, Al^{3+}

The group 3 cations that we will consider are Fe^{3+}, Ni^{2+}, Mn^{2+}, and Al^{3+}. These cations do not precipitate as insoluble chlorides (as do those of group 1) or as sulfides in acidic solution (like those of group 2). These ions can be separated from those of group 4 by precipitation as insoluble hydroxides or sulfides under slightly alkaline conditions. The separation is shown in the flow chart of Figure 30.4. As in group 2, the pH of the solution controls the sulfide-ion concentration. The slightly alkaline conditions employed here favor a higher sulfide-ion concentration than that used in the group 2 separation. FeS, NiS, and MnS are more soluble than the sulfides of group 2 and therefore require a higher sulfide-ion concentration for their precipitation. Making the solution slightly alkaline reduces the hydrogen-ion concentration. We see that decreasing the hydrogen-ion concentration causes the following equilibrium to shift to the right:

$$H_2S(aq) \rightleftharpoons 2H^+(aq) + S^{2-}(aq)$$

This results in an increase in the sulfide-ion concentration.

Aqueous ammonia is a weak base:

$$NH_3(aq) + H_2O(l) \rightleftharpoons NH_4^+(aq) + OH^-(aq)$$

A mixture of aqueous NH_3 and NH_4Cl makes a buffer solution whose pH allows the precipitation of FeS, MnS, and NiS. Moreover, in this slightly alka-

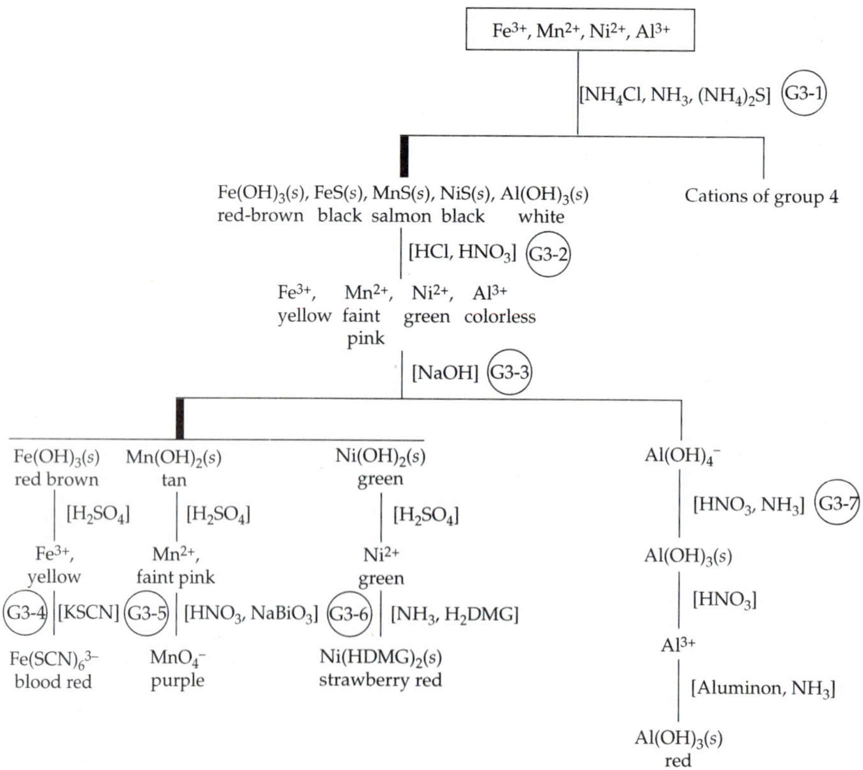

▲ **FIGURE 30.4** Group 3 flow chart.

line solution, the insoluble hydroxides $Fe(OH)_3$ and $Al(OH)_3$ also precipitate. Although $Mg(OH)_2$ is insoluble, it does not precipitate with the group 3 hydroxides because the OH^- ion concentration is too small. The common ion NH_4^+, from NH_4Cl, controls the OH^- concentration and keeps it sufficiently low, preventing the $Mg(OH)_2$ from precipitating.

As you analyze the group 3 ions, pay particular attention to the colors of the solutions and the precipitates. Aqueous solutions of Al^{3+} ions are colorless, while those of Fe^{3+} appear yellow to reddish-brown; Ni^{2+} solutions are green, and Mn^{2+} *very* faint pink.

The reactions involved in the precipitation of the group 3 cations are:

$$Fe^{3+}(aq) + 3OH^-(aq) \longrightarrow \mathbf{Fe(OH)_3}(s) \quad \text{red-brown} \quad [1]$$

$$Al^{3+}(aq) + 3OH^-(aq) \longrightarrow \mathbf{Al(OH)_3}(s) \quad \text{colorless} \quad [2]$$

$$Ni(NH_3)_6^{2+}(aq) + S^{2-}(aq) \longrightarrow 6NH_3(aq) + \mathbf{NiS}(s) \quad \text{black} \quad [3]$$

$$Mn^{2+}(aq) + S^{2-}(aq) \longrightarrow \mathbf{MnS}(s) \quad \text{salmon} \quad [4]$$

In aqueous NH_3 solution, Ni^{2+} ions exist as the ammine complex $Ni(NH_3)_6^{2+}$, and addition of $(NH_4)_2S$ results in the precipitation of NiS. Some $Fe(OH)_3$ is reduced by sulfide:

$$2\mathbf{Fe(OH)_3}(s) + 3S^{2-}(aq) \longrightarrow 6OH^-(aq) + 2\mathbf{FeS}(s) + S(s) \quad \text{black} \quad [5]$$

The group 3 precipitates are dissolved using HCl. Nitric acid is also added to help dissolve NiS by oxidizing S^{2-} ions to elemental sulfur. At the same time, nitric acid oxidizes Fe^{2+} to Fe^{3+}:

$$\mathbf{Fe(OH)_3}(s) + 3H^+(aq) \longrightarrow Fe^{3+}(aq) + 3H_2O(l) \quad [6]$$

$$\mathbf{FeS}(s) + 2H^+(aq) \longrightarrow Fe^{2+}(aq) + H_2S(g) \quad [7]$$

$$3Fe^{2+}(aq) + NO_3^-(aq) + 4H^+(aq) \longrightarrow 3Fe^{3+}(aq) + NO(g) + 2H_2O(l) \quad [8]$$

$$\mathbf{Al(OH)_3}(s) + 3H^+(aq) \longrightarrow Al^{3+}(aq) + 3H_2O(l) \quad [9]$$

$$3\mathbf{NiS}(s) + 2NO_3^-(aq) + 8H^+(aq) \longrightarrow 3Ni^{2+}(aq) + 3S(s) + 2NO(g) + 4H_2O(l) \quad [10]$$

$$\mathbf{MnS}(s) + 2H^+(aq) \longrightarrow Mn^{2+}(aq) + H_2S(g) \quad [11]$$

After all ions are in solution, addition of excess strong base allows the separation of Fe^{3+}, Ni^{2+}, and Mn^{2+} ions from Al^{3+} ions. $Fe(OH)_3$, $Ni(OH)_2$, and $Mn(OH)_2$ precipitate, while $Al(OH)_3$, being amphoteric, redissolves in excess base forming aluminate ions:

$$Fe^{3+}(aq) + 3OH^-(aq) \longrightarrow \mathbf{Fe(OH)_3}(s) \quad \text{red-brown} \quad [12]$$

$$Ni^{2+}(aq) + 2OH^-(aq) \longrightarrow \mathbf{Ni(OH)_2}(s) \quad \text{green} \quad [13]$$

$$Mn^{2+}(aq) + 2OH^-(aq) \longrightarrow \mathbf{Mn(OH)_2}(s) \quad \text{tan} \quad [14]$$

$$Al^{3+}(aq) + 4OH^-(aq) \longrightarrow Al(OH)_4^-(aq) \quad \text{colorless} \quad [15]$$

The hydroxide precipitates are dissolved by the addition of H_2SO_4 to give a solution of Fe^{3+}, Ni^{2+}, and Mn^{2+} ions. After the solution is divided into three equal parts, tests for the individual ions are made as described below.

Iron A very sensitive test for Fe^{3+} uses the thiocyanate ion, SCN^-. If Fe^{3+} is present, a blood-red solution results when SCN^- is added:

$$Fe^{3+}(aq) + 6SCN^-(aq) \longrightarrow Fe(SCN)_6^{3-}(aq) \quad \text{red} \quad [16]$$

Manganese Because of its intense purple color, the permanganate ion, MnO_4^-, affords a suitable confirmatory test for Mn^{2+}. When a solution of Mn^{2+} is acidified with HNO_3 and then treated with sodium bismuthate, $NaBiO_3$, Mn^{2+} is oxidized to MnO_4^-:

$$2Mn^{2+}(aq) + \mathbf{5NaBiO_3}(s) + 14H^+(aq) \longrightarrow 5Bi^{3+}(aq) + 5Na^+(aq) + 7H_2O(l) + 2MnO_4^-(aq) \quad \text{purple} \quad [17]$$

Nickel The presence of Ni^{2+} is confirmed by the formation of a bright red precipitate when an organic compound called dimethylglyoxime (abbreviated as H_2DMG) is added to an ammoniacal solution:

$$Ni(NH_3)_6^{2+}(aq) + 2H_2DMG(aq) \longrightarrow 4NH_3(aq) + 2NH_4^+(aq) + \mathbf{Ni(HDMG)_2}(s) \quad \text{red} \quad [18]$$

Aluminum When a solution that contains $Al(OH)_4^-$ is acidified and then made slightly alkaline with the weak base NH_3, $Al(OH)_3$ precipitates:

$$Al(OH)_4^- + 4H^+(aq) \longrightarrow Al^{3+}(aq) + 4H_2O(l) \quad [19]$$

$$Al^{3+}(aq) + 3NH_3(aq) + 3H_2O(l) \longrightarrow 3NH_4^+(aq) + \mathbf{Al(OH)_3}(s) \quad [20]$$

Aluminum hydroxide is not easily seen, however, for it is a gelatinous, translucent substance. To help see the hydroxide, it is precipitated in the presence of a red dye. The dye is adsorbed on the $Al(OH)_3$, giving it a cherry-red color.

PROCEDURE

G3-1 Precipitation of Group 3 Cations

Place the decantate from Procedure G2-1 or seven drops of known or unknown solution in a small (10 mm × 75 mm) test tube. If the decantate from Procedure G2-1 has a precipitate, centrifuge, decant, and discard the precipitate. Add five drops of 2 *M* NH_4Cl and stir; add 15 *M* NH_3 dropwise with stirring until the solution is just basic to litmus paper. Usually this requires only a few drops of the NH_3. Then add two additional drops of the 15 *M* NH_3 and 1 mL (20 drops) of water. Stir thoroughly. Next add 10 drops of $(NH_4)_2S$ and mix thoroughly. Heat the test tube in a boiling-water bath for about 5 min. If excessive frothing occurs, temporarily remove the test tube from the hot-water bath. Centrifuge and test for completeness of precipitation using one drop of $(NH_4)_2S$. Note the color of the precipitate. Decant and save the decantate for group 4 analysis, or discard into a designated receptacle, as appropriate.

Wash the precipitate two times with 20 drops of a solution made by mixing equal portions of water and 1 *M* $NH_4C_2H_3O_2$. For each washing, stir the precipitate with the wash solution and heat the mixture in the water bath before centrifuging. Discard the supernatant wash liquid.

G3-2 Dissolution of Group 3 Precipitate

Treat the precipitate from Procedure G3-1 with 12 drops of 12 *M* HCl, then cautiously add five drops of 16 *M* HNO_3 and carefully mix the solution. **(CAUTION: HNO_3 *can cause severe burns. If you come in contact with the acid, wash the area with copious amounts of water.*)** Heat the test tube in the hot-water bath until the precipitate dissolves and a clear, but not necessarily colorless, solution is obtained. Add 10 drops of water, centrifuge to remove any sulfur that has precipitated, and decant into an evaporating dish. Note the color of the decantate.

G3-3 Separation of Iron, Nickel, and Manganese from Aluminum

Make the solution in the evaporating dish from Procedure G3-2 strongly basic, using 6 *M* NaOH, mixing thoroughly. If the precipitate is pasty and nonfluid, add 12 drops of water. Note the color of the precipitate. Transfer to a test tube and centrifuge. Decant, saving the decantate, which may contain aluminum, for Procedure G3-7. To the precipitate add 20 drops of water and 10 drops of 6 *M* H_2SO_4. Stir and heat in a water bath for 3 min or until the precipitate dissolves. Add 12 drops of water and divide the solution into three approximately equal volumes.

G3-4 Test for Fe^{3+}

To one of the three samples from Procedure G3-3 add two drops of 0.2 *M* KSCN (potassium thiocyanate). A blood-red *solution* confirms the presence of iron as $Fe(SCN)_6^{3-}$. Traces of iron that have been introduced as impurities along the way will give a weak test result. If you are in doubt about your results, perform this test on 10 drops of your original sample.

G3-5 Test for Mn^{2+}

To the second portion of a solution from Procedure G3-3 add an equal volume of water and four drops of 3 *M* HNO_3. Mix and then add a few grains of solid sodium bismuthate, $NaBiO_3$. Mix thoroughly with a stirring rod and centrifuge. A pink or purple color is due to MnO_4^- and confirms the presence of manganese.

G3-6 Test for Ni^{2+}

To a third portion of the sample from Procedure G3-3 add 6 *M* NH_3 until the solution is basic. If a precipitate forms, remove it by centrifuging and decanting, keeping the decantate. Add about four drops of dimethylglyoxime reagent, mix, and allow to stand. The formation of a strawberry-red *precipitate* indicates the presence of nickel.

G3-7 Test for Al^{3+}

Treat only half of the decantate from Procedure G3-3 with 16 *M* HNO_3 until the solution is slightly acidic. **(CAUTION: HNO_3 *can cause severe burns. Wash with water immediately if you come in contact with the acid.*)** Then add 15 *M* NH_3, while stirring, until the solution is distinctly alkaline. Allow at least 1 min for the formation of $Al(OH)_3$. Centrifuge and carefully remove the supernatant liquid with a capillary pipet without disturbing the gelatinous $Al(OH)_3$. Discard the supernate. Wash the precipitate two times with 20 drops of hot water, discarding the decantate. Dissolve the precipitate in seven drops of 3 *M* HNO_3. Add three drops of Aluminon reagent, which colors the solution; stir; and add 6 *M* NH_3 dropwise until the solution is just alkaline to litmus paper (avoid an excess). Stir and centrifuge. The formation of a cherry-red *precipitate,* not solution, confirms the presence of aluminum.

REVIEW QUESTIONS

Before beginning Part III of this experiment in the laboratory, you should be able to answer the following questions:

1. What are the symbols and charges of the group 3 cations?
2. What are the colors of the following ions: Cu^{2+}, Fe^{3+}, Al^{3+}, Ni^{2+}?
3. What are the colors of the following solids: $Fe(OH)_3$, MnS, $Al(OH)_3$, $Ni(OH)_2$?
4. How can Fe^{3+} be separated from Al^{3+}?
5. How can $Ni(OH)_2$ be separated from $Al(OH)_3$?
6. Complete and balance the following equations:
   (a) $Fe^{3+}(aq) + OH^-(aq) \longrightarrow$
   (b) $Al(OH)_3(s) + H^+(aq) \longrightarrow$
   (c) $FeS(s) + H^+(aq) \longrightarrow$
   (d) $NiS(s) + NO_3^-(aq) + H^+(aq) \longrightarrow$
7. Give the formula for a reagent that precipitates
   (a) Pb^{2+} but not Ni^{2+}.
   (b) Fe^{3+} but not Al^{3+}.
8. What cation forms a
   (a) blood-red solution with thiocyanate ion?
   (b) bright red precipitate with dimethylglyoxime?
9. If solid NH_4Cl is added to 3 *M* NH_3, does the pH increase, decrease, or remain the same?

PART IV: Chemistry of Group 4 Cations Ba^{2+}, Ca^{2+}, NH_4^+, Na^+

In addition to the ammonium ion, the cations of group 4 consist of ions of the alkali and alkaline earth metals. The cations we will consider in this group are Ba^{2+}, Ca^{2+}, NH_4^+, and Na^+. Because their chlorides and sulfides are soluble, these ions do not precipitate with groups 1, 2, or 3.

Sodium ions are a common impurity and were even introduced (as was ammonium ion) in some of the reagents that were used in the analysis of groups 1, 2, and 3. Hence, in the analysis of a general unknown mixture, tests for these ions must be made on the original sample even before performing the group analysis. The flow chart for group 4 is shown in Figure 30.5.

Barium Because barium chromate, $BaCrO_4$ ($K_{sp} = 1.2 \times 10^{-10}$), is less soluble than calcium chromate, $CaCrO_4$ ($K_{sp} = 7.1 \times 10^{-4}$), Ba^{2+} can be separated from Ca^{2+} by precipitation as the insoluble yellow chromate salt:

$$Ba^{2+}(aq) + CrO_4^{2-}(aq) \longrightarrow \mathbf{BaCrO_4}(s) \quad \text{yellow} \qquad [1]$$

$BaCrO_4$ is insoluble in the weak acid $HC_2H_3O_2$, but it is soluble in the presence of the strong acid HCl because it is the salt of a weak acid, H_2CrO_4. After $BaCrO_4$ is dissolved in HCl, a flame test is performed on the resulting

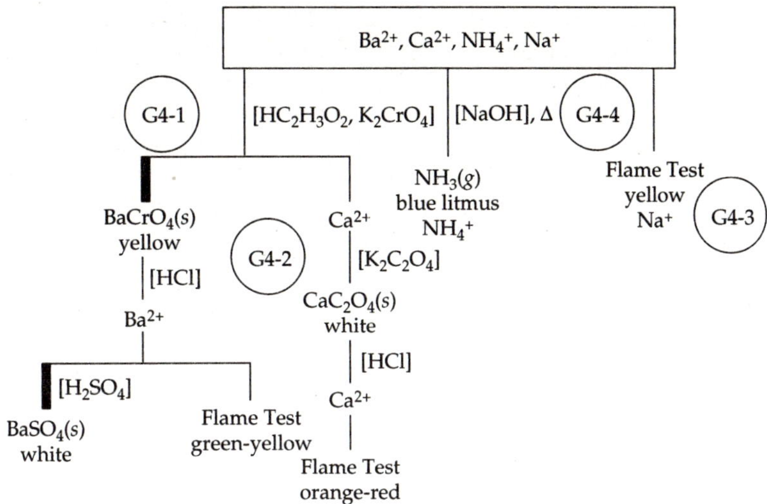

▲ **FIGURE 30.5** Group 4 flow chart.

solution. A green-yellow flame is indicative of Ba^{2+}. Further confirmation of Ba^{2+} is precipitation of $BaSO_4$, which is white:

$$Ba^{2+}(aq) + SO_4^{2-}(aq) \longrightarrow \mathbf{BaSO_4}(s) \qquad \text{white} \qquad [2]$$

Calcium Calcium oxalate, CaC_2O_4, is very insoluble ($K_{sp} = 4.0 \times 10^{-9}$). The formation of a white precipitate when oxalate ion is added to a slightly alkaline solution confirms the presence of Ca^{2+}:

$$Ca^{2+}(aq) + C_2O_4^{2-}(aq) \longrightarrow \mathbf{CaC_2O_4}(s) \qquad \text{white} \qquad [3]$$

Additional evidence for the calcium ion is obtained from a flame test. Dissolution of CaC_2O_4 with HCl, followed by a flame test, produces a transient orange-red flame that is characteristic of calcium ions.

Sodium Most sodium salts are soluble. The simplest test for sodium ion is a flame test. Sodium salts impart a characteristic yellow color to a flame; the test is very sensitive, and because of the prevalence of sodium ions, much care must be exercised to keep equipment clean and free from contamination by these ions.

Ammonium The ammonium ion, NH_4^+, is the conjugate acid of the base ammonia, NH_3. The test for NH_4^+ takes advantage of the following equilibrium:

$$NH_4^+(aq) + OH^-(aq) \rightleftharpoons NH_3(aq) + H_2O(l) \qquad [4]$$

Thus, when a strong base is added to a solution of an ammonium salt and this solution is heated, NH_3 gas is evolved. The NH_3 can easily be detected by its effect upon red litmus.

PROCEDURE

G4-1 Separation and Identification of Ba^{2+}

If the solution, "known" or "unknown," contains only cations of group 4, place seven drops of the solution in a small test tube; otherwise, use the supernate from group 3 analysis, Procedure G3-1. Add eight drops of 6 *M* acetic acid, $HC_2H_3O_2$, and one drop of 1 *M* K_2CrO_4 and mix. The formation of a yellow precipitate indicates the presence of Ba^{2+}. Centrifuge and save the decantate for Procedure G4-2 to test for calcium. Dissolve the precipitate with 6 *M* HCl and perform a flame test as described below.

To perform the flame test, obtain a piece of platinum or Nichrome wire that has been sealed in a piece of glass tubing. Clean the wire by dipping it in 12 *M* HCl that is contained in a small test tube, and heat the wire in the hottest part of your Bunsen burner flame. Repeat this operation until no color is seen when the wire is placed in the flame. Several cleanings will be required before this is achieved. Then dip the wire into the solution to be tested, and place the wire in the flame. A pale green flame confirms the presence of Ba^{2+}. If the concentration of Ba^{2+} is very low, you may not detect the green color.

As further confirmation of barium, add 10 drops of 6 *M* H_2SO_4 to the solution on which the flame test was performed. A white precipitate confirms the presence of Ba^{2+}.

G4-2 Test for Ca^{2+}

Make the decantate from Procedure G4-1 alkaline to litmus with 15 *M* NH_3. If a precipitate forms, centrifuge and discard the precipitate. Add seven drops of 1 *M* $K_2C_2O_4$ (potassium oxalate) and stir. The formation of a white precipitate indicates the presence of calcium ion. Should no precipitate form immediately, warm the test tube briefly in the hot-water bath and then cool.

Additional evidence for Ca^{2+} is obtained from a flame test. Dissolve the precipitate in 6 *M* HCl and then perform a flame test. A transitory red-orange color that appears when the wire is first placed in the flame and later reappears somewhat more red as the wire is heated is characteristic of the calcium ion. If the concentration of Ca^{2+} is very low, you may not observe the red color.

G4-3 Test for Na^+

The flame test for sodium is very sensitive, and traces of sodium ion will impart a characteristic yellow color to the flame. Just about every solution has a trace of sodium and thus will give a positive test. On the basis of the intensity and duration of the yellow color, you can decide whether Na^+ is merely a contaminant or present in substantial quantity. Using a clean wire, perform a flame test on your original (untreated) unknown. To help you make a decision as to the presence of sodium, run a flame test on distilled water and then on a 0.2 *M* NaCl solution. Compare the tests.

G4-4 Test for NH_4^+

Place 2 mL of the original (untreated) unknown or known in a 100-mL beaker (or evaporating dish) and add 2 mL of 6 *M* NaOH. Moisten a piece of red litmus paper with water and stick it to the convex side of a small watch glass. Cover the beaker or evaporating dish with the watch glass, convex side down. (The litmus paper must not come into contact with any NaOH.) Gently warm the beaker (or evaporating dish) with a small burner flame; do not boil. Allow the covered beaker (or evaporating dish) to stand for 3 min. A change in the color of the litmus paper from red to blue confirms the presence of ammonium ion.

REVIEW QUESTIONS

Before beginning Part IV of this experiment in the laboratory, you should be able to answer the following questions:

1. What are the symbols and charges of the group 4 cations?
2. Which is less soluble: $BaCrO_4$ or $CaCrO_4$?
3. What is the color of
   (a) $BaSO_4$?
   (b) $BaCrO_4$?
   (c) CaC_2O_4?
4. What color do the following ions impart to a flame?
   (a) Ba^{2+}
   (b) Ca^{2+}
   (c) Na^+
5. What reagent will precipitate
   (a) Cu^{2+} but not Ba^{2+}?
   (b) Ag^+ but not Ca^{2+}?
   (c) Ba^{2+} but not NH_4^+?
6. Complete and balance the following equations:
   (a) $Ba^{2+}(aq) + CrO_4^{2-}(aq)$
   (b) $NH_4^+(aq) + OH^-(aq)$

PART V: CHEMISTRY OF ANIONS SO_4^{2-}, NO_3^-, CO_3^{2-}, Cl^-, Br^-, I^-, CrO_4^{2-}, PO_4^{3-}, S^{2-}, SO_3^{2-}

DISCUSSION

A systematic scheme based on the kinds of principles involved in cation analysis can be designed for the analysis of anions. This would involve separation of the anions followed by their identification. However, it is generally easier to take another approach to the identification of anions. An effort is made either to eliminate or verify the presence of certain anions on the basis of the color and solubility of the samples; then the material being analyzed is subjected to a series of preliminary tests. From the results of the preliminary tests and observations, certain anions may definitely be shown to be present or absent; then specific tests are performed for those anions not definitely eliminated in the preliminary tests and observations. The preliminary tests include treating the solid with concentrated sulfuric acid and using silver nitrate and $BaCl_2$ as precipitating agents. The behavior of the anions in these tests is conveniently summarized in Tables 30.1 and 30.2.

In this experiment, characteristics of the following 10 anions are considered: sulfate (SO_4^{2-}), nitrate (NO_3^-), carbonate (CO_3^{2-}), chloride (Cl^-), bromide (Br^-), iodide (I^-), chromate (CrO_4^{2-}), phosphate (PO_4^{3-}), sulfide (S^{2-}), and sulfite (SO_3^{2-}).

As in the case of cation identification, the physical and chemical properties of the compounds formed by the anions provide the basis for their identification. We will proceed by using the following general procedures.

TABLE 30.1 Summary of Preliminary Tests

Anion	Concentrated H_2SO_4 solid salt	$AgNO_3$ Neutral	$AgNO_3$ Acidic	$BaCl_2$ Neutral	$BaCl_2$ Acidic
NO_3^-	Cold: no observable reaction Hot: brown gas	No reaction		No reaction	
Cl^-	HCl(*g*), colorless; pungent	AgCl(*s*) / white	Insol.	No reaction	
Br^-	Br_2, red-brown; pungent	AgBr(*s*) / cream	Insol.	No reaction	
I^-	Solids turn dark brown, violet vapors; H_2S odor	AgI(*s*) / yellow	Insol.	No reaction	
S^{2-}	H_2S(g), odor of rotten eggs free S deposited	Ag_2S(*s*) / black	Insol.	No reaction	
SO_4^{2-}	No evidence of reaction	No reaction		$BaSO_4$(*s*) / white	Insol.
SO_3^{2-}	SO_2(*g*), colorless; choking odor	Ag_2SO_3(*s*) / white	Sol.	$BaSO_3$(*s*) / white	Sol.
CO_3^{2-}	CO_2(*g*), colorless; odorless	Ag_2CO_3(*s*) / white → black	Sol.	$BaCO_3$(*s*) / white	Sol.
PO_4^{3-}	No evidence of reaction	Ag_3PO_4(*s*) / yellow	Sol.	$Ba_3(PO_4)_2$(*s*) / white	Sol.
CrO_4^{2-}	Solid changes from yellow to orange-red	Ag_2CrO_4(*s*) / red-brown	Sol.	$BaCrO_4$(*s*) / yellow	Sol.

PROCEDURE

Examination of the Solid

The color of a substance offers a clue to its constituents. For example, many transition metal salts are colored. Those of nickel, Ni^{2+}, are generally green; iron(III), Fe^{3+}, reddish-brown to yellow; of iron(II), Fe^{2+}, grayish-green; of chromium(III), Cr^{3+}, green to bluish-gray to black; of copper(II), Cu^{2+}, blue to green to black; of cobalt(II), Co^{2+}, wine-red to blue; and of manganese(II), Mn^{2+}, pink to tan. By contrast, only a few anions are colored, and these contain transition metals as well. The colored anions are chromate, CrO_4^{2-}, yellow; dichromate, $Cr_2O_7^{2-}$, orange-red; and permanganate, MnO_4^{2-}, violet-purple. Hence, the color of the solid may be used as a very good indicator of the presence of these ions. Observe the color of your unknown and note it on the report sheet.

Solubility of Salt

The solubility of a substance in water and in acidic or basic solutions, together with a knowledge of the cations present, often permits us to narrow down the choice of anions present. For example, a white substance containing Ag^+ that is insoluble in acid solution (HNO_3) but soluble in aqueous ammonia (NH_3) is most probably AgCl. There is some ambiguity, however, for AgBr behaves similarly. Thus tests for both chloride and bromide would be necessary to confirm which anion is present. Test the solubility of your solid unknown anion sample in water, 6 *M* HNO_3, 6 *M* HCl, and 6 *M* NH_3,

TABLE 30.2 Behavior of Anions with Concentrated Sulfuric Acid, H_2SO_4

A. Cold H_2SO_4	
SO_4^{2-}	No reaction.
NO_3^-	No reaction.
PO_4^{3-}	No reaction.
CO_3^{2-}	A colorless, odorless gas forms. $CO_3^{2-} + 2H^+ \longrightarrow H_2O + CO_2$
Cl^-	A colorless gas forms. It has a sharp, pungent odor, gives an acid test result with litmus, and fumes in moist air. $Cl^- + H^+ \longrightarrow HCl$
Br^-	A brownish-red gas forms. It has a sharp odor, gives an acid test result with litmus, and fumes in moist air. The odor of SO_2 may be detected. $Br^- + H_2SO_4 \longrightarrow HSO_4^- + HBr$; $H_2SO_4 + 2HBr \longrightarrow 2H_2O + SO_2 + Br_2$
I^-	Solid turns dark brown immediately, with the slight formation of violet fumes. The gas has the odor of rotten eggs, gives an acidic test result with litmus, and fumes in moist air. $I^- + H_2SO_4 \longrightarrow HSO_4^- + HI$; $H_2SO_4 + 8HI \longrightarrow H_2S + 4H_2O + 4I_2$; $H_2SO_4 + 2HI \longrightarrow 2H_2O + SO_2 + I_2$
S^{2-}	A colorless gas with the odor of rotten eggs forms, and some free sulfur is deposited. $S^{2-} + H_2SO_4 \longrightarrow SO_4^{2-} + H_2S$; $H_2SO_4 + H_2S \longrightarrow 2H_2O + SO_2 + S$
SO_3^{2-}	A colorless gas with a sharp, choking odor forms. $SO_3^{2-} + H_2SO_4 \longrightarrow SO_4^{2-} + H_2O + SO_2$
CrO_4^{2-}	Color changes from yellow to orange-red. $2K_2CrO_4 + H_2SO_4 \longrightarrow K_2Cr_2O_7 + H_2O + K_2SO_4$
B. Hot Concentrated H_2SO_4	
There are no additional reactions with any of the anions except NO_3^-, which forms brown fumes of NO_2 gas. $4NO_3^- + 4H^+ \longrightarrow 4NO_2 + O_2 + 2H_2O$	

and note your results on your report sheet. Compare your results with the solubility properties of ions and solids given in Appendix C.

Reactions with $AgNO_3$ and $BaCl_2$

If your unknown substance is soluble in water, you may obtain additional clues as to the presence of certain anions by treating a solution of your unknown separately with solutions of silver nitrate ($AgNO_3$) and barium chloride ($BaCl_2$). The silver salts of all your possible anions except nitrate and sulfate are insoluble in water and have the following colors (see Table 30.1):

AgCl: white	Ag_2CrO_4: red-brown
AgBr: cream	Ag_2S: black
AgI: yellow	Ag_2CO_3: white
Ag_3PO_4: yellow	Ag_2SO_3: white

All these solids except AgCl, AgBr, AgI, and Ag_2S are soluble in dilute nitric acid. Dissolve some of your solid unknown, about the size of a small pea, in water and add four drops of 0.1 *M* $AgNO_3$ solution. Note the color of any precipitate that forms, and enter your observation on your report sheet. Dissolve another small pea–size portion of your sample in water, add a few drops of 6 *M* HNO_3 to make the solution acidic, and then add four drops of 0.1 *M* $AgNO_3$ and stir the mixture. Record your observation on your report sheet.

The barium salts $BaCl_2$, $BaBr_2$, BaI_2, BaS, and $Ba(NO_3)_2$ are soluble in water and in slightly basic solution, but $BaSO_4$, $BaSO_3$, $BaCO_3$, $BaCrO_4$, and $Ba_3(PO_4)_2$ are insoluble (see Table 30.1). Only $BaSO_4$ is insoluble in acidic solution, whereas the other barium salts that are insoluble in water dissolve in acidic solution. All the barium salts are white except $BaCrO_4$, which is yellow. Dissolve a small pea–size portion of your unknown in water, then add a few drops of 0.2 *M* barium chloride and stir. Record your observations on your report sheet. Acidify the mixture with a few drops of 6 *M* HNO_3 and stir thoroughly. Record your observations on your report sheet.

Reactions of the Solid Unknown with Concentrated H_2SO_4

Several of the anions form volatile weak acids with, or are oxidized by, concentrated sulfuric acid. Careful observation of the reaction of concentrated sulfuric acid with your solid unknown can give you further clues as to the presence of specific anions. A summary of the pertinent reactions of concentrated sulfuric acid with various anions is given in Table 30.2.

Place a small pea–size amount of the solid in a dry, small test tube. Add one or two drops of 18 *M* H_2SO_4 and observe everything that occurs, especially the color and odor of gas formed. **(CAUTION:** ***Concentrated*** H_2SO_4 ***causes severe burns. Do not get it on your skin or clothing. If you come in contact with it, immediately wash the area with copious amounts of water. You must wear your eye protection as you should at all times when in the laboratory!*)** DO NOT place your nose directly over the mouth of the test tube, but carefully fan gases toward your nose. Then carefully heat the test tube, but not so strongly as to boil the H_2SO_4. **(CAUTION:** ***If you heat the acid too strongly, it could come shooting out!*)** Note whether or not brown fumes of NO_2 are produced. **(CAUTION:** ***Do not look down into the test tube. Do not point the test tube at yourself or at your neighbors.*)** SAFETY GLASSES MUST BE WORN.

Specific Tests for Anions

When an anion is indicated by the preliminary tests with $AgNO_3$, $BaCl_2$, or concentrated H_2SO_4, it is confirmed using the appropriate specific test. Make an aqueous solution of the solid unknown and perform the following tests on portions of this solution:

Sulfate Place 10 drops of a solution of the anion unknown in a test tube, acidify with 6 *M* HCl, and add a drop of $BaCl_2$ solution. A white precipitate of $BaSO_4$ confirms SO_4^{2-} ions (see Note 1).

$$Ba^{2+}(aq) + SO_4^{2-}(aq) \longrightarrow \mathbf{BaSO_4}(s)$$

Note 1 Sulfites are slowly oxidized to sulfates by atmospheric oxygen. Consequently, sulfites commonly show a positive test for sulfates.

$$2SO_3^{2-}(aq) + O_2(g) \longrightarrow 2SO_4^{2-}(aq)$$

$$SO_4^{2-}(aq) + Ba^{2+}(aq) \longrightarrow \mathbf{BaSO_4}(s)$$

Sulfite Place 10 drops of a solution of the anion unknown in a test tube, acidify with 6 *M* HCl, add two to three drops of 0.2 *M* $BaCl_2$, and mix thoroughly. If a precipitate ($BaSO_4$) forms, remove it by centrifuging and decanting. To the clear decantate, add a drop of 3% H_2O_2 (see Note 2). The formation of a white precipitate of $BaSO_4$ confirms SO_3^{2-} ions.

Note 2 H_2O_2 oxidizes sulfite to sulfate:

$$SO_3^{2-}(aq) + H_2O_2(aq) \longrightarrow SO_4^{2-}(aq) + H_2O(l)$$

$$Ba^{2+}(aq) + SO_4^{2-}(aq) \longrightarrow \mathbf{BaSO_4}(l)$$

Any SO_4^{2-} originally present was previously removed as $BaSO_4$ by centrifugation.

Chromate Place two drops of a solution of the anion unknown in a test tube, add 10 drops of water, and make the solution just acidic with 3 *M* HNO_3. Add five to six drops of ether (KEEP THE ETHER AWAY FROM FLAMES) and 1 drop of 3% H_2O_2, stir well, and then allow the precipitate to settle. A blue coloration of the top ether layer (see Note 3) confirms CrO_4^{2-} ions (see Note 4).

Note 3 The blue coloration in the ether layer is due to the presence of chromium peroxide, CrO_5.

$$2CrO_4^{2-}(aq) + 2H^+(aq) \longrightarrow Cr_2O_7^{2-}(aq) + H_2O(l)$$

$$Cr_2O_7^{2-}(aq) + 4H_2O_2(aq) + 2H^+(aq) \longrightarrow 2CrO_5(aq) + 5H_2O(l)$$

Note 4 If your anion unknown is not colored or did not indicate CrO_4^{2-} in the H_2SO_4 reaction, or if both conditions apply, this test may be omitted.

Iodide Place five drops of a solution of the anion unknown in a test tube, add five drops of 6 *M* $HC_2H_3O_2$, and then add two drops of 0.2 *M* KNO_2. A reddish-brown coloration due to the presence of I_2 confirms I^-. If the brown color is very faint, add a few drops of mineral oil and shake well. A violet color in the top (mineral oil) layer confirms I^-.

$$NO_2^-(aq) + H^+(aq) \longrightarrow HNO_2(aq)$$

$$2HNO_2(aq) + 2I^-(aq) + 2H^+(aq) \longrightarrow 2NO(g) + I_2(aq) + 2H_2O(l)$$

Bromide Iodides will interfere with this test (see Note 5). Place five drops of a solution of the anion unknown in a test tube and add five drops of chlorine water. A brown coloration due to the liberation of Br_2 confirms Br^-. If the solution is shaken with a few drops of mineral oil, the brown color will concentrate in the top layer, which is mineral oil. Allow about 20 s for the layers to separate.

$$Cl_2(aq) + 2Br^-(aq) \longrightarrow 2Cl^-(aq) + Br_2(aq)$$

Note 5 Iodide ions, if present, must be removed before testing for bromide. To remove iodide, acidify the solution with 3 *M* HNO_3 and add 2 *M* KNO_2, dropwise, with constant stirring, until there is no further increase in the depth of the brown color. Extract once by shaking with five drops of mineral oil. Discard the mineral-oil layer into a designated receptacle. Boil the water layer carefully until the iodine has been largely driven off. Test the colorless, or nearly colorless, solution for bromide as directed above.

Nitrate Iodides, bromides, and chromates interfere with this test and must be removed (see Note 6) if they are present. Place 10 drops of a solution of the anion salt in a small test tube, add five drops of $FeSO_4$ solution, and mix the solution. Carefully, without agitation, pour concentrated H_2SO_4 down the inside of the test tube so as to form two layers. Allow to stand for 1 or 2 min. The formation of a brown ring between the two layers confirms NO_3^- ions. **(CAUTION:** ***Do not get the*** $\boldsymbol{H_2SO_4}$ ***on yourself or on your clothing. If you do, wash immediately with copious amounts of water.*****)**

(1) $3Fe^{2+}(aq) + NO_3^-(aq) + 4H^+(aq) \longrightarrow 3Fe^{3+}(aq) + NO(aq) + 2H_2O(l)$

(2) $NO(aq) + Fe^{2+}(aq)\ (\text{excess}) \longrightarrow Fe(NO)^{2+}(aq)$ (brown)

Note 6 Iodides and bromides react with concentrated H_2SO_4 to liberate I_2 and Br_2.

$$SO_4^{2-}(aq) + 8I^-(aq) + 10H^+(aq) \longrightarrow H_2S(g) + 4I_2(aq) + 4H_2O(l)$$

$$SO_4^{2-}(aq) + 2I^-(aq) + 4H^+(aq) \longrightarrow SO_2(g) + I_2(aq) + 2H_2O(l)$$

$$SO_4^{2-}(aq) + 2Br^-(aq) + 4H^+(aq) \longrightarrow SO_2(g) + Br_2(aq) + 2H_2O(l)$$

Chromate ions, if present, will be reduced by Fe^{2+} to green Cr^{3+}.

$$2CrO_4^{2-}(aq) + 2H^+(aq) \longrightarrow Cr_2O_7^{2-}(aq) + H_2O(l)$$

$$Cr_2O_7^{2-}(aq) + 6Fe^{2+}(aq) + 14H^+(aq) \longrightarrow 2Cr^{3+}(aq) + 6Fe^{3+}(aq) + 7H_2O(l)$$

The colors of I_2, Br_2, and Cr^{3+} will interfere with detection of the brown color of $Fe(NO)^{2+}$. Consequently, I^-, Br^-, and CrO_4^{2-} must be removed as follows: Place four drops of the unknown anion solution in a test tube and add 0.2 *M* $Pb(C_2H_3O_2)_2$ until precipitation is complete. Centrifuge and decant, discarding the precipitate ($PbCrO_4$). Treat the decantate with 0.1 *M* $Hg(C_2H_3O_2)_2$ until precipitation is complete. Centrifuge the solution to separate the precipitate (HgI_2, $HgBr_2$) and treat the decantate as above to test for nitrate.

Carbonate Sulfites will interfere with this test for carbonates.

A When sulfites are absent. Place a small amount of the solid anion unknown in a small test tube and add a few drops of 6 *M* H_2SO_4. If a colorless, odorless gas evolves, hold a drop of $Ba(OH)_2$ solution over the mouth of the test tube, using either an eyedropper or a Nichrome wire loop (see Figure 30.6). CO_3^{2-} ions are confirmed if the drop turns milky.

(1) $2H^+(aq) + CO_3^{2-}(aq) \longrightarrow CO_2(g) + H_2O(l)$

(2) $CO_2(g) + Ba(OH)_2(aq) \longrightarrow \mathbf{BaCO_3}(s) + H_2O(l)$

B When sulfites are present. Place a small amount of the solid anion unknown in a test tube, and add an equal amount of solid Na_2O_2. **(CAUTION:** ***Sodium peroxide is a very strong oxidizing agent, and all contact with skin should be avoided. If you do get some on yourself, immediately wash it off with large volumes of water.*****)** Add three to four drops of water, mix thoroughly, then proceed as in **A** above to test for carbonate.

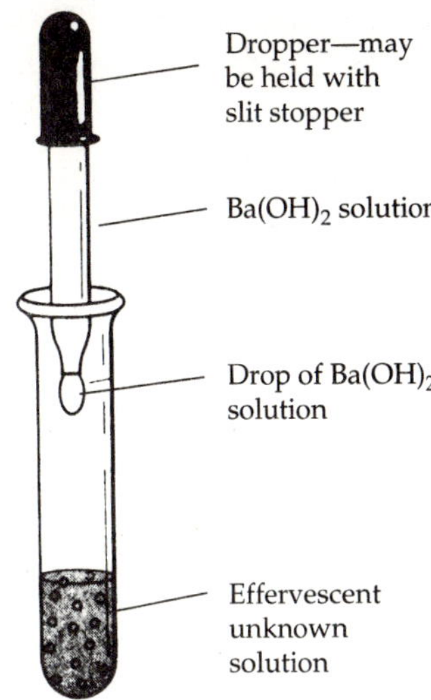

▲ **FIGURE 30.6** Test for carbonate ion.

Sulfide Place a small quantity of the solid unknown anion in a test tube and (UNDER THE HOOD) add 10 drops of 6 *M* HCl. Hold a piece of filter paper that has been moistened with 0.2 *M* $Pb(C_2H_3O_2)_2$ over the mouth of the test tube so that any gas that escapes comes into contact with the paper. A brownish or silvery black stain (PbS) on the paper confirms the presence of S^{2-}. If no blackening of the lead acetate occurs after 1 min, heat the tube gently. If still no reaction occurs, add a small amount of granulated zinc to the contents of the tube. If the lead acetate is not darkened, S^{2-} is absent.

$$S^{2-}(aq) + 2H^+(aq) \longrightarrow H_2S(g)$$

$$Pb^{2+}(aq) + H_2S(aq) \longrightarrow \mathbf{PbS}(s) + 2H^+(aq)$$

$$\mathbf{Zn}(s) + HgS(aq) + 2H^+(aq) \longrightarrow Zn^{2+}(aq) + \mathbf{Hg}(l) + H_2S(aq)$$

Chloride Sulfides, bromides, and iodides will interfere with this test (see Note 7). Place 10 drops of a solution of the anion unknown in a test tube, and add a drop of $AgNO_3$ solution. A white, curdy precipitate confirms Cl^- ions.

$$Ag^+(aq) + Cl^-(aq) \longrightarrow \mathbf{AgCl}(s)$$

Note 7 Since Ag_2S, AgBr, and AgI are also insoluble in acid solution, this test is not conclusive unless S^{2-}, Br^-, and I^- are definitely shown to be absent. Chromate ions, if present in high concentrations, may also interfere.

Interference from chromate can be eliminated by dilution with 3 *M* HNO_3. Sulfide ions can be removed by boiling the solution after adding two drops of 6 *M* H_2SO_4 until the escaping vapors give no test for H_2S with lead acetate paper. If you suspect the presence of chloride, bromide, iodide, or any combination of these, they may all be confirmed as follows:

Place 10 drops of a solution of the anion unknown in a test tube, and add five drops of $AgNO_3$ solution. Centrifuge and discard the supernate. To the precipitate add 10 drops of concentrated NH_3 solution and three drops of yellow ammonium sulfide solution. Stir the mixture with a glass rod and warm gently until the black Ag_2S coagulates. Centrifuge and discard the precipitate. Transfer the solution to a 50-mL beaker and boil to expel NH_3 and to decompose the ammonium sulfide. When the solution becomes cloudy, add five or six drops of 6 *M* HNO_3 and continue heating until H_2S is completely removed. Then carry out the following tests:

A Place one drop of the solution on a piece of filter paper; add one drop of iodide-free starch solution and one drop of 0.2 *M* KNO_2. A blue color confirms iodide.

$$\mathbf{AgI}(s) + 2NH_3(aq) \longrightarrow Ag(NH_3)_2^+(aq) + I^-(aq)$$

$$2Ag(NH_3)_2^+(aq) + (NH_4)_2S(aq) \longrightarrow \mathbf{Ag_2S}(s) + 2NH_4^+(aq) + 4NH_3(g)$$

$$I^-(aq) + HNO_2(aq) + 2H^+(aq) \longrightarrow 2NO(g) + I_2(aq) + 2H_2O(aq)$$

$$I_2 + \text{starch} \longrightarrow \text{starch} \cdot I_2 \text{ complex (blue)}$$

B If iodide is present, add three or four drops of 0.2 *M* KNO_2 to the solution in the beaker and boil until no more brown fumes are evolved. If iodide is absent, proceed directly with the test for bromide. Cool the beaker to room temperature by running cold water over its outer surface. To the room-temperature beaker, add four or five drops of 3 *M* HNO_3, then a small pea-size piece of solid Na_2O_2. A brown coloration due to the presence of Br_2 confirms Br^-.

$$4H^+(aq) + O_2^{2-}(aq) + 2Br^-(aq) \longrightarrow 2H_2O(l) + Br_2(aq)$$

C If bromide is present, boil the uncovered contents of the beaker for 30 s to expel the remainder of the bromine. Allow to stand for 30 s and decant the solution into a test tube. Centrifuge if necessary. Add five-drops of 0.1 *M* $AgNO_3$. A white precipitate of AgCl confirms chloride.

$$Ag^+(aq) + Cl^-(aq) \longrightarrow \mathbf{AgCl}(s)$$

Phosphate

A When iodides are absent. Place four or five drops of a solution of the anion unknown in a test tube, then add two drops of 6 *M* HNO_3 and three or four drops of $(NH_4)_2MoO_4$ (ammonium molybdate). Mix thoroughly and heat almost to boiling for 2 min. Formation, sometimes very slow, of a finely divided yellow precipitate confirms phosphate.

$$PO_4^{3-}(aq) + 12MoO_4^{2-}(aq) + 24H^+(aq) + 3NH_4^+(aq) \longrightarrow \mathbf{(NH_4)_3PO_4 \cdot 12MoO_3}(s) + 12H_2O(l)$$

B When iodides are present. The phosphate test gives a green solution when iodide is present. If iodide is known to be present, acidify the solution with 6 *M* HNO_3, add four drops of 0.1 *M* $AgNO_3$, centrifuge to remove AgI, and test the decantate for phosphate as in **A** above.

Waste Disposal Instructions All waste from this experiment should be placed in appropriate containers in the laboratory.

REVIEW QUESTIONS

Before beginning Part V of this experiment in the laboratory, you should be able to answer the following questions:

1. What are the names and formulas of the anions to be identified?
2. Describe the behavior of each solid containing the anions toward concentrated H_2SO_4.

3. If you had a mixture of NaCl and Na_2CO_3, would the action of concentrated H_2SO_4 allow you to decide whether both Cl^- and CO_3^{2-} were present?

4. Identify each of the following anions from the information given (consult Appendix C):
   (a) Its barium salt is insoluble in water, but its copper salt is soluble.
   (b) Its copper salt is insoluble in water, but its sodium salt is soluble.
   (c) Its mercury(II) salt is soluble in water, but its mercury(I) salt is insoluble.

5. A mixture of barium and silver salts is water-soluble. What anions may not be present in this mixture?

6. A white solid unknown is readily soluble in water, and upon treatment of this solution with HCl, a colorless, odorless gas is evolved. This gas reacts with $Ba(OH)_2$ to give a white precipitate. What anion is indicated?

7. How would treatment with concentrated H_2SO_4 allow you to distinguish between the following:
   (a) $BaSO_4$ and $Hg(NO_3)_2$?
   (b) $CaBr_2$ and Na_3PO_4?
   (c) ZnS and $BaSO_4$?
   (d) HgI_2 and K_2CrO_4?

8. Five different salts each imparted a yellow color to a flame and reacted with cold H_2SO_4 as follows. Identify each salt.
   (a) Effervescence was observed, and the evolved gas was colorless and odorless and did not fume in moist air.
   (b) Effervescence was observed, and the evolved gas was pale reddish-brown, had a sharp odor, and fumed strongly in moist air.
   (c) No effervescence was observed, and the solid changed color from yellow to orange.
   (d) Effervescence was observed, and the evolved gas was colorless, had a sharp odor, and fumed in moist air.
   (e) Effervescence was observed, and the evolved colorless gas had a very sharp odor but did not fume in moist air and did not discolor a piece of filter paper that had been moistened with a solution of lead acetate.

9. How would you test for NO_3^- in the presence of I^-?

10. How would you test for CO_3^{2-} in the presence of SO_3^{2-}?

PART VI: ANALYSIS OF A SIMPLE SALT

PROCEDURE

Cation Identification

Note the color of your solid and eliminate any cations on the basis of the color. Place a small pea-size portion of your unkown in a test tube and add just enough distilled water to dissolve it. Follow the procedures for the identification of group 1, 3, and 4 cations and identify the cation. Once the cation has been identified, proceed to Anion Identification.

Anion Identification

Test the solubility of your solid in distilled water and note its reactions with 18 *M* H_2SO_4, $AgNO_3$, and $BaCl_2$ solutions. Compare your results with the information in Table 30.1. Once the anion has been identified, perform the specific test for that anion to confirm its presence.

Name ______________________ Desk ______________________

Date ______________ Laboratory Instructor ______________________

Unknown no. ______________________

REPORT SHEET | EXPERIMENT

Abbreviated Qualitative-Analysis Scheme | 30

PART I: GROUP 1 CATIONS

Record the reagent used in each step, your observations, and the equations for each precipitation reaction.

Procedure	*Reagent*	*Observations*	*Equations*	*Mark (+) if observed in unknown*
G1-1	HCl	white ppt	$Ag^+ + Cl^- \longrightarrow AgCl(s)$ $Hg_2^{2+} + 2Cl^- \longrightarrow Hg_2Cl_2(s)$ $Pb^{2+} + 2Cl^- \longrightarrow PbCl_2(s)$	
G1-2	K_2CrO_4	yellow ppt.	$Pb^{2+} + CrO_4^{2-} \longrightarrow PbCrO_4(s)$	
G1-3	NH_3	ppt. turned dark gray	$Hg_2Cl_2 + 2NH_3 \longrightarrow HgNH_2Cl + Hg + NH_4Cl$ $AgCl + 2NH_3 \longrightarrow Ag(NH_3)_2^+ + Cl^-$ $Ag(NH_3)_2^+ + 2H^+ + Cl^- \longrightarrow AgCl(s) + 2NH_4^+$	

Cations in group 1 unknown Depends upon unknown

NOTES AND CALCULATIONS

Name ______________________ Desk ______________________

Date ______________ Laboratory Instructor ______________________

Unknown no. ______________

REPORT SHEET | EXPERIMENT

Abbreviated Qualitative-Analysis Scheme | 30

PART II: GROUP 2 CATIONS

Record the reagent used in each step, your observations, and the equations for each precipitation reaction.

Procedure	*Reagent*	*Observations*	*Equations*	*Mark (+) if observed in unknown*
G2-1	H_2S	black ppt. formed	$Pb^{2+} + S^{2-} \longrightarrow PbS(s)$ $Cu^{2+} + S^{2-} \longrightarrow CuS(s)$ $2Bi^{3+} + 3S^{2-} \longrightarrow Bi_2S_3(s)$ $Sn^{4+} + 2S^{2-} \longrightarrow SnS_2(s)$	
G2-2	$(NH_4)_2S$	no change	$SnS_2 + S^{2-} \longrightarrow SnS_3^{2-}$	
G2-3	HNO_3	black ppt. dissolved	$3PbS + 8H^+ + 2NO_3^- \longrightarrow 3Pb^{2+} + 3S + 2NO + 4H_2O$ $3CuS + 8H^+ + 2NO_3^- \longrightarrow 3Cu^{2+} + 3S + 2NO + 4H_2O$ $Bi_2S_3 + 8H^+ + 2NO_3^- \longrightarrow 2Bi^{3+} + 3S + 2NO + 4H_2O$	
G2-4	H_2SO_4 NH_3 $Sn(OH)_4^{2-}$	white ppt. forms, solution turned blue, black ppt. formed	$Pb^{2+} + SO_4^{2-} \longrightarrow PbSO_4$ $Cu^{2+} + 4NH_3 \longrightarrow Cu(NH_3)_4^{2+}$ $2Bi(OH)_3 + 3Sn(OH)_4^{2-} \longrightarrow 3Sn(OH)_6^{2-} + 2Bi$	
G2-5	$Sn(OH)_4^{2-}$ $HgCl_2$	black ppt. formed dark gray ppt.	$2Bi(OH)_3 + 3Sn(OH)_4^{2-} \longrightarrow 3Sn(OH)_6^{2-} + 2Bi$ $2SnCl_4^{2-} + 3HgCl_2 \longrightarrow 2SnCl_6^{2-} + Hg_2Cl_2 + Hg$	

Cations in group 2 unknown ___Depends upon unknown___

NOTES AND CALCULATIONS

Name ______________________ Desk ______________________

Date ______________ Laboratory Instructor ______________________

Unknown no. ______________

REPORT SHEET | EXPERIMENT

Abbreviated Qualitative-Analysis Scheme | 30

PART III: GROUP 3 CATIONS

Record the reagent used in each step, your observations, and the equations for each precipitation reaction.

Procedure	*Reagent*	*Observations*	*Equations*	*Mark (+) if observed in unknown*
G3-1	NH_3, $(NH_4)_2S$	soln. turned blue; brown ppt. formed and then turned black	$Fe^{3+} + 3OH^- \longrightarrow Fe(OH)_3(s)$ $Al^{3+} + 3OH^- \longrightarrow Al(OH)_3(s)$ $Ni^{2+} + S^{2-} \longrightarrow NiS(s)$ $Mn^{2+} + S^{2-} \longrightarrow MnS(s)$	
G3-2	HCl, HNO_3	ppt. dissolved and green soln. forms	$Fe(OH)_3(s) + 3H^+ \longrightarrow Fe^{3+} + 3H_2O$ $Al(OH)_3(s) + 3H^+ \longrightarrow Al^{3+} + 3H_2O$ $MnS(s) + 2H^+ \longrightarrow Mn^{2+} + H_2S(g)$ $3NiS(s) + 2NO_3^- + 8H^+ \longrightarrow 3Ni^{2+} + 3S(s) + 2NO(g) + 4H_2O$	
G3-3	NaOH	a brown ppt. formed	$Fe^{3+} + 3OH^- \longrightarrow Fe(OH)_3(s) \xrightarrow{H^+} Fe^{3+}$ $Ni^{2+} + 2OH^- \longrightarrow Ni(OH)_2(s) \xrightarrow{H^+} Ni^{2+}$ $Mn^{2+} + 2OH^- \longrightarrow Mn(OH)_2(s) \xrightarrow{H^+} Mn^{2+}$ $Al^{3+} + 4OH^- \longrightarrow Al(OH)_4^-$	
G3-4	KSCN	soln. turned blood red	$Fe^{3+} + 6SCN^- \longrightarrow Fe(SCN)_6^{3-}$	
G3-5	$NaBiO_3$	soln. turned purple	$2Mn^{2+} + 5NaBiO_3 + 14H^+ \longrightarrow 5Bi^{3+} + 2MnO_4^- + 5Na^+ + 7H_2O$	

Procedure	*Reagent*	*Observations*	*Equations*	*Mark (+) if observed in unknown*
G3-6	H_2DMG	a red ppt. formed	$Ni(NH_3)_6^{2+} + 2H_2DMG \longrightarrow$ $Ni(HDMG)_2(s) + 4NH_3 + 2NH_4^+$	
G3-7	NH_3, Aluminon reagent	a red ppt. formed	$Al(OH)_4^- + 4H^+ \longrightarrow Al^{3+} + 3H_2O$ $Al^{3+} + 3NH_3 + 3H_2O \longrightarrow$ $Al(OH)_3(s) + 3NH_4^+$	

Cations in group 3 unknown Depends upon unknown

Name ______________________ Desk ______________________

Date ______________ Laboratory Instructor ______________________

Unknown no. ______________________

REPORT SHEET | EXPERIMENT

Abbreviated Qualitative-Analysis Scheme | 30

PART IV: GROUP 4 CATIONS

Record the reagent used in each step, your observations, and the equations for each precipitation reaction.

Procedure	*Reagent*	*Observations*	*Equations*	*Mark (+) if observed in unknown*
G4-1	K_2CrO_4 H_2SO_4	yellow ppt. formed white ppt. formed no green flame	$Ba^{2+} + CrO_4^{2-} \longrightarrow BaCrO_4(s)$ $Ba^{2+} + SO_4^{2-} \longrightarrow BaSO_4(s)$	
G4-2	$K_2C_2O_4$	white ppt. formed no red flame	$Ca^{2+} + C_2O_4^{2-} \longrightarrow CaC_2O_4(s)$	
G4-3		fluffy yellow flame		
G4-4	NaOH	litmus turned blue	$NH_4^+ + OH^- \longrightarrow NH_3 + H_2O$	

Cations in group 4 unknown Depends upon unknown

NOTES AND CALCULATIONS

Name ______________________ Desk ______________

Date ______________ Laboratory Instructor ______________________

Unknown no. ______________

REPORT SHEET | EXPERIMENT

Abbreviated Qualitative-Analysis Scheme | 30

PARTS I-IV: GENERAL-CATION UNKNOWN

Record the reagent used in each step, your observations, and the equations for each precipitation reaction.

Procedure	*Reagent*	*Observations*	*Equations*

Cations in general unknown Unknown Dependent

NOTES AND CALCULATIONS

Name ______________________ Desk ______________________

Date ______________ Laboratory Instructor ______________________

Unknown no. ______________

REPORT SHEET | EXPERIMENT

Abbreviated Qualitative-Analysis Scheme | 30

PART V: ANIONS

(Note to Instructor: We have the students perform the preliminary tests ($AgNO_3$, $BaCl_2$ and H_2SO_4) on the ten individual anion knowns and have the results reported on a separate sheet of paper.)

1. EXAMINATION OF THE SOLID

Color? Unknown Dependent Homogeneous? Unknown Dependent

2. SOLUBILITY

Water	6 *M* HNO_3	6 *M* HCl	6 *M* NH_3
Soluble	Soluble (gas forms)	Soluble (gas forms)	Soluble

3. REACTIONS WITH $AgNO_3$ AND $BaCl_2$ SOLUTIONS

$AgNO_3$ *Observations*

$AgNO_3$ + 6 *M* HNO_3 white ppt. that partially dissolves on acidification

Anions indicated Cl^-

Anions eliminated Br^-, I^-, PO_4^{3-}, CrO_4^{2-}, S^{2-}, SO_3^{2-}

$BaCl_2$ *Observations*

$BaCl_2$ + 6 *M* HNO_3 white ppt. that dissolves on acidification with the evolution of a gas

Anions indicated CO_3^{2-}, SO_3^{2-}

Anions eliminated SO_4^{2-}, CrO_4^{2-}

4. REACTION OF SOLID WITH CONCENTRATED H_2SO_4

Observations

Anions indicated pungent colorless gas; Cl^-, CO_3^{2-}

Anions eliminated Br^-, I^-, S^{2-}, NO_3^-, CrO_4^{2-}

5. SPECIFIC TEST RESULTS

Test made	*Observation and equation*	*Anion confirmed*
Cl^-	$Ag^+ + Cl^- \longrightarrow AgCl(s)$ white	Cl^-
SO_3^{2-}	No white ppt.	absent
CO_3^{2-}	White ppt. formed $2H^+ + CO_3^{2-} \longrightarrow CO_2(g) + H_2O(l)$ $CO_2(g) + Ba(OH)_2 \longrightarrow BaCO_3(s) + H_2O(l)$	CO_3^{2-}

6. ANIONS CONFIRMED IN UNKNOWN SOLID Cl^-, CO_3^{2-}

NOTES AND CALCULATIONS

Name ________________________________ Desk ________________

Date ________________ Laboratory Instructor ________________________

REPORT SHEET | EXPERIMENT

Abbreviated Qualitative-Analysis Scheme

30

PART VI: SIMPLE SALTS

Cation in Unknown Solid (Results are unknown dependent)

1. Color of the solid: ________________

   Cations indicated: ________________

   Cations eliminated: ________________

2. For each step, record the procedure number, the reagent(s) used, your observations, and the corresponding equation(s) for each reaction.

Procedure	*Reagent*	*Observation(s)*	*Equation(s)*

Anion in Unknown Solid

1. Solubility of solid (see Appendix C):

Water ________ 6 *M* HNO_3 ________ 6 *M* HCl ________ 6 *M* NH_3 ________

2. Reactions of solid with concentrated (18 *M*) H_2SO_4 (observations):
With cold acid: ____________________
With hot acid (only if necessary): ____________________
Anions indicated: ____________________
Anions eliminated: ____________________

3. Reactions of unknown solution with $AgNO_3$ and $BaCl_2$ solutions (observations):
With 0.1 *M* $AgNO_3$: (obs) ____________________
With 0.1 *M* $AgNO_3$ + HNO_3: (obs) ____________________
Anions indicated: ____________________
Anions eliminated: ____________________
With 0.2 *M* $BaCl_2$: ____________________
With 0.2 *M* $BaCl_2$ + HNO_3: ____________________
Anions indicated: ____________________
Anions eliminated: ____________________

4. Specific test results
Test made for (anion): ____________________
Observation(s) and equation(s): ____________________
____
____
____

The anion in unknown solid: ____________________
The chemical formula of unknown solid is: ____________________

QUESTIONS

1. Write balanced chemical equations for the reactions occurring in each of the following mixtures.
   (a) HCl is added to solid $Mg_3(PO_4)_2$.
   Note: $Mg_3(PO_4)_2$ is insoluble
   $6HCl(aq) + Mg_3(PO_4)(s) \longrightarrow 3MgCl_2(aq) + 2H_3PO_4(aq)$
   $6H^+(aq) + Mg_3(PO_4)_2(s) \longrightarrow H_3PO_4(aq) + 3Mg^{2+}$

   (b) HBr is added to solid FeS.
   $2HBr(aq) + FeS(s) \longrightarrow H_2S(s) + FeBr_2$
   $2H^+(aq) + FeS(s) \longrightarrow H_2S(g) + Fe^{2+}$

   (c) H_2O_2 is added to a solution containing SO_3^{2-}.
   $H_2O_2(aq) + SO_3^{2-}(aq) \longrightarrow SO_4^{2-} + H_2O(l)$

   (d) $(NH_4)_2S$ solution is added to solid AgBr.
   $(NH_4)_2S(aq) + AgBr(s) \longrightarrow$ No apparent reaction

   (e) HCl is added to a solution of Na_2CrO_4.
   $2H^+(aq) + 2CrO_4^{2-}(aq) \longrightarrow H_2O(l) + Cr_2O_7^{2-}(aq)$

2. What conclusions can be drawn from the following observations?
   (a) An acidic solution of unknown anions forms no precipitate when $BaCl_2$ solution is added.
   SO_4^{2-} is absent

   (b) A solution of unknown anions forms a pale yellow precipitate when $AgNO_3$ solution is added.
   If the solution is acidic, S^{2-} is absent, I^- is present, and Cl^-, Br^-, and NO_3^- may be present.
   If solution is neutral, the last statement applies, but also CrO_4^{2-} is absent and SO_3^-, CO_3^{2-}, and PO_4^{3-} may be present.

   (c) Addition of 6 *M* NH_3 to a pale blue solution of an unknown gave a deep blue solution and a white precipitate.
   Cu^{2+} and possibly Al^{3+} and Mg^{2+} are present

3. Why is it more difficult to identify all of the components of a mixture than to identify the cation and anion in a simple salt?
   One reaction may mask another. For example, black Ag_2S may mask or hide AgCl, white, AgBr, cream, etc.

4. The observation that a solid is colorless allows you to suggest that a broad category of elements is probably not present in this solid. What is the general name for this category of elements?
   Transition series elements

5. A white compound dissolves in water to give an acidic solution. Addition of either NaOH or NH_3 to this colorless solution produces a flocculent white precipitate that is soluble in acid. If an excess of NaOH is added to a water solution of this compound, a colorless precipitate forms that later dissolves as more NaOH is added. Addition of $AgNO_3$ to a water solution of the compound gives no reaction. The solid shows no reaction with concentrated H_2SO_4, and addition of a barium chloride solution to a solution of the compound gives a white precipitate that is insoluble in both acid and base. Identify the compound.
   $Al_2(SO_4)_3$

6. A white salt dissolves in water to give a neutral solution. No precipitate is formed when the aqueous solution is treated with the buffer NH_4Cl, NH_3. No precipitate forms when $(NH_4)_2S$ is added to the buffer solution. However, a yellow precipitate forms when K_2CrO_4 is added to a solution of the salt that is acidified with $HC_2H_3O_2$, and the precipitate dissolves in concentrated HCl. When $AgNO_3$ is added to a solution of the salt that is acidified with HNO_3, a white precipitate forms. What is the salt?
   $BaCl_2$

NOTES AND CALCULATIONS

Experiment

Colorimetric Determination of Iron

OBJECTIVE

To become acquainted with the principles of colorimetric analysis.

APPARATUS AND CHEMICALS

Apparatus

balance	cuvettes
125-mL Erlenmeyer flask	1-, 2- and 5-mL pipets
spectrophotometer	50-mL volumetric flasks (6)

Chemicals

standard iron solution, $Fe(NO_3)_3$ + HNO_3 (1 mL = 0.050 mg Fe)	1 *M* $NH_4C_2H_3O_2$
0.30% *o*-phenanthroline	10% hydroxylamine hydrochloride
unknown iron sample	6 *M* H_2SO_4

PREPARE YOUR CALIBRATION CURVE IN TEAMS OF THREE, BUT ANALYZE YOUR UNKNOWN INDIVIDUALLY.

DISCUSSION

The basis for what the chemist calls *colorimetric analysis* is the variation in the intensity of the color of a solution with changes in concentration. The color may be due to an inherent property of the constituent itself—for example, MnO_4^- is purple—or it may be due to the formation of a colored compound as the result of the addition of a suitable reagent. By comparing the intensity of the color of a solution of unknown concentration with the intensities of solutions of known concentrations, the concentration of an unknown solution may be determined.

You will analyze for iron in this experiment by allowing iron(II) to react with an organic compound (*o*-phenanthroline) to form an orange-red complex ion. Note in its structure (shown below) that *o*-phenanthroline has two pairs of unshared electrons that can be used to form coordinate covalent bonds.

The equation for the formation of the complex ion is

$$3C_{12}H_8N_2 + Fe^{2+} \longrightarrow \underset{\text{Orange-red}}{[(C_{12}H_8N_2)_3Fe]^{2+}}$$

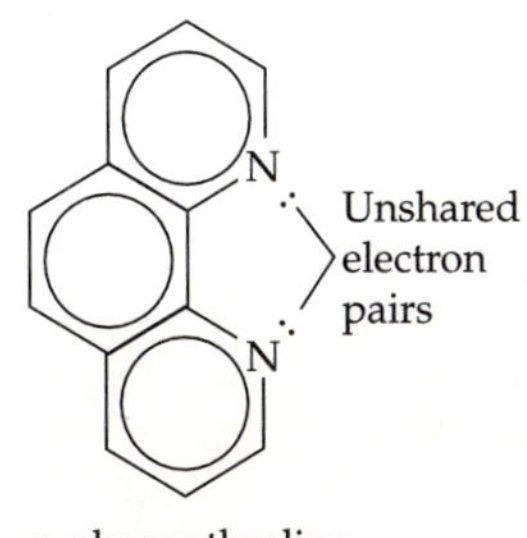

o-phenanthroline

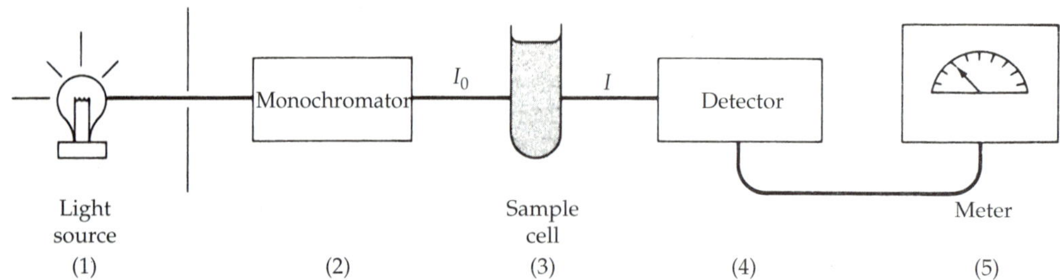

▲ **FIGURE 31.1** Schematic representation of a spectrophotometer.

Before the colored iron(II) complex is formed, however, all the Fe^{3+} present must be reduced to Fe^{2+}. This reduction is accomplished by the use of an excess of hydroxylamine hydrochloride:

$$4Fe^{3+} + \underset{\text{Hydroxylamine}}{2NH_2OH} \longrightarrow 4Fe^{2+} + N_2O + 6H^+ + H_2O$$

Although the eye can discern differences in color intensity with reasonable accuracy, an instrument known as a *spectrophotometer,* which eliminates the "human" error, is commonly used for this purpose. Basically, it is an instrument that measures the fraction I/I_0 of an incident beam of light of a particular wavelength and of intensity I_0 that is transmitted by a sample. (Here, I is the intensity of the light transmitted by the sample.) A schematic representation of a spectrophotometer is shown in Figure 31.1. The instrument has these five fundamental components:

1. A light source that produces light with a wavelength range from about 375 to 650 nm
2. A monochromator, which *selects* a particular wavelength of light and sends it to the sample cell with an intensity of I_0
3. The sample cell, which contains the solution being analyzed
4. A detector that measures the intensity, I, of the light transmitted from the sample cell; if the intensity of the incident light is I_0 and the solution absorbs light, the intensity of the transmitted light, I, is less than I_0
5. A meter that indicates the intensity of the transmitted light

For a given substance, the amount of light absorbed depends on the

1. concentration
2. cell or path length
3. wavelength of light
4. solvent

Plots of the amount of light absorbed versus wavelength are called *absorption spectra.* There are two common ways of expressing the amount of light absorbed. One is in terms of *percent transmittance,* %*T*, which is defined as

$$\%T = \frac{I}{I_0} \times 100 \qquad [1]$$

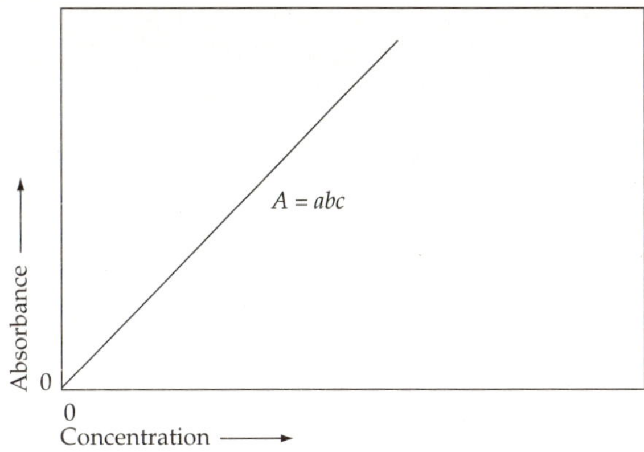

▲ **FIGURE 31.2** Relationship between absorbance and concentration according to the Beer-Lambert law.

As the term implies, percent transmittance corresponds to the percentage of light transmitted. When the sample in the cell is a solution, I is the intensity of light transmitted by the solution, and I_0 is intensity of light transmitted when the cell only contains solvent. Another method of expressing the amount of light absorbed is in terms of *absorbance, A,* which is defined by

$$A = \log \frac{I_0}{I} \qquad [2]$$

The term *optical density,* OD, is synonymous with absorbance. If there is no absorption of light by a sample at a given wavelength, the percent transmittance is 100, and the absorbance is 0. On the other hand, if the sample absorbs all of the light, $\%T = 0$ and $A = \infty$.

Absorbance is related to concentration by the Beer-Lambert law:

$$A = abc$$

where A is absorbance, b is solution path length, c is concentration in moles per liter, and a is molar absorptivity or molar extinction coefficient. There is a linear relationship between absorbance and concentration when the Beer-Lambert law is obeyed, as illustrated in Figure 31.2. However, since deviations from this law occasionally occur, it is wise to construct a calibration curve of absorbance versus concentration.

PROCEDURE

A. Preparation of the Calibration Curve

Accurately pipet 1.00 mL of standard iron solution (1.00 mL = 0.050 mg Fe) into a 50-mL volumetric flask. Add 1 mL of 1 *M* ammonium acetate, 1 mL of 10% hydroxylamine hydrochloride, and 10 mL of 0.30% *o*-phenanthroline solution. Dilute to exactly 50.0 mL with distilled water. Mix well to develop the characteristic orange-red color of the iron(II)-phenanthroline complex. Allow the color to develop for 45 min. Fill halfway a clean, dry cuvette (colorimeter tube) with the colored solution and determine the absorbance at 510 nm using a Spectronic 20 or other colorimeter. Operating instructions for the Spectronic 20 are given below.

Repeat using 2.0-mL, 3.0-mL, 4.0-mL, and 5.0-mL portions of the standard solution. Plot your results with milligrams of iron along the abscissa

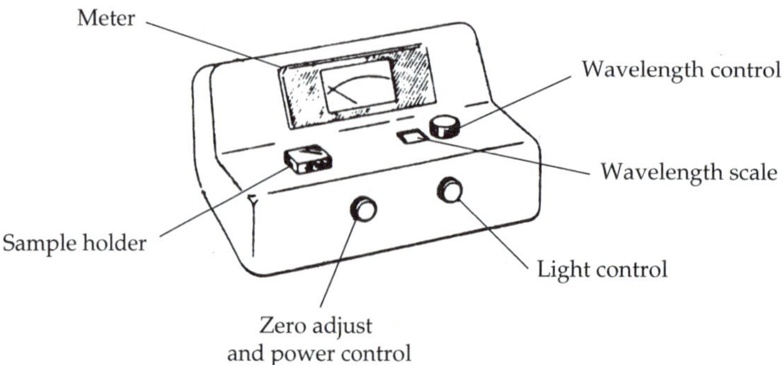

▲ FIGURE 31.3 Spectrophotometer controls.

(horizontal axis) and absorbance along the ordinate (vertical axis). This curve should be turned in with your report sheet. (HINT: Time may be saved if these solutions are all made at the same time.)

Operating Instructions for Spectronic 20

Instructor: Explain the operation of the spectrophotometer that the students will use.

1. Turn the wavelength-control knob (Figure 31.3) to the desired wavelength.
2. Turn on the instrument by rotating the power control clockwise and allow the instrument to warm up about 5 min. With no sample in the holder but with the cover closed, turn the zero adjust to bring the meter needle to zero on the "percent transmittance" scale.
3. Fill the cuvette about halfway with distilled water (or solvent blank) and insert it in the sample holder, aligning the line on the cuvette with that of the sample holder; close the cover and rotate the light-control knob until the meter reads 100% transmittance.
4. Remove the blank from the sample holder and replace it with the cuvette containing the sample whose absorbance is to be measured. Align the lines on the cuvette with the holder and close the cover. Read percent transmittance or optical density from the meter.

B. Determination of Iron

Accurately weigh out about 0.1 g to four significiant figures (0.05 g for 12% to 15% iron samples) of the iron unknown into a 50-mL volumetric flask, add five drops of 6 *M* sulfuric acid, and dilute to exactly 50 mL. Mix thoroughly and then transfer this solution to a clean 125-mL Erlenmeyer flask. Pipet exactly 1 mL of this solution into a thoroughly rinsed 50-mL volumetric flask and repeat the procedure described above in Part A (do not add standard iron solution). Repeat this procedure on two additional 1-mL aliquots of unknown solution.

Using the observed absorbance and the calibration curve, calculate the milligrams of iron in 1 mL of solution and the percentage of iron in the sample. Calculate the mean and standard deviation of your results.

Waste Disposal Instructions Sulfuric acid and the organic solutions are hazardous. Do not pour the solutions down the drain of the sink. Use the appropriate containers in the laboratory for disposal.

REVIEW QUESTIONS

Before beginning this experiment in the laboratory, you should be able to answer the following questions:

1. For a given substance, the amount of light absorbed depends on what four factors?
2. How are percent transmittance and absorbance related algebraically?
3. What are the five fundamental components of a spectrophotometer?
4. State the Beer-Lambert law, and define all terms in it.
5. What is the purpose of preparing a calibration curve?
6. Why is hydroxylamine hydrochloride used in this experiment?
7. If the percent transmittance for a sample is 80 at 350 nm, what is the value of A?
8. Suppose your experimental absorbance is greater than 1. How would you modify your procedure?
9. If 4.0 mL of a standard iron solution (1 mL = 0.050 mg Fe) is diluted to 50 mL, what is the final iron concentration in mg Fe/mL?
10. If aqueous $Co(NO_3)_2$ has an extinction coefficient of 5.1 L/mol-cm at 505 nm, show that a 0.0400 M $Co(NO_3)_2$ solution will give an absorbance of 0.20.

NOTES AND CALCULATIONS

Name ______________________ Desk ______________________

Date ______________________ Laboratory Instructor ______________________

Unknown no. ______________________

REPORT SHEET | EXPERIMENT

Colorimetric Determination of Iron | 31

A. Calibration Curve

(show calculations)

Note: ppm $= \frac{\text{mg}}{\text{L}} = \frac{10^{-3}\text{g}}{10^{3}\text{mL}} \times 10^{6}$

	Concentration of Fe (mg Fe/mL)	*Absorbance*
1.	0.0010	0.17
2.	0.0020	0.33
3.	0.0030	0.52
4.	0.0040	0.66
5.	0.0050	0.83

From $A = abc$

$$a = \frac{A}{bc}$$

$$= \frac{0.17}{(1.0\text{ cm})(1.0\text{ ppm})}$$

$$= 0.17\text{ cm}^{-1}\text{ppm}^{-1}$$

Do the above data obey the Beer-Lambert law? Yes

Why? There is a linear relationship between absorbance and the concentration of Fe.

B. Unknown Determination

1. Sample mass 0.0943 g Volume of solution 50.0 mL

	Absorbance or %T	*Concentration of Fe (mg Fe/mL)*
Trial 1	1.22; % T = 6.0	0.0072
Trial 2	1.23; % T = 5.9	0.0072
Trial 3	1.21; % T = 6.1	0.0071
Mean	1.22; % T = 6.0	0.0072

$$c = \frac{A}{ab}$$

$$= \frac{1.22}{(0.17\text{ cm}^{-1}\text{ppm}^{-1})(1.0\text{ cm})}$$

$$= 7.2\text{ ppm}$$

2. Concentration of Fe in mg/mL 0.0072 ± 0.00006

$$\frac{0.0072 + 0.0072 + 0.0071}{3} = 0.0072$$

3. Standard deviation (show calculations)

$$SD = \sqrt{\frac{2(0.00)^2 + (0.0001)^2}{2}} = 7 \times 10^{-5}$$

4. Percent Fe in original sample (show calculations) 19.0% ± 0.2

   0.0943 g diluted to 50 mL ⟶ 0.001886 g/mL

   1 mL of this diluted to 50 mL ⟶ 0.0000377 g/mL or 0.0377 mg/mL

$$\frac{0.00718 \text{ mg/mL}}{0.0377 \text{ mg/mL}} \times 10^2 = 19.0\% \text{ Fe} \qquad \frac{0.00724 \text{ mg/mL}}{0.0377 \text{ mg/mL}} \times 10^2 = 19.2\% \text{ Fe}$$

$$\frac{0.00712}{0.00377} = 10^2 = 18.9\% \text{ Fe} \qquad s = \sqrt{\frac{(0.0)^2 + (0.2)^2 + (0.1)^2}{2}} = 0.2$$

$$\frac{19.0 + 19.2 + 18.9}{3} = 19.0$$

QUESTIONS

1. Why is the line on the cuvette always aligned with that of the sample holder?

   To minimize any irregularities in the glass; they will have a constant effect.

2. Why was hydroxylamine hydrochloride added to your sample?

   To reduce Fe^{3+} to Fe^{2+}.

3. Iron(II) reacts with water by a hydrolysis reaction. In order to prevent this hydrolysis, acid has been added to the standard iron solution. How would your final results change if no acid had been added to the standard iron solution?

   The concentration of $[Fe(C_{12}H_8N_2)_3]^{2+}$ would be less than it should be and, therefore, the absorbance would have been less. Hence, % Fe would be low.

4. Suppose a solution of $Co(NO_3)_2$ has an extinction coefficient of 5.1 L/mol-cm at 505 nm. On the graph paper provided, plot a graph of *A* versus C (mol/L) for solutions of 0.020, 0.040, 0.060, 0.080, and 0.100 *M* $Co(NO_3)_2$ in a 1-cm cell. On the same graph, plot the percent transmittance, %*T*, of each solution versus concentration.

   $\log \%T = 2.00 - A$

   $A = abc$

   $= 5.1 \text{ L/mol-cm} \times 1 \text{ cm} \times C$

   $= 5.1 \text{ L/mol} \times C$

Co(mol/L)	A	%T
0.020	0.102	79.07
0.040	0.204	62.52
0.060	0.306	49.43
0.080	0.408	39.08
0.100	0.510	30.90

5. An 8.64 ppm (1 ppm = 1 mg/L) solution of $FeSCN^{2+}$ has a transmittance of 0.295 when measured in a 1.00-cm cell at 580 nm. Calculate the extinction coefficient at this wavelength.

   $A = 2.00 - \log \%T = 2.00 - \log 29.5$

   $= 2.00 - 1.47 = 0.53$

$$\text{and } a = \frac{A}{bc} = \frac{0.53}{(1.00 \text{ cm})\left(8.64 \times 10^{-3} \frac{\text{g}}{\text{L}} / 113.9 \text{ g/mol}\right)}$$

$$= 7.0 \times 10^3 \text{ L mol}^{-1} \text{ cm}^{-1}.$$

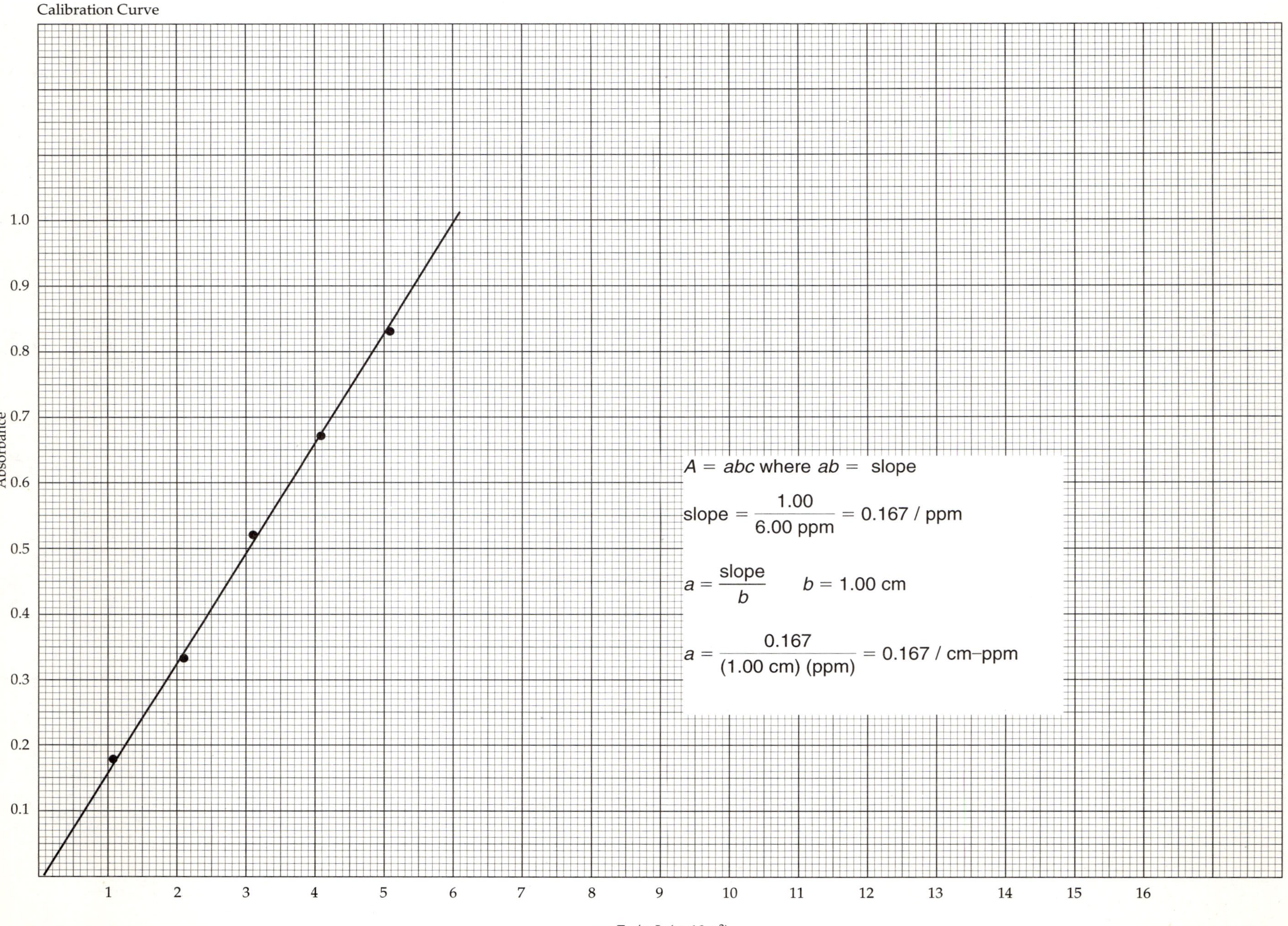
Calibration Curve
Absorbance
1.0
0.9
0.8
0.7
0.6
0.5
0.4
0.3
0.2
0.1
1
2
3
4
5
6
7
8
9
10
11
12
13
14
15
16
mg Fe/mL (× 10 + 3)
$A = abc$ where $ab =$ slope
slope $= \frac{1.00}{6.00 \text{ ppm}} = 0.167$ / ppm
$a = \frac{\text{slope}}{b}$ $b = 1.00$ cm
$a = \frac{0.167}{(1.00 \text{ cm})(\text{ppm})} = 0.167$ / cm–ppm

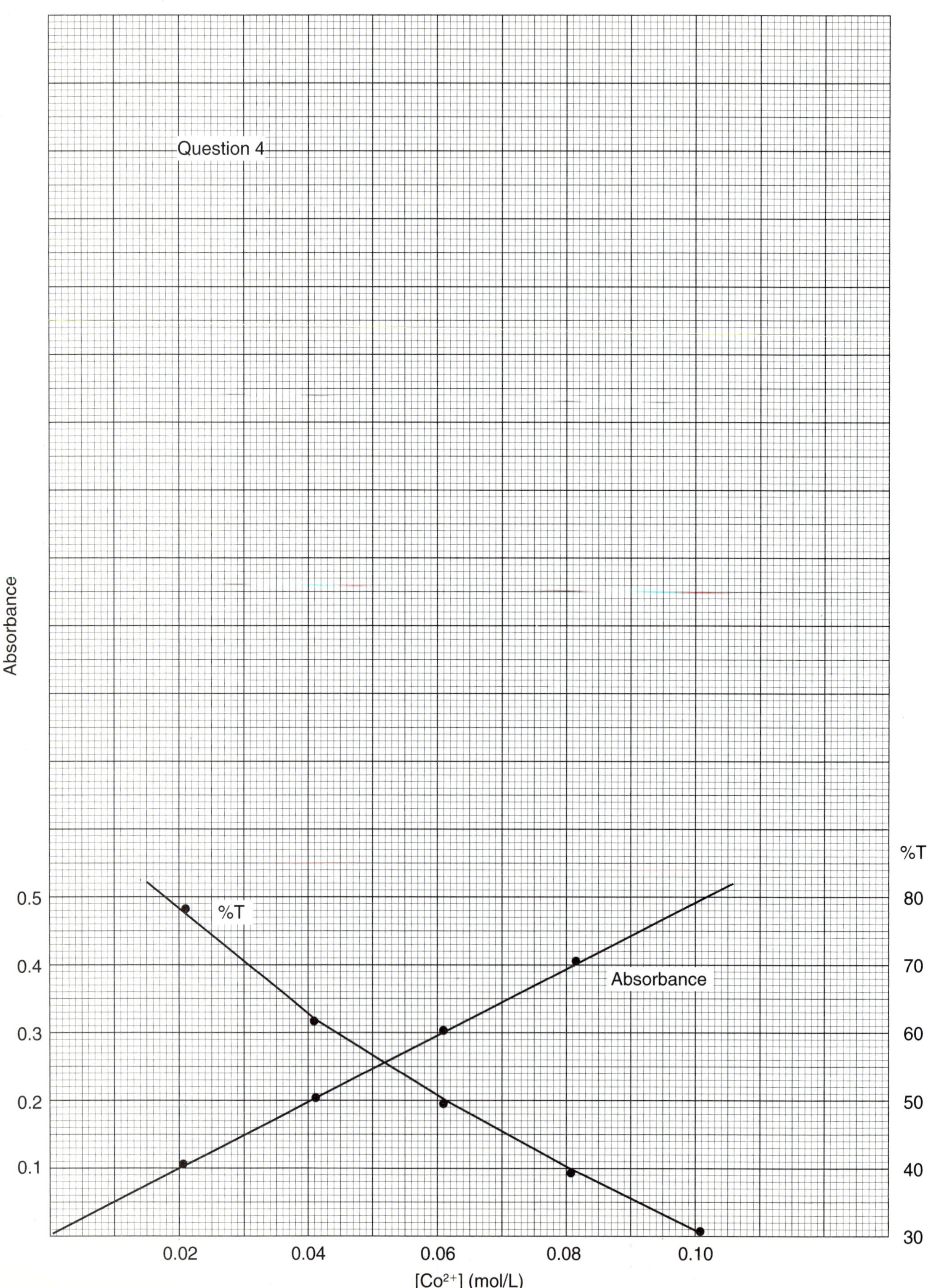

Question 4
Absorbance
%T
0.5
0.4
0.3
0.2
0.1
80
70
60
50
40
30
%T
Absorbance
0.02
0.04
0.06
0.08
0.10
[Co2+] (mol/L)

Experiment 32

Determination of Orthophosphate in Water

OBJECTIVE

To gain some familiarity with the techniques of spectrophotometric analysis by analyzing a water solution for its phosphate content.

APPARATUS AND CHEMICALS

Apparatus

balance
1-, 5-, and 10-mL pipets
cuvettes
100-mL volumetric flasks (6)
spectrophotometer
5-mL graduated pipet
1-L volumetric flask

Chemicals

ammonium vanadomolybdate solution
KH_2PO_4 (oven-dried)
$CHCl_3$ (chloroform)
conc. H_2SO_4
water sample

PREPARE YOUR CALIBRATION CURVE IN TEAMS OF THREE, BUT EVALUATE YOUR UNKNOWN INDIVIDUALLY.

DISCUSSION

Tripolyphosphates have been found to be extremely effective in enhancing the cleansing ability of detergents; they also are very inexpensive. Their aid in cleaning is probably due, in part, to the stable complexes that they form with Ca^{2+} and Mg^{2+}, thus softening the water. Their extensive use, however, has very serious side effects on the environment.

The accelerated eutrophication, or overfertilization, of our lakes has aroused a great deal of ecological concern.* Nutrient enrichment enhances the growth of algae and other microscopic organisms. This produces the green scum of an algal bloom on the water surface, masses of water weeds, and a depletion of dissolved oxygen; it also kills fish and other aquatic organisms and produces malodorous water systems.

When the photosynthetically active algae population near a lake's surface rapidly expands, most of the oxygen produced escapes to the atmosphere. After the algae die, they sink to the lake bottom, where they are biochemically oxidized. This depletes the dissolved oxygen needed to support aquatic life. When oxygen is removed, anaerobic decomposition of the algae continues, producing foul odors.

Although many factors affect algal growth, the only one that is readily subject to preventive control is the supply of nutrients. The many nutrients important to the growth of algae include phosphorus, carbon, nitrogen, sulfur, potassium, calcium, and magnesium. Many environmentalists have

*Eutrophication is the process whereby a body of water becomes rich in plant nutrient minerals and organisms but often deficient in oxygen.

accepted the idea that phosphorus is generally the key nutrient that limits the plant growth that a body of water can support.

There are at least four major sources of phosphorus associated with human activity: human and food wastes, fertilizers, industrial wastes, and detergents. Although detergent products contribute only about one-third of the phosphates entering our water systems, curtailing this particular source is a logical place to begin to combat eutrophication.

The phosphate found in natural waters is present as othophosphate, PO_4^{3-}, as well as the polyphosphates $P_2O_7^{-4}$ and $P_3O_{10}^{5-}$. The species present, PO_4^{3-}-, HPO_4^{2-}, $H_2PO_4^-$, or H_3PO_4, depend on the pH. Trace amounts are also present as organophosphorus compounds. Detergents usually contain triphosphate, $P_3O_{10}^{5-}$, which slowly hydrolyzes to produce orthophosphate, PO_4^{3-}, according to the following reaction:

$$P_3O_{10}^{5-} + 2H_2O \longrightarrow 3PO_4^{3-} + 4H^+$$

Instructor: Give the students the source of the sample.

In this experiment, you will determine the amount of orthophosphate present in a sample from a natural body of water, whose source will be given to you by the laboratory instructor.

Analytical Method

In dilute phosphate solutions ammonium metavanadate, (NH_4VO_3), molybdate (MoO_4^{2-}), and phosphate (PO_4^{3-}) condense to form an intensely yellow-colored compound called a *heteropoly acid,* whose formula is thought to be $(NH_4)_3PO_4 \cdot NH_4VO_3 \cdot 16MoO_3$. The intensity of the yellow color is directly proportional to the concentration of phosphate. The relative amount of color developed is measured with a spectrophotometer. The amount of light absorbed by the sample is directly proportional to the concentration of the colored substance. This is stated by the Beer-Lambert law,

$$A = abc$$

where A is absorbance, b is solution path length, c is concentration, and a is absorptivity or extinction coefficient.

The colored solutions that you study in this experiment have been found to obey the Beer-Lambert law in the region of wavelengths ranging from 350 nm to 410 nm. It is convenient to run this experiment at 400 nm. The amount of phosphate in the unknown sample of interest is determined by comparison with a calibration curve constructed by using a distilled-water reference solution and solutions of known phosphate concentration. The minimum detectable concentration of phosphate is about 0.01 mg/L (10 ppb). The usual experimental precision will lie within about ±1% of the result obtained by an experienced analyst.

Comparative Phosphate Levels in Water Systems

Limiting nutrients and their critical concentrations are likely to differ in different bodies of water. Analysis of the waters of 17 Wisconsin lakes has led to the suggestion that an annual average concentration of 0.015 mg/L of inorganic phosphorus (0.05 mg phosphate/L) is the critical level above which algal blooms can be expected if other nutrients, such as nitrogen, are in sufficient supply. During the 1968–69 period, Lake Tahoe in Nevada had an average phosphate level of 0.006 mg/L, while its tributaries averaged 0.018 mg/L. Lake Tahoe is one of the two purest lakes in the world.

The other is Lake Baikal, in Russia. This figure thus represents the lowest natural-water value of phosphate one is likely to find. In July of 1969 the phosphate level of Lahontan Reservoir (about 55 km east of Reno, Nevada) was 0.52 mg/L. By comparison, Lake Erie's phosphate level increased from 0.014 mg/L in 1942 to 0.40 mg/L in 1967-68. The U.S. Public Health Service has set 0.1 mg/L of phosphorus (0.3 mg phosphate/L) as the maximum value allowable for drinking water. Raw sewage contains an average of about 30 mg/L of orthophosphate, of which about 25% is removed by most secondary sewage-treatment plants.

PROCEDURE

A. Preparation of Calibration Curve

Dissolve about 136 mg of oven-dried KH_2PO_4 (weigh accurately) in about 500 mL of water. Quantitatively transfer this solution to a 1-L volumetric flask, add 0.5 mL of 98% H_2SO_4, and dilute to the mark with distilled water. This yields a stock solution that is about 1×10^{-3} *M* in various phosphate species. From this stock solution, prepare a series of six solutions with phosphate concentrations of 2×10^{-5}, 5×10^{-5}, 1×10^{-4}, 2×10^{-4}, 5×10^{-4}, and 7.5×10^{-4} *M* by appropriate dilution of the stock solution. You must know the precise concentrations of these solutions.

Each point on the calibration curve is obtained by mixing 10 mL of the phosphate solution with 5 mL of the ammonium vanadomolybdate solution (see Note 1, below) and measuring the absorbance on the Spectronic 20 (Figure 32.1) or equivalent spectrophotometer at 400 nm. Your curve is constructed by plotting absorbance as the ordinate (vertical axis) versus concentration of phosphate as the abscissa (horizonal axis) as in Figure 32.2. A straight line passing through the origin should be obtained. Your calibration curve should be handed in with your report sheet.

Operating Instructions for Spectronic 20

Instructor: Explain the operation of the spectrophotometer that the students will use.

1. Turn the wavelength-control knob (Figure 32.1) to the desired wavelength.
2. Turn on the instrument by rotating the power control clockwise and allow the instrument to warm up about 5 min. With no sample in the holder but with the cover closed, turn the zero adjust to bring the meter needle to zero on the "percent transmittance" scale.

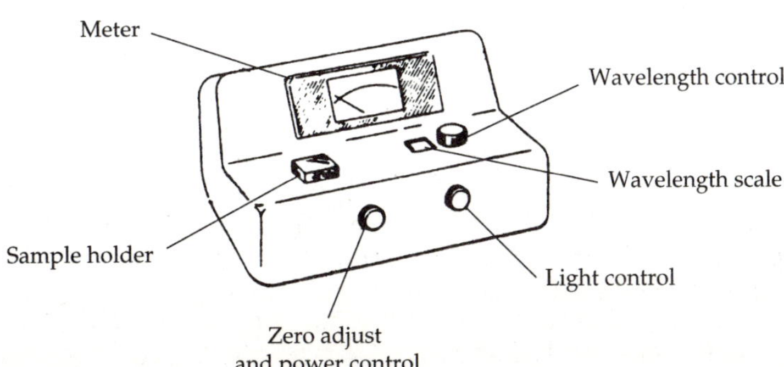

▲ **FIGURE 32.1** Spectrophotometer controls for a Spectronic 20.

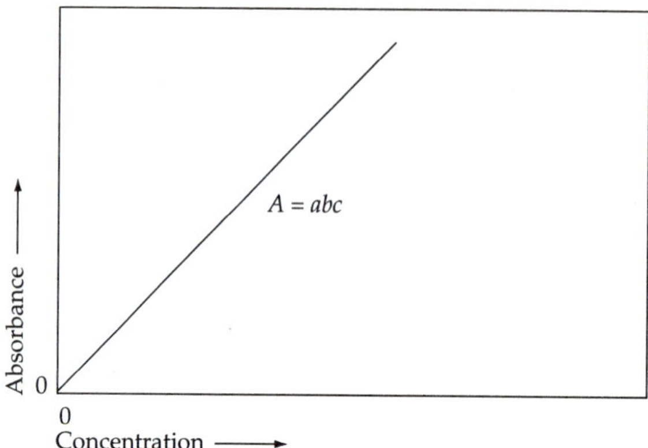

▲ FIGURE 32.2 Relationship between absorbance and concentration according to the Beer-Lambert law.

3. Fill the cuvette about halfway with distilled water (or solvent blank) and insert it in the sample holder, aligning the line on the cuvette with that of the sample holder; close the cover and rotate the light-control knob until the meter reads 100% transmittance.
4. Remove the blank from the sample holder and replace it with the cuvette containing the sample whose absorbance is to be measured. Align the lines on the cuvette with the holder and close the cover. Read the percent transmittance or optical density from the meter.

B. Analysis of Water Sample

Instructor: Note sample preparation and preservation.

The sample to be analyzed should be taken in a clean bottle, either glass or plastic, which has been rinsed with dilute HCl and distilled water. The bottle should be filled to the top and tightly stoppered. If the sample is particularly turbid (cloudy or thick with sediment), it should be filtered before the analysis is attempted. If the sample is stored for a long period before analysis, it should be preserved by the addition of 5 mL of chloroform per liter of sample. This will arrest the loss of phosphate by microbiological activity. Care must be taken in handling chloroform, since it is very toxic. If you collect and analyze your sample on the same day, chloroform is not needed.

Just prior to the analysis, add 0.5 mL of 98% H_2SO_4 to your sample to hydrolyze any polyphosphate present. **(CAUTION: *Concentrated H_2SO_4 causes severe burns. Do not get any on your skin. If you come in contact with it, wash the area immediately with copious amounts of water.*)** Transfer 2.0 mL of the hydrolyzed phosphate unknown solution to a clean 100-mL volumetric flask and dilute to the mark with distilled water. Add 5 mL of the ammonium vanadomolybdate solution to 10 mL of this solution and measure the absorbance at 400 nm. If the absorbance is not within your calibration range, prepare either a more dilute or more concentrated sample, whichever is necessary. Measure the absorbance of three separate samples and report the phosphate concentration with its standard deviation. When you make the calculations, *remember* that you have diluted your sample by a factor of 50.

Instructor: Students often forget the dilution in their calculations.

Waste Disposal Instructions The solutions used in this experiment are not harmful and should be disposed of according to your instructor's directions.

Note 1 Ammonium vanadomolybdate solution: Dissolve 40 g of ammonium molybdate (molybdic acid, 85% MoO_3) in about 400 mL of distilled water. Dissolve 1 g of ammonium metavanadate, NH_4VO_2, in about 300 mL of distilled water and add 200 mL of concentrated nitric acid. Mix the two solutions and dilute to 1 L. This solution is stable for about 90 days and will be provided for your analysis.

REVIEW QUESTIONS

Before beginning this experiment in the laboratory, you should be able to answer the following questions:

1. What volume of 1.0×10^{-3} *M* solution is required to make 50.0 mL of solution with the following concentrations: 2.0×10^{-5}, 5.0×10^{-5}, 1.0×10^{-4}, 2.0×10^{-4}, 5.0×10^{-4}, and 7.5×10^{-4} *M*?
2. Which is thought to be the light-absorbing species in this experiment?
3. Write a balanced chemical equation for the formation of the light-absorbing species that results from reaction of phosphate and ammonium vanadomolybdate.
4. State the Beer-Lambert law, and define all terms in it.
5. Why is a calibration curve constructed? How?
6. How do you know whether to measure the absorbance of a more *dilute* or more *concentrated* solution if the absorbance of your unknown solution is not within the limits of your calibration curve?
7. A 0.1500 *M* sample of $Co(NO_3)_2$ gave an absorbance of 0.76 at 505 nm in a 1-cm cell. What is the cobalt concentration of a solution giving an absorbance of 0.52 in the same cell at the same wavelength?
8. A 8.64-ppm (1 ppm = 1 mg/L) solution of $FeSCN^{2+}$ has a transmittance of 0.30 when measured in a 1.00-cm cell at 580 nm. Calculate the extinction coefficient for $FeSCN^{2+}$ at this wavelength.
9. Define the term *eutrophication.*
10. If raw sewage contains 30 mg/L phosphate, and a secondary sewage-treatment plant removes 25% of the phosphate, would a secondary treatment plant provide potable water if 0.3 mg/L is the maximum phosphate concentration allowable in drinking water?

12. Write a balanced chemical equation for the hydrolysis of triphosphate, $P_3O_{10}^{5-}$, to orthophosphate, PO_4^{3-}.

NOTES AND CALCULATIONS

Name ______________________ Desk ______________________

Date ______________ Laboratory Instructor ______________________

Unknown no. ______________

REPORT SHEET | EXPERIMENT

Determination of Orthophosphate in Water | 32

A. Preparation of Calibration Curve

1. Mass of KH_2PO_4 0.1368 g (MW = 136.09 amu)
2. Concentration of phosphate stock solution 1.006×10^{-3} mol/L PO_4^{3-}
3. Calibration solutions

For literature data see: R. E. Kitson and M. G. Mellon, *Ind. Eng. Chem. Anal. Ed., 16,* 379 (1944), C. Wadlin and M. G. Mellon, *Anal. Chem., 25,* 1668 (1953); C. J. Barton, *Anal. Chem., 20,* 1068 (1948); K. P. Quinlan and M. A. Desesa, *Anal. Chem., 27,* 1626 (1955).

	Concentration, mol/L PO_4^{3-}	*Absorbance*
A	2.0×10^{-5}	0.05
B	5.0×10^{-5}	0.13
C	1.0×10^{-4}	0.20
D	2.0×10^{-4}	0.45
E	5.0×10^{-4}	1.25
F	7.5×10^{-4}	1.40

B. Analysis of Water Sample

1. Water-sample source stream (source dependent)

	Absorbance	*Concentration of PO_4^{3-} in aliquot*	*Concentration of PO_4^{3-} in sample*
A	0.090	3.4×10^{-5}	1.7×10^{-3} *M* or 230 ppm
B	0.070	2.4×10^{-5}	1.2×10^{-3} *M* or 160 ppm
C	0.070	2.4×10^{-5}	1.2×10^{-3} *M* or 160 ppm

2. Concentration of original solution
(show calculations) ___183 ppm___ ± ___55 ppm___

$$M = \frac{(3.4 \times 10^{-5}\,M)(100\text{ mL})}{2\text{ mL}} = 1.7 \times 10^{-3}\,M$$

$$\frac{(160 + 160 + 230)\text{ ppm}}{3} = 183\text{ ppm}$$

$$\frac{1.7 \times 10^{-3}\text{ mol}}{\text{L}} \times \frac{136\text{ g}}{\text{mol}} \times \frac{10^3\text{ mg}}{\text{g}} = 230\text{ mg/L} = 230\text{ ppm}$$

3. Standard deviation (show calculation)

$$SD = \sqrt{\frac{(70)^2 + 2(23)^2}{2}} = 55\text{ ppm}$$

4. Is the sample suitable as drinking water in terms of phosphate concentration? Will algal blooms form in it?
Since $PO_4^{3-} > 0.3$ mg/L or 0.30 ppm, this water is not suitable to drink. Since $PO_4^{3-} > 0.5$ mg/L or 0.5 ppm, algal blooms will occur if other nutrients are in high enough concentration.

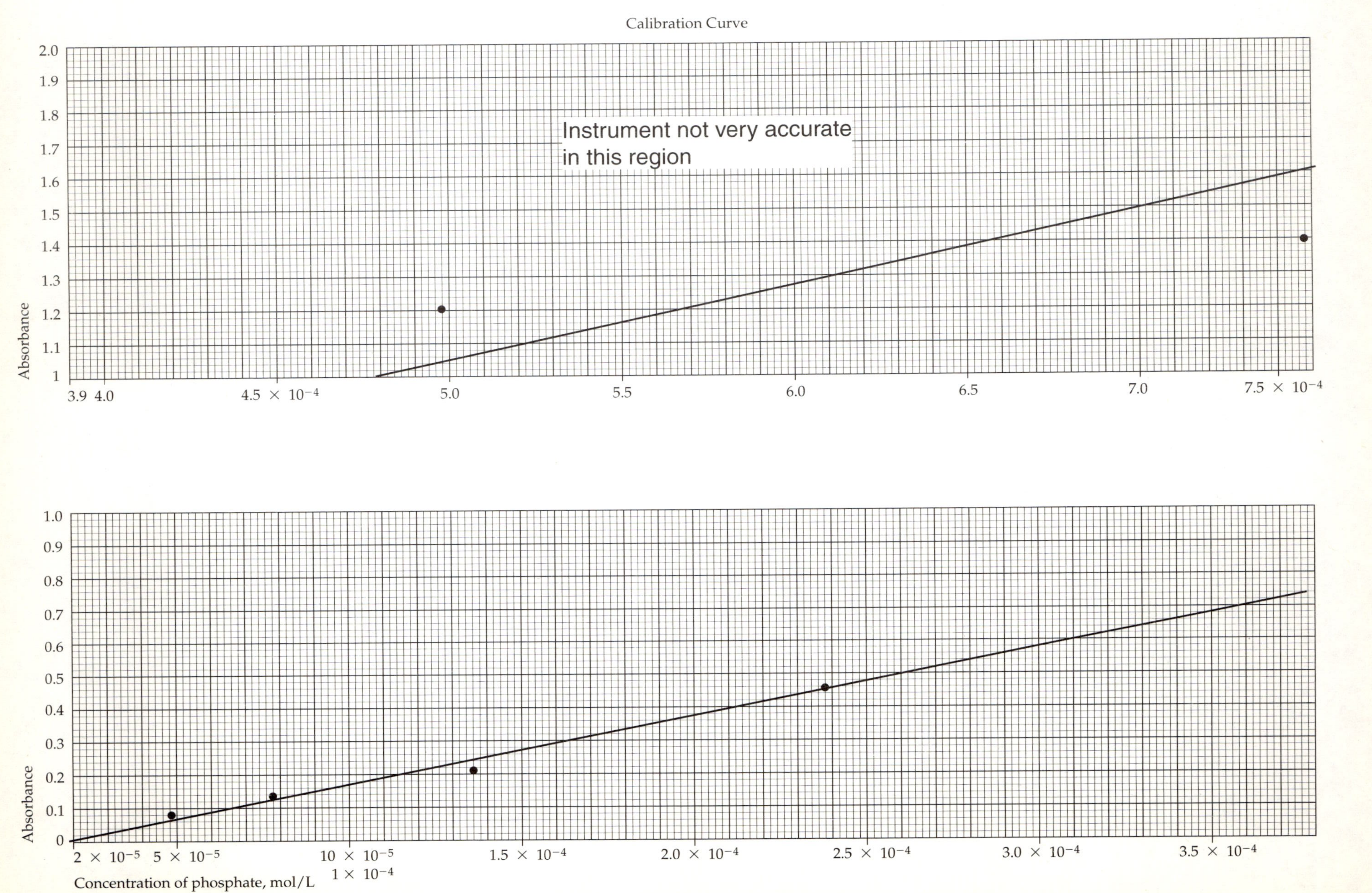

Calibration Curve
Instrument not very accurate in this region
Absorbance
2.0
1.9
1.8
1.7
1.6
1.5
1.4
1.3
1.2
1.1
1
3.9 4.0
4.5 × 10⁻⁴
5.0
5.5
6.0
6.5
7.0
7.5 × 10⁻⁴
Absorbance
1.0
0.9
0.8
0.7
0.6
0.5
0.4
0.3
0.2
0.1
0
2 × 10⁻⁵
5 × 10⁻⁵
10 × 10⁻⁵
1 × 10⁻⁴
1.5 × 10⁻⁴
2.0 × 10⁻⁴
2.5 × 10⁻⁴
3.0 × 10⁻⁴
3.5 × 10⁻⁴
Concentration of phosphate, mol/L

NOTES AND CALCULATIONS

Experiment

Analysis of Water for Dissolved Oxygen

OBJECTIVE

To gain a basic understanding of quantitative techniques of volumetric analysis by determining the dissolved-oxygen content of a water sample.

APPARATUS AND CHEMICALS

Apparatus

balance
50-mL buret
500-mL Erlenmeyer flask
250-mL Erlenmeyer flasks (3)
100-mL graduated cylinder
250-mL narrow-mouth, glass-stoppered bottle or 1-pint bottle
buret clamp and ring stand
barometer
1-L beaker
2-mL pipets (2)
25-mL pipet
thermometer
1-L volumetric flask
Bunsen burner and flask

Chemicals

alkaline iodide-azide reagent*
chloroform
2.15 *M* $MnSO_4$ (freshly prepared)
1% boiled starch solution
water sample (unknown)
1 *M* H_2SO_4
conc. H_2SO_4
NaN_3 solution*
$Na_2S_2O_3 \cdot 5H_2O$

DISCUSSION

The oxygen normally dissolved in water is indispensable to fish and other water-dwelling organisms. Certain pollutants deplete the dissolved oxygen during the course of their decomposition. This is particularly true of many organic compounds that are present in sewage or dead algae. These are decomposed by the aerobic metabolism of microorganisms, which use these organic compounds for food. The metabolic process is an oxidation of the organic compounds—the dissolved oxygen is the oxidizing agent. Thus, while these microorganisms are removing the pollutants, they are also removing the dissolved oxygen that otherwise would be present to support aquatic life. Since the solubility of most gases in solution decreases as the temperature of the solution increases, thermal pollution also decreases the dissolved oxygen content.

As a logical consequence of this, one empirical standard for determining water quality is the dissolved-oxygen content (DO). The survival of aquatic life depends on the water's ability to maintain certain minimum concentrations of the vital dissolved oxygen. Fish require the highest levels, invertebrates lower levels, and bacteria the least. For a diversified warm-water

*The alkaline iodide-azide solution is prepared by dissolving 500 g of NaOH and 135 g of NaI in 1 L of distilled water and adding a solution of 10 g NaN_3 in 40 mL of distilled water.

biota, including game fish, the DO concentration should be at least 5 mg/L (5 ppm). Another water-quality standard is the biological oxygen demand (BOD). The BOD is the amount of oxygen needed by the microorganisms to remove the pollutant. To determine BOD, a sample containing organic pollutants is incubated with its microorganisms for a definite time, usually 5 days, and the amount of oxygen removed is measured. The BOD is taken as the difference in DO before and after incubation. A BOD of 1 ppm is characteristic of nearly pure water. Water is considered fairly pure with a BOD of 3 ppm, and of doubtful purity when the BOD level reaches 5 ppm. Monitoring of water quality therefore logically includes analysis for dissolved oxygen.

This experiment outlines the analysis of water samples for their total dissolved oxygen (DO) content using the azide modification of the iodometric (Winkler) method. This is the procedure most commonly used for analysis of sewage, effluents, and streams. It is based on the use of Mn (II) compounds that are oxidized to Mn(IV) compounds by the oxygen in the water sample. The Mn(IV) compound in turn reacts with NaI to produce iodine, I_2. The released I_2 is then titrated with standardized sodium thiosulfate, $Na_2S_2O_3$, using starch as an indicator. The chemical reactions involved are as follows:

$$MnSO_4(aq) + 2KOH(aq) \longrightarrow Mn(OH)_2(s) + K_2SO_4(aq)$$

$$2Mn(OH)_2(s) + O_2(aq) \longrightarrow 2MnO(OH)_2(s)$$

$$MnO(OH)_2(s) + 2H_2SO_4(aq) \longrightarrow Mn(SO_4)_2(aq) + 3H_2O(l)$$

$$Mn(SO_4)_2(aq) + 2NaI(aq) \longrightarrow MnSO_4(aq) + Na_2SO_4(aq) + I_2(aq)$$

$$2Na_2S_2O_3(aq) + I_2(aq) \longrightarrow Na_2S_4O_6(aq) + 2NaI(aq)$$

The net overall chemical equation for this sequence of reactions is:

$$\tfrac{1}{2}O_2(g) + 2S_2O_3^{2-}(aq) + 2H^+(aq) \longrightarrow S_4O_6^{2-}(aq) + H_2O(l)$$

PROCEDURE

Instructor: Have students start boiling water at the beginning of the laboratory period.

A. Standard Sodium Thiosulfate Solution

First prepare and standardize a 0.025 *M* sodium thiosulfate solution as follows: Begin boiling about 800 mL of distilled water. Weigh out 6.205 g of $Na_2S_2O_3 \cdot 5H_2O$, and dissolve it in 500 mL of water that has been boiled to remove CO_2. Add 5 mL of chloroform and 0.4 g of NaOH to retard bacterial decomposition and dilute to 1 L in a 1-L volumetric flask.

Accurately weigh out about 0.223 g of dry potassium iodate. Dissolve it in 250 mL of previously boiled water in a 250-mL volumetric flask. To each of three 25-mL aliquots of this solution in labeled 250-mL Erlenmeyer flasks add 0.5 g of potassium iodide and about 2 mL of 1 *M* sulfuric acid. Titrate each of these solutions with your thiosulfate solution, stirring constantly. When the color of the solution becomes a pale yellow, dilute to approximately 200 mL with distilled water, add about 2 mL of 1% starch solution, and continue the titration until the color changes from blue to colorless for the first time. Ignore any return of color. Record the final buret reading and subtract that value from the initial reading to give the amount of thiosulfate used. The reactions involved in this standardization are

$$IO_3^-(aq) + 5I^-(aq) + 6H^+(aq) \rightleftharpoons 3I_2(aq) + 3H_2O(l)$$

$$2Na_2S_2O_3(aq) + I_2(aq) \rightleftharpoons Na_2S_4O_6(aq) + 2NaI(aq)$$

You are actually titrating with your thiosulfate the iodine formed by the first reaction. Calculate the molarity of your thiosulfate from the following equations:

$$\text{moles } Na_2S_2O_3 = 6 \times (\text{moles } KIO_3)$$

$$\text{moles } Na_2S_2O_3 = V_{Na_2S_2O_3} \times M_{Na_2S_2O_3}$$

$$\text{moles } KIO_3 = \frac{(\text{g } KIO_3)}{(214.02 \text{ g/mol})} \times \frac{25.00 \text{ mL}}{250.0 \text{ mL}}$$

Thus,

$$M_{Na_2S_2O_3} = \frac{(6)(\text{g } KIO_3)(25.00 \text{ mL})}{(214.02 \text{ g/mol})(\text{volume } Na_2S_2O_3 \text{ in liters})(250.0 \text{ mL})}$$

EXAMPLE 33.1

A 0.2264-g sample of KIO_3 was dissolved in 250.0 mL of water. To each 25.00-mL aliquot of this solution 0.500 g of KI and 2.00 mL of 1.00 M H_2SO_4 were added. Titration of the liberated iodine required 25.30 mL of $Na_2S_2O_3$ solution. Calculate the molarity of the $Na_2S_2O_3$ solution.

SOLUTION: From the analytical reactions

$$IO_3^-(aq) + 5I^-(aq) + 6H^+(aq) \longrightarrow 3I_2(aq) + 3H_2O(aq)$$

$$2Na_2S_2O_3(aq) + I_2(aq) \longrightarrow Na_2S_4O_6(aq) + 2NaI(aq)$$

it is seen that exactly three moles of iodine are liberated for each mole of iodate, and one mole of iodine reacts with two moles $S_2O_3^{2-}$ whence, in a 25.00 mL aliquot,

$$\text{moles } KIO_3 = \left(\frac{0.2264 \text{ g}}{214.02 \text{ g/mol}}\right)\left(\frac{25.00 \text{ mL}}{250.0 \text{ mL}}\right)$$

$$= 1.058 \times 10^{-4} \text{ mol}$$

and since

$$\text{moles } Na_2S_2O_3 = 6 \times (\text{moles } KIO_3) = 6(1.058 \times 10^{-4} \text{ mol})$$

$$= 6.347 \times 10^{-4} \text{ mol}$$

then

$$M_{Na_2S_2O_3} = \frac{\text{moles } Na_2S_2O_3}{\text{liters of solution}}$$

$$= \frac{6.347 \times 10^{-4} \text{ mol}}{0.02530 \text{ L}}$$

$$= 0.02509\ M$$

B. Water-Sample Analysis

Collection of Sample Collect the sample in a narrow-mouthed, glass-stoppered bottle (250-to-300 mL capacity). Avoid entrapment or dissolution of atmospheric oxygen. Allow the bottle to overflow its volume and replace the stopper so that no air bubbles are entrained; avoid excessive agitation, which will dissolve atmospheric oxygen. Record the temperature of the water sample in degrees Celsius. The sample should be analyzed as soon as possible. If the method outlined below is used, the sample may be preserved for four to eight hours by adding 0.7 mL of concentrated H_2SO_4 and 1 mL of sodium azide solution (2 g NaN_3/100 mL H_2O) to the DO bottle. This will arrest biologic activity and maintain the DO if the bottle is stored at the temperature of collection or at 10° to 20°C while tightly sealed.

Release of Iodine Open the sample bottle with great care to avoid aeration and add 2 mL of 2.15 *M* manganous sulfate solution from a volumetric pipet. Similarly, add 2 mL of alkaline iodide-azide reagent using another pipet. The neck of the bottle will again have excess liquid, so replace the stopper carefully to avoid splashing. Thoroughly mix the contents of the bottle by inverting the bottle several times. A milky precipitate will form and will gradually change to a yellowish brown color. Allow the precipitate to settle so that the clear solution occupies the top third of the bottle. Carefully remove the stopper and immediately add 2 mL of concentrated sulfuric acid. This addition should be made by bringing the pipet tip against the neck of the bottle just slightly below the surface of the liquid. Stopper the bottle and then mix the contents by gentle inversion until the precipitate dissolves. At this point the yellowish-brown color due to liberated iodine should appear. The sample need not be titrated immediately, but if titration is delayed, the sample should be stored in darkness. The titration should be done within several hours.

Titration Accurately measure 200 mL of the sample into a 500-mL Erlenmeyer flask. Titrate the sample with the standardized thiosulfate solution with constant stirring. When the color of the solution becomes a pale yellow, add about 2 mL of 1% starch solution and continue titrating until the color changes from blue to colorless for the first time. Record the volume of titrant necessary. If time permits, analyze another sample.

Calculation of Dissolved Oxygen Content According to the above equations, 1 mol of O_2 (32 g) requires 4 mol of $Na_2S_2O_3$ in reaching an end point. The number of moles of $Na_2S_2O_3$ is equal to the volume of $Na_2S_2O_3$ (in liters) times the concentration of the $Na_2S_2O_3$ (in molarity):

$$\text{moles } Na_2S_2O_3 = \text{volume}_{Na_2S_2O_3} \times \text{molarity}_{Na_2S_2O_3}$$

From the information given above, calculate the number of grams of O_2 in your 200-mL sample. From this, calculate the number of milligrams of O_2 per liter of solution. Since a liter of solution will weigh approximately 1000 g (the bulk of the solution is water), 1 mg/L is equivalent to 1 mg in 10^6 mg, or 1 million mg, of solution. Therefore, the number of milligrams of O_2 per liter is often referred to as parts per million (ppm).

EXAMPLE 33.2

To a 200.0-mL water sample 0.5000 g of KI, 2.000 mL of 1.000 *M* H_2SO_4, and 2.000 mL of starch solution were added. The liberated iodine required 7.88 mL of 0.0251 *M* $Na_2S_2O_3$. Calculate the O_2 concentration in the sample in ppm.

SOLUTION: From the analytical reactions

$$MnSO_4 + 2KOH \longrightarrow Mn(OH)_2 + K_2SO_4$$

$$2Mn(OH)_2 + O_2 \longrightarrow 2MnO(OH)_2$$

$$MnO(OH)_2 + 2H_2SO_4 \longrightarrow Mn(SO_4)_2 + 3H_2O$$

$$Mn(SO_4)_2 + 2KI \longrightarrow MnSO_4 + K_2SO_4 + I_2$$

$$2Na_2S_2O_3 + I_2 \longrightarrow Na_2S_4O_6 + 2NaI$$

it is seen that each mole of O_2 requires 4 moles of $Na_2S_2O_3$.

$$
\begin{aligned}
\text{moles } Na_2S_2O_3 &= (\text{volume } Na_2S_2O_3) \times (\text{molarity } Na_2S_2O_3) \\
&= (0.00788 \text{ L})(0.0251 \text{ mol/L}) \\
&= 1.98 \times 10^{-4} \text{ mol} \\
\text{moles } O_2 &= \tfrac{1}{4} \text{ mol } Na_2S_2O_3 \\
&= 4.95 \times 10^{-5} \text{ mol} \\
\text{weight } O_2 &= (4.95 \times 10^{-5} \text{ mol}) \times (32.0 \text{ g/mol}) \\
&= 1.58 \times 10^{-3} \text{ g} \\
\text{concentration } O_2 &= \frac{1.58 \text{ mg}}{0.200 \text{ L}} = 7.90 \text{ mg/L} \\
&= 7.90 \text{ ppm}
\end{aligned}
$$

A correction factor may be applied to your answer to correct for solution loss during the addition of manganous sulfate and sulfuric acid. This amounts to multiplication by 204/200 if these reagents were added to a 200-mL bottle.

Comparisons in Dissolved Oxygen Contents The amount of oxygen dissolved in water depends not only on the amount of chemical pollution but also on such factors as water temperature and the atmospheric pressure above the water. At temperatures between 0°C and 39°C, the amount of O_2 that will be present in oxygen-saturated distilled water is given by the equation

$$\text{ppm dissolved } O_2 = \frac{(P - p) \times 0.678}{35 + T} = \text{SLDO}$$

where P is the barometric pressure in mm Hg, T is the temperature of the water in °C, and p is the vapor pressure of water at the temperature of the water. Calculate the saturation level (SL) for your water sample. The percent saturation, %SL, is given by

$$\%\text{SL} = \frac{100(\text{DO in ppm})}{(\text{SLDO in ppm})}$$

Calculate the %SL for your sample.

EXAMPLE 33.3

A water sample at 12°C and 652 mm Hg was found to contain 7.90 ppm O_2. Calculate the percent saturation of this sample.

SOLUTION:

$$
\begin{aligned}
\text{SLDO} &= \frac{(652 \text{ mm} - 10.5 \text{ mm})(0.678 \text{ ppm}-°\text{C/mm})}{(35 + 12)°\text{C}} \\
&= 9.3 \text{ ppm} \\
\%\ \text{SL} &= \frac{(100)(7.90 \text{ ppm})}{9.3 \text{ ppm}} = 85\%
\end{aligned}
$$

Waste Disposal Instructions All solutions should be disposed of in the appropriately labeled containers.

REVIEW QUESTIONS

Before beginning this experiment in the laboratory, you should be able to answer the following questions:

1. What is the molarity of a thiosulfate solution if 27.35 mL of it reacts with 0.1309 g of KIO_3?
2. Suggest how chloroform helps to preserve the water sample.
3. Which species is being oxidized, and which is reduced in the reaction

$$Mn(SO_4)_2 + 2KI \rightleftharpoons MnSO_4 + K_2SO_4 + I_2$$

4. During the course of the standardization of $S_2O_3^{2-}$ with KIO_3, the text tells you to ignore the return of the blue color with time after the end point has been reached in the reaction. Suggest a reason for the return of the blue color. Write a chemical reaction that is consistent with your suggestion.
5. Calculate the concentration in mg/L (ppm) of a solution that is 1×10^{-3} M in O_2, assuming a density of 1 g/mL for the solution.
6. Some characteristic BOD levels are given below:

Source	BOD range (ppm)
Untreated municipal sewage	100–400
Runoff from barnyards and feed lots	100–10,000
Food-processing wastes	100–10,000

Assuming the minimum values above, by what factor must these waters be diluted with pure water to reduce the BOD to a value sufficiently low to support aquatic life (BOD $\leqslant$ 5 ppm)?

7. A 0.150-g sample of KIO_3 required 10.6 mL of $Na_2S_2O_3$ solution for its standardization. What is the molarity of the $Na_2S_2O_3$ solution?
8. In the determination of the BOD of a water sample, it was found that the DO level was 8.4 ppm before incubation and 5.6 ppm after five days of incubation. What is the BOD of this sample?
9. Would the water sample in question 8 support aquatic life?

Name ______________________ Desk ______________________

Date ______________ Laboratory Instructor ______________________

Unknown no. ______________

REPORT SHEET | EXPERIMENT

Analysis of Water for Dissolved Oxygen | 33

A. Standard Sodium Thiosulfate Solution

1. Mass in grams of KIO_3 in 250 mL ___0.2264___ g

	Sample 1	*Sample 2*	*Sample 3*
2. Final buret reading	26.31 mL	26.87 mL	27.50 mL
3. Initial buret reading	1.01 mL	1.52 mL	2.22 mL
4. Volume of $Na_2S_2O_3$ solution	25.30 mL	25.35 mL	25.28 mL
	0.02530 L	0.02535 L	0.02528 L
5. Molarity of $Na_2S_2O_3$ (show calculations)	0.02509 *M*	0.02504 *M*	0.02511 *M*

Avg. *M* ___0.02508___ ± ___0.00004___

$$M = \frac{(6)(0.2264)(0.100)}{(214.02)(0.02530)} = 0.02509$$

$$M = \frac{(6)(0.2264)(0.100)}{(214.02)(0.02535)} = 0.02504 \qquad M = \frac{(6)(0.2264)(0.100)}{(214.02)(0.02528)} = 0.02511$$

6. Standard deviation (show calculations) ___4×10^{-5}___

$$SD = \sqrt{\frac{(1 \times 10^{-5})^2 + (4 \times 10^{-5})^2 + (3 \times 10^{-5})^2}{2}} = 4 \times 10^{-5}$$

B. Water-Sample Analysis

Water temperature ___12.0___ °C   Barometric pressure = 652 mm Hg

	Sample 1	*Sample 2*
1. Volume of $Na_2S_2O_3$ required to titrate water sample		
Final buret reading	7.98 mL	18.23 mL
Initial buret reading	0.10 mL	10.00 mL
Volume of $Na_2S_2O_3$	7.88 mL	8.23 mL

(used standard $Na_2S_2O_3$)

	Sample 1	*Sample 2*
2. Moles of $Na_2S_2O_3$ required to titrate water sample (show calculations)	1.98×10^{-4} mol	2.06×10^{-4} mol

$(0.02508 \text{ mol/L})(7.88 \times 10^{-3} \text{ L}) = 1.98 \times 10^{-4}$ mol; $(0.02508 \text{ mol/L})(8.23 \times 10^{-3} \text{ L}) = 2.06 \times 10^{-4}$ mol

	Sample 1	*Sample 2*
3. Moles of O_2 present in 200 mL of water sample (show calculations)	4.95×10^{-5} mol	5.15×10^{-5} mol

mole O_2 = (1/4)(mole $Na_2S_2O_3$)

$(1/4)(1.98 \times 10^{-4}) = 4.95 \times 10^{-5}$ mol

$(4.94 \times 10^{-5} \text{ mol})(32 \text{ g/mol}) = 1.58 \times 10^{-3}$ g

	Sample 1	*Sample 2*
4. Grams of O_2 present in 200 mL of water sample	1.58×10^{-3} g	1.65×10^{-3} g
5. Milligrams of O_2 present in 200 mL of water sample	1.58 mg	1.65 mg
6. Milligrams of O_2 present in 1000 mL of water sample	7.90 ppm	8.25 ppm
7. Saturation level for your water sample (show calculations)	9.3 ppm	

$$\frac{(652 - 10.5)(0.678)}{35 + 12} = \frac{435}{47} = 9.3 \text{ ppm}$$

8. Percent saturation level of your water sample 87%
(show calculations)

$$\frac{7.90}{9.3} \times 10^2 = 85\% \qquad \frac{8.25}{9.3} \times 10^2 = 89\% \qquad \frac{85 + 89}{2} = 87$$

9. Do you think that this water has sufficient DO to sustain aquatic life?
Yes

10. Calculate the SLDO for water at 20°, 30°, 40°, 50°, and 100°C and 700 mm Hg, and plot the data on the graph paper provided. How would you expect the solubility of a gas in water to change with temperature? Are your calculations in agreement with your expectations?

$$\text{At } 20°\text{C} = 8.4 = \frac{(700 \text{ mm Hg} - 17.5 \text{ mm Hg})(0.678)}{35 + 20}$$

30°C = 7.0
40°C = 5.8
50°C = 4.9
100°C = 0

It should decrease with increasing temperature. Yes, calculations agree with expectations.

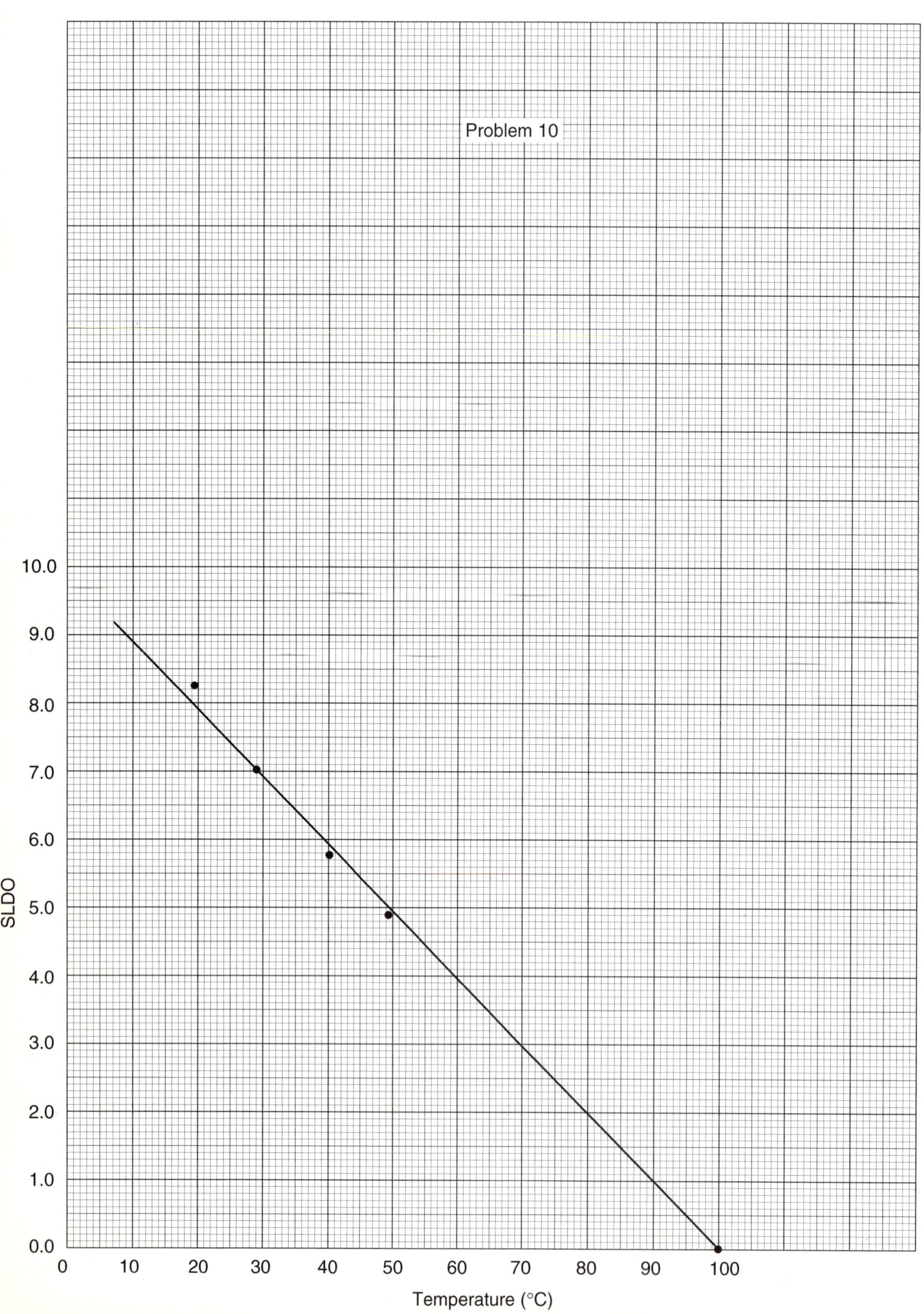

Problem 10
10.0
9.0
8.0
7.0
6.0
5.0
4.0
3.0
2.0
1.0
0.0
SLDO
0
10
20
30
40
50
60
70
80
90
100
Temperature (°C)

NOTES AND CALCULATIONS

Preparation and Reactions of Coordination Compounds: Oxalate Complexes

Experiment

OBJECTIVE

To gain some familiarity with coordination compounds by preparing a representative compound and witnessing some typical reactions.

APPARATUS AND CHEMICALS

Apparatus

balance
Bunsen burner and hose
8-oz wide-mouth bottle
No. 6 two-hole rubber stopper
glass stirring rod
9-cm filter paper
wire gauze
9-cm Büchner funnel
250-mL suction flask
aspirator
100-mL and 250-mL beakers
thermometer
ring stand and iron ring
glass wool

Chemicals

cis- and *trans-*$K[Cr(C_2O_4)_2(H_2O)_2]$ (prep given)
6 *M* NH_3
$Na_2C_2O_4$ (sodium oxalate)
$K_2C_2O_4 \cdot H_2O$ (potassium oxalate monohydrate)
acetone
ferrous ammonium sulfate
6 *M* H_2SO_4
aluminum powder
6 *M* KOH
$H_2C_2O_4$ (oxalic acid)
$K_2Cr_2O_7$ (potassium dichromate)
50% ethanol
95% ethanol
$CuSO_4 \cdot 5(H_2O)$
6% H_2O_2
6 *M* HCl

DISCUSSION

When gaseous boron trifluoride, BF_3, is passed into liquid trimethylamine, $(CH_3)_3N$, a highly exothermic reaction occurs, and a creamy white solid, $(CH_3)_3N{:}\,BF_3$, separates. This solid, which is an adduct of trimethylamine and boron trifluoride, is a coordination compound. It contains a coordinate covalent, or dative, bond uniting the Lewis acid BF_3 with the Lewis base trimethylamine. Numerous coordination compounds are known, and in fact nearly all compounds of the transition elements are coordination compounds wherein the metal is a Lewis acid and the atoms or molecules joined to the metal are Lewis bases. These Lewis bases are called *ligands,* and the coordination compounds are usually denoted by square brackets when their formulas are written. The metal and the ligands bound to it constitute what is termed the coordination sphere. In writing chemical formulas for coordination compounds, we use square brackets to set off the coordination sphere from other parts of the compound. For example, the salt $NiCl_2 \cdot 6H_2O$ is in reality the coordination compound $[Ni(H_2O)_6]Cl_2$, with an octahedral geometry, as shown in Figure 34.1.

Instructor: Do a demonstration.

The apexes of a *regular* octahedron are all equivalent positions. Thus, each of the monodentate (one donor site) H_2O molecules in the $[Ni(H_2O)_6]^{2+}$ ion

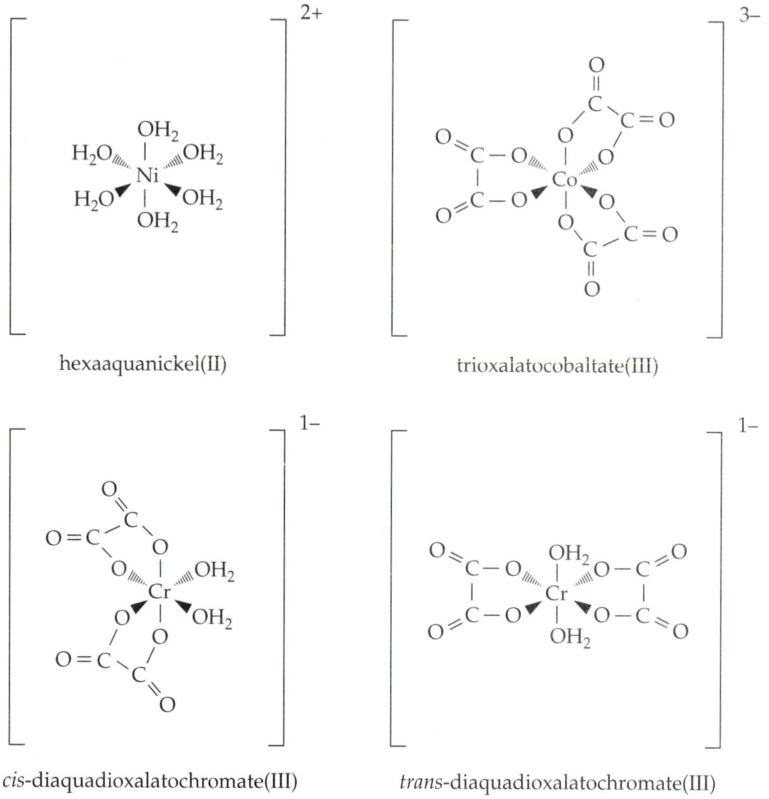

hexaaquanickel(II)

trioxalatocobaltate(III)

cis-diaquadioxalatochromate(III)

trans-diaquadioxalatochromate(III)

▲ **FIGURE 34.1** Typical octahedral (6-coordinate) coordination compounds.

and the three bidentate (two donor sites) oxalate ions, $C_2O_4^{2-}$, in $[Co(C_2O_4)_3]^{2-}$ are in identical environments. The water molecules in the two isomeric compounds, *cis*- and *trans*-$[Cr(C_2O_4)_2(H_2O)_2]^-$, are in equivalent environments within each complex ion (coordination compound), but the two isomeric ions are not equivalent to one another. The two water molecules are adjacent in the *cis* isomer and opposite one another in the *trans* isomer. These two isomers are termed *geometric isomers*, and although they have identical empirical and molecular formulas, their geometrical arrangements in space are different. Consequently, they have different chemical and physical properties, as your laboratory instructor will demonstrate through the reactions shown in Figure 34.2.

Your goal in this experiment is to prepare an oxalate-containing coordination compound. In Experiment 35 you will analyze it by titration for its oxalate content. You will prepare *one* of the following compounds:

1. $K_3 [Cr(C_2O_4)_3] \cdot 3H_2O$
2. $K_2 [Cu(C_2O_4)_2] \cdot 2H_2O$
3. $K_3 [Fe(C_2O_4)_3] \cdot 3H_2O$
4. $K_3 [Al(C_2O_4)_3] \cdot 3H_2O$

Your laboratory instructor will tell you which one to prepare. Each of these compounds will be prepared by someone in your lab section so that you can compare their properties. **(CAUTION:** ***Oxalic acid is a toxic compound and***

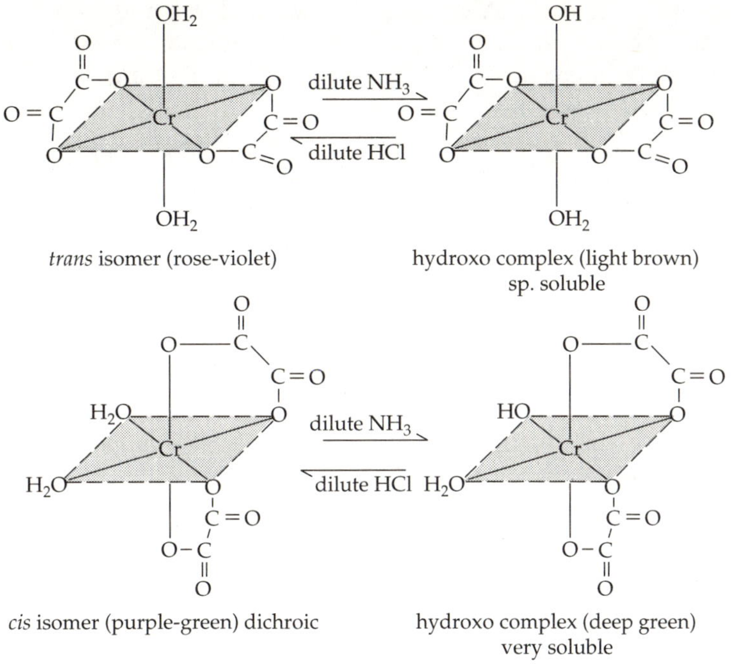

▲ **FIGURE 34.2** Reaction of geometric isomers: *cis-* and *trans-* $[Cr(C_2O_4)_2(H_2O)_2]^-$ ions. The procedure is to place a small amount of an aqueous solution of each complex on separate pieces of filter paper on a watch glass. Let them dry, and then add a drop of reagent. Observe the results.

Instructor: Do this demonstration.

is absorbed through the skin. Should any come in contact with your skin, wash it off immediately with copious amounts of water.)

PROCEDURE

Prepare one of the complexes whose synthesis is given below.

1. Preparation of $K_3[Cr(C_2O_4)_3] \cdot 3H_2O$

$$K_2Cr_2O_7 + 7H_2C_2O_4 \cdot 2H_2O + 2K_2C_2O_4 \cdot H_2O \longrightarrow 2K_3[Cr(C_2O_4)_3] \cdot 3H_2O + 6CO_2 + 17H_2O$$

Slowly add 3.6 g of potassium dichromate to a suspension of 10 g of oxalic acid in 20 mL of H_2O in a 250-mL beaker. The orange-colored mixture should spontaneously warm up almost to boiling as a vigorous evolution of gas commences. When the reaction has subsided (about 15 min), dissolve 4.2 g of potassium oxalate monohydrate in the hot, green-black liquid and heat to boiling for 10 min. Allow the beaker and its contents to cool to room temperature. Add about 10 mL of 95% ethanol, with stirring, into the cooled solution in the beaker. Further cool the beaker and its contents in ice. The cooled liquid should thicken with crystals. After cooling in ice for 15 to 20 min, the crystals should be collected by filtration with suction using a Büchner funnel and filter flask. Wash the crystals on the funnel with three 10-mL portions of 50% aqueous ethanol followed by 25 mL of 95% ethanol and dry

Instructor: If desired, all these preparations can be scaled down by a factor of four or five.

the product in air. Weigh the air-dried material and store it in a vial. You should obtain about 9 g of product. Calculate the theoretical yield and determine your percentage yield. Reactions of chromium(III) are slow, and your yield will be low if you work too fast.

$$\% \text{ yield} = 100 \times \frac{\text{actual yield in grams}}{\text{theoretical yield in grams}}$$

SAVE your sample for analysis in Experiment 35.

EXAMPLE 34.1

In the preparation of *cis*-$K[Cr(C_2O_4)_2(H_2O)_2] \cdot 2H_2O$, 12.0 g of oxalic acid was allowed to react with 4.00 g of potassium dichromate, and 8.20 g of *cis*-$K[Cr(C_2O_4)_2(H_2O)_2] \cdot 2H_2O$ was isolated. What is the percent yield in this synthesis?

SOLUTION:

$$K_2Cr_2O_7 + 7H_2C_2O_4 \cdot 2H_2O \longrightarrow 2K[Cr(C_2O_4)_2(H_2O)_2] \cdot 2H_2O + 6CO_2 + 13H_2O$$

From the above reaction, we see that 1 mole of $K_2Cr_2O_7$ reacts with 7 moles of $H_2C_2O_4$ to produce 2 moles of $K[Cr(C_2O_4)_2(H_2O)_2] \cdot 2H_2O$. In our synthesis we have used the following:

$$\text{moles } K_2Cr_2O_7 = \frac{4.00 \text{ g}}{294.19 \text{ g / mol}} = 0.0136 \text{ mol}$$

$$\text{moles } H_2C_2O_4 \cdot 2H_2O = \frac{12.0 \text{ g}}{126.07 \text{ g / mol}} = 0.0952 \text{ mol}$$

Our reaction requires a 7:1 molar ratio of oxalic acid to $K_2Cr_2O_7$, and we have actually used a 6.999 molar ratio or, within experimental error, the stoichiometric amount of each reagent. Hence, the number of moles of $K[Cr(C_2O_4)_2(H_2O)_2] \cdot 2H_2O$ formed should be twice the number of moles of $K_2Cr_2O_7$ reacted.

$$\text{moles } K[Cr(C_2O_4)_2(H_2O)_2] \cdot 2H_2O \text{ (expected)} = (2)(0.0136 \text{ mol}) = 0.0272 \text{ mol}$$

The theoretical yield of $K[Cr(C_2O_4)_2(H_2O)_2] \cdot 2H_2O$ is

$$(0.0272 \text{ mol})(339.2 \text{ g/mol}) = 9.23 \text{ g}$$

Our percentage yield is then

$$\% \text{ yield} = \frac{(100)(8.20 \text{ g})}{(9.23 \text{ g})} = 88.8\%$$

2. Preparation of $K_2[Cu(C_2O_4)_2] \cdot 2H_2O$

$$CuSO_4 \cdot 5H_2O + 2K_2C_2O_4 \cdot H_2O \longrightarrow K_2[Cu(C_2O_4)_2] \cdot 2H_2O + K_2SO_4 + 5H_2O$$

Heat a solution of 6.2 g of copper sulfate pentahydrate in 12 mL of water to about 90°C and add it rapidly, with vigorous stirring, to a hot (~90°C) solution of 10.0 g of potassium oxalate monohydrate ($K_2C_2O_4 \cdot H_2O$) in 50 mL of water contained in a 100-mL beaker. Cool the mixture by setting the beaker in an ice bath for 15 to 30 min. Suction-filter the resultant crystals using a Büchner funnel and filter flask and wash the crystals successively with about 12 mL of cold water, then 10 mL of absolute ethanol, and finally 10 mL of acetone, and air dry. Weigh the air-dried material and store it in a vial. You should obtain about 7 g of product. Calculate the theoretical yield and determine your percentage yield. SAVE your sample for analysis in Experiment 35.

3. Preparation of $K_3[Fe(C_2O_4)_3] \cdot 3H_2O$

$$(NH_4)_2[Fe(H_2O)_2(SO_4)_2] \cdot 6H_2O + H_2C_2O_4 \cdot 2H_2O \longrightarrow FeC_2O_4 + H_2SO_4 + (NH_4)_2SO_4 + 10H_2O$$

$$H_2C_2O_4 \cdot 2H_2O + 2FeC_2O_4 + 3K_2C_2O_4 \cdot H_2O + H_2O_2 \longrightarrow 2K_3[Fe(C_2O_4)_3] \cdot 3H_2O + H_2O$$

This preparation contains two separate parts. Iron(II) oxalate is prepared first and then converted to $K_3[Fe(C_2O_4)_3] \cdot 3H_2O$ by oxidation with hydrogen peroxide, H_2O_2, in the presence of potassium oxalate.

Prepare a solution of 10 g of ferrous ammonium sulfate hexahydrate in 30 mL of water containing a few drops of 6 M H_2SO_4 (to prevent premature oxidation of Fe^{2+} to Fe^{3+} by O_2 in the air). Then add, with stirring, a solution of 6 g of oxalic acid in 50 mL of H_2O. Yellow iron(II) oxalate forms. Carefully heat the mixture to boiling while stirring constantly to prevent bumping. Decant and discard the supernatant liquid and wash the precipitate several times by adding about 30 mL of hot water, stirring, and decanting the liquid. Filtration is not necessary at this point.

To the wet iron(II) oxalate, add a solution of 6.6 g of $K_2C_2O_4 \cdot H_2O$ in 18 mL of water and heat the mixture to about 40°C. SLOWLY AND CAUTIOUSLY add 17 mL of 6% H_2O_2, while stirring constantly and maintaining the temperature at 40°C. After the addition of H_2O_2 is complete, heat the mixture to boiling and add a solution containing 1.7 g of oxalic acid in 15 mL of water. When adding the oxalic acid solution, add the first 8 mL all at once and the remaining 5 mL dropwise, keeping the temperature near boiling. Filter off any solid by gravity and add 20 mL of 95% ethanol to the filtrate. Cover the beaker with a watch glass and store it in your lab desk until the next period. Filter by suction using a Büchner funnel and filter flask (Figure 34.3) and wash the green crystals with a 50% aqueous ethanol solution, then with acetone, and air-dry. Weigh the product and store it in a vial

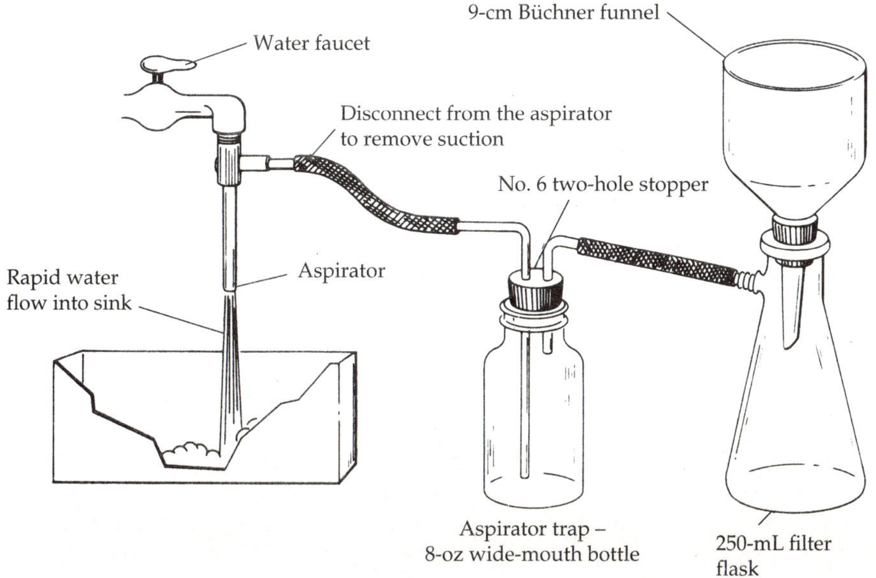

▲ **FIGURE 34.3** Suction filtration assembly.

in the dark. This complex is photosensitive and reacts with light according to the following:

$$[Fe(C_2O_4)_3]^{3-} \xrightarrow{h\nu} [Fe(C_2O_4)_2]^{2-} + 2CO_2 + e^-$$

In order to demonstrate this, place a small specimen on a watch glass near the window and observe any changes that occur during the lab period. You should obtain about 8 g of product. Calculate the theoretical yield and determine your percent yield. SAVE your sample for analysis in Experiment 35. Keep it away from light by wrapping the vial with aluminium foil.

4. Preparation of $K_3[Al(C_2O_4)_3] \cdot 3H_2O$

$$Al + 3KOH + 3H_2C_2O_4 \cdot 2H_2O \longrightarrow K_3[Al(C_2O_4)_3] \cdot 3H_2O + 6H_2O + \tfrac{3}{2}H_2$$

Place 1 g of aluminum powder in a 200-mL beaker and cover with 10 mL of hot water. Add 20 mL of 6 *M* KOH solution in small portions to regulate the vigorous evolution of hydrogen. Finally, heat the liquid almost to boiling in order to dissolve any residual metal. Maintain the heating and add a solution of 13 g of oxalic acid in 100 mL of water in small portions. During the neutralization, hydrated alumina will precipitate, but it will redissolve at the end of the addition after gentle boiling. Cool the solution in an ice bath and add 50 mL of 95% ethanol. If oily material separates, stir the solution and scratch the sides of the beaker with your glass rod to induce crystallization. Suction-filter the product using the Büchner funnel and suction flask and wash with a 20-mL portion of ice-cold 50% aqueous ethanol and finally with small portions of absolute ethanol. Dry the product in air, weigh it, and store it in a stoppered bottle. You should obtain about 11 g of product. Calculate the theoretical yield and determine your percent yield. SAVE your sample for analysis in Experiment 35.

Instructor: Demonstration materials

Preparation of Materials for the Demonstration

(These preparations should be done a week before the laboratory period.)

cis-$K[Cr(C_2O_4)_2(H_2O)_2] \cdot 2H_2O$

$$K_2Cr_2O_7 + 7H_2C_2O_4 \cdot 2H_2O \rightleftharpoons 2K[Cr(C_2O_4)_2(H_2O)_2] \cdot 2H_2O + 6CO_2 + 13H_2O$$

Separately powder in a *dry* mortar 12 g of oxalic acid dihydrate and 4 g of potassium dichromate. Mix the powders as intimately as possible by gentle grinding in the mortar. Moisten a large evaporating dish (10 cm) with water and pour off all the water but do not wipe dry. Place the powdered mixture in the evaporating dish as a *compact heap*; it will become moistened by the water that remains in the evaporating dish. Cover the evaporating dish with a large watch glass and warm it gently on a hot plate. A vigorous spontaneous reaction will soon occur and will be accompanied by frothing, as steam and CO_2 escape. The mixture should then liquefy to a deep-colored syrup. Pour about 20 mL of ethanol on the hot liquid and continue to gently warm it on the hot plate. Triturate (grind or crush) the product with a spatula until it solidifies. If complete solidification cannot be effected with one portion of alcohol, decant the liquid, add another 20 mL of alcohol, warm gently, and resume the trituration until the product is entirely crystalline and granular. The yield is essentially quantitative at about 9 g. This compound is intensely dichroic, appearing in the solid state as almost black in diffuse daylight and deep purple in artificial light.

***trans*-K[Cr(C$_2$O$_4$)$_2$(H$_2$O)$_2$] · 3H$_2$O** Dissolve 12 g of oxalic acid dihydrate in a minimum of boiling water in a 300-mL (or larger) beaker. Add to this in small portions a solution of 4 g of potassium dichromate in a minimum of hot water and cover the beaker with a watch glass while the violent reaction proceeds. After the addition is complete, cool the contents of the beaker and allow spontaneous evaporation at room temperature to occur so that the solution reduces to about one-third its original volume (this takes 36 to 48 hours). Collect the deposited crystals by suction filtration and wash several times with cold water and alcohol and air-dry. The yield is about 6.5 g. The complex is rose-colored with a violet tinge and is not dichroic.

Waste Disposal Instructions All oxalate- and metal-containing solutions should be disposed of in appropriate containers.

REVIEW QUESTIONS

Before beginning this experiment in the laboratory, you should be able to answer the following questions:

1. Define the terms *Lewis acid* and *Lewis base.*
2. Define the terms *ligand* and *coordination sphere.*
3. Define and give an example of a coordination compound.
4. Define the term *geometric isomer.*
5. Draw structures for all possible isomers of the six-coordinate compounds $[Co(NH_3)_4Cl_2]$ and $[Co(NH_3)_3Cl_3]$.
6. Are the chlorine atoms in equivalent environments in each of the compounds $[Co(NH_3)_4Cl_2]$ and $[Co(NH_3)_3Cl_3]$?
7. What is the meaning of *dichroism?*
8. What is the meaning of *trituration?*
9. Look up the preparation of an oxalate complex of Ni, Mn, or Co. Cite your reference and state whether this preparation would be suitable to add to this experiment. Why or why not?
10. Find an analytical method to determine the amount of Fe, Cu, Cr, or Al in your oxalate complex. Cite the reference to the method. Could you do the determination with the chemicals and equipment available in your laboratory? Why or why not?
11. Oxalic acid is used to remove rust and corrosion from automobile radiators. How do you think it works?

NOTES AND CALCULATIONS

Name ______________________ Desk ______________________

Date ______________ Laboratory Instructor ______________________

REPORT SHEET | EXPERIMENT

Preparation and Reactions of Coordination Compounds: Oxalate Complexes

34

1. Complex prepared $K_3[Fe(C_2O_4)_3] \cdot 3H_2O$
2. Chemical reaction for its preparation ______________________
$(NH_4)_2[Fe(H_2O)_2(SO_4)_2] \cdot 6H_2O + H_2C_2O_4 \cdot 2H_2O \longrightarrow FeC_2O_4 + H_2SO_4 + (NH_4)_2SO_4 + 10H_2O;$
$H_2O_2 + H_2C_2O_4 \cdot 2H_2O + 2FeC_2O_4 + 3K_2C_2O_4 \cdot H_2O \longrightarrow 2K_3[Fe(C_2O_4)_3] \cdot 2H_2O + H_2O$

3. Theoretical yield of oxalate complex (show calculations):

$$\frac{10 \text{ g}}{428 \text{ g/mol}} = 0.023 \text{ mol}$$

0.023 mol $(NH_4)_2[Fe(H_2O)_2(SO_4)_2] \cdot 6H_2O \longrightarrow$ 0.023 mol FeC_2O_4
0.023 mol $FeC_2O_4 \longrightarrow$ 0.023 mol $K_3[Fe(C_2O_4)_3] \cdot 3H_2O$
(0.023 mol $K_3[Fe(C_2O_4)_3] \cdot 3H_2O$)(491 g/mol) = 11 g

4. Experimental yield of oxalate complex 7.64 g

5. Percent yield of oxalate complex (show calculations):

$$\% \text{ yield} = \frac{\text{experimental yield}}{\text{theoretical yield}} \times 100 = \frac{7.64 \text{ g}}{11 \text{ g}} \times 100 = 69\%$$

Student yields range from 30% to 95%, depending on the particular complex prepared and its solubility. The chromium reactions are slow, and yields of the chromium complex are low if students work too rapidly.

6. Color and general appearance of complex:
green crystalline material

7. Describe the reactions of *cis*- and *trans*-$K[Cr(C_2O_4)_2(H_2O)_2]$ with NH_3 and the reverse reactions with HCl using chemical equations. List any observations, such as color changes or apparent solubilities.
See Figure 34.2 of Laboratory Manual.

8. Is your complex soluble in H_2O? Yes Alcohol? No Acetone? Yes

QUESTIONS

1. Sodium trioxalatocobaltate(III) trihydrate is prepared by the following reactions:

$$[Co(H_2O)_6]Cl_2 + K_2C_2O_4 \cdot H_2O \longrightarrow CoC_2O_4 + 2KCl + 7H_2O$$

$$2CoC_2O_4 + 4H_2O + H_2O_2 + 4Na_2C_2O_4 \longrightarrow 2Na_3[Co(C_2O_4)_3] \cdot 3H_2O + 2NaOH$$

What is the percent yield of $Na_3[Co(C_2O_4)_3] \cdot 3H_2O$ if 7.6 g is obtained from 12.5 g of $[Co(H_2O)_6]Cl_2$?

$$\frac{12.5 \text{ g } [Co(H_2O)_6]Cl_2}{237.93 \text{ g/mol}} = 0.0525 \text{ mol}$$

$$\text{mol } Na_3[Co(C_2O_4)_3] \cdot 3H_2O = \text{mol } [Co(H_2O)_6]Cl_2$$

$$\text{Theoretical yield} = (0.0525 \text{ mol}) \times \left(\frac{446.01 \text{ g}}{\text{mol}}\right) = 23.4 \text{ g}$$

$$\% \text{ yield} = \frac{7.6 \text{ g}}{23.4 \text{ g}} \times 100 = 32\%$$

2. Why are $K_3[Cr(C_2O_4)_3] \cdot 3H_2O$, $K_2[Cu(C_2O_4)_2] \cdot 2H_2O$, and $K_3[Fe(C_2O_4)_3] \cdot 3H_2O$ colored, whereas $K_3[Al(C_2O_4)_3] \cdot 3H_2O$ is colorless?

   Cr^{3+}, Cu^{2+}, and Fe^{3+} all have incomplete *d* orbitals characteristic of transition metals. A feature of incomplete *d* orbitals is that electronic transitions among them occur in the visible region of the spectrum. Al^{3+} has no *d* electrons and, as a consequence, does not absorb visible light.

3. What are the names of the following compounds?

$K_3[Cr(C_2O_4)_3] \cdot 3H_2O$	Potassium trioxalatochromate(III) trihydrate
$K_2[Cu(C_2O_4)_2] \cdot 2H_2O$	Potassium dioxalatocuprate(II) dihydrate
$K_3[Fe(C_2O_4)_3] \cdot 3H_2O$	Potassium trioxalatoferrate(III) trihydrate
$K_3[Al(C_2O_4)_3] \cdot 3H_2O$	Potassium trioxalatoaluminate trihydrate

4. What is the percent oxalate in each of the following compounds?

   A. $K_3[Cr(C_2O_4)_3] \cdot 3H_2O$ – FW = 487.42 amu
   B. $K_2[Cu(C_2O_4)_2] \cdot 2H_2O$ – FW = 342.28 amu
   C. $K_3[Fe(C_2O_4)_3] \cdot 3H_2O$ – FW = 491.27 amu
   D. $K_3[Al(C_2O_4)_3] \cdot 3H_2O$ – FW = 462.40 amu

A. $\%C_2O_4 = \frac{264.1}{487.4} \times 100 = 54.18\%$

C. $\%C_2O_4 = \frac{264.1}{491.3} \times 100 = 53.76\%$

B. $\%C_2O_4 = \frac{176.0}{342.3} \times 100 = 51.42\%$

D. $\%C_2O_4 = \frac{264.1}{462.4} \times 100 = 57.12\%$

Experiment 35

Oxidation-Reduction Titrations I: Determination of Oxalate

OBJECTIVE

To gain some familiarity with redox chemistry through analysis of an oxalate sample.

APPARATUS AND CHEMICALS

Apparatus

- 50-mL buret
- buret clamp
- 400-mL beakers (3)
- glass stirring rods
- ring stand
- balance
- weighing bottle
- thermometer
- Bunsen burner and hose

Chemicals

- an oxalate sample (either an unknown or one of the complexes from Experiment 34)
- 1.0 *M* H_2SO_4
- $Na_2C_2O_4$ (primary standard)
- ~0.02 *M* $KMnO_4$

DISCUSSION

Potassium permanganate reacts with oxalate ions to produce carbon dioxide and water in an acidic solution, and the permanganate ion is reduced to manganese(II) as follows:

$$5C_2O_4^{2-}(aq) + 2MnO_4^-(aq) + 16H^+(aq) \longrightarrow 10CO_2(g) + 8H_2O(l) + 2Mn^{2+}(aq) \quad [1]$$

Because this reaction proceeds slowly at room temperature, it is necessary to heat the solution gently to obtain satisfactory reaction rates. No indicators are necessary in permanganate titrations, since the end points are easily observed. The permanganate ion is intensely purple, whereas the manganese(II) ion is nearly colorless. The first slight excess of permanganate imparts a pink color to the solution, signaling that all of the oxalate has been consumed.

The balanced half-reactions associated with Equation [1] are:

Oxidation: $$C_2O_4^{2-}(aq) \longrightarrow 2CO_2(g) + 2e^- \quad [2]$$

Reduction: $$MnO_4^-(aq) + 8H^+(aq) + 5e^- \longrightarrow Mn^{2+}(aq) + 4H_2O(l) \quad [3]$$

In this experiment, you will standardize a $KMnO_4$ solution—that is, you will determine its molarity—by titrating it against a very pure sample of sodium oxalate, $Na_2C_2O_4$. You will then use your standardized $KMnO_4$ to determine the percentage of oxalate ion, $C_2O_4^{2-}$, either in an unknown sample or in the complex you prepared in Experiment 34. The basis of the determination is that the reagents react in a molar ratio of 5:2

$$5 \times (\text{mol } MnO_4^-) = 2 \times (\text{mol } C_2O_4^{2-}) \quad [4]$$

Measuring the volume of $KMnO_4$ that reacts with a known mass of sodium oxalate, $Na_2C_2O_4$, allows us to calculate the molarity of the $KMnO_4$ solution. Recall that the product of the volume in liters and molarity of a solution is moles, Equation [5]

$$M \times V_L = \text{moles} = \frac{mass}{molar\ mass} \quad [5]$$

Because volumes are measured in mL, a more convenient form of Equation [5] is

$$M \times V_{mL} = \text{mmol} = \frac{mass(10^3)}{molar\ mass} \quad [6]$$

where, V_{mL} is the volume in mL and mmol is millimoles (that is, 10^{-3} moles). Example 35.1 illustrates the calculation of molarity for a standardization; Example 35.2 illustrates analysis for oxalate.

EXAMPLE 35.1

What is the molarity of a $KMnO_4$ solution if 40.41 mL is required to titrate 0.2538 g of $Na_2C_2O_4$?

SOLUTION: The reaction proceeds according to Equation [1]. At the equivalence point:

$$\text{mmol } KMnO_4 = 2/5 \text{ mmol } Na_2C_2O_4$$

The number of mmoles $Na_2C_2O_4$ is

$$\frac{0.2538 \text{ g}}{134.0 \text{ g/mol}} \times \frac{10^3 \text{ mmol}}{1 \text{ mol}} = 1.894 \text{ mmol } Na_2C_2O_4$$

Now $M \times V_{mL}$ = the number of mmoles for $KMnO_4$. Thus,

$$M \times 40.41 \text{ mL} = 2/5\ (1.894 \text{ mmol})\ KMnO_4$$

and

$$M = \frac{0.7576 \text{ mmol}}{40.41 \text{ mL}} = 0.01875\ M\ KMnO_4$$

EXAMPLE 35.2

What is the percent oxalate, $C_2O_4^{2-}$, in a 1.429-g sample if 34.21 mL of 0.02000 M $KMnO_4$ solution is required for titration?

SOLUTION: We want to find:

$$\%\ C_2O_4^{2-} = \frac{\text{mass of } C_2O_4^{2-}}{\text{mass of sample}} \times 100$$

Thus, we need to know the mass of $C_2O_4^{2-}$ in the sample with a mass of 1.429 g. At the equivalence point,

$$\begin{aligned}\text{mmol } C_2O_4^{2-} &= 5/2 \text{ mmol } KMnO_4 \\ &= 5/2\ (0.0200 \text{ mmol/mL})(34.21 \text{ mL}) \\ &= 1.711 \text{ mmol } C_2O_4^{2-}\end{aligned}$$

The corresponding mass of $C_2O_4^{2-}$ is:

$$\begin{aligned}m &= 1.711 \text{ mmol } C_2O_4^{2-} \times \frac{1 \text{ mol } C_2O_4^{2-}}{10^3 \text{ mmol } C_2O_4^{2-}} \times \frac{88.02 \text{ g } C_2O_4^{2-}}{\text{mol } C_2O_4^{2-}} \\ &= 0.1506 \text{ g } C_2O_4^{2-}\end{aligned}$$

and

$$\%\ C_2O_4^{2-} = \frac{0.1506 \text{ g } C_2O_4^{2-}}{1.429 \text{ g}} \times 100 = 10.54\%$$

PROCEDURE

A. Preparation of $KMnO_4$ Solution

Potassium permanganate solutions are not stable over long periods of time, because they react slowly with the organic matter present even in most distilled water. The solution should be protected from heat and light as far as possible, because both induce decomposition, producing manganese dioxide, which seems to act as a catalyst for further decomposition. Because commercial samples of the permanganate usually contain some manganese dioxide, it is advisable to boil and filter the permanganate solutions before standardizing. Almost any form of organic matter will reduce permanganate solutions, so care must be taken to keep the solutions out of contact with rubber, filter paper, dust particles, and so on. For this reason permanganate solutions are filtered through sintered glass.

For this determination you will be provided with a potassium permanganate solution that is approximately 0.02 *M*, which will have to be standardized.

Instructor: Time may be saved if the students are provided with a standardized permaganate solution.

B. Standardization of Permanganate Solution—Sodium Oxalate Method

Several reducing agents can be used as primary standards for permanganate solutions, but sodium oxalate is very commonly used.

The oxidation of the oxalate ion is carried out in an acid solution that is maintained at a temperature between 80° and 90°C. This oxidation is catalyzed by the Mn^{2+} ion, which is a product of the oxidation-reduction reaction. The intense pink color of the permanganate ion may persist until several milliliters of the permanganate reagent have been added; however, when sufficient Mn^{2+} ions have been formed to catalyze the reaction, the pink color will suddenly disappear and will continue to do so with each drop of reagent until the equivalence point is reached, at which point one drop in excess will turn the solution pink.

The titration can be accomplished more quickly if we know the approximate volume of $KMnO_4$ required. Assume a 0.20-g sample of $Na_2C_2O_4$ is weighed out. This corresponds to:

$$\text{mmol } Na_2C_2O_4 = 0.20 \text{ g} \times \frac{1 \text{ mol}}{134.0 \text{ g}} \times \frac{10^3 \text{ mmol}}{1 \text{ mol}} = 1.5 \text{ mmol}$$

The reaction requires 2/5 mmol $KMnO_4$ for each mmol $Na_2C_2O_4$. If a 0.020 *M* $KMnO_4$ solution is used, the number of mL required in a titration is (see Equations [4] and [5]):

$$2/5 \text{ (1.5 mmol)} = V_{\text{mL}} \times 0.020 \text{ mmol/mL}$$

$$V_m = 30 \text{ mL}$$

A volume of $KMnO_4$ solution up to within about 5.0 mL of this amount, that is, up to 25 mL, can be rapidly added to the hot, acidified oxalate solution without much danger of overrunning the end point. The permanganate solution is then added dropwise until one drop in excess turns the solution pink.

Procedure Weigh out from a weighing bottle, to the nearest 0.1 mg, triplicate portons of about 0.20 g each of pure sodium oxalate into 400-mL beakers. Run each trial as follows: Add about 250 mL of 1.0 *M* sulfuric acid;

Instructor: Caution the students to be careful with the thermometers.

then stir the solution (not with the thermometer), and warm the mixture until the oxalate has dissolved and the temperature has been brought to 80° to 90°C. After taking the initial reading of the permanganate in the buret (see Note 1), titrate with the permanganate, stirring constantly, while keeping the solution above 70°C at all times. Add the permanganate dropwise when near the end point, allowing each drop to decolorize before adding the next. The end point is reached when the faintest visible shade of pink remains even after the solution has been allowed to stand for 15 s (see Note 2). Again read the buret and record the volume of permanganate solution used.

From the data obtained, calculate the molarity of the potassium permanganate solution. The three results should agree within 3 ppt (parts per thousand). If they do not agree, titrate another sample.

Note 1 A 0.02 *M* permanganate solution is so deeply colored that the bottom of the meniscus is difficult to read. Thus, it is necessary to read the top of the surface of the solution, taking the usual precautions to have the eyes level with the solution surface when the reading is made. Remember, the buret reading can be estimated to ±0.01 mL.

Note 2 If there is any doubt whether or not an end point has been reached, it is good practice to take the buret reading and then add another drop of the permanganate. The development of an intense color indicates that the reading did correspond to the true end point.

Ordinarily, a permanganate end point is not permanent. Due to the action of dissolved reducing species in the water, the permanganate is slowly reduced and the color fades. A pink color that remains after stirring the solution for 15 s should be taken as the true end point.

C. The Determination of Oxalate in Your Oxalate Complex or Unknown

The unknown for this determination may be either one that your instructor provides, or the oxalate complex you prepared in Experiment 34, in which case read Note 3 before starting the procedure.

Procedure Into three separate 400-mL beakers, weigh out from a weighing bottle, to the nearest 0.1 mg, triplicate portions of about 0.2 to 0.3 g of the oxalate complex or about 0.5 g of unknown. Add about 250 mL of 1.0 *M* sulfuric acid. Titrate with the standardized permanganate solution as described above in Procedure B. Calculate the weight percent of $C_2O_4^{2-}$. A standard deviation of 0.3 or better represents acceptable results.

Note 3 Chromium oxalate complexes are difficult to analyze by titration with MnO_4^- because of their dark colors and slow rates of decomposition. In order to analyze them, first decompose the complex by boiling about 0.6 to 1 g of the complex in 10 mL of water and 10 mL of 3 *M* KOH for 15 min. Filter off the green solid $Cr(OH)_3$ and wash with distilled water. Combine the filtrate and washings, dilute to 25 mL in a volumetric flask, pipet 5 mL of this solution into a 400 mL beaker and titrate with MnO_4^- as above.

Waste Disposal Instructions Dispose of all oxalate- and metal-containing solutions in appropriate containers in the laboratory.

REVIEW QUESTIONS

Before beginning this experiment in the laboratory, you should be able to answer the following questions:

1. If 0.4586 g of sodium oxalate, $Na_2C_2O_4$, requires 34.53 mL of a $KMnO_4$ solution to reach the end point, what is the molarity of the $KMnO_4$ solution?

2. Titration of an oxalate sample gave the following percentages: 15.35%, 15.55%, and 15.65%. Calculate the average and the standard deviation.

3. Why does the solution decolorize on standing after the equivalence point has been reached?

4. Why is the $KMnO_4$ solution filtered and why should it not be stored in a rubber-stoppered bottle?

5. What volume of 0.100 *M* $KMnO_4$ would be required to titrate 0.36 g of $K_2[Cu(C_2O_4)_2] \cdot 2H_2O$?

6. Calculate the percent $C_2O_4^{2-}$ in each of the following: $H_2C_2O_4$, $Na_2C_2O_4$, $K_2C_2O_4$, and $K_3[Al(C_2O_4)_3] \cdot 3H_2O$.

7. If 28.70 mL of 0.0200 *M* $KMnO_4$ was required to titrate a 0.250-g sample of $K_3[Fe(C_2O_4)_3] \cdot 3H_2O$, what is the percent $C_2O_4^{2-}$ in the complex?

8. What is the percent purity of the complex in question 7?

NOTES AND CALCULATIONS

Name ______________________ Desk ______________________

Date ______________ Laboratory Instructor ______________________

Unknown no. or complex ______________________

REPORT SHEET | EXPERIMENT

Oxidation-Reduction Titrations I: Determination of Oxalate

35

B. Standardization of $KMnO_4$

Mass of $Na_2C_2O_4$	*Trial 1*	*Trial 2*	*Trial 3*
Weighing bottle initial mass	1.0872 g	1.0301 g	1.0063 g
Weighing bottle final mass	0.7670 g	0.7681 g	0.7690 g
Mass of $Na_2C_2O_4$	0.3202 g	0.2620 g	0.2373 g
Titration volume of $KMnO_4$			
Final reading	47.97 mL	39.72 mL	35.83 mL
Initial reading	0.06 mL	1.12 mL	0.41 mL
Volume of $KMnO_4$	47.91 mL	38.60 mL	35.42 mL
Calculations			
mMoles of $Na_2C_2O_4$	2.390×10^{-3}	1.955×10^{-3}	1.771×10^{-3}
Volume of $KMnO_4$	0.04791 L	0.03860 L	0.03542 L
mMoles of $KMnO_4$	9.560×10^{-4}	7.820×10^{-4}	7.084×10^{-4}
Molarity of $KMnO_4$	0.01995 *M*	0.02026 *M*	0.02000 *M*

Average molarity 0.02007

Standard deviation (show calculations) 0.0001

$$\frac{0.01995 + 0.02026 + 0.02000}{3} = 0.02007$$

$$SD = \sqrt{\frac{(1.2 \times 10^{-4})^2 + (9.0 \times 10^{-5})^2 + (7 \times 10^{-5})^2}{2}} = 1.2 \times 10^{-4}$$

C. Analysis of Oxalate Complex or Unknown

Mass of sample	*Trial 1*	*Trial 2*	*Trial 3*
Weighing bottle initial mass	2.1536 g	2.4870 g	2.3830 g
Weighing bottle final mass	0.7243 g	0.7250 g	0.7350 g
Mass of sample	1.4293 g	1.7620 g	1.6480 g
Titration			
Final reading	30.11 mL	36.89 mL	33.89 mL
Initial reading	0.31 mL	0.24 mL	0.06 mL
Volume of $KMnO_4$	29.80 mL	36.65 mL	33.83 mL

Calculations	*Trial 1*	*Trial 2*	*Trial 3*
Millimoles of oxalate, $C_2O_4^{2-}$	1.495	1.839	1.697
Mass of oxalate, $C_2O_4^{2-}$	0.1316 g	0.1619 g	0.1494 g
Percent oxalate, $C_2O_4^{2-}$	9.21%	9.19%	9.07%

Average percent oxalate 9.16% Standard deviation 0.065

If you analyzed your oxalate complex from Experiment 34, complete the following.

Theoretical percent oxalate in your complex (show calculations) not applicable

Determine the purity of your complex as $\frac{\text{Experimental \% oxalate}}{\text{Theoretical \% oxalate}} \times 100 = $ % purity.

QUESTIONS

1. Balance the following reactions:
   (a) $MnO_4^- + e^- + H^+ \longrightarrow Mn^{2+} + H_2O$; $MnO_4^- + 5e^- + 8H^+ \longrightarrow Mn^{2+} + 4H_2O$
   (b) $MnO_4^- + e^- + H^+ \longrightarrow MnO_2 + H_2O$; $MnO_4^- + 3e^- + 4H^+ \longrightarrow MnO_2 + 2H_2O$
   (c) $MnO_4^- + e^- + H^+ \longrightarrow Mn^{3+} + H_2O$; $MnO_4^- + 4e^- + 8H^+ \longrightarrow Mn^{3+} + 4H_2O$

2. How many grams of $KMnO_4$ are required to prepare 2.000 L of 0.0400 *M* $KMnO_4$ solution that is used in titrating $Na_2C_2O_4$?

   (0.0400 mol/L)(158.0 g/mol)(2.000 L) = 12.6 g $KMnO_4$

3. MnO_4^- reacts with Fe^{2+} in acid solution to produce Fe^{3+} and Mn^{2+}. Write a balanced equation for this reaction.

   $8H^+(aq) + MnO_4^-(aq) + 5Fe^{2+}(aq) \longrightarrow 5Fe^{3+}(aq) + Mn^{2+}(aq) + 4H_2O(l)$

Preparation of Sodium Bicarbonate and Sodium Carbonate

OBJECTIVE

To become acquainted with the chemistry involved in the production of $NaHCO_3$ and its conversion to Na_2CO_3.

APPARATUS AND CHEMICALS

Apparatus

pH paper
8-oz wide-mouthed bottles (2)
rubber tubing
ring stand
thistle tube or long-stem funnel
100-mL beaker
test tubes (3)
wood splints
glass tubing
Bunsen burner
utility clamp
250-mL Erlenmeyer flask
two-hole rubber stoppers (2)
funnel and filter paper

Chemicals

marble chips
$Ca(H_2PO_4)_2$ (solid)
6 *M* NH_3
50% ethanol
glycerol
6 *M* HCl
$(NH_4)_2CO_3$-NaCl saturated solution*
ice

DISCUSSION

Sodium bicarbonate ($NaHCO_3$), more properly called sodium hydrogen carbonate, and sodium carbonate (Na_2CO_3) are both important commercial chemicals. Sodium bicarbonate is commonly called baking soda, because it is used in cooking. It is the main ingredient in baking powders, where it serves as a source of carbon dioxide, which causes dough to rise. Most of the sodium bicarbonate manufactured is converted to sodium carbonate by mild heating:

$$2NaHCO_3(s) \xrightarrow{\Delta} Na_2CO_3(s) + H_2O(g) + CO_2(g) \quad [1]$$

Anhydrous sodium carbonate, commonly known as soda ash, is used in large amounts in the manufacture of glass, enamels, soap, and paper. It is also used in water softening. More than 7 million tons of sodium carbonate is produced annually in the United States alone.

Both sodium bicarbonate and sodium carbonate are prepared industrially by the Solvay process. The Solvay process uses the readily available and relatively inexpensive raw materials salt and limestone; it also uses ammonia, which is not inexpensive. Limestone, a rock composed mainly of $CaCO_3$, is heated to make carbon dioxide:

$$CaCO_3(s) \xrightarrow{\Delta} CaO(s) + CO_2(g) \quad [2]$$

*Prepared by adding 360 mL of water to a mixture of 55 g of $NaHCO_3$, 205 g of NaCl, and 240 mL of 7.4 *M* NH_3, stirring magnetically overnight and filtering by gravity.

Ammonia and the carbon dioxide from limestone are passed through a saturated sodium chloride solution at about 0°C. Under these conditions, sodium bicarbonate precipitates out as finely divided crystals that can be separated by filtration from the ammonium chloride solution. The overall reaction can be written as follows:

$$CO_2(g) + NH_3(g) + H_2O(l) + Na^+(aq) + Cl^-(aq) \longrightarrow NaHCO_3(s) + NH_4^+(aq) + Cl^-(aq) \quad [3]$$

To a large measure, the Solvay process is economically feasible because the ammonia can be recovered from the ammonium chloride solution and recycled. The calcium oxide (quicklime) produced from the thermal decomposition of limestone according to Equation [1] is used to regenerate the ammonia as follows:

$$2NH_4^+(aq) + CaO(s) \longrightarrow 2NH_3(g) + H_2O(l) + Ca^{2+}(aq) \quad [4]$$

Although it is not practical to carry out the entire process in your laboratory, the process may be illustrated in part by starting with a solution that is saturated with both NaCl and $(NH_4)_2CO_3$ and bubbling in carbon dioxide. The formation of sodium bicarbonate under these conditions may be represented by the following equation:

$$2Na^+(aq) + CO_2(g) + CO_3^{2-}(aq) + H_2O(l) \longrightarrow 2NaHCO_3(s) \quad [5]$$

In this experiment, you will make sodium bicarbonate and sodium carbonate and study some of the properties of these two substances.

PROCEDURE

Instructor: Illustrate how to safely insert the thistle tube through the rubber stopper to avoid injury.

A. Preparation of Sodium Bicarbonate

Set up the apparatus shown in Figure 36.1. **(CAUTION!** ***Take care to avoid injury while assembling the apparatus. Use glycerol to lubricate the thistle tube, glass tubing, and holes in the rubber stopper before inserting the glass into the rubber stoppers. Protect your hands with a towel.)*** Place 55 g of marble chips into the bottle that is fitted with the thistle tube and place 30 mL of water into the bottle that is clamped to the ring stand. BE CERTAIN *that the glass tubing extends down beneath the surface of the water.* Place 50 mL of a saturated solution of $(NH_4)_2CO_3$ that is also saturated with NaCl into the 250-mL Erlenmeyer flask. Ask your instructor to inspect your setup.

Through the thistle tube, add enough 6 *M* HCl to the bottle so that the lower end of the thistle tube extends beneath the surface of the acid. As the carbon dioxide forms and bubbles through the solution, shake the Erlenmeyer flask vigorously, splashing the solution so that it is thoroughly mixed with the gas. Allow the reaction to proceed until a large amount of solid is formed in the flask (about 90 min). Disconnect the flask, add 3 mL of 6 *M* NH_3, and filter the contents. The solid can be transferred completely to the filter funnel by rinsing the flask with the filtrate. Wash the solid with 20 mL of 50% ethanol and air dry. Divide your product into four equal portions for later use, as described below.

B. Properties of Sodium Bicarbonate

On the report sheet, describe the appearance of the sodium bicarbonate that you have prepared (1).

Dissolve one portion of your product in 5 mL of water. Measure the pH of the solution with pH paper and record its value (2). Account for the pH of the solution by means of an ionic equation (3).

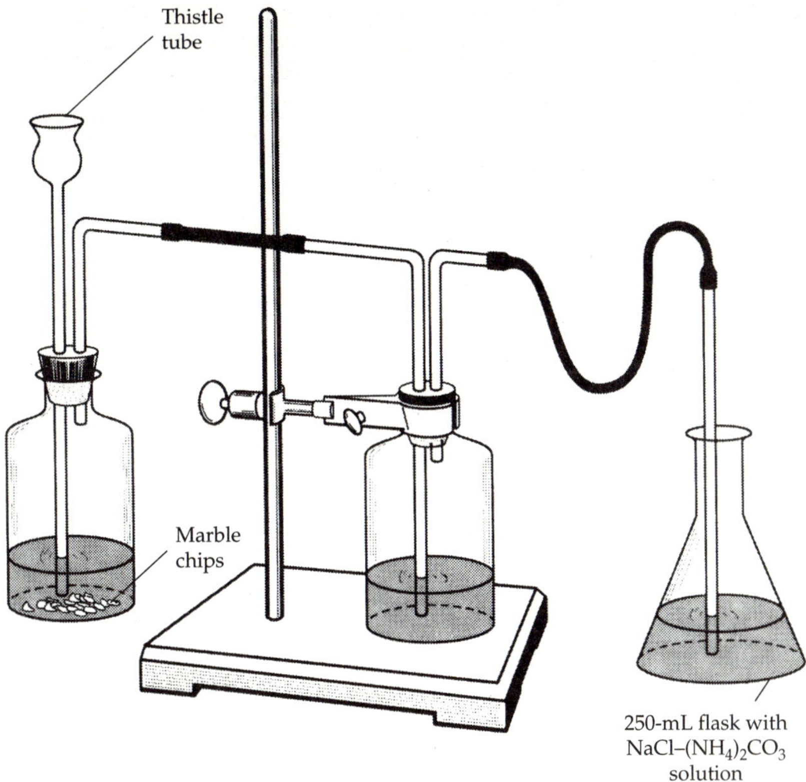

▲ **FIGURE 36.1** Preparation of sodium bicarbonate

Place a second portion of the sodium bicarbonate in the test tube and cautiously add about 2 mL of 6 *M* HCl. Test the gas with a burning splint. What is the gas? (4) Write an equation for the reaction (5). Based upon these observations, do you think that sodium bicarbonate would be suitable to neutralize acid that was spilled on your hand? Justify your answer (6).

C. Preparation of Sodium Carbonate

Place another portion of the sodium bicarbonate in a test tube and heat it. Insert a burning splint in the test tube. Write an equation for the reaction that occurs when sodium bicarbonate is heated (7). After the test tube and contents have cooled, add about 5 mL of water. Measure the pH of the solution and record the value (8). How does the pH of this solution compare with that of the sodium bicarbonate of part B? Which is more alkaline (9)? Write an ionic equation that accounts for the pH of the solution (10).

D. Baking Powders

Baking powders are used because they liberate carbon dioxide, which causes dough to "rise" or to be leavened. Baking powders are mixtures of the following three essential ingredients: (1) baking soda, which furnishes carbon dioxide; (2) an acid substance that furnishes hydrogen ions to react with sodium bicarbonate when water is added; and (3) an inert substance such as starch, to prevent intimate contact between ingredients (1) and (2) and to keep the mixture dry. The various brands of baking powders differ only in the nature of the acid ingredient. The most common acid sources are calcium dihydrogen phosphate, $Ca(H_2PO_4)_2$; potassium hydrogen tartrate (cream of tartar), $KHC_4H_4O_6$; and sodium alum, $NaAl(SO_4)_2 \cdot 12H_2O$.

Dissolve about 0.5 g of $Ca(H_2PO_4)_2$ in water. Measure and record the pH of the solution (11). Write an equation that accounts for the formation of the hydrogen ions from this substance (12).

Mix about 1.5 g of finely powdered $Ca(H_2PO_4)_2$ with about 0.5 g of your *dry* sodium bicarbonate in a 100-mL beaker. Add 5 mL of water to the mixture. Is there any evidence of a reaction? Record your observations (13). Write an equation that accounts for the reaction that occured (14).

REVIEW QUESTIONS

Before beginning this experiment in the laboratory, you should be able to answer the following questions:

1. Complete and balance the following equations:
   (a) $NaHCO_3(s) \xrightarrow{\Delta}$
   (b) $CaCO_3(s) \xrightarrow{\Delta}$
   (c) $CaO(s) + H_2O(l) \longrightarrow$
   (d) $NH_3(g) + H_2O(l) \longrightarrow$
   (e) $NH_4^+(aq) + CaO(s) \longrightarrow$
   (f) $H^+(aq) + HCO_3^-(aq) \longrightarrow$
2. How is the solubility of a gas related to its partial pressure above the solution?
3. How could you simply test for the presence of CO_3^{2-} in an unknown?
4. How could you test for the ammonium ion?
5. HCl and NaCl are both very soluble in water. The solubility of NaCl in aqueous HCl is much less than in pure water. Why?
6. How much $NaHCO_3$ could be prepared by the addition of excess CO_2 to a solution containing 15 g of NaCl and 10 g of $(NH_4)_2CO_3$ according to the reaction

$$2NaCl(aq) + CO_2(g) + (NH_4)_2CO_3(aq) + H_2O(l) \longrightarrow 2NaHCO_3(s) + 2NH_4Cl(aq)$$

7. Why should the bottom of the thistle tube extend beneath the surface of the hydrochloric acid?

Name ______________________ Desk ______________________

Date ______________ Laboratory Instructor ______________________

REPORT SHEET | EXPERIMENT

Preparation of Sodium Bicarbonate and Sodium Carbonate

36

B. Properties of Sodium Bicarbonate

1. Appearance of $NaHCO_3$ white, fine, crystalline powder
2. pH of $NaHCO_3$ solution 9.5
3. Ionic equation $HCO_3^-(aq) + H_2O(aq) \rightleftharpoons H_2CO_3(aq) + OH^-(aq)$
4. Identity of gas CO_2
5. $NaHCO_3(s) + HCl(aq) \longrightarrow NaCl(aq) + H_2O(l) + CO_2(g)$
6. Yes; it is mildly alkaline (pH = 9.5) and reacts with acids as in (5)

C. Preparation of Sodium Carbonate

7. $2NaHCO_3(s) \xrightarrow{\Delta} Na_2CO_3(s) + H_2O(g) + CO_2(g)$
8. pH of Na_2CO_3 11.5
9. Na_2CO_3 is more alkaline
10. $CO_3^{2-} + H_2O \rightleftharpoons HCO_3^- + OH^-$; $HCO_3^- + H_2O \rightleftharpoons H_2CO_3 + OH^-$; $H_2CO_3 \rightleftharpoons H_2O + CO_2$

D. Baking Powders

11. pH of $Ca(H_2PO_4)_2$ solution 3.0
12. $H_2PO_4^-(aq) \rightleftharpoons H^+(aq) + HPO_4^{2-}(aq)$
13. Evidence of reaction Bubbles formed; a gas, CO_2, evolved
14. $H^+(aq) + HCO_3^-(aq) \rightleftharpoons H_2O(l) + CO_2(g)$; because $H_2PO_4^-$ is a weak acid, the more correct answer is: $H_2PO_4^-(aq) + HCO_3^- \longrightarrow HPO_4^{2-}(aq) + H_2O(l) + CO_2(g)$

QUESTIONS

1. Dry-chemical fire extinguishers contain $NaHCO_3$. How might powered $NaHCO_3$ help extinguish a fire?
It smothers the fire by keeping air (oxygen) physically away, as well as forming a layer of CO_2. CO_2 is more dense than air and does not support combustion.

2. Complete and balance the following equations:
$KHC_4H_4O_6 + NaHCO_3 \longrightarrow NaKC_4H_4O_6 + H_2O + CO_2(g)$
$HC_2H_3O_2 + NaHCO_3 \longrightarrow NaC_2H_3O_2 + H_2O + CO_2(g)$

3. Why is a sodium carbonate solution more alkaline than a sodium bicarbonate solution?
CO_3^{2-} is a stronger Brønsted base than HCO_3^-; the conjugate acid of CO_3^{2-}, HCO_3^-, is a weaker acid than the conjugate acid, H_2CO_3, of HCO_3^-.

4. Consult a handbook and determine the solubility of the following in mol/L at 20°C:
(a) NH_4Cl ____5.6 *M*____ cold water and ____14.2 *M*____ hot water
(b) Na_2CO_3 ____0.67 *M*____ cold water and ____4.3 *M*____ hot water

Oxidation-Reduction Titrations II: Analysis of Bleach

OBJECTIVE

To show how redox reactions can be used to determine the amount of hypochlorite in household bleach.

APPARATUS AND CHEMICALS

Apparatus

50-mL burets (2)	balance
250-mL Erlenmeyer flasks (3)	6-in. test tubes (3)
500-mL Erlenmeyer flask	10- and 100-mL graduated cylinders
100-mL beaker	wash bottle
buret clamp	ring stand

Chemicals

3 *M* H_2SO_4	3 *M* KI
0.0100 *M* KIO_3	1 *M* $Na_2S_2O_3$
starch indicator (freshly prepared)	3% ammonium molybdate

DISCUSSION

Laundering white clothes to remove dirt and stains is an everyday chore and is usually accomplished with the aid of bleach. The effectiveness of a bleach to whiten and remove stains is related to its oxidizing (bleaching) strength. The use of detergents helps to remove grease by an emulsification action, while agitation helps loosen grime and dirt. Most liquid "chlorine" bleaches, such as Clorox or Purex, contain the hypochlorite ion, OCl^- as the oxidizing agent (note the last syllable in Clorox). Hypochlorite is generally present as the sodium salt, NaOCl or the calcium salt, $Ca(OCl)_2$. Nonchlorine bleaches that are used in washing colored as well as white fabrics utilize the oxidative properties of hydrogen peroxide.

This experiment illustrates how redox reactions can be used to quantitatively determine the amount of oxidizing agent in liquid hypochlorite bleaches.

Two redox reactions are involved in this experiment in the analysis of the oxidizing capacity of a liquid bleach. Initially you will add an excess of potassium iodide solution to the bleach. The iodide ions, I^-, are oxidized to iodine I_2 after the solution has been acidified:

$$HOCl(aq) + 2I^-(aq) + H^+(aq) \longrightarrow I_2(aq) + Cl^-(aq) + H_2O(l) \qquad [1]$$

(NOTE: In the presence of a strong acid, the hypochlorite ion is converted to hypochlorous acid. Why?) The iodine that is formed is then titrated with a standardized sodium thiosulfate, $Na_2S_2O_3$, solution that quantitatively reduces the iodine to iodide as follows:

$$2S_2O_3^{2-}(aq) + I_2(aq) \longrightarrow 2I^-(aq) + S_4O_6^{2-}(aq) \qquad [2]$$

Starch is used as the indicator for this reaction; the starch solution is not added until the dark brownish color, due to the iodine, has changed to a pale yellow. When the starch is added, the yellow color changes to a blue-black. The end point in the titration is reached when a drop of the thiosulfate causes the solution to become colorless. Starch and iodine (actually the triiodide ion, I_3^-) form a blue-black complex. If the starch is added too soon in the titration, the formation of the blue-black complex is not easily reversed making the end point very slow and difficult to detect.

$$I_3^- + \text{starch} \rightleftharpoons \underset{\text{blue-black}}{\text{starch} \cdot I_3^-\text{(complex)}} \qquad [3]$$

Adding Equations [1] and [2] yields Equation [4],

$$HOCl(aq) + 2S_2O_3^{2-}(aq) + H^+ \longrightarrow Cl^-(aq) + S_4O_6^{2-}(aq) + H_2O(l) \qquad [4]$$

which shows that for every mole of hypochlorite, two moles of thiosulfate are required. Thus, from the volume of standardized thiosulfate that is required to react with the liberated iodine and from the weight of bleach we can calculate the percentage oxidizing agent, by mass. We will assume the oxidizing agent to be NaOCl. This is illustrated in Example 37.1.

EXAMPLE 37.1

A 0.501-g sample of bleach was treated with an excess of KI. The iodine liberated required 10.21 mL of 0.0692 *M* $Na_2S_2O_3$ for titration. What is the percentage NaOCl in the bleach?

SOLUTION: The number of moles of $Na_2S_2O_3$ used in the titration is twice the number of moles of hypochlorite; or alternatively, the number of moles of hypochlorite that react is half the number of moles as $Na_2S_2O_3$.

$$\text{moles NaOCl} = {}^1/_2 \text{ moles } Na_2S_2O_3$$

$$= {}^1/_2 (0.0692 \text{ moles } Na_2S_2O_3/L)(0.01021\ L) = 3.53 \times 10^{-4} \text{ mol}$$

Changing this number of moles to grams:

$$\text{grams NaOCl} = 3.53 \times 10^{-4} \text{ mol NaOCl} \times \frac{74.5 \text{ g NaOCl}}{\text{mol NaOCl}} = 0.0263 \text{ g NaOCl}$$

And the percentage NaOCl is:

$$\%\text{NaOCl} = \frac{0.0263 \text{ g}}{0.501 \text{ g}} \times 100 = 5.25\%$$

Your $Na_2S_2O_3$ solution used to titrate the I_2 formed according to Equation [1] will be standardized using potassium iodate KIO_3 as the primary standard. As in the analysis of the bleach solution described above, two redox reactions, Equations [5] and [6] are involved in the standardization. Notice in Equation [5] that IO_3^- plays an analogous oxidative role to HOCl in Equation [1].

$$IO_3^-(aq) + 5I^-(aq) + 6H^+(aq) \longrightarrow 3I_2(aq) + 3H_2O(l) \qquad [5]$$

$$3I_2(aq) + 6S_2O_3^{2-}(aq) \longrightarrow 6I^-(aq) + 3S_4O_6^{2-}(aq) \qquad [6]$$

The standardization procedure is similar to the procedure described above for the analysis of bleach. An excess of KI is allowed to react with a known amount of KIO_3 in acidic solution. The iodine formed according to Equation [5] will be titrated with your $Na_2S_2O_3$ solution using starch as the indicator. Once again, the starch is not added until the iodine solution has turned a pale yellow. The end point in the titration is signaled by the disappearance of the blue-black color.

The stoichiometric relation between the primary standard KIO_3 and $Na_2S_2O_3$ can easily be seen by adding Equations [5] and [6] to give Equation [7]. (NOTE: Equation [6] is obtained by multiplying Equation [2] by the factor 3):

$$IO_3^-(aq) + 6S_2O_3^{2-}(aq) + 6H^+(aq) \longrightarrow I^-(aq) + 3S_4O_6^{2-}(aq) + 3H_2O(l) \quad [7]$$

Equation [7] shows that for each mole of KIO_3 used in the titration, six moles of $Na_2S_2O_3$ are required. Example 37.2 illustrates how the concentration of a $Na_2S_2O_3$ solution can be determined using KIO_3 as the primary standard.

EXAMPLE 37.2

What is the concentration of a $Na_2S_2O_3$ solution if 21.21 mL of the solution was required to titrate the iodine formed from 20.95 mL of 0.0100 *M* KIO_3 and excess KI?

SOLUTION: First determine the number of moles of KIO_3 that react:

$$\text{moles } KIO_3 = (0.02095 \text{ L})\,(0.0100 \text{ mol } KIO_3/\text{L}) = 2.10 \times 10^{-4} \text{ mol}$$

According to Equation [7], the number of moles of $Na_2S_2O_3$ that react is six times the number of moles of KIO_3:

$$\text{moles } Na_2S_2O_3 = 6 \text{ (moles } KIO_3)$$
$$= 6 \times 2.10 \times 10^{-4} \text{ mol} = 1.26 \times 10^{-3} \text{ mol}$$

Hence, the molarity of the $Na_2S_2O_3$ is:

$$\text{molarity } Na_2S_2O_3 = \frac{1.26 \times 10^{-3} \text{ mol}}{0.02121 \text{ L}} = 0.0594\ M$$

In this experiment, you will first standardize a sodium thiosulfate solution and then use it to analyze a liquid bleach.

PROCEDURE

You can perform this experiment most expeditiously if you procure the following solutions at the beginning of the laboratory period:

3 *M* H_2SO_4	35 mL	1/4 of test tube of 3% ammonium molybdate
0.0100 *M* KIO_3	50 mL	
1 *M* $Na_2S_2O_3$	15 mL	1/2 test tube starch solution
distilled water	400 mL	
3 *M* KI	15 mL	

The flask or test tube for each solution *must* be clean and labeled. Each clean vessel should be rinsed with a small amount of the required solution before filling except for the flask for the $Na_2S_2O_3$. It should be rinsed with distilled water. Avoid using the 250-mL Erlenmeyer flask, for it will be needed as a titration flask. Dispose of the chemicals as instructed.

Instructor: Give waste-disposal instructions.

A. Standardization of 0.05 *M* $Na_2S_2O_3$ Solution

Prepare about 300 mL of 0.05 *M* $Na_2S_2O_3$ solution by diluting 15 mL of 1 *M* $Na_2S_2O_3$ solution to about 300 mL with distilled water. Be careful to thoroughly mix the solution. The approximately 0.05 *M* $Na_2S_2O_3$ will be standardized with potassium iodate. Rinse and fill a buret with the standard

KIO_3. Rinse and fill a second buret with the 0.05 *M* sodium thiosulfate to be standardized.

To carry out the standardization, first accurately dispense about 15 mL of the KIO_3 solution into a 250 mL Erlenmeyer flask. Record the initial and final reading of the buret to the nearest 0.02 mL. Add 25 mL of distilled water and 3 mL of 3 *M* KI to the flask and swirl the flask. Add 2 mL of 3 *M* H_2SO_4 and swirl the flask. A deep brown color should appear, indicating the presence of iodine. After recording the initial buret reading, immediately* begin titration with the sodium thiosulfate solution. As titrant is added, the color of the solution in the flask will fade to light brown and then light yellow. When the solution reaches a light yellow, add 0.5 mL of starch indicator and mix. The solution will turn a dark blue-black color. Slowly continue to add titrant drop by drop. The dark blue-black color will fade and when the solution becomes a transparent bright blue the end point is generally a drop or two away. The end point is the transition from a blue to a colorless solution. Record the final buret reading. Repeat the titration two more times. Calculate the molarity of the $Na_2S_2O_3$ for the three trials and then calculate the average molarity and standard deviation.

(NOTE: Solid iodine crystals may form during each titration, as the iodine concentration exceeds the solubility product. All of the iodine must be redissolved and reacted before the titration is complete. As the titration proceeds and the solution becomes pale yellow, the flask should be swirled until all of the iodine is dissolved. Over time, the iodine will clump together and become more and more difficult to dissolve. For this reason, the titrations should be performed as rapidly as practical.)

B. Determination of the Oxidizing Capacity of an Unknown Liquid Bleach

The oxidizing capacity of an unknown liquid bleach will be determined in the following way: Place a clean test tube into a 100-mL beaker, weigh, and record the mass. Add about 0.5 mL of bleach to the test tube. **(CAUTION: *Liquid bleaches containing sodium hypochlorite are corrosive to the skin. Take care when handling the liquid bleach not to get any on yourself. If you do, immediately wash the area with a copious amount of water.*)** Re-weigh and record the mass. The total mass of bleach should be 0.4 to 0.6 g. Pour the bleach into a 250-mL titration flask and rinse the test tube several times with distilled water to ensure that all of the bleach is transferred (the total volume of water added to the flask should be about 25 mL, including the water used to rinse the test tube). Add 3 mL of 3 *M* KI to the flask and swirl the flask. Add 2 mL of 3 *M* H_2SO_4 and swirl the flask. Add 5 drops of 3% ammonium molybdate catalyst immediately after the acid is added. The molybdate ion catalyzes the reaction between the iodide and oxidizing agent. Proceed with the titration as in the standardization of the $Na_2S_2O_3$ solution above.

Perform two more titrations. From the concentration and volume of the added sodium thiosulfate solution used to titrate the bleach, calculate the

---

*Iodide is oxidized by oxygen in air:

$$4I^-(aq) + O_2(g) + 4H^+(aq) \rightarrow 2I_2(aq) + 2H_2O(l)$$

The reaction is slow in neutral solution but is faster in acid and is accelerated by sunlight. After the solution has been acidified it must be titrated immediately. Moreover, after reaching the end point the solution may darken on standing for an extended length of time.

mass of sodium hypochlorite, NaOCl, present. For each titration, calculate the strength of the bleach as the effective percentage mass of the bleach that is sodium hypochlorite. Report the average percentage mass of sodium hypochlorite in the bleach. This is the oxidizing capacity of the bleach. Calculate the standard deviation.

REVIEW QUESTIONS

Before beginning this experiment in the laboratory, you should be able to answer the following questions:

1. Write the formula for the salts that are oxidizing agents in chlorine bleaches.
2. Complete and balance the following equations
   (a) $HOCl + I^- + H^+ \longrightarrow$
   (b) $S_2O_3^{2-} + I_2 \longrightarrow$
3. Is the iodate ion, IO_3^-, an oxidizing or reducing agent?
4. In the standardization process, when the iodate-iodide solution is acidified with H_2SO_4, it should be titrated immediately. Why?
5. What is the color of the iodate-iodide solution being titrated with $Na_2S_2O_3$
   (a) before the starch solution is added?
   (b) after the starch solution is added?
   (c) at the end point?
6. How many moles of $Na_2S_2O_3$ react with each mole of KIO_3 used in the standardization?
7. If 16.92 mL of a $Na_2S_2O_3$ solution were required to titrate the iodine formed from 20.95 mL of a 0.0111 *M* KIO_3 solution and excess KI, what is the molarity of the $Na_2S_2O_3$ solution?
8. What is meant by the term *standard solution*?
9. What are the oxidation states of
   (a) chlorine in OCl^-?
   (b) sulfur in $Na_2S_2O_3$?

NOTES AND CALCULATIONS

Name ______________________ Desk ______________

Date ______________ Laboratory Instructor ______________

Unknown no. ______________

REPORT SHEET | EXPERIMENT

Oxidation-Reduction Titrations II: Analysis of Bleach

37

A. Standardization of 0.05 *M* $Na_2S_2O_3$ Solution

	Trial 1	*Trial 2*	*Trial 3*
Volume of KIO_3			
Final buret reading	15.43 mL	30.20 mL	48.25 mL
Initial buret reading	0.31 mL	15.45 mL	32.96 mL
mL KIO_3 used	15.12 mL	14.75 mL	15.29 mL
Volume of $Na_2S_2O_3$			
Final buret reading	17.05 mL	32.96 mL	49.09 mL
Initial buret reading	0.94 mL	17.05 mL	32.96 mL
mL $Na_2S_2O_3$ used	16.11 mL	15.91 mL	16.13 mL
Molarity of $Na_2S_2O_3$	0.0563 *M*	0.0556 *M*	0.0569 *M*

Average molarity 0.0563 Standard deviation ±0.00047

Show Calculations

$$\text{moles } KIO_3 = (0.01512 \text{ L})(0.0100 \text{ mol } KIO_3 / \text{L}) = 1.51 \times 10^{-4} \text{ mol}$$

$$\text{moles } Na_2S_2O_3 = 6 \times \text{moles } KIO_3 = 6 \times 1.51 \times 10^{-4} \text{ mol} = 9.07 \times 10^{-4} \text{ mol}$$

$$M\ Na_2S_2O_3 = 9.07 \times 10^{-4} \text{ mol } Na_2S_2O_3 / 0.01611 \text{ L} = 0.0563\ M$$

$$\text{Average } M = (0.0563 + 0.0556 + 0.0569) / 3 = 0.0563\ M$$

$$SD = \sqrt{\frac{(0.0)^2 + (3 \times 10^{-4})^2 + (6 \times 10^{-4})^2}{2}} = 4.7 \times 10^{-4}$$

B. Determination of the Oxidizing Capacity of an Unknown Liquid Bleach

	Trial 1	*Trial 2*	*Trial 3*
Mass of beaker, test tube, and bleach	56.143 g	56.567 g	55.655 g
Mass of beaker and test tube	55.690 g	56.111 g	55.125 g
Mass of unknown bleach	0.453 g	0.456 g	0.530 g
Volume of $Na_2S_2O_3$			
Final buret reading	5.52 mL	10.95 mL	17.05 mL
Initial buret reading	0.40 mL	5.52 mL	10.95 mL
mL $Na_2S_2O_3$	5.12 mL	5.43 mL	6.10 mL
Mass NaOCl	0.0107 g	0.0114 g	0.0128 g
Percent NaOCl	2.37%	2.49%	2.41%

Show Calculations

$$\text{moles NaOCl} = \tfrac{1}{2}\text{ moles }Na_2S_2O_3$$

$$= \tfrac{1}{2}\,(0.0563\text{ mol / L})\,(0.00512\text{ L}) = 1.44 \times 10^{-4}\text{ mol NaCl}$$

$$\text{mass NaOCl} = (1.44 \times 10^{-4}\text{ mol})\,(74.5\text{ g NaOCl / mol}) = 0.0107\text{ g NaOCl}$$

$$\%\text{NaOCl} = (0.0107\text{ g / }0.453\text{ g})\,100 = 2.37\%$$

$$\text{Average}\% = (2.37 + 2.49 + 2.41) / 3 = 2.42\%$$

$$\text{SD} = \sqrt{\frac{(0.05)^2 + (0.07)^2 + (0.02)^2}{2}} = 0.062$$

QUESTIONS

1. In the standardization of the $Na_2S_2O_3$ solution, if you did not titrate the liberated iodine immediately, how would this likely affect the value of the molarity you calculate? Explain.

   In addition to the I_2 formed by the reaction of KIO_3 and KI, additional I_2 may be formed due to air oxidation of I^- in the acidic solution. More $Na_2S_2O_3$ solution would be used in the titration than theoretically should be. Thus, the calculated molarity of the $Na_2S_2O_3$ would be less than the true value.

2. Some cleansers contain bromate salts as the oxidizing agent. The bromate ion, BrO_3^-, oxidizes the iodide ion as follows:

$$BrO_3^- + 6I^- + 6H^+ \longrightarrow 3I_2 + Br^- + 3H_2O$$

   On a molar basis, which is a more effective bleaching agent, $KBrO_3$ or NaOCl?

   On a molar basis, $KBrO_3$ is more effective, because one mole of $KBrO_3$ oxidizes six moles of I^-, whereas one mole of NaOCl only oxidizes two moles of I^-.

NOTES AND CALCULATIONS

Experiment 38

Molecular Geometry: Experience with Models

OBJECTIVE

To become familiar with the three-dimensional aspects of organic molecules.

MATERIALS

Prentice-Hall Molecular Model Set for General and Organic Chemistry

DISCUSSION

Organic compounds are extremely numerous—in fact, there are approximately 2×10^6 known organic compounds. The chemical and physical properties of these compounds depend on what elements are present, how many atoms of each element are present, and how these atoms are arranged in the molecule. Molecular formulas often, but not always, permit one to distinguish between two compounds. For example, even though there are eight atoms in both C_2H_6 and C_2H_5Cl, we know immediately that these are different substances on the basis of their molecular formulas. Similarly, inspection of the molecular formulas C_2H_6 and C_3H_8 reveals that these are different compounds. However, there are many substances that have identical molecular formulas but are completely different compounds. Consider the molecular formula C_2H_6O. There are two compounds that correspond to this formula: ethyl alcohol and dimethyl ether. While the molecular formula gives no clue as to which compound one may be referring, examination of the *structural formula* immediately reveals a different arrangement of atoms for these substances:

```
   H   H                  H         H
   |   |                  |         |
H—C—C—O—H            H—C—O—C—H
   |   |                  |         |
   H   H                  H         H
```

Ethyl alcohol   Dimethyl ether

In addition, when molecular models (ball-and-stick type) are used, trial and error will show that there are just two ways that two carbons, six hydrogens, and one oxygen can be combined to form molecules. Compounds that have the same molecular formula but different structural formulas are termed *isomers*. This difference in molecular structure results in differences in chemical and physical properties of isomers. In the case of ethyl alcohol and dimethyl ether, whose molecular formulas are C_2H_6O, these differences are very pronounced (Table 38.1). In other cases, the differences may be more subtle.

TABLE 38.1 **Properties of Ethyl Alcohol and Dimethyl Ether**

Property	Ethyl Alcohol	Dimethyl ether
Boiling point	78.5°C	−24°C
Melting point	−117°C	−139°C
Solubility in H_2O	Infinite	Slight
Behavior toward sodium	Reacts vigorously, liberating hydrogen	No reaction

The importance of the use of structural formulas in organic chemistry becomes evident when we consider the fact that there are 35 known isomers corresponding to the formula C_9H_{20}! For the sake of convenience, *condensed structural formulas* are often used. The structural and condensed structural formulas for ethyl alcohol and dimethyl ether are

```
                          H   H               H         H
                          |   |               |         |
Structural formulas:  H — C — C — O — H   H — C — O — C — H
                          |   |               |         |
                          H   H               H         H
```

Condensed structural formulas: CH_3CH_2OH (Ethyl alcohol) CH_3OCH_3 (Dimethyl ether)

A short glance at these formulas readily reveals their difference. The compounds differ in their **connectivity.** The atoms in ethyl alcohol are connected or bonded in a different sequence from those in dimethyl ether.

You must learn to translate these condensed formulas into three-dimensional mental structures and to translate structures represented by molecular models to condensed structural formulas.

PROCEDURE

Construct models of the molecules as directed and determine the number of isomers by trial and error as directed below. Use a black ball with four holes for carbon, a white ball with one hole for hydrogen, and a green ball with one hole for chlorine. Answer all questions on the report sheet at the end of this experiment.

A. Methane

Construct a model of methane, CH_4. Place the model on the desk top and note the symmetry of the molecule. Note that the molecule looks the same regardless of which three hydrogens are resting on the desk. All four hydrogens are said to be *equivalent*. Now, grasp the top hydrogen and tilt the molecule so that only two hydrogens rest on the desk and the other two are in a plane parallel to the desk top (Figure 38.1). Now imagine pressing this methane model flat onto the desk top. The resulting imaginary projection in the plane of the desk is the conventional representation of the structural formula of methane:

```
    H
    |
H — C — H
    |
    H
```

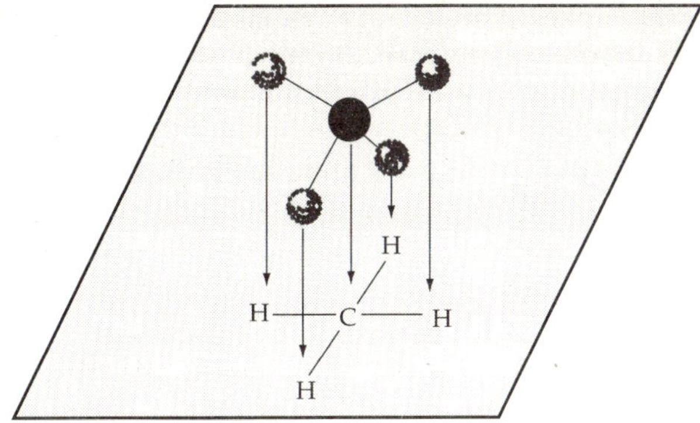

▲ **FIGURE 38.1** Model of methane.

Replace one of the hydrogen atoms with a chlorine atom to construct a model of chloromethane (or methyl chloride), CH_3Cl. Replace a second hydrogen atom by a chlorine atom to make dichloromethane, CH_2Cl_2. Convince yourself that the two formulas

$$\begin{array}{ccc} & H & \\ & | & \\ Cl- & C & -Cl \\ & | & \\ & H & \end{array} \quad \text{and} \quad \begin{array}{ccc} & Cl & \\ & | & \\ Cl- & C & -H \\ & | & \\ & H & \end{array}$$

represent the same three-dimensional structure and are not isomers. Replace another hydrogen to make $CHCl_3$, chloroform (or trichloromethane). Finally, make CCl_4, carbon tetrachloride.

B. Ethane

Make a model of ethane, C_2H_6, from your model of CH_4 by replacing one of the hydrogens by a CH_3 unit; the $—CH_3$ unit is called the *methyl group*. Note that the hydrogens of ethane are all equivalent. Replace one of the hydrogens in your ethane model with a chlorine. Does it matter which hydrogen you replace? How many compounds result? Now examine your model of C_2H_5Cl and note how many different hydrogen atoms are present. If another hydrogen of C_2H_5Cl is replaced with a chlorine atom to yield $C_2H_4Cl_2$, how many isomers would result? Does the formula $C_2H_4Cl_2$ distinguish the possible molecules corresponding to this formula? Write the structural and condensed structural formulas for all isomers of $C_2H_4Cl_2$ and assign the IUPAC names to each.

C. Propane

From your model of ethane, construct a molecular model of propane, C_3H_8, by replacing one of the hydrogen atoms with a methyl group, $—CH_3$. Examine your model of propane and note how many different hydrogen atoms are present in propane. If one of the hydrogens of propane is replaced with chlorine, how many isomers of C_3H_7Cl are possible? Write them and give their names. How many isomers are there corresponding to the formula $C_3H_6Cl_2$? Write their formulas and name them. Determine for yourself that there are five isomers with the formula $C_3H_5Cl_3$. Write both the structural and condensed structural formulas and IUPAC names for these isomers. By this stage in this experiment, you should realize that a systematic approach is most useful in determining the number of isomers for a given formula.

D. Butane

The formula of butane is C_4H_{10}. From your model of propane, C_3H_8, construct all of the possible isomers of butane by replacing a hydrogen atom with the methyl group, —CH_3. How many isomers of butane are there? List their structural formulas and IUPAC names. There are four isomers corresponding to the formula C_4H_9Cl. Write their structural formulas and name them. How many isomers of $C_4H_8Cl_2$ are there? Use your models to help answer this question. Write and name all of the isomers of $C_4H_8Cl_2$.

E. Pentane

Use your models in a systematic manner to determine how many isomers there are for the formula C_5H_{12}. Write their structural formulas and name them. Write and name all isomers for the formula $C_5H_{11}Cl$.

F. Cycloalkanes

Cycloalkanes corresponding to the formula C_nH_{2n} exist. Without using too much force, try to construct models of cyclopropane, C_3H_6; cyclobutane, C_4H_8; cyclopentane, C_5H_{10}; and cyclohexane, C_6H_{12}. Although cyclopropane and cyclobutane exist, would you anticipate these to be highly stable molecules? How many isomers of 1,2-dichlorocyclopentane are there? The answer is not obvious. There are three. If you cannot determine for yourself that there are three, check with your laboratory instructor. This is another aspect of isomerization that is most significant in biochemical systems. Although you may think this is trivial, such differences are of utmost importance in nature!

G. Alkenes

Using two of the longer, narrower, and more flexible bonds, construct a model of ethene (ethylene), C_2H_4. Note the rigidity of the molecule; note that there is no rotation about the carbon-carbon double bond as there is in the case of carbon-carbon single bonds such as in ethane or propane. How many isomers are there corresponding to the formulas C_2H_3Cl? $C_2H_2Cl_2$? Draw the formulas and name them.

REVIEW QUESTIONS

Before beginning this experiment in the laboratory, you should be able to answer the following questions:

1. Distinguish between molecular and structural formulas.
2. What is a condensed structural formula? Give an example.
3. What is the meaning of the term *isomer*?
4. Why should the properties of structural isomers differ?
5. Draw the structural formulas for ethane and propane.

6. Distinguish between molecular and empirical formulas.
7. The molecular formula of benzene is C_6H_6. What is the empirical formula of benzene?
8. Distinguish between geometric and structural isomers.
9. Carbon has a valence of 4, oxygen 2, and hydrogen 1. How many compounds of C, H, and O containing only one carbon and one oxygen can you make? Draw their structures.
10. Draw Lewis electron-dot formulas for the compounds in question 9 and name them.

NOTES AND CALCULATIONS

Name ____________________ Desk ____________________

Date ____________________ Laboratory Instructor ____________________

REPORT SHEET | EXPERIMENT

Molecular Geometry: Experience with Models | 38

A. Methane

Write the structure for and name each of the chloromethanes.

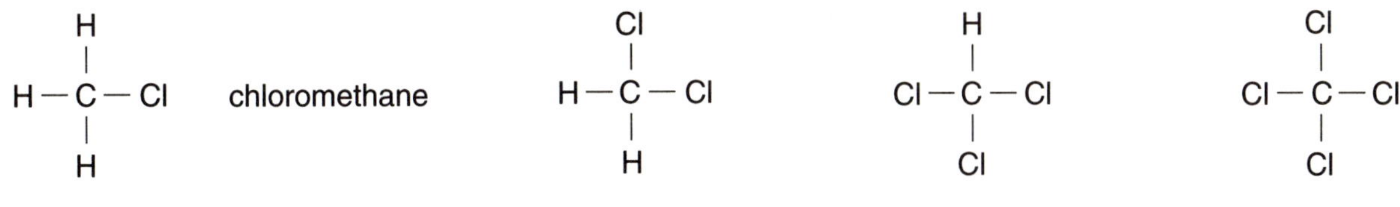

dichloromethane

trichloromethane
chloroform

tetrachloromethane
carbon tetrachloride

B. Ethane

Write the structures for C_2H_5Cl and $C_2H_4Cl_2$, and name each compound.

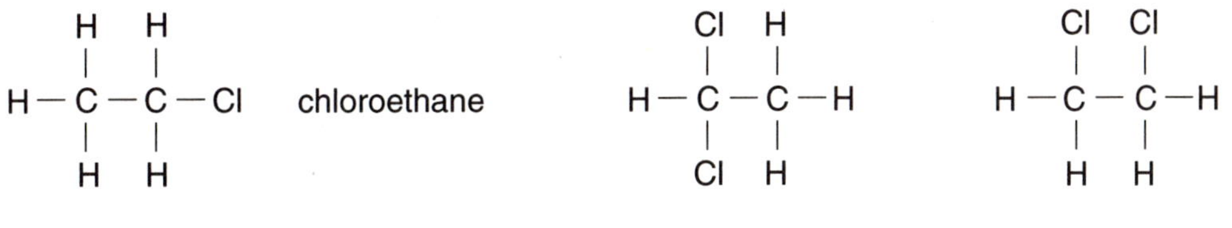

1,1-dichloroethane

1,2-dichloroethane

C. Propane

1. Write the structural and condensed formulas as well as the names for all isomers of C_3H_7Cl and $C_3H_6Cl_2$.

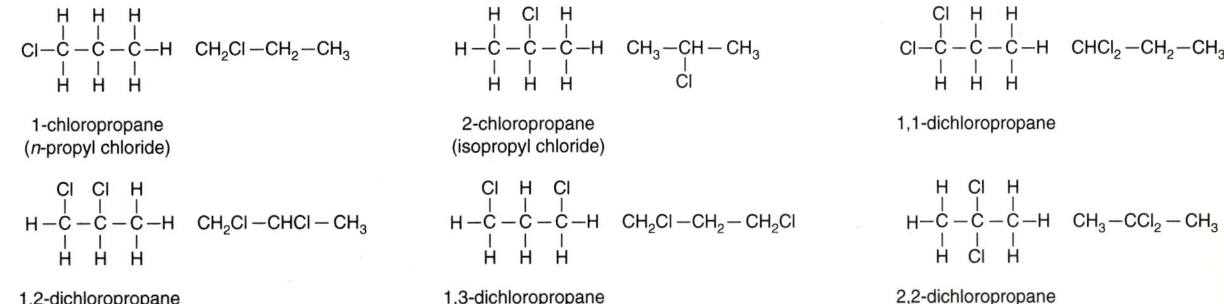

2. Write the condensed and structural formulas, as well as the names for all the isomers of $C_3H_5Cl_3$.

```
    H   H   Cl                                     H   Cl  Cl
    |   |   |                                      |   |   |
H - C - C - C - Cl   CH3-CH2-CCl3              H - C - C - C - Cl   CH3-CH-CHCl2
    |   |   |                                      |   |   |               |
    H   H   Cl                                     H   H   H               Cl
```

1,1,1-trichloropropane — $CH_3{-}CH_2{-}CCl_3$

1,1,2-trichloropropane — $CH_3{-}CH(Cl){-}CHCl_2$

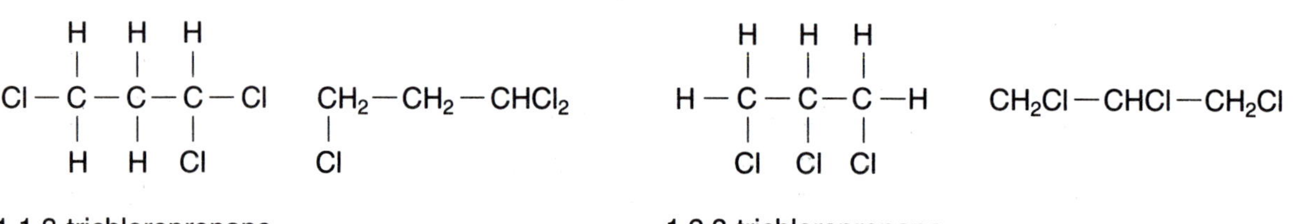


```
    H   Cl  H
    |   |   |
H - C - C - C - H      CH2Cl-CCl2-CH3
    |   |   |
    Cl  Cl  H
```

$CH_2Cl{-}CCl_2{-}CH_3$

1,2,2-trichloropropane

D. Butane

1. Write the structural formulas and names for all the butanes.

```
    H   H   H   H
    |   |   |   |
H - C - C - C - C - H
    |   |   |   |
    H   H   H   H
```

butane

```
    H   H   H
    |   |   |
H - C - C - C - H
    |   |   |
    H   |   H
    H - C - H
        |
        H
```

2-methylpropane

2. Give the structural formulas and names for all the isomers of
   (a) C_4H_9Cl

```
    Cl  H   H   H
    |   |   |   |
H - C - C - C - C - H
    |   |   |   |
    H   H   H   H
```

1-chlorobutane

```
    H   Cl  H   H
    |   |   |   |
H - C - C - C - C - H
    |   |   |   |
    H   H   H   H
```

2-chlorobutane

```
    H   Cl  H
    |   |   |
H - C - C - C - H
    |   |   |
    H   |   H
    H - C - H
        |
        H
```

2-chloro-2-methyl-propane

```
     H   H   H
     |   |   |
Cl - C - C - C - H
     |   |   |
     H   |   H
     H - C - H
         |
         H
```

1-chloro-2-methyl-propane

(b) $C_4H_8Cl_2$

- $Cl-CHCl-CH_2-CH_2CH_3$ — 1,1-dichlorobutane
- $CH_2Cl-CHCl-CH_2-CH_3$ — 1,2-dichlorobutane
- $CH_2Cl-CH_2-CHCl-CH_3$ — 1,3-dichlorobutane
- $CH_2Cl-CH_2-CH_2-CH_2Cl$ — 1,4-dichlorobutane
- $CH_3-CCl_2-CH_2-CH_3$ — 2,2-dichlorobutane
- $CH_3-CHCl-CHCl-CH_3$ — 2,3-dichlorobutane
- $Cl-CHCl-CH(CH_3)-CH_3$ — 1,1-dichloro-2-methylpropane
- $Cl-CH_2-CCl(CH_3)-CH_3$ — 1,2-dichloro-2-methylpropane
- $Cl-CH_2-CH(CH_3)-CH_2Cl$ — 1,3-dichloro-2-methylpropane

E. Pentane

1. Write the structural formulas and names for all the isomers of C_5H_{12}.

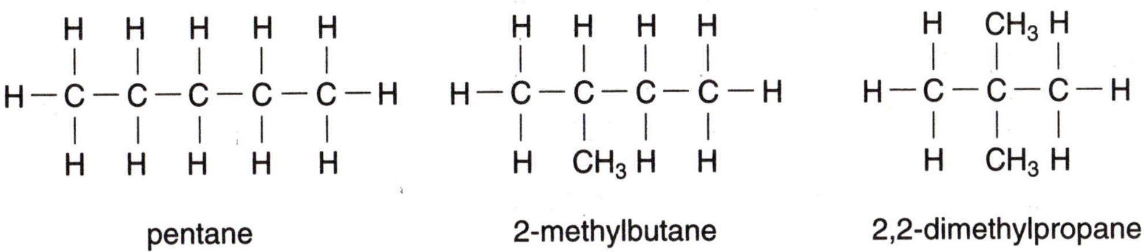

2. Give the structural formula and names for all the isomers of $C_5H_{11}Cl$.

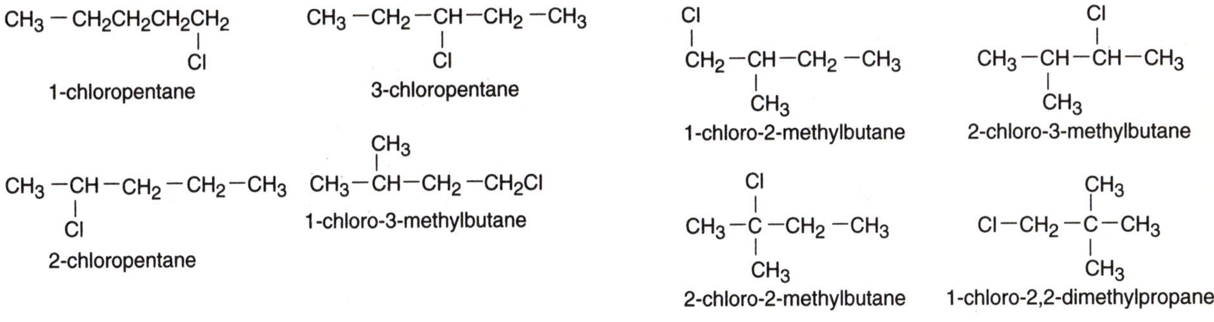

F. Cycloalkanes

Using the following format for the structure of cyclopentane, give the structures of all isomers of 1,2-dichlorocyclopentane and 1,3-dichlorocyclopentane:

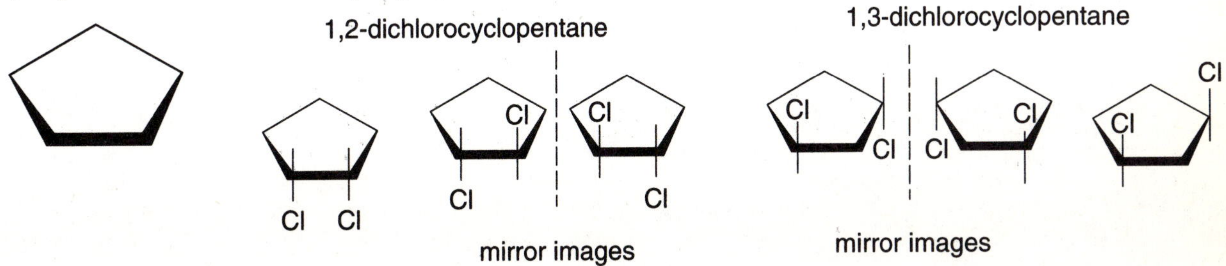

G. Alkenes

Give the structures and names of all isomers of $C_2H_2Cl_2$.

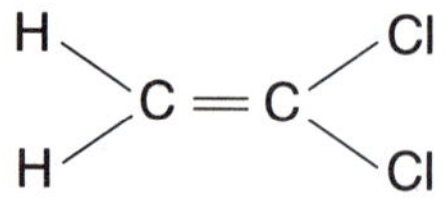

$$\mathrm{H_2C{=}CCl_2}$$

1,1-dichloroethene

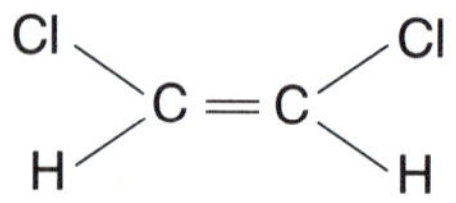

$$\mathrm{ClHC{=}CHCl}\ (\text{Cl, Cl same side})$$

cis-1,2-dichloroethene

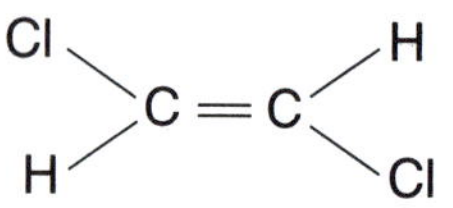

$$\mathrm{ClHC{=}CHCl}\ (\text{Cl, Cl opposite sides})$$

trans-1,2-dichloroethene

QUESTIONS

1. Write and name all isomers for the alkane C_6H_{14}.

$CH_3CH_2CH_2CH_2CH_2CH_3$ — hexane

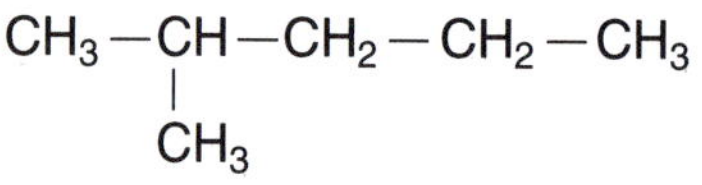

$CH_3-CH(CH_3)-CH_2-CH_2-CH_3$ — 2-methylpentane

$CH_3-CH_2-CH(CH_3)-CH_2-CH_3$ — 3-methylpentane

$CH_3-C(CH_3)_2-CH_2-CH_3$ — 2,2-dimethylbutane

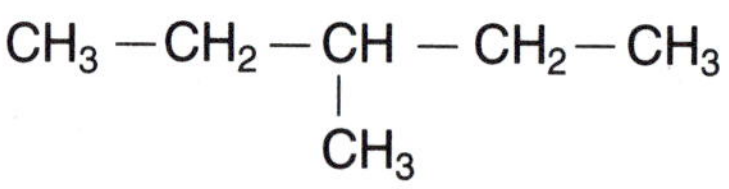

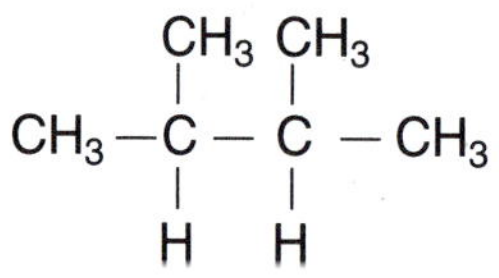

$CH_3-CH(CH_3)-CH(CH_3)-CH_3$ — 2,3-dimethylbutane

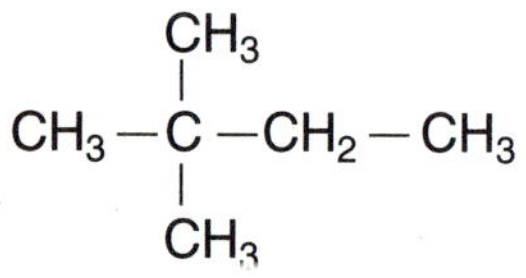

2. Name the following four compounds:

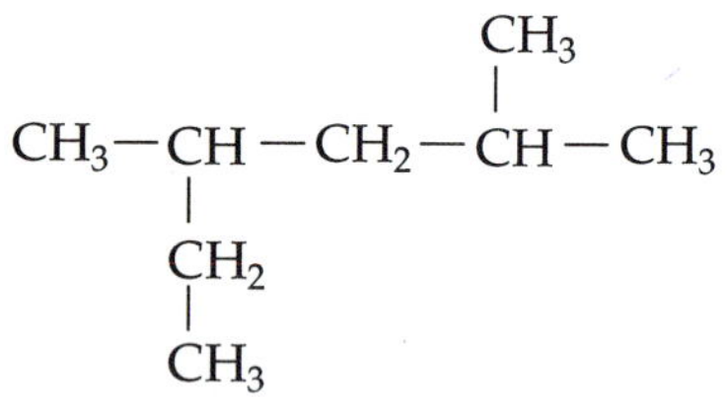

$CH_3-CH(CH_2CH_3)-CH_2-CH(CH_3)-CH_3$

2,4-dimethylhexane

$CH_3-C(Cl)(CH_3)-CH_2-CH_3$

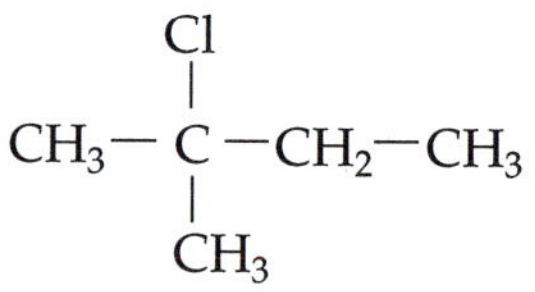

2-chloro-2-methylbutane

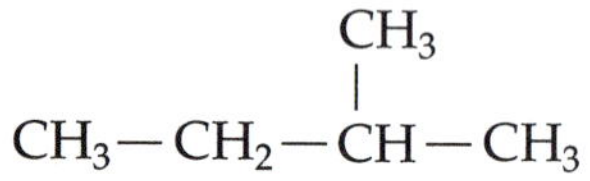

$CH_3-CH_2-CH(CH_3)-CH_3$

2-methylbutane

$CH_3-CH_2-CH(C_2H_5)-CH(CH(CH_3)_2)CH_3$

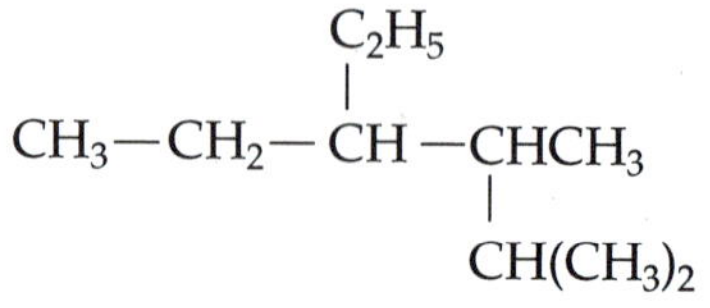

4-ethyl-2,3-dimethylhexane

3. Give the structural formulas for the following.

(a) 2-chloropentane

$$\begin{array}{ccccccccc} CH_3 & - & CH & - & CH_2 & - & CH_2 & - & CH_3 \\ & & | & & & & & & \\ & & Cl & & & & & & \end{array}$$

(b) 3-chloro-3-methylpentane

$$\begin{array}{ccccccccc} & & & & CH_3 & & & & \\ & & & & | & & & & \\ CH_3 & - & CH_2 & - & C & - & CH_2 & - & CH_3 \\ & & & & | & & & & \\ & & & & Cl & & & & \end{array}$$

(c) 2,2,3-trimethylhexane

$$\begin{array}{ccccccccccc} & & CH_3 & & CH_3 & & & & & & \\ & & | & & | & & & & & & \\ CH_3 & - & C & - & CH & - & CH_2 & - & CH_2 & - & CH_3 \\ & & | & & & & & & & & \\ & & CH_3 & & & & & & & & \end{array}$$

(d) 4-ethyl-2,2-dimethylhexane

$$\begin{array}{ccccccccccc} & & CH_3 & & & & C_2H_5 & & & & \\ & & | & & & & | & & & & \\ CH_3 & - & C & - & CH_2 & - & CH & - & CH_2 & - & CH_3 \\ & & | & & & & & & & & \\ & & CH_3 & & & & & & & & \end{array}$$

NOTES AND CALCULATIONS

Experiment

Preparation of Aspirin and Oil of Wintergreen

OBJECTIVE

To illustrate the synthesis of organic compounds.

APPARATUS AND CHEMICALS

Apparatus

125-mL Erlenmeyer flask	600-mL beaker
10-mL graduated cylinder	large test tube
small watch glass	Bunsen burner and hose
thermometer	funnel
filter paper	ring stand and ring
wire gauze	

Chemicals

acetic anhydride	absolute ethanol
1% $FeCl_3$ solution	methyl alcohol
salicylic acid	concentrated H_2SO_4
ice	

DISCUSSION

Esters are derivatives of organic acids. They can be prepared by the reaction of carboxylic acids with alcohols in the presence of a catalytic amount of mineral acid. The reaction

$$\underset{\text{Acetic acid}}{CH_3\overset{\overset{O}{\|}}{C}-OH} + \underset{\text{Methyl alcohol}}{CH_3OH} \rightleftharpoons \underset{\text{Methyl acetate}}{CH_3\overset{\overset{O}{\|}}{C}-OCH_3} + H_2O$$

is termed *esterification.* Esters usually have pleasant, fruitlike odors and are responsible for the flavors and fragrances of many fruits and flowers. For example, benzyl acetate has been found to be a principal ingredient of jasmine. Some common esters and their odors are octyl acetate, oranges; *n*-pentyl acetate, bananas; and butyl butyrate, pineapples. Generally speaking, the odor of natural products is due to more than one substance. For example, the volatile oil from pineapples contains at least six compounds.

In this experiment, you will prepare two esters of *o*-hydroxybenzoic acid, which is also called salicylic acid. One of the esters is acetylsalicylic acid, or, as it is commonly called, aspirin. It can be prepared by heating salicylic acid and acetic acid for an extended period of time. However, instead of using acetic acid, you will use acetic anhydride, because it is more reactive and will accomplish the esterification more quickly. The reaction is

$$C_6H_4(OH)(CO_2H) + (CH_3CO)_2O \xrightarrow{H^+} C_6H_4(OCOCH_3)(CO_2H) + CH_3CO_2H$$

Salicylic acid Acetic anhydride Acetylsalicylic acid

Aspirin is one of the oldest and generally most useful drugs known. It is both an analgesic (painkiller) and an antipyretic (reduces fever). Over 30 billion tablets are made in the United States annually. This is more than 200 tablets per person per year!

The second ester you will prepare is methyl salicylate, which is a component of oil of wintergreen. It is prepared by esterification of the carboxylic acid group, —CO_2H, of salicylic acid with methyl alcohol. The reaction is

$$C_6H_4(OH)(COOH) + CH_3OH \rightleftharpoons C_6H_4(OH)(COOCH_3) + H_2O$$

Salicylic acid Methyl alcohol Methyl salicylate

It is used quite extensively as a flavoring agent and in rubbing liniments for sore muscles.

Phenols are a class of compounds in which the —OH group is attached to an aromatic ring. Note that this group is present in both salicylic acid and methyl salicylate. Many phenols, but not all, form colored complexes with ferric chloride. The colors range from green through blue and red through violet. Hence, a 1% $FeCl_3$ solution is employed as a test for the presence of phenols.

Waste Disposal Instructions Place all waste from this experiment in designated containers.

PROCEDURE

A. Synthesis of Aspirin

Place about 3 grams of salicylic acid in a 125-mL Erlenmeyer flask. On the report sheet, record to the nearest 0.01 g the mass of salicylic acid used. Cautiously add 6 mL of acetic anhydride and then five drops of concentrated sulfuric acid. **(CAUTION:** ***Acetic anhydride and concentrated*** H_2SO_4 ***can cause severe burns if they come in contact with skin. If you get any of these reagents on you, immediately wash the area with copious amounts of water.*****)** Thoroughly mix the reagents by swirling the flask. Place the flask in a beaker of water warmed to 80° to 90°C. Heat the flask and its contents for 20 min. Remove the Erlenmeyer flask from the hot water and allow it to cool to room temperature. Add 40 mL of distilled water to the mixture in the Erlenmeyer flask. Thoroughly mix the contents by swirling. Cool the mixture in an ice bath to complete the crystallization. Filter and wash the crystals on the filter with a little ice-cold water. Allow the crystals to drain and then press them between several sheets of paper toweling or filter paper. Allow the crystals to air dry.

Your product is not likely to be pure, and certainly it is *not* suitable for human ingestion. Dissolve a few crystals of salicylic acid and a few crystals of your aspirin in 5 mL of water in separate test tubes. Add a drop of 1% ferric chloride ($FeCl_3$) solution to each test tube and note the color. Does your aspirin contain any unreacted salicylic acid?

Most solids may be purified by *recrystallization.* This is usually achieved by dissolving the substance in a suitable solvent at the boiling point, filtering the hot solution by gravity to remove any suspended insoluble particles, and letting crystalization proceed as the solution cools.The following is a conventional procedure for recrystallizing aspirin: Dissolve about 6 g in about 20 mL of absolute ethanol (calculate the volume of ethanol needed for your mass of aspirin) in a 125-mL Erlenmeyer flask and warm the alcohol in a water bath (NOT OVER A FLAME, SINCE ETHANOL IS VERY FLAMMABLE). When the aspirin has dissolved, add 50 mL of warm (50°C) distilled water. (If any crystals form at this point, heat the solution over the water bath again until they all dissolve.) Let the solution cool slowly, with the mouth of the flask covered with a watch glass. After crystallization is complete, isolate the crystals by gravity filtration, wash them with a little ice-cold distilled water, and air dry. Test the recrystallized material with 1% $FeCl_3$ solution for phenolic impurities. Did the recrystallization procedure remove the phenolic impurities? Your asprin is still probably not pure and is *not* suitable for human ingestion.

After your aspirin has dried, obtain and record its mass. Based on the mass of salicylic acid used and the mass of your product, calculate the percent yield of aspirin. Turn in your product to your laboratory instructor in a sample bottle or test tube labeled with your name or save it for Experiment 40.

B. Synthesis of Methyl Salicylate

Place 1 g of salicylic acid and 5 mL of methyl alcohol in a large test tube. Add 3 drops of concentrated sulfuric acid and then place the test tube in a water bath at 70°C for about 15 min. Note the odor. Add a drop of 1% $FeCl_3$ solution to the test tube and note any color change. Would you expect a color change? Dispose of chemicals used in the test in the designated receptacles.

REVIEW QUESTIONS

Before beginning this experiment in the laboratory, you should be able to answer the following questions:

1. What are esters? What is their general structure?
2. What is the role of the mineral acids in the esterification process?
3. From what substances can esters be prepared?
4. What is one general physical property of esters?
5. Describe a test for phenols.
6. What is the carboxylic acid group?
7. What is the purpose of recrystallization?
8. 1.02 g of acetylsalicylic acid was obtained from 1.00 g of salicylic acid by reaction with excess acetic anhydride. Calculate the percent yield of acetylsalicylic acid.
9. Two esters with the empirical formula $C_3H_6O_2$ may be prepared. Write structural formulas for these esters and name them.
10. How would you prepare the two esters in question 9?

11. Esters may be saponified by reaction with a strong base. Predict the products of the reaction.

$$CH_3\overset{\overset{\displaystyle O}{\|}}{C}OCH_3 + NaOH \xrightarrow{H_2O}$$

12. Give the structural formulas for isopropyl acetate and ethyl salicylate.

Name ______________________ Desk ______________________

Date ______________ Laboratory Instructor ______________________

REPORT SHEET | EXPERIMENT 39

Preparation of Aspirin and Oil of Wintergreen

A. Synthesis of Aspirin

Mass of salicylic acid ___2.98___ g

Mass of *dry* recrystallized aspirin ___2.69___ g

Percent yield

Moles of salicylic acid used (mol wt of salicylic acid = 138 amu) ___0.0216___ mol $\frac{2.98\text{ g}}{138\text{ g/mol}} = 0.0216\text{ mol}$

Theoretical number of moles of aspirin ___0.0216___ mol

Theoretical grams of aspirin (mol wt of aspirin = 180 amu) ___3.89___ g

$$\text{Percent yield} = \frac{\text{grams aspirin obtained}}{\text{theoretical grams aspirin}} \times 10^2$$

(0.0216 mol)(180 g/mol) = 3.89 g

$$= \frac{2.69\text{ g}}{3.89\text{ g}} \times 10^2$$

___69.2___ % yield

Color of $FeCl_3$ plus salicylic acid ___purple___

Color of $FeCl_3$ plus aspirin ___pale yellow___

Color of $FeCl_3$ plus recrystallized aspirin ___colorless___

Did the recrystallization remove phenolic impurities? ___yes___

B. Synthesis of Methyl Salicylate

Odor ___wintergreen mint___

Color of $FeCl_3$ plus product ___purple___

QUESTIONS

1. Why was *cold* as opposed to *warm* water used to wash the aspirin that you prepared?

   To minimize the loss of product by solubility.

2. Explain why you would or would not expect to observe a color change if $FeCl_3$ were added to the following.

   (a) Pure aspirin

   No color change because no phenol —OH group is present if the aspirin is pure.

   (b) Oil of wintergreen

   Yes, there should be a color change because of the presence of the phenol group.

3. Write the structure of the products that you would expect from the following reactions:

   (a)

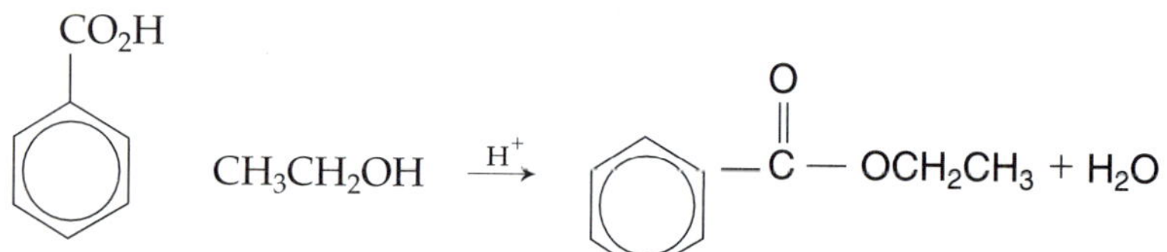

$$C_6H_5CO_2H + CH_3CH_2OH \xrightarrow{H^+} C_6H_5\text{—}C(=O)\text{—}OCH_2CH_3 + H_2O$$

   (b) CH_3CH_2OH + acetic anhydride $\xrightarrow{H^+}$ $CH_3\text{—}C(=O)\text{—}O\text{—}CH_2CH_3 + CH_3CO_2H$

   (c) $CH_3CH_2CO_2H + CH_3CH_2CH_2OH \xrightarrow{H^+} CH_3CH_2C(=O)OCH_2CH_2CH_3 + H_2O$

4. How would you prepare each of the following?

   (a) Ethyl acetate (two ways)

$$CH_3\text{—}C(=O)\text{—}OH + CH_3CH_2OH \longrightarrow CH_3\text{—}C(=O)\text{—}OCH_2CH_3 + H_2O$$

$$(CH_3\text{—}C(=O))_2O + CH_3CH_2OH \longrightarrow CH_3\text{—}C(=O)\text{—}OCH_2CH_3 + CH_3CO_2H$$

   (b) Methyl acetate

   $CH_3CO_2H + CH_3OH \longrightarrow CH_3CO_2CH_3 + H_2O$

   $(CH_3CO)_2O + CH_3OH \longrightarrow CH_3CO_2CH_3 + CH_3CO_2H$

5. If your experimental yield of aspirin is greater than 100%, how could this occur?

   Product was not pure: It may not be dried sufficiently and/or it may be contaminated with acetic acid.

Analysis of Aspirin

OBJECTIVE

To determine the purity of aspirin by acid-base titrations; to become acquainted with the concept of back-titration analyses.

APPARATUS AND CHEMICALS

Apparatus

50-mL burets (2)
600-mL beaker
utility clamp
ring stand and ring
wire gauze
250-mL Erlenmeyer flasks (3)
buret clamp
Bunsen burner and hose
balance
boiling chips

Chemicals

aspirin (student preparation or nonbuffered commercial tablets)
phenolphthalein solution
95% ethyl alcohol
0.1 *M* NaOH, standardized
0.1 *M* HCl, standardized

DISCUSSION

The aspirin you prepared in Experiment 39 is not likely to be pure. The most likely impurities are acids, either acetic or salicylic. Even most commercial aspirin tablets are not 100% acetylsalicylic acid. Most aspirin tablets contain a small amount of "binder," which helps prevent the tablets from crumbling. Even though the binder is chemically inert and was deliberately added by the manufacturer, its presence means that aspirin tablets are not 100% acetylsalicylic acid. Moreover, moisture can hydrolyze aspirin; thus, aspirin that is not kept dry can decompose. You may be able to detect a vinegarlike odor in aspirin if it has been exposed to moisture for an extended period of time. The hydrolysis product responsible for this odor is, in fact, acetic acid. It is formed in the following way:

$$C_6H_4(COOH)(OCOCH_3) + H_2O \xrightarrow{H^+ \text{ catalyst}} C_6H_4(COOH)(OH) + CH_3COOH$$

In this experiment, you will determine the purity of either the aspirin you prepared or commercial tablets. In particular, you will determine the percentage of acetylsalicylic acid in the material you analyze. The basis of the analysis utilizes acid-base titrations.

A *titration* is a process for determining the amount of analyte* present in a solution by the incremental addition of known volumes of a standard solution until the reaction between the analyte and the titrant is judged to be complete. Occasionally, it is convenient or necessary to add an excess of the titrant and then titrate the excess with another reagent. This process is called *back-titration.* In this technique, a measured amount of the reagent, which would normally be the titrant, is added to the sample so that there is a slight excess. After the reaction with the analyte is allowed to go to completion, the amount of excess (unreacted) reagent is determined by titration with another standard solution. Hence, by knowing the number of millimoles (mmol) of reagent taken and measuring the number in excess, we can calculate the number of millimoles of analyte by difference:

$$\text{millimoles reagent reacted} = \text{total millimoles} - \text{millimoles back-titrated}$$

Example 40.1

When nitrogen is analyzed by the Kjeldahl method, all the nitrogen in the sample is converted into NH_3. The NH_3 is distilled into a solution containing excess acid, and the excess acid is titrated with a standard base solution. If the nitrogen from a 1.325-g fertilizer sample is converted into NH_3 and distilled into 50.00 mL of 0.2030 *M* HCl, and if 25.32 mL of 0.1980 *M* KOH is required to back-titrate the excess HCl, how much NH_3 has been liberated from the fertilizer? Calculate the percentage of nitrogen in the fertilizer.

SOLUTION:

$$\text{mmol } NH_3 = \text{mmol HCl} - \text{mmol KOH}$$

$$\begin{aligned}\text{mmol } NH_3 &= \text{mmol N}\\ &= (50.00\text{ mL})(0.2030\text{ mmol/mL}) - (25.32\text{ mL})(0.1980\text{ mmol/mL})\\ &= 10.15 - 5.013\\ &= 5.14\text{ mmol}\end{aligned}$$

$$\begin{aligned}\text{mass } NH_3 &= (5.14 \times 10^{-3}\text{ mol})(17.0\text{ g/mol})\\ &= 0.0874\text{ g}\end{aligned}$$

$$\begin{aligned}\%\,\text{N} &= \frac{(5.14\text{ mmol})(14.00\text{ mg/mmol})}{1325\text{ mg}} \times 100\\ &= 5.43\%\end{aligned}$$

At low temperature, acetylsalicylic acid can be neutralized with base according to Equation [1]:

$$C_6H_4(OCOCH_3)(CO_2H) + OH^- \longrightarrow C_6H_4(OCOCH_3)(CO_2^-) + H_2O \qquad [1]$$

*An analyte is the substance being determined in any analytical procedure.

If no acidic impurities were present, you could determine the purity of the aspirin by titration with a standardized base solution. However, if acid impurities are present, titration of the aspirin will neutralize not only the acetylsalicylic acid (Equation [1]) but the acidic impurities as well. Thus, from such a titration the *total* number of millimoles of acid present in the aspirin can be calculated by measuring the volume of standard NaOH required to reach the phenolphthalein end point. The total number of millimoles (mmol) of acid may be calculated according to Equation [2]:

$$\text{total millimoles acid} = \text{milliliters NaOH} \times \text{molarity NaOH} \qquad [2]$$

In order to determine the amount of acetylsalicylic acid present in your material, we will take advantage of the fact that this substance will react with additional base reasonably rapidly at elevated temperatures, according to Equation [3]:

$$C_6H_4(OCOCH_3)(CO_2^-) + OH^- \xrightarrow{\Delta} C_6H_4(OH)(CO_2^-) + CH_3CO_2^- \qquad [3]$$

This reaction represents what is termed a base-promoted hydrolysis or saponification of esters. The reaction is the reverse of the esterification process. After you have neutralized all acidic material in the aspirin, you will add a known excess amount of millimoles of base to cause this reaction to occur. The excess base that is not consumed in the hydrolysis will be determined by a *back-titration* with standard HCl. From these data you can calculate the grams of acetylsalicylic acid in your material. Example 40.2 illustrates such a calculation.

EXAMPLE 40.2

A 0.5130-g sample of aspirin prepared by a student required 27.98 mL of 0.1000 M NaOH for neutralization. An additional 42.78 mL of 0.1000 M NaOH was added, and the sample was heated to hydrolyze the acetylsalicylic acid. After the reaction mixture cooled, the excess base was back-titrated with 14.29 mL of 0.1056 M HCl. How many grams of acetylsalicylic acid are in the sample? What is the percentage of acetylsalicylic acid (or the purity)?

Instructor: Students often have difficulty with the concepts involved in back-titration. Discuss it with them.

SOLUTION: First recognize that the 27.98 mL of base was used to neutralize all acidic material present in the sample. The total number of millimoles of base added for the hydrolysis reaction is

$$\begin{aligned}\text{mmol NaOH} &= 42.78 \text{ mL} \times 0.1000\ M \\ &= 4.278 \text{ mmol}\end{aligned}$$

The millimoles of HCl used in the titration corresponds to the number of millimoles of base that were *not* consumed in the hydrolysis reaction, Equation [3]:

$$\begin{aligned}\text{mmol HCl} &= \text{mmol excess NaOH} \\ &= 0.1056\ M \times 14.29 \text{ mL} \\ &= 1.509 \text{ mmol}\end{aligned}$$

The difference between the millimoles of base added for the hydrolysis and those which were not consumed equals the number of millimoles of base that brought about hydrolysis. From Equation [3] we see that this is exactly equal to the number of millimoles of acetylsalicylic acid:

$$4.278 \text{ mmol} - 1.509 \text{ mmol} = 2.769 \text{ mmol}$$

The molar mass of acetylsalicylic acid is 180.2 g. Hence, the number of grams of this acid can be found as follows:

$$\text{grams} = 2.769 \text{ mmol} \times \frac{180.2 \text{ g}}{10^3 \text{ mmol}}$$

$$= 0.4990 \text{ g acid}$$

Thus,

$$\% \text{ purity} = \frac{0.4990 \text{ g}}{0.5130 \text{ g}} \times 100 = 97.27\%$$

PROCEDURE

Instructor: The students should be provided with standardized NaOH and HCl solutions.

Weigh to the nearest milligram about 0.5 g of the aspirin you prepared in Experiment 39 or commercial nonbuffered tablets into a clean, dry 250-mL Erlenmeyer flask. Prepare two burets, one for acid and the other for base, by rinsing them with standard acid and base. After you fill the burets, record the molarities on your report sheet and also record the initial buret readings. Remember, the buret reading can be estimated to ±0.02 mL.

Add about 25 mL of 95% ethyl alcohol that has been cooled to about 15°C to the flask and swirl the flask to dissolve the aspirin. **(CAUTION:** ***Ethyl alcohol is flammable!*****).** Add two drops of phenolphthalein and rapidly titrate the sample with standard 0.1 *M* NaOH to a faint pink end point. Record the volume of NaOH used. This volume of base corresponds to that which is required to neutralize *all* acids present in your sample, that is, impurities as well as the acetylsalicylic acid.

To hydrolyze the aspirin, you will add additional NaOH to the flask from your buret. This may require your refilling the buret with more standard base. Record the initial buret reading. The amount of base to use for the hydrolysis is determined as follows: Add 15 mL to the volume of base required in the previous titration. Add *about* this volume of NaOH to the Erlenmeyer flask from the buret. Record the buret reading; you will know the exact volume of NaOH used for the hydrolysis from these buret readings. Heat the mixture for 15 min in a bath of boiling water in a 600-mL beaker, as shown in Figure 40.1, to hydrolyze the aspirin. **(CAUTION:** ***Remember that ethyl alcohol is flammable!*****)** Swirl the flask occasionally. Cool the flask to room temperature with cold tap water or an ice bath. If the solution is not pink, add two more drops of phenolphthalein indicator. Record the initial volume of HCl and back-titrate the excess base with the standard HCl until the pink color disappears. Record the volume of HCl used. Repeat this procedure with your other two samples.

From these data calculate the grams of acetylsalicylic acid in your aspirin samples, the percentage of aspirin, the average percentage of aspirin, and the standard deviation.

Waste Disposal Instructions Dispose of all waste as instructed by your instructor.

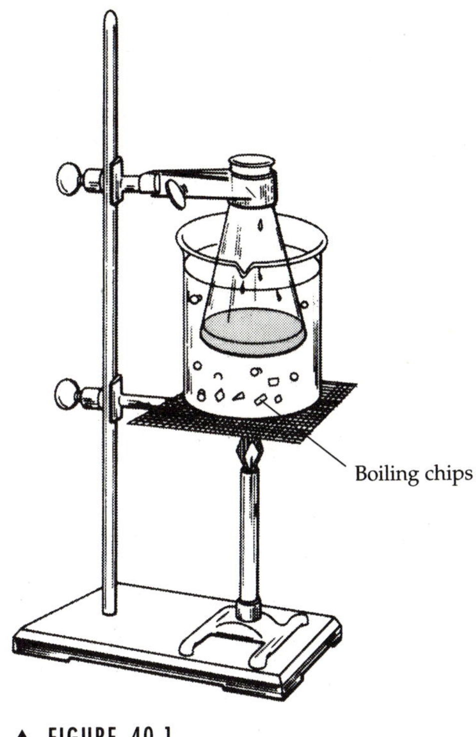

▲ FIGURE 40.1

REVIEW QUESTIONS

Before beginning this experiment in the laboratory, you should be able to answer the following questions:

1. Define the following terms: *analyte, standard solution, titration, titrant,* and *standardization.*
2. Write balanced equations for the cold neutralization of acetylsalicylic acid. If the mixture is hot, what other reaction may occur?
3. Illustrate the base-promoted hydrolysis of an ester.
4. If 33.5 mL of 0.25 M NaOH exactly neutralize 35.8 mL of HCl, what is the molarity of the HCl?
5. How many millimoles of NaOH are present in 35.8 mL of 0.125 M NaOH? How many grams of NaOH are present?
6. What volume of 0.15 M NaOH would be required to exactly neutralize 0.65 g of pure aspirin?
7. Account for the fact that old aspirin may smell like vinegar.
8. What is a buffer?
9. Why might a buffer be added to aspirin?
10. Why would you not use this procedure to analyze buffered aspirin?

NOTES AND CALCULATIONS

Name ______________________ Desk ______________________

Date ______________________ Laboratory Instructor ______________________

REPORT SHEET | EXPERIMENT

Analysis of Aspirin | 40

Check one: My preparation ( ); commercial tablets (x)
Brand name Skaggs

Molarity of NaOH 0.1007 *M*
Molarity of HCl 0.0992 *M*

Mass of Aspirin

	Trial 1	*Trial 2*	*Trial 3*
Weighing bottle initial mass	20.0845 g	19.4567 g	18.7402 g
Weighing bottle final mass	19.4567 g	18.7402 g	18.1201 g
Mass of aspirin	0.6278 g	0.7165 g	0.6201 g

Volume of* NaOH *Required to Neutralize All Acid Material

	Trial 1	*Trial 2*	*Trial 3*
Final reading	33.93 mL	39.19 mL	33.86 mL
Initial reading	0.19 mL	0.69 mL	0.34 mL
Volume of NaOH	33.74 mL	38.50 mL	33.52 mL
Millimoles of NaOH	3.398	3.877	3.375

Assuming that the only acid present is acetylsalicylic acid, calculate the grams of acetylsalicylic acid present in the aspirin.

(3.398 mmol)(0.1802 g/mmol) = 0.6123 g
(3.877 mmol)(0.1802 g/mmol) = 0.6986 g
(3.375 mmol)(0.1802 g/mmol) = 0.6082 g

Volume of* NaOH *Used in Hydrolysis

	Trial 1	*Trial 2*	*Trial 3*
Final reading	49.35 mL	55.00 mL	49.03 mL
Initial reading	0.21 mL	0.64 mL	0.11 mL
Volume of NaOH	49.14 mL	54.36 mL	48.92 mL
Millimoles of NaOH	4.948	5.474	4.926

Volume of* HCl *in Back-Titration

	Trial 1	*Trial 2*	*Trial 3*
Final reading	19.34 mL	31.41 mL	48.31 mL
Initial reading	2.64 mL	13.65 mL	31.41 mL
Volume of HCl	16.70 mL	17.76 mL	16.90 mL
Millimoles of HCl	1.657	1.762	1.676

Analysis

	Trial 1	*Trial 2*	*Trial 3*
Grams of acetylsalicylic acid in sample	0.5930 g	0.6689 g	0.5857 g
Percent of aspirin	94.46%	93.36%	94.45%

Average percentage of aspirin 94.09% Standard deviation 0.63

Show your calculations

mmol aspirin = 4.948 − 1.657 = 3.291 mmol

g aspirin = 3.291 mmol × 0.1802 g/mmol = 0.5930 g

% aspirin = 0.5930 g/0.6278 g × 100 = 94.46%

$$SD = \sqrt{\frac{(0.37)^2 + (0.73)^2 + (0.36)^2}{2}} = 0.63$$

From the above data, calculate the number of millimoles of acidic impurities in each of your three analyses.

TRIAL 1: 4.948 − 1.657 = 3.291 mmol; mmol impurity = 3.398 − 3.291 = 0.107 mmol

TRIAL 2: 5.474 − 1.762 = 3.712 mmol; mmol impurity = 3.877 − 3.712 = 0.165 mmol

TRIAL 3: 4.926 − 1.676 = 3.250 mmol; mmol impurity = 3.375 − 3.250 = 0.125 mmol

For each of your three analyses, what is the ratio of acetylsalicylic acid to other acid impurities?

TRIAL 1: $\frac{3.291 \text{ mmol}}{0.107 \text{ mmol}} = 30.8$

TRIAL 2: $\frac{3.712 \text{ mmol}}{0.165 \text{ mmol}} = 22.5$

TRIAL 3: $\frac{3.250 \text{ mmol}}{0.125 \text{ mmol}} = 26.0$

Ion-Exchange Resins: Analysis of a Calcium, Magnesium, or Zinc Salt

Experiment

OBJECTIVE

To become acquainted with the technique of ion-exchange chromatography and some of its applications.

APPARATUS AND CHEMICALS

Apparatus

balance	250-mL Erlenmeyer flasks (3)
100- and 250-mL beakers	50-mL burets (2)
buret clamp and ring stand	glass wool
glass rod (long)	polyethylene bottle
pH paper	weighing paper

Chemicals

cation-exchange resin (Dowex 50W-X-8)	deionized water
6 *M* HCl	phenolphthalein indicator solution
~0.1 *M* NaOH	potassium hydrogen phthalate (KHP)
unknown calcium, magnesium, or zinc salt	

DISCUSSION

I. Ion-Exchange Theory

Chromatography is one of the more important and versatile techniques available to separate the components of a mixture. *Chromatography* is a physical means of *separation* in which the components of a mixture become distributed between two different phases. These two phases are commonly called the *stationary phase* and the *mobile phase*. The mobile phase is usually a liquid or a gas, whereas the stationary phase is usually a solid. If the stationary phase is a polymer that possesses exchangeable ions, such as those illustrated in Figure 41.1, then the chromatographic technique is called *ion-exchange chromatography*. Polymers with exchangeable ionic functional groups have become known as *ion-exchange resins*. These ion-exchange resins are long-chain polymers that contain polar functional groups and have no appreciable water solubility. They come in the form of spherical beads that swell when placed in contact with water, so that the water solution can enter the interior of the bead and come into intimate contact with these functional groups. The polar functional groups are strong electrolytes and are thus completely ionized when in contact with water solutions. Ions of like charge equilibrate between the solution and the resin according to Equations [1] and [2]:

$$R_z^-H^+(s) + M^+(aq) \rightleftharpoons R_z^-M^+(s) + H^+(aq) \quad [1]$$

$$R_z^+OH^-(s) + X^-(aq) \rightleftharpoons R_z^+X^-(s) + OH^-(aq) \quad [2]$$

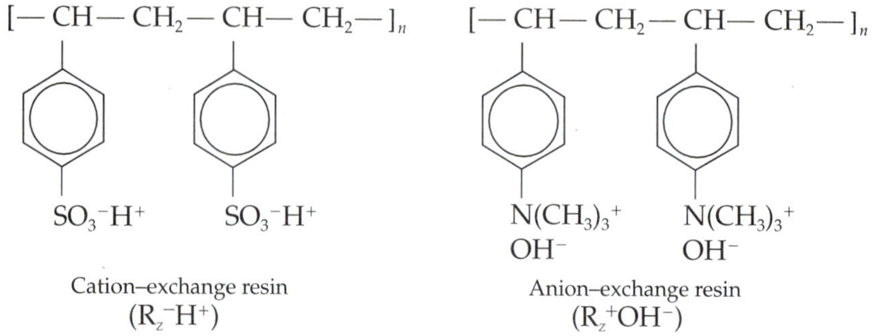

▲ **FIGURE 41.1** Ion-exchange chromatographic resins.

where R_z symbolizes the resin. A resin that behaves according to Equation [1] is called a cation-exchange resin, and one that behaves according to Equation [2] is called an anion-exchange resin. In general, the positions of these equilibria are influenced both by the nature and the concentrations of the ions in solution. Ions with a greater positive charge and a smaller size tend to have a greater affinity for the cation-exchange resin than do larger ions with smaller charges. As shown in Equation [1], one H^+ ion is liberated for each M^+ cation. Two H^+ ions would be liberated for each M^{2+} cation, three H^+ ions for each M^{3+} cation, and so forth. Thus, the extent of exchange of cations with H^+ on a resin would be expected to decrease in the orders: $Th^{4+} > Al^{3+} > Ca^{2+} > Na^+$ and $Cs^+ > Rb^+ > K^+ > Na^+ > Li^+$. In the latter series there appears to be a discrepancy because we expect the Cs^+ ion to be larger than the Li^+ ion and thus to have a smaller charge density. However, the Li^+ ion is hydrated and is tightly bound to six H_2O molecules, whereas the Cs^+ ion is bound to only four H_2O molecules, making the $[Li(H_2O)_6]^+$ ion larger than the $[Cs(H_2O)_4]^+$ ion.

In the same way, when small, highly negative anions come into contact with an anion-exchange resin, they liberate one OH^- for each negative charge. Thus we would expect the position of equilibrium [2] to be influenced by the charge density of the anion also. As a consequence, since anions are not so tightly bound to H_2O molecules as are cations, we expect the extent of exchange of anions in solutions with OH^- ions on the resin to decrease in the orders: $F^- > Cl^- > Br^- > I^-$ and $PO_4^{3-} > SO_4^{2-} > ClO_4^-$, for example. Because ion-exchange resins possess the ability to remove cations and anions from water and replace them with H^+ or OH^- ions, respectively, ion-exchange resins are commonly used to *deionize* water and to soften water. You may have a water softener in your home; if you do, you can see that it is just a large container of cation-exchange resin.

II. Hard Water

The term *hard water* has its origin in the fact that mineral impurities such as Ca^{2+}, Mg^{2+}, and Fe^{3+}, which are frequently present in natural water, react with soap to form gummy precipitates. They also react with CO_3^{2-} and SO_4^{2-} in the water to leave what is called a *boiler scale* ($CaCO_3$, $MgCO_3$, or $CaSO_4$) on walls of the vessels in which hard water is boiled. Ordinary soap is a water-soluble sodium (solid soaps) or potassium (liquid soaps) salt of a long-chain organic acid, RCO_2H, such as sodium stearate, $C_{17}H_{35}CO_2Na$. Sodium stearate reacts with Mg^{2+} or Ca^{2+} ions to form a sticky, insoluble precipitate according to Equation [3].

$$2C_{17}H_{35}CO_2^-(aq) + Ca^{2+}(aq) \rightleftharpoons Ca(C_{17}H_{35}CO_2)_2(s) \qquad [3]$$

This results in waste of the soap and the formation of an undesirable film on items that the soap is used to clean. You may have observed the formation of such a precipitate and the fact that soap does not form much of a foam or lather in hard water.

Suppose, then, that a solution containing calcium chloride, $CaCl_2$, was slowly passed through a bed of cation-exchange resin contained in a column. In this case equilibrium [4] would occur:

$$2R_z^-H^+(s) + Ca^{2+}(aq) + 2Cl^-(aq) \rightleftharpoons [(R_z^-)_2Ca^{2+}](s) + 2H^+(aq) + 2Cl^-(aq) \qquad [4]$$

The *eluate*, or solution, coming out of the column (Figure 41.2) would contain H^+ and Cl^- ions, and the Ca^{2+} ions would be retained by the column because

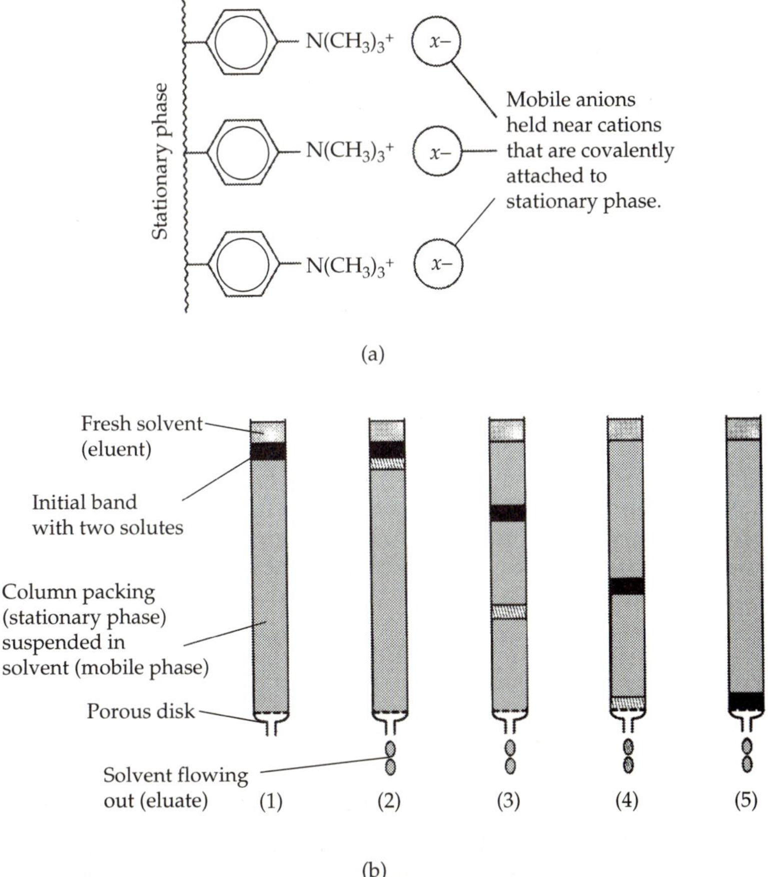

▲ **FIGURE 41.2** (a) Ion-exchange chromatography; this is an example of an anion-exchange resin, since only anions can be attracted to it. (b) Schematic representation of a chromatographic separation. The solute with a greater affinity for the stationary phase remains on the column longer.

their affinity for the resin is greater than that of the H^+ ions. If this eluate were then passed through an anion-exchange resin in the OH^- form, then equilibrium [5] would occur on the column:

$$2H^+(aq) + 2Cl^-(aq) + 2R_z^+OH^-(s) \rightleftharpoons 2R_z^+Cl^-(s) + 2H^+(aq) + 2OH^-(aq) \qquad [5]$$

Thus, passing a $CaCl_2$ solution successively through a cation- and then an anion-exchange resin would produce a solution containing neither Ca^{2+} nor Cl^- ions in the eluate. Since OH^- ions react with H^+ ions according to Equation [6], the eluate would be pure water:

$$H^+(aq) + OH^-(aq) \longrightarrow H_2O(l) \qquad [6]$$

Water treated in this way is said to be *deionized.*

In order to carry out such a process in actuality, it is necessary to have a large volume of resin with respect to the volume of solution, because ion-exchange resins typically have an exchange capacity of only about 5 mmol/g on a dry basis. Also, it is necessary to have a long contact time between the solution and the resin so that all the ionic sites on the resin may contact the ions in solution. Thus we generally use a column such as illustrated in Figure 41.2 and slowly pass through it the solution to be exchanged.

Ion-exchange resins are useful not only for separation and purification purposes, but also for analytical purposes. Suppose you passed a solution containing a known mass of a pure soluble calcium salt (which contained an unknown amount of calcium) through a cation-exchange resin. Equilibrium [7] would occur on the column:

$$2R_z^-H^+(s) + Ca^{2+}(aq) \longrightarrow [(R_z^-)_2Ca^{2+}](s) + 2H^+(aq) \qquad [7]$$

The liberated H^+ in the eluate from the column could be determined by a simple acid-base titration with a standard NaOH solution. Assuming that all the Ca^{2+} was captured by the column, the calcium content of the sample can be easily calculated.

Example 41.1

A 0.2940-g sample of a soluble calcium salt was passed through a cation-exchange column in the H^+ form. The H^+ in the eluate required 26.22 mL of 0.1025 *M* NaOH to reach a phenolphthalein end point. Calculate the percentage calcium in the original sample.

SOLUTION: The acid-base reaction is $H^+(aq) + OH^-(aq) \longrightarrow H_2O(l)$.

$$\text{moles } H^+ = (26.22 \text{ mL})\left(\frac{1\text{L}}{1000 \text{ mL}}\right)\left(\frac{0.1025 \text{ mol}}{\text{L}}\right)$$

$$= 2.688 \times 10^{-3} \text{ mol}$$

$$\text{moles } Ca^{2+} = \tfrac{1}{2} \text{ mol } H^+ \text{ (see Equation [7])}$$

$$= \frac{2.688 \times 10^{-3} \text{ mol}}{2}$$

$$= 1.344 \times 10^{-3} \text{ mol}$$

$$\text{mass } Ca^{2+} = (1.344 \times 10^{-3} \text{ mol})(40.08 \text{ g/mol})$$

$$= 53.87 \times 10^{-3} \text{ g}$$

$$\% \ Ca^{2+} = \left(\frac{53.87 \times 10^{-3} \text{ g } Ca^{2+}}{0.2940 \text{ g sample}}\right)(100)$$

$$= 18.32 \ \% \ Ca^{2+}$$

Record all data directly onto the report sheets.

PROCEDURE

In this experiment, you are to determine the percent calcium, magnesium, or zinc in an unknown by performing a cation exchange followed by an acid-base titration of the eluate. Your instructor will tell you which cation you are determining. You will first standardize your base solution (NaOH, the titrant).

A. Standardization of NaOH

Place about 400 mL of approximately 0.1 *M* NaOH solution into a polyethylene bottle. Standardize the NaOH solution by titrating against KHP, using phenolphthalein indicator. Remember, the buret reading can be estimated to ±0.01 mL.

Instructor: Time may be saved if the students are provided with standardized NaOH solution.

B. Preparation of the Ion-Exchange Column and Determination of Cation

Weigh out about 10 g of the cation ion-exchange resin to the nearest gram. The resin must be in the acid form. In order to ensure that it is, slurry the resin in a 250-mL Erlenmeyer flask with a mixture of 50 mL of 6 *M* HCl and 100 mL of deionized water. Allow the slurry to stand, with occasional swirling, for 15 min. Then decant the acid solution and wash the resin with five 50-mL portions of deionized water, decanting the wash liquid after each washing. Finally add 50 mL of deionized water to the resin.

Using a long wire or glass rod, push a small plug of glass wool to the bottom of a 50-mL buret. Then put 5 mL of deionized water on top of the glass wool in the buret, clamp the buret vertically in a buret clamp on a ring stand, and transfer the resin slurry to the buret, with the aid of more deionized water, if necessary. Drain off excess water occasionally from the bottom of the buret. Never allow the liquid level in the buret to fall below the level of the resin, because this usually causes channels to develop in the resin column and greatly reduces its efficiency. Wash the resin by allowing deionized water to flow through it until the pH of the effluent is the same as the pH of the deionized water (use pH paper to check the pH of the deionized water and the eluate). Then drain water from the buret until the water level is just barely above the top of the resin.

Obtain a sample of a soluble calcium, magnesium, or zinc salt from your laboratory instructor. With the aid of weighing paper, weigh out to the nearest milligram approximately 0.25 to 0.30 g of this salt. Transfer the weighed salt to a 100-mL beaker and dissolve it in about 25 mL of deionized water. Transfer the salt solution to the ion-exchange column, rinse the beaker twice with 5-mL portions of deionized water, and add the rinsings to the ion-exchange column. As the liquid level begins to reach the top of the column, drain liquid (the eluate) from the bottom of the column and collect the eluate in a 250-mL Erlenmeyer flask. The buret stopcock should be adjusted so that the eluate flow rate is about 2 to 3 mL/min. Continue to drain solution from the column while periodically adding deionized water to the top of the column until you have collected about 125 mL of solution in the Erlenmeyer flask. At all times maintain the liquid level in the column above the level of the resin bed. When about 125 mL of eluate has been collected, take a small drop of eluate on the end of a glass rod as it emerges from the buret and test it with pH paper to ensure that the pH of the eluate is the same as the pH of the deionized-water wash. Continue to elute the column with deionized water until the two pH values are the same. Close the buret and set the flask and eluate aside and label the flask "Trial 1."

Repeat the above procedure with two more approximately 0.30-g samples of your unknown and label the two Erlenmeyer flasks "Trial 2" and "Trial 3."

Add two to three drops of phenolphthalein indicator solution to each of the three Erlenmeyer flasks and titrate each of the eluate samples with your standardized NaOH solution to a faint pink end point.

Remove the resin from the buret, regenerate it by slurrying it with 50 mL of 6 *M* HCl, and wash five times with deionized water as described for the preparation of the column. Return the regenerated resin to the resin stock bottle.

From your data, calculate the number of moles of cation in your sample and the percentage of cation in your sample.

REVIEW QUESTIONS

Before beginning this experiment in the laboratory, you should be able to answer the following questions:

1. Describe how an ion-exchange column works.
2. Define *stationary* and *mobile phases.*
3. Define the terms *eluant* and *eluate.*
4. Why do smaller, highly charged ions have a greater affinity for an ion-exchange column than do larger ions with smaller charges?
5. What are hard water and deionized water?
6. Would you expect Na^+ or Mg^{2+} to be more slowly eluted from an ion-exchange column?
7. A 0.300-g sample of pure $ZnCl_2 \cdot nH_2O$ was placed on an ion-exchange column in the H^+ form. The H^+ released in the eluate required 33.95 mL of 0.1024 *M* NaOH to reach a phenolphthalein end point. Calculate the percent zinc in the sample.
8. Calculate the value of *n* for the data in question 7.
9. If a 0.135-g sample of NaCl was placed on an ion-exchange column in the H^+ form and the eluate titrated with 0.1231 *M* NaOH, what volume of the NaOH solution would be required to reach a phenolphthalein end point?

Name ____________________ Desk ____________________

Date ____________________ Laboratory Instructor ____________________

Salt analyzed ____________________

REPORT SHEET | EXPERIMENT 41

Ion-Exchange Resins: Analysis of a Calcium, Magnesium, or Zinc Salt

A. Standardization of NaOH

	Trial 1	*Trial 2*	*Trial 3*
Mass of bottle + KHP	28.7185 g	28.4008 g	28.0849 g
Mass of bottle	28.4008 g	28.0849 g	27.7393 g
Mass of KHP used	0.3177 g	0.3159 g	0.3456 g
Final buret reading	13.97 mL	26.35 mL	39.72 mL
Initial buret reading	1.42 mL	13.97 mL	26.35 mL
mL of NaOH used	12.55 mL	12.38 mL	13.37 mL
Molarity of NaOH	0.1241 *M*	0.1250 *M*	0.1266 *M*

Average molarity 0.1252 *M*

Standard deviation ±0.0012

(show calculations and standard deviation)

$$M = \frac{0.3177 \text{ g} \times 10^3 \text{ mmol/mol}}{12.54 \text{ mL} \times 204.2 \text{ g/mol}}$$

$$= 0.1241\ M$$

$$SD = \sqrt{\frac{(0.0012)^2 + (0.0002)^2 + (0.0014)^2}{2}}$$

$$= \pm 0.0012$$

B. Preparation of the Ion-Exchange Column and Determination of Cation

(Unknown dependent)

	Trial 1	*Trial 2*	*Trial 3*
Mass of sample + paper	0.6772 g	0.6831 g	0.7031 g
Mass of paper	0.4138 g	0.3955 g	0.3972 g
Mass of sample	0.2634 g	0.2876 g	0.3059 g
Final buret reading	18.81 mL	39.69 mL	22.62 mL
Initial buret reading	0.00 mL	18.81 mL	0.05 mL
mL of NaOH used	18.81 mL	20.88 mL	22.57 mL
Moles H^+	2.355×10^{-3}	2.614×10^{-3}	2.826×10^{-3}
Moles M^{2+}	1.177×10^{-3}	1.307×10^{-3}	1.413×10^{-3}

Mass M^{2+}	(Mg)	28.61×10^{-3} g	31.77×10^{-3} g	34.34×10^{-3} g
Percent M^{2+}		10.86%	11.05%	11.22%

Average 11.04% Standard deviation ±0.14

(Show calculations)

$$\text{Ave} = \frac{10.86 + 11.05 + 11.22}{3} = 11.04$$

$$SD = \sqrt{\frac{(0.18)^2 + (0.01)^2 + (0.08)^2}{2}} = 0.14$$

QUESTIONS

1. Assuming that your salt was a hydrated form of a sulfate, $MSO_4 \cdot nH_2O$, calculate n. What values are quoted in handbooks for common hydrates of your cation?

$MgSO_4 \cdot nH_2O$ Handbook lists: $MgSO_4$ and $MgSO_4 \cdot 7H_2O$

$$\%\text{Mg} = \frac{24.31}{\text{molar mass}} \times 10^2$$

$$\text{molar mass} = \frac{24.31 \times 10^2}{11.04}$$

$$= 220.2$$

$$\text{molar mass of } MgSO_4 = 120.4$$

$$\text{molar mass of } H_2O = 18.0$$

$$220.2 - 120.4 = 99.8$$

$$\frac{99.8}{18.0} = 5.54$$

$$\therefore n = 6$$

2. A 0.300-g sample of a compound of empirical formula $CrCl_3 \cdot 6H_2O$ was dissolved in water and passed through a cation-exchange column in the H^+ form. The eluate required 33.45 mL of 0.100 M NaOH to reach a phenolphthalein end point. Explain this result in terms of each of the following.
(a) The charge on Cr in the compound

$$33.45 \text{ mL} \times 0.100 \frac{\text{mmol}}{\text{mL}} \text{ NaOH} = 3.35 \text{ mmol NaOH} = 3.35 \text{ mmol } H^+$$

If the charge on the Cr compound is 1+, 2+, or 3+, the corresponding number of mmol of Cr compound is, respectively, 3.35/1, 3.35/2, or 3.35/3 mmol. (If charge is 0, no acid is exchanged.) See below* to decide that the compound is $[Cr(H_2O)_6]^{3+} Cl_3$.

(b) What is the structure of the compound, $[Cr(H_2O)_6]^{3+}Cl_3$, $[Cr(H_2O)_5Cl]^{2+}Cl_2$, $[Cr(H_2O)_4Cl_2]^{+}Cl$, or $[Cr(H_2O)_3Cl_3]$

if 0.300 g = 3.35×10^{-3} mol, 1 mol = 89.55 g
or if 0.300 g = 3.35×10^{-3} mol/2, 1 mol = 179.1 g
or if 0.300 g = 3.35×10^{-3} mol/3, 1 mol = 268.7 g

These do not correspond exactly to any molar mass below* but are close to that of $[Cr(H_2O)_6]Cl_3$.

(c) The percentage of Cr in the compound

Atomic wt. Cr = 51.996 $\quad \%Cr = \dfrac{51.996}{266.45} \times 100 = 19.514\%$

Molar mass $CrCl_3 \cdot 6H_2O$ = 266.45

*Compound	Molar mass	Charge
$[Cr(H_2O)_6]Cl_3$	266.45	3+
$[Cr(H_2O)_5Cl]Cl_2$	248.43	2+
$[Cr(H_2O)_4Cl_2]Cl$	230.41	1+
$[Cr(H_2O)_3Cl_3]$	212.39	0

NOTES AND CALCULATIONS

Appendices

Appendix

A

Chemical Arithmetic

Elementary mathematics is frequently used in the study of general chemistry. Exponential arithmetic, significant figures, and logarithms are of particular importance and widespread application in these calculations. These are discussed in turn in this appendix.

EXPONENTIAL ARITHMETIC

Many quantities that we measure in chemistry are either very large or very small. Because of this, it becomes convenient to express numbers in scientific notation. Scientific notation is a way of expressing all numbers as a product. The two members of the product are, first, a number between 1 and 10 and, second, the power of 10 that places the decimal point. This second number is called the *exponential term* and is written as 10 with a right-hand superscript (the exponent). The exponent denotes the power of 10, that is, how many times 10 is multiplied by 10. Some examples of the exponential method of expressing numbers are given below:

$$1000 = 1 \times 10^3 \qquad 0.1 = 1 \times 10^{-1}$$
$$100 = 1 \times 10^2 \qquad 0.01 = 1 \times 10^{-2}$$
$$10 = 1 \times 10^1 \qquad 0.001 = 1 \times 10^{-3}$$
$$1 = 1 \times 10^0 \qquad 2386 = 2.386 \times 1000 = 2.386 \times 10^3$$
$$0.123 = 1.23 \times 0.1 = 1.23 \times 10^{-1}$$

As should be evident from the above examples, the power (exponent) of 10 is equal to the number of places the decimal is shifted to give the digit number. The efficacy of using exponential numbers becomes readily apparent when one compares writing 1,230,000,000 with writing 1.23×10^9, or 0.000,000,000,36 with 3.6×10^{-10}.

Once numbers have been expressed as exponentials, the question arises: How does one perform mathematical operations such as addition or multiplication with exponentials? The answers to this question are illustrated in the following examples.

Addition of Numbers in Scientific Notation

Convert all the numbers to the same power of 10 and add the digit terms of the numbers.

EXAMPLE

$5.0 \times 10^{-2} + 3 \times 10^{-3}$

$$\begin{array}{r} 50 \times 10^{-3} \\ +\quad 3 \times 10^{-3} \\ \hline 53 \times 10^{-3} = 5.3 \times 10^{-2} \end{array} \qquad \text{or} \qquad \begin{array}{r} 5.0 \times 10^{-2} \\ +\ 0.3 \times 10^{-2} \\ \hline 5.3 \times 10^{-2} \end{array}$$

Subtraction of Numbers in Scientific Notation

Convert all the numbers to the same power of 10 and subtract the digit terms of the numbers.

EXAMPLE

$5.0 \times 10^{-6} - 4 \times 10^{-7}$

$$\begin{array}{r} 5.0 \times 10^{-6} \\ -\ 0.4 \times 10^{-6} \\ \hline 4.6 \times 10^{-6} \end{array} \quad \text{or} \quad \begin{array}{r} 50 \times 10^{-7} \\ -\ 4 \times 10^{-7} \\ \hline 46 \times 10^{-7} = 4.6 \times 10^{-6} \end{array}$$

Multiplication of Numbers in Scientific Notation

Multiply the digit terms in the usual way and add the exponents of the exponential terms (that is, $10^a \times 10^b = 10^{a+b}$).

EXAMPLES

$$(4.2 \times 10^{-8})(2.0 \times 10^{3}) = 8.4 \times 10^{(-8+3)} = 8.4 \times 10^{-5}$$

$$(4.2 \times 10^{-8})(2.0 \times 10^{-3}) = 8.4 \times 10^{-11}$$

$$(4.2 \times 10^{8})(2.0 \times 10^{-3}) = 8.4 \times 10^{[8+(-3)]} = 8.4 \times 10^{5}$$

Division of Numbers in Scientific Notation

Divide the digit terms of the numerator by the digit term of the denominator and subtract the exponents of the exponential terms (that is, $10^a/10^b = 10^{a-b}$).

EXAMPLES

$$\frac{4.2 \times 10^{-8}}{2.0 \times 10^{3}} = 2.1 \times 10^{-11}$$

$$\frac{4.2 \times 10^{-8}}{2.0 \times 10^{-3}} = 2.1 \times 10^{-5}$$

$$\frac{4.2 \times 10^{8}}{2.0 \times 10^{-3}} = 2.1 \times 10^{[8-(-3)]} = 2.1 \times 10^{11}$$

Squaring of Exponentials

Square the digit term in the usual way and multiply the exponent of the exponential term by 2 (that is, $(10^a)^2 = 10^{2a}$).

EXAMPLES

$$(4.0 \times 10^{-2})^2 = 16 \times 10^{-2 \times 2} = 16 \times 10^{-4} = 1.6 \times 10^{-3}$$

$$(5.0 \times 10^{4})^2 = 25 \times 10^{4 \times 2} = 25 \times 10^{8} = 2.5 \times 10^{9}$$

$$(1.20 \times 10^{3})^2 = 1.44 \times 10^{3 \times 2} = 1.44 \times 10^{6}$$

Raising Numbers in Scientific Notation to a General Power

Raise the digit term to the power in the usual way, and multiply the exponent of the exponential term by the power (that is, $(10^a)^b = 10^{ab}$).

EXAMPLES

$$(2 \times 10^{-2})^3 = 8 \times 10^{-2 \times 3} = 8 \times 10^{-6}$$

$$(1 \times 10^{3})^5 = 1 \times 10^{3 \times 5} = 1 \times 10^{15}$$

Extraction of Square Roots of Numbers in Scientific Notation

Decrease or increase the exponential term so that the power of 10 is evenly divisible by 2. Extract the square root of the digit term by inspection, by logarithms, or by calculator, and divide the exponential term by 2 (that is, $\sqrt{10^a} = 10^{a/2}$).

EXAMPLE

Find the square root of 1.6×10^{-7}:

$$1.6 \times 10^{-7} = 16 \times 10^{-8}$$

$$\sqrt{16 \times 10^{-8}} = \sqrt{16} \times \sqrt{10^{-8}} = 4.0 \times 10^{-8/2} = 4.0 \times 10^{-4}$$

Combined Operations

Any combination of the above operations is performed by doing each operation individually and combining the results.

EXAMPLE

Solve the equation: $a/(2.5 \times 10^{-3})=(7.4 \times 10^{8})/(3.9 \times 10^{-6})$

$$a = \frac{(2.5 \times 10^{-3})(7.4 \times 10^{8})}{(3.9 \times 10^{-6})} = 4.7 \times 10^{11}$$

Because the operations of multiplication and division are commutative, the order of operations is immaterial.

$$a = \frac{(2.5 \times 10^{-3})(7.4 \times 10^{8})}{(3.9 \times 10^{-6})} = \frac{1.9 \times 10^{6}}{3.9 \times 10^{-6}} = 4.7 \times 10^{11}$$

SIGNIFICANT FIGURES

Many operations in the chemistry laboratory, such as weighing a compound or measuring the volume of a liquid, involve measurements of some kind. It is important to record these data properly so that the value reported will correctly represent the accuracy of the measurement. The following is a brief guide for calculating and reporting numerical results.

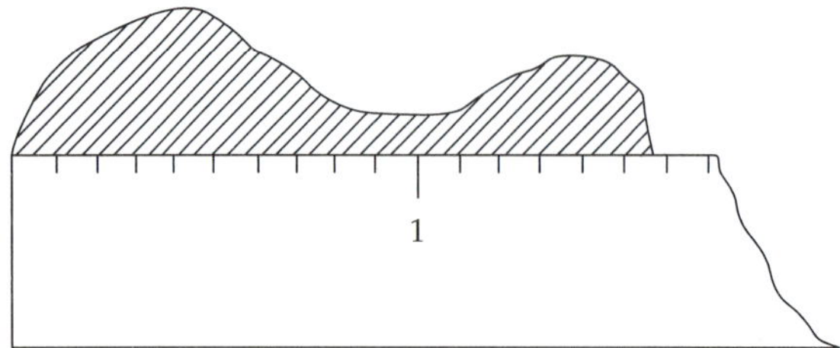

▲ **FIGURE A.1**

Every observed measurement that is made is really an approximation. For example, the length of the object in Figure A.1 is between 1.5 and 1.6 units. Its length is seen to be approximately 1.56 units. There is uncertainty in the last digit, 6; it is estimated. Consider the recorded mass of an object as 2.6 g. This means that the object was weighed to the nearest tenth (0.1) of a gram and that its exact mass is between 2.55 g and 2.65 g. In recording a result, it is the last digit that represents a degree of uncertainty; for example, in 2.6 g it is the number 6 that represents some degree of uncertainty. We say the number 2.6 g contains two significant figures, the numbers 2 and 6 being the significant figures. If the recorded mass of the object were 2.63$\underline{3}$ g, there would be four significant figures (2, 6, 3, and 3), and this would mean that the object was weighed to the nearest thousandth (0.001 g) of a gram. Thus it is the last (underscored) 3 which has been estimated. *Significant figures* refer to those digits we know with certainty plus the first doubtful or estimated digit.

Zeros

A zero may or may not represent a significant figure.* The number 5.00 represents three significant figures, the zeros being significant; this result implies that the measurement was made to the nearest one-hundredth part. Similarly, the zeros in 5.034, 7.206, 9.310, 10.20, and 2.001 are all regarded as significant figures. Each of the five preceding numbers contains four significant figures.

It is a common error to omit the zeros that indicate the reliability of a number. An object whose mass is 2 g when weighed to the nearest 0.01 g on a triple-beam balance should be recorded as 2.00 g, *not* 2 g.

When the zero is used to locate a decimal point, as in 0.023, the zero is *not* a significant figure. There are only two significant figures in each of the following: 0.027, 0.00078, and 0.0010. These numbers could be written 2.7×10^{-2}, 7.8×10^{-4}, and 1.0×10^{-3}. In numbers such as 10, 100, or 1000, the number of significant figures is uncertain. These should be written as 1.0×10^{1}, 1.0×10^{2}, or 1.0×10^{3} if two significant figures are used.

Exact Numbers

Some numbers by their very nature are exact numbers. When you say you are holding three test tubes in your hand, you mean exactly three test tubes, not two or four. Thus if this figure 3 is used in a calculation, you may regard it as containing as many significant figures as desired: 3.0000. . . .

Another example of exact numbers is a number used to relate quantities within the same system of units:

*See *Journal of Chemical Education*, *54*, 578 (1977); *57*, 646 (1980) for further discussion.

$$1 \text{ ft} = 12 \text{ in.}$$

$$1 \text{ cm} = 10 \text{ mm}$$

This quality of exactness follows from the definition of the equality. Note, however, that the relationship between units in two different systems is not exact. For example:

$$1 \text{ kg} = 2.205 \text{ lb}$$

One kilogram does not exactly equal 2.205 lb, because the two systems have been defined independently of one another. Thus, 1 kg is approximately equal to 2.205 lb.

Rounding Off Numbers

A number may be rounded off to the desired number of significant figures by dropping one or more digits at the end of the number. The following rules should be observed when rounding off a number:

1. When the first digit dropped is less than 5, the last digit retained should remain unchanged (this is called rounding down). Examples: 17.373 becomes 17.37; 1.5324 becomes 1.53 if rounded off to three significant figures and 1.5 if rounded off to two significant figures.
2. When the first digit dropped is equal to or greater than 5, the last digit retained is increased by 1. Statistically, 5's should be rounded up only half the time. Round up if the number preceding the 5 is even, round down if the number preceding the 5 is odd. Examples: 19.765 becomes 19.77; 8.1525 becomes 8.153 if rounded off to four significant figures and 8.2 if rounded off to two significant figures.

Significant Figures and Calculations

In any calculation in which experimental results are used, the final result should contain only as many significant figures as are justified by the experiment. Thus the least precise measurement dictates the number of significant figures that should be present in the final answer.

Addition and subtraction In addition and subtraction, retain only as many decimal places in the result as there are in that component with the least number of decimal places. For example:

$$\begin{array}{l} 21.1 \\ \ \ 2.035 \\ \underline{\ \ 6.12\ \ \ } \\ 29.255 \text{ becomes } 29.3 \end{array}$$

Multiplication and division In multiplication and division, the answer should contain only as many significant figures as are contained in the factor with the least number of significant figures. For example:

$$21.71 \times 0.029 \times 89.2 = 56.159428$$

The number 0.029 contains only two significant figures; therefore, according to the rule above, the answer should be rounded off to contain two significant figures: 56.

Problems

1. How many significant figures are present in each of the following numbers: (a) 454 g; (b) 22.01 cm; (c) 18.00 mL; (d) 0.020 g?
2. Add each of the following: (a) 311 cm + 10.1 cm + 1.21 cm; (b) 18.00 mL + 1.2 mL + 0.71 mL; (c) 3.286 ft + 7.01 ft + 0.001 ft.
3. Multiply each of the following: (a) 3.70 × 1.11; (b) 3.70 × 2.2; (c) 3.70 × 0.022.
4. Divide each of the following: (a) $\frac{98.98}{4.90}$; (b) $\frac{75.24}{1.1}$; (c) $\frac{37.1}{2.5312}$

Answers: **1.** (a) 3; (b) 4; (c) 4; (d) 2
2. (a) 322 cm; (b) 19.9 mL; (c) 10.30 ft.
3. (a) 4.11; (b) 8.1 (c) 0.081
4. (a) 20.2; (b) 68; (c) 14.7

THE USE OF LOGARITHMS AND EXPONENTIAL NUMBERS

The *common logarithm* of a number is the power to which the number 10 must be raised to equal that number. For example, the logarithm of 100 is 2 because the number 10 must be raised to the second power to be equal to 100, that is, $\log_{10} 100 = 2$, or $10^2 = 100$. Additional examples are given in Table A.1.

What is the logarithm of 60? Because 60 lies between 10 and 100, which have logarithms of 1 and 2, respectively, the logarithm of 60 must lie between 1 and 2. The logarithm of 60 is 1.78, that is, $60 = 10^{1.78}$, or $\log_{10} 60 = 1.78$.

Every logarithm is made up of two parts, called the *characteristic* and the *mantissa.* The characteristic is that part of the logarithm which lies to the left of the decimal point; thus the characteristic of the logarithm of 60 is 1. The mantissa is that part of the logarithm which lies to the right of the decimal point; thus the mantissa of the logarithm of 60 is .78. The characteristic of the logarithm of a number greater than 1 is 1 less than the number of digits to the left of the decimal point of the number, as shown by the following table:

TABLE A.1 Some Examples of Common Logarithms

Number	Number expressed exponentially	Logarithm
10,000	10^4	4
1,000	10^3	3
10	10^1	1
1	10^0	0
0.1	10^{-1}	−1
0.01	10^{-2}	−2
0.001	10^{-3}	−3
0.0001	10^{-4}	−4

Number	Characteristic	Number	Characteristic
60	1	2.340	0
600	2	23.40	1
6,000	3	234.0	2
52,840	4	2340.0	3

The mantissa of the logarithm of a number is independent of the position of the decimal point. Thus the logarithms of 2.340, 23.40, 234.0, and 2340.0 all have the same mantissa, and the logarithms of these numbers are 0.3692, 1.3692, 2.3692, and 3.3692, respectively. You may verify this on your calculator.

The meaning of the mantissa and characteristic can be better understood from a consideration of their relationship to exponential numbers. For example, 2340 may be written as 2.340×10^3. The logarithm of (2.340×10^3) = the logarithm of 2.340 plus the logarithm of 10^3. Recall that $(10^a) \times (10^b) = 10^{a+b}$. To find the logarithm of a product, $\log 10^{a+b} = a + b = \log 10^a + \log 10^b$. The logarithm of 2.340 is 0.3692 and the logarithm of 10^3 is 3. Thus the logarithm of $2340 = 3 + 0.3692 = 3.3692$.

The logarithm of a number less than 1 has a negative value, and a convenient method of obtaining the logarithm of such a number is given below.

Example

Obtain the logarithm of 0.00234. When expressed exponentially, $0.00234 = 2.34 \times 10^{-3}$. The logarithm of 2.34×10^{-3} equals the logarithm of 2.34 plus the logarithm of 10^{-3}. The logarithm of 2.34 is 0.369, and the logarithm of 10^{-3} is -3. Thus the logarithm of $0.00234 = 0.369 + (-3) = -2.631$.

To multiply two numbers, we add the logarithms of the numbers.

Example

Find 412×353:

$$\begin{array}{rl} \text{Logarithm of 412} & = 2.615 \\ +\ \text{Logarithm of 353} & = 2.548 \\ \hline \text{Logarithm of product} & = 5.163 \end{array}$$

The number whose logarithm is 5.163 is called the *antilogarithm* of 5.163, and it is 1.45×10^5. Thus, $412 \times 353 = (4.12 \times 10^2)(3.53 \times 10^2) = 14.5 \times 10^4 = 1.45 \times 10^5$.

To divide two numbers we subtract the logarithms of the numbers.

Example

Find 412/353:

$$\begin{array}{rl} \text{Logarithm of 412} & = 2.615 \\ -\ \text{Logarithm of 353} & = 2.548 \\ \hline \text{Logarithm of quotient} & = 0.067 \end{array}$$

The antilogarithm of 0.0671 is 1.17. Thus

$$\frac{412}{353} = 1.17 \quad \text{or} \quad \frac{4.12 \times 10^2}{3.53 \times 10^2} = 1.17$$

Combined operations are performed in precisely the same manner.

Significant Figures and Common Logarithms

For the common logarithm of a measured quantity, the number of digits after the decimal point (the number of digits in the mantissa) equals the number of significant figures in the original number. For example, if 23.5 is a measured quantity (three significant figures) then log 23.5 = 1.371 (three significant figures in the mantissa). The characteristic, 1, just places the decimal point and is not a significant figure.

EXAMPLE

Find $\dfrac{(353)(295)}{(412)}$:

$$\begin{array}{lr}
\text{Logarithm of } 353 & = 2.548 \\
+\ \text{Logarithm of } 295 & = 2.470 \\
\hline
\text{Logarithm of } 353 \times 295 & = 5.018 \\
-\ \text{Logarithm of } 412 & = 2.615 \\
\hline
\text{Logarithm of quotient} & = 2.403
\end{array}$$

The antilogarithm of 2.403 is 253. Thus (353)(296)/(412) = 253.

The extraction of roots of numbers by means of logarithms is a simple procedure.

EXAMPLE

Find $\sqrt[3]{7235} = (7235)^{1/3}$:

$$\begin{array}{c}
\text{Logarithm of } 7235 = 3.8594 \\
\tfrac{1}{3} \text{ logarithm of } 7235 = 1.2865 \\
\text{Antilogarithm } 1.2865 = 1.934 \times 10^1 = 19.34
\end{array}$$

Thus $19.34 = (7235)^{1/3}$.

Powers are found in the same fashion.

EXAMPLE

$(353)^3 = ?$

$$\begin{array}{ll}
\text{Logarithm } 353 & = 2.548 \\
3 \text{ logarithm } 353 & = 7.644 \\
\text{Antilogarithm } 7.644 & = 4.40 \times 10^7
\end{array}$$

Thus $(3.53 \times 10^2)^3 = 4.40 \times 10^7$.

Finding the antilogarithm of a negative logarithm is best illustrated by example.

Example

Find the antilogarithm of −7.1594. First convert the mantissa to a positive value, since there are no tables of negative logarithms. This is done as follows:

$$-7.1594 = -8 + 0.8406$$

Then find the antilogarithm of this logarithm as follows:

$$\text{Antilog } -7.1594 = \text{antilog } (-8) \times \text{antilog } (0.8406)$$
$$= 10^{-8} \times 6.930$$

Thus antilog of $-7.1594 = 6.930 \times 10^{-8}$.

After a little practice you should find that logarithms simplify the mathematical operations involving very large or very small numbers. Operations involving roots and powers are most easily performed using logarithms. Some basic rules to remember are the following:

$$10^a \times 10^b = 10^{(a+b)} \qquad \log (10^a \times 10^b) = \log 10^a + \log 10^b = a + b$$

$$\frac{10^a}{10^b} = 10^{(a-b)} \qquad \log (10^a/10^b) = \log 10^a - \log 10^b = a - b$$

$$(10^a)^b = 10^{ab} \qquad \log (10^a)^b = b \log 10^a = ba$$

$$\sqrt{10^a} = (10^a)^{1/2} = 10^{a/2} \qquad \log (10^a)^{1/2} = \tfrac{1}{2} \log 10^a = a/2$$

Appendix

Graphical Interpretation of Data: Calibration Curves and Least-Squares Analysis

GRAPHICAL INTERPRETATION OF DATA

TABLE B.1 Relationship Between Mass and Volume of Mercury

Volume (mL)	Mass (g)
2.00	27.0
2.50	33.75
3.00	40.5
3.50	47.25
4.00	54.0
4.50	60.75

Relationships between two or more variables can easily be visualized when the data are plotted or graphed. In such a graph the horizontal axis (x-axis) represents the experimentally varied variable, called the *independent variable.* The vertical axis (y-axis) represents the *dependent variable,* which responds to a change in the independent variable. For example, consider the data in Table B.1 for the mass of mercury as a function of volume. In this case volume is the independent variable and mass the dependent variable. A graph of these data is shown in Figure B.1. Clearly, the relationship between mass and volume of mercury is linear. The equation for a linear relationship is of the form $y = ax + b$, where a is the slope of the line ($\Delta y/\Delta x$) and b is the intercept with the y-axis. That is, b is the value of y when $x = 0$. Therefore, the equation for the data in Table B.1 is

$$\text{mass} = (a)\text{volume} + b$$

where $a = 13.5$ g/mL and $b = 0$.

or

$$\text{mass} = \left(\frac{13.5\text{ g}}{\text{mL}}\right)\text{volume} + 0$$

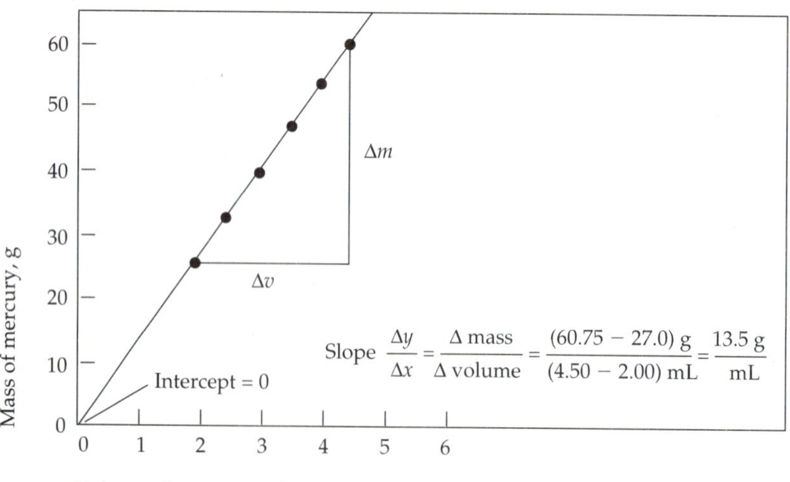

▲ **FIGURE B.1**

In this case, the slope, 13.5 g/mL, has the units of mass per unit volume, which is density. In practice, experimental data have uncertainties, and all points, therefore, usually do not lie precisely on the line. In these cases, draw the best straight line or calculate the best straight line from a least-squares analysis and then determine the slope and intercept of this line, as above.

CALIBRATION CURVES: LEAST-SQUARES ANALYSIS

Many analytical methods require a calibration step in which standards containing known amounts of analyte, x, are analyzed in the same way as the unknown. The experimental result, y, is then plotted versus x to give a calibration curve such as that shown in Figure B.2. These plots often are straight lines or linear relations. Seldom, however, do the experimental data fall exactly on the line, usually because of the experimental errors. The experimenter is then obligated to draw the "best" straight line through the experimental points. Statistical methods are available that allow for an objective determination of such a line and also for an estimation of the uncertainties associated with this line. Statisticians call this technique *linear-regression analyses.* We will use the simplest of these techniques, called the *method of least squares.*

We assume that a linear relation of the form

$$y = ax + b$$

does in fact exist.

Method

The line generated by a least-squares method is the one that minimizes the squares of the individual vertical displacements, or *residuals* (Figure B.2), from that line. In addition to providing the best fit of the experimental points to the line, the method calculates the slope, a, and intercept, b, of the line.

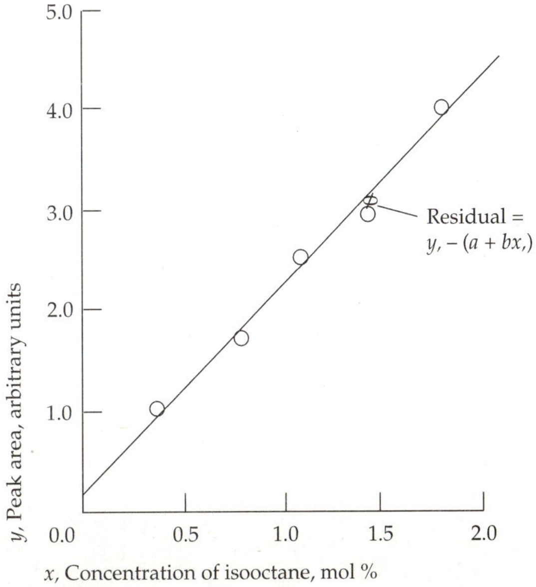

▲ **FIGURE B.2**

We define:

$$S_{xx} = \Sigma(x_i - \bar{x})^2 = \Sigma x_i^2 - \frac{(\Sigma x_i)^2}{n} \qquad [1]$$

$$S_{yy} = \Sigma(y_i - \bar{y})^2 = \Sigma y_i^2 - \frac{(\Sigma y_i)^2}{n} \qquad [2]$$

$$S_{xy} = \Sigma(x_i - \bar{x})(y_i - \bar{y}) = \Sigma x_i y_i - \frac{\Sigma x_i \Sigma yi}{n} \qquad [3]$$

where x_i and y_i are individual pairs of values for x and y defining each of the points to be plotted. The quantity n is the number of points, and $\bar{x}$ and $\bar{y}$ are the average values of x and y, that is,

$$\bar{x} = \frac{\Sigma x_i}{n} \qquad \bar{y} = \frac{\Sigma y_i}{n}$$

Note that S_{xx} and S_{yy} are simply the sums of the squares of the deviations from the means for individual values of x and y.

From S_{xx}, S_{yy}, and S_{xy} we may calculate:

1. The slope of the line, a:

$$a = \frac{S_{xy}}{S_{xx}}$$

2. The intercept of the line, b:

$$b = \bar{y} - a\bar{x}$$

3. The standard deviation about the regression, Sr, which is based upon deviations of the individual points from the line:

$$S_r = \sqrt{\frac{S_{yy} - a^2 S_{xx}}{n - 2}}$$

Example B.1

Given the following data, carry out a least-squares analysis.

x_i	y_i	x_i^2	y_i^2	$x_i y_i$
0.352	1.09	0.12390	1.1881	0.38368
0.803	1.78	0.64481	3.1684	1.42934
1.08	2.60	1.16640	6.7600	2.80800
1.38	3.03	1.90140	9.1809	4.18140
1.75	4.01	3.06250	16.0801	7.01750
5.365	12.51	6.90201	36.3775	15.81942

SOLUTION:

$$S_{xx} = \Sigma x_i^2 - \frac{(\Sigma x_i)^2}{n} = 6.90201 - \frac{(5.365)^2}{5} = 1.145365$$

$$S_{yy} = \Sigma y_i^2 - \frac{(\Sigma y_i)^2}{n} = 36.3755 - \frac{(12.51)^2}{5} = 5.07748$$

$$S_{xy} = \Sigma x_i y_i - \frac{\Sigma x_i \Sigma y_i}{n} = 15.81992 - \frac{(5.365)(12.51)}{5} = 2.39669$$

$$a = \frac{2.39669}{1.145365} = 2.0925 = 2.09$$

$$b = \bar{y} - a\bar{x} = \frac{12.51}{5} - 2.09\frac{(5.365)}{5} = 0.259$$

Thus the best linear equation is

$$y = 2.09x + 0.259$$

and $$S_r = \sqrt{\frac{S_{yy} - b^2 S_{xx}}{n-2}} = \sqrt{\frac{5.07748 - (2.0925)^2(1.145365)}{5-2}}$$

$$= \pm 0.144$$

$$= \pm 0.14$$

Appendix

Summary of Solubility Properties of Ions and Solids

	Cl^-	SO_4^{2-}	CO_3^{2-} PO_4^{3-}	CrO_4^{2-}	OH^- O^{2-}	H_2S pH = 0.5	S^{2-}, pH = 9
Li^+, Na^+, K^+, NH_4^+	S	S	S	S	S	S	S
Ba^{2+}	S	I	A	A	S^-	S	S
Ca^{2+}	S	S^-	A	S	S^-	S	S
Mg^{2+}	S	S	A	S	A	S	S
Fe^{3+}	S	S	A	A	A	S	A
Cr^{3+}	S	S	A	A	A	S	A
Al^{3+}	S	S	A, B	A, B	A, B	S	A, B
Ni^{2+}	S	S	A, N	A, N	A, N	S	A^+, O^+
Co^{2+}	S	S	A	A	A	S	A^+, O^+
Zn^{2+}	S	S	A, B, N	A, B, N	A, B, N	S	A
Mn^{2+}	S	S	A	A	A	S	A
Cu^{2+}	S	S	A, N	A, N	A, N	O	O
Cd^{2+}	S	S	A, N	A, N	A, N	A^+, O	A^+, O
Bi^{3+}	A	A	A	A	A	O	O
Hg^{2+}	S	S	A	A	A	O^+, C	O^+, C
Sn^{2+}, Sn^{4+}	A, B	A, B	A, B	A, B	A, B	A^+, C	A^+, C
Sb^{3+}	A, B	A, B	A, B	A, B	A, B	A^+, C	A^+, C
Ag^+	A^+, N	S^-, N	A, N	A, N	A, N	O	O
Pb^{2+}	HW, B, A^+	B	A, B	B	A, B	O	O
Hg_2^{2+}	O^+	S^-, A	A	A	A	O^+	O^+

Key: S, soluble in water.
A, soluble in acid (6 *M* HCl or other nonprecipitating, nonoxidizing acid).
B, soluble in 6 *M* NaOH.
O, soluble in hot 6 *M* HNO_3.
N, soluble in 6 *M* NH_3.
I, insoluble in any common reagent.
S^-, slightly soluble in water.
A^+, soluble in 12 *M* HCl.
O^+, soluble in aqua regia.
C, soluble in 6 *M* NaOH containing excess S^{2-}.
HW, soluble in hot water.

Example: For Cd^{2+} and OH^- the entry is A, N. This means that $Cd(OH)_2(s)$, the product obtained when solutions containing Cd^{2+} and OH^- are mixed, will dissolve to the extent of at least 0.1 mol/L when treated with 6 *M* HCl or 6 *M* NH_3. Since 6 *M* HNO_3, 12 *M* HCl, and aqua regia are at least as strongly acidic as 6 *M* HCl, $Cd(OH)_2(s)$ would also be soluble in those reagents.

Solubility Rules

Water-soluble salts	
Na^+, K^+, NH_4^+	All sodium, potassium, and ammonium salts are soluble.
NO_3^-, ClO_3^-, $C_2H_3O_2^-$	All nitrates, chlorates, and acetates are soluble.
Cl^-	All chlorides are soluble except AgCl, Hg_2Cl_2, and $PbCl_2$.*
Br^-	All bromides are soluble except AgBr, Hg_2Br_2, $PbBr_2$,* and $HgBr_2$.*
I^-	All iodides are soluble except AgI, Hg_2I_2, PbI_2, and HgI_2.
SO_4^{2-}	All sulfates are soluble except $CaSO_4$,* $SrSO_4$, $BaSO_4$, Hg_2SO_4, $PbSO_4$, and Ag_2SO_4.
Water-insoluble salts	
CO_3^{2-}, SO_3^{2-}, PO_4^{3-}, CrO_4^{2-}	All carbonates, sulfites, phosphates, and chromates are insoluble except those of alkali metals and NH_4^+.
OH^-	All hydroxides are insoluble except those of alkali metals and $Ca(OH)_2$,* $Sr(OH)_2$,* and $Ba(OH)_2$.
S^{2-}	All sulfides are insoluble except those of the alkali metals, alkaline earths, and NH_4^+.

*Slightly soluble.

Another way of stating the above rule is as follows:

Solubility Guidelines for Common Ionic Compounds in Water

Soluble compounds		**Important exceptions**
Compounds containing	NO_3^-	None
	$C_2H_3O_2^-$	None
	Cl^-	Salts of Ag^+, Hg_2^{2+}, and Pb^{2+}
	Br^-	Salts of Ag^+, Hg_2^{2+}, and Pb^{2+}
	I^-	Salts of Ag^+, Hg_2^{2+}, and Pb^{2+}
	SO_4^{2-}	Salts of Ca^{2+}, Sr^{2+}, Ba^{2+}, Hg_2^{2+}, and Pb^{2+}
Insoluble compounds		**Important exceptions**
Compounds containing	S^{2-}	Salts of NH_4^+, the alkali metal cations, and Ca^{2+}, Sr^{2+}, and Ba^{2+}
	CO_3^{2-}	Salts of NH_4^+ and the alkali metal cations
	PO_4^{3-}	Salts of NH_4^+ and the alkali metal cations
	OH^-	Compounds of the alkali metal cations, Ca^{2+}, Sr^{2+}, and Ba^{2+}

Appendix

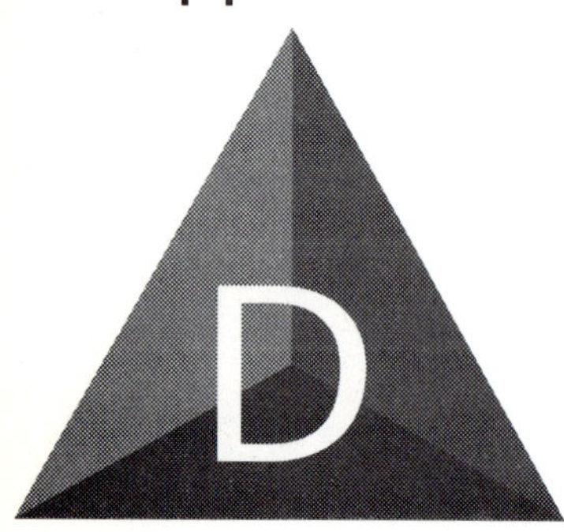

Solubility-Product Constants for Compounds at 25°C

Name	Formula	K_{sp}
Barium carbonate	$BaCO_3$	5.1×10^{-9}
Barium chromate	$BaCrO_4$	1.2×10^{-10}
Barium fluoride	BaF_2	1.0×10^{-6}
Barium hydroxide	$Ba(OH)_2$	5×10^{-3}
Barium oxalate	BaC_2O_4	1.6×10^{-7}
Barium phosphate	$Ba_3(PO_4)_2$	3.4×10^{-23}
Barium sulfate	$BaSO_4$	1.1×10^{-10}
Cadmium carbonate	$CdCO_3$	5.2×10^{-12}
Cadmium hydroxide	$Cd(OH)_2$	2.5×10^{-14}
Cadmium sulfide*	CdS	8×10^{-28}
Calcium carbonate	$CaCO_3$	2.8×10^{-9}
Calcium chromate	$CaCrO_4$	7.1×10^{-4}
Calcium fluoride	CaF_2	3.9×10^{-11}
Calcium hydroxide	$Ca(OH)_2$	5.5×10^{-6}
Calcium oxalate	CaC_2O_4	1×10^{-9}
Calcium phosphate	$Ca_3(PO_4)_2$	2.0×10^{-29}
Calcium sulfate	$CaSO_4$	9.1×10^{-6}
Cerium(III) fluoride	CeF_3	8×10^{-16}
Chromium(III) fluoride	CrF_3	6.6×10^{-11}
Chromium(III) hydroxide	$Cr(OH)_3$	6.3×10^{-31}
Cobalt(II) carbonate	$CoCO_3$	1.4×10^{-13}
Cobalt(II) hydroxide	$Co(OH)_2$	1.6×10^{-15}
Cobalt(III) hydroxide	$Co(OH)_3$	1.6×10^{-44}
Cobalt(II) sulfide*	CoS	5.0×10^{-22}
Copper(I) bromide	$CuBr$	5.3×10^{-9}
Copper(I) chloride	$CuCl$	1.2×10^{-6}
Copper(I) sulfide	Cu_2S	2.5×10^{-48}
Copper(II) carbonate	$CuCO_3$	1.4×10^{-10}
Copper(II) chromate	$CuCrO_4$	3.6×10^{-6}
Copper(II) hydroxide	$Cu(OH)_2$	2.2×10^{-20}
Copper(II) phosphate	$Cu_3(PO_4)_2$	1.3×10^{-37}
Copper(II) sulfide	CuS	6×10^{-37}
Gold(I) chloride	$AuCl$	2.0×10^{-13}
Gold(III) chloride	$AuCl_3$	3.2×10^{-25}
Iron(II) carbonate	$FeCO_3$	3.2×10^{-11}
Iron(II) hydroxide	$Fe(OH)_2$	8.0×10^{-16}
Iron(II) sulfide*	FeS	6×10^{-19}
Iron(III) hydroxide	$Fe(OH)_3$	4×10^{-38}
Lanthanum fluoride	LaF_3	7×10^{-17}
Lanthanum iodate	$La(IO_3)_3$	6.1×10^{-12}
Lead carbonate	$PbCO_3$	7.4×10^{-14}
Lead chloride	$PbCl_2$	1.6×10^{-5}
Lead chromate	$PbCrO_4$	2.8×10^{-13}
Lead fluoride	PbF_2	2.7×10^{-8}
Lead hydroxide	$Pb(OH)_2$	1.2×10^{-15}
Lead sulfate	$PbSO_4$	1.6×10^{-8}
Lead sulfide*	PbS	8.0×10^{-28}
Magnesium hydroxide	$Mg(OH)_2$	1.8×10^{-11}
Magnesium oxalate	MgC_2O_4	8.6×10^{-5}
Manganese carbonate	$MnCO_3$	1.8×10^{-11}
Manganese hydroxide	$Mn(OH)_2$	1.9×10^{-13}
Manganese(II) sulfide*	MnS	3.0×10^{-11}
Mercury(I) chloride	Hg_2Cl_2	1.3×10^{-18}
Mercury(I) oxalate	$Hg_2C_2O_4$	2.0×10^{-13}
Mercury(I) sulfide	Hg_2S	1×10^{-47}
Mercury(II) hydroxide	$Hg(OH)_2$	3.0×10^{-26}
Mercury(II) sulfide*	HgS	4×10^{-53}
Nickel carbonate	$NiCO_3$	6.6×10^{-9}
Nickel hydroxide	$Ni(OH)_2$	1.6×10^{-14}
Nickel oxalate	NiC_2O_4	4×10^{-10}
Nickel sulfide*	NiS	3×10^{-20}
Silver arsenate	Ag_3AsO_4	1.0×10^{-22}
Silver bromide	$AgBr$	5.0×10^{-13}
Silver carbonate	Ag_2CO_3	8.1×10^{-12}
Silver chloride	$AgCl$	1.8×10^{-10}
Silver chromate	Ag_2CrO_4	1.1×10^{-12}
Silver cyanide	$AgCN$	1.2×10^{-16}
Silver iodide	AgI	8.3×10^{-17}
Silver sulfate	Ag_2SO_4	1.4×10^{-5}
Silver sulfide	Ag_2S	6×10^{-51}
Strontium carbonate	$SrCO_3$	1.1×10^{-10}
Tin(II) hydroxide	$Sn(OH)_2$	1.4×10^{-28}
Tin(II) sulfide	SnS	1×10^{-26}
Zinc carbonate	$ZnCO_3$	1.4×10^{-11}
Zinc hydroxide	$Zn(OH)_2$	1.2×10^{-17}
Zinc oxalate	ZnC_2O_4	2.7×10^{-8}
Zinc sulfide*	ZnS	2×10^{-25}

*For a solubility equilibrium of the type $MS(s) + H_2O(l) \rightleftharpoons M^{2+}(aq) + HS^-(aq) + OH^-(aq)$

Appendix

Dissociation Constants for Acids at 25°C

Name	Formula	K_{a1}	K_{a2}	K_{a3}
Acetic	$HC_2H_3O_2$	1.8×10^{-5}		
Arsenic	H_3AsO_4	5.6×10^{-3}	1.0×10^{-7}	3.0×10^{-12}
Arsenous	H_3AsO_3	5.1×10^{-10}		
Ascorbic	$HC_6H_7O_6$	8.0×10^{-5}	1.6×10^{-12}	
Benzoic	$HC_7H_5O_2$	6.5×10^{-5}		
Boric	H_3BO_3	5.8×10^{-10}		
Butanoic	$HC_4H_7O_2$	1.5×10^{-5}		
Carbonic	H_2CO_3	4.3×10^{-7}	5.6×10^{-11}	
Chloroacetic	$HC_2H_2O_2Cl$	1.4×10^{-3}		
Chlorous	$HClO_2$	1.1×10^{-2}		
Citric	$H_3C_6H_5O_7$	7.4×10^{-4}	1.7×10^{-5}	4.0×10^{-7}
Cyanic	HCNO	3.5×10^{-4}		
Formic	$HCHO_2$	1.8×10^{-4}		
Hydrazoic	HN_3	1.9×10^{-5}		
Hydrocyanic	HCN	4.9×10^{-10}		
Hydrofluoric	HF	6.8×10^{-4}		
Hydrogen chromate ion	$HCrO_4^-$	3.0×10^{-7}		
Hydrogen peroxide	H_2O_2	2.4×10^{-12}		
Hydrogen selenate ion	$HSeO_4^-$	2.2×10^{-2}		
Hydrosulfuric acid	H_2S	9.5×10^{-8}	1×10^{-19}	
Hypobromous	HBrO	2.5×10^{-9}		
Hypochlorous	HClO	3.0×10^{-8}		
Hypoiodous	HIO	2.3×10^{-11}		
Iodic	HIO_3	1.7×10^{-1}		
Lactic	$HC_3H_5O_3$	1.4×10^{-4}		
Malonic	$H_2C_3H_2O_4$	1.5×10^{-3}	2.0×10^{-6}	
Nitrous	HNO_2	4.5×10^{-4}		
Oxalic	$H_2C_2O_4$	5.9×10^{-2}	6.4×10^{-5}	
Paraperiodic	H_5IO_6	2.8×10^{-2}	5.3×10^{-9}	
Phenol	HC_6H_5O	1.3×10^{-10}		
Phosphoric	H_3PO_4	7.5×10^{-3}	6.2×10^{-8}	4.2×10^{-13}
Propionic	$HC_3H_5O_2$	1.3×10^{-5}		
Pyrophosphoric	$H_4P_2O_7$	3.0×10^{-2}	4.4×10^{-3}	
Selenous	H_2SeO_3	2.3×10^{-3}	5.3×10^{-9}	
Sulfuric	H_2SO_4	Strong acid	1.2×10^{-2}	
Sulfurous	H_2SO_3	1.7×10^{-2}	6.4×10^{-8}	
Tartaric	$H_2C_4H_4O_6$	1.0×10^{-3}	4.6×10^{-5}	

Appendix

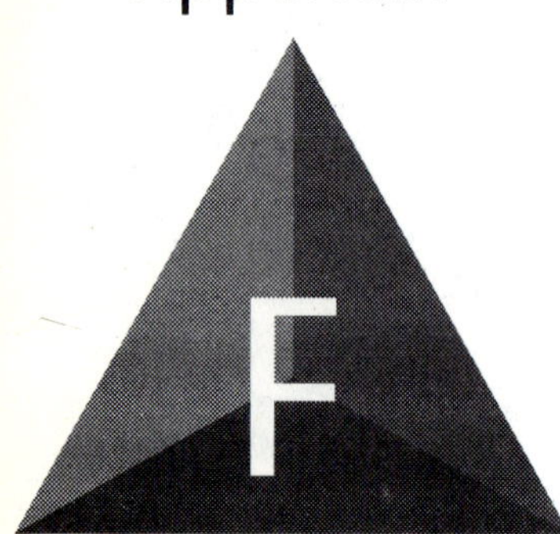

Dissociation Constants for Bases at 25° C

Name	Formula	K_b
Ammonia	NH_3	1.8×10^{-5}
Aniline	$C_6H_5NH_2$	4.3×10^{-10}
Dimethylamine	$(CH_3)_2NH$	5.4×10^{-4}
Ethylamine	$C_2H_5NH_2$	6.4×10^{-4}
Hydrazine	H_2NNH_2	1.3×10^{-6}
Hydroxylamine	$HONH_2$	1.1×10^{-8}
Methylamine	CH_3NH_2	4.4×10^{-4}
Pyridine	C_5H_5N	1.7×10^{-9}
Trimethylamine	$(CH_3)_3N$	6.4×10^{-5}

Appendix

Selected Standard-Reduction Potentials at 25°C

Half-reaction	***E*° (V)**
$Ag^+(aq) + e^- \longrightarrow Ag(s)$	+0.799
$AgBr(s) + e^- \longrightarrow Ag(s) + Br^-(aq)$	+0.095
$AgCl(s) + e^- \longrightarrow Ag(s) + Cl^-(aq)$	+0.222
$Ag(CN)_2^-(aq) + e^- \longrightarrow Ag(s) + 2CN^-(aq)$	−0.31
$Ag_2CrO_4(s) + 2e^- \longrightarrow 2Ag(s) + CrO_4^{2-}(aq)$	+0.446
$AgI(s) + e^- \longrightarrow Ag(s) + I^-(aq)$	−0.151
$Ag(S_2O_3)_2^{3-} + e^- \longrightarrow Ag(s) + 2S_2O_3^{2-}(aq)$	+0.01
$Al^{3+}(aq) + 3e^- \longrightarrow Al(s)$	−1.66
$H_3AsO_4(aq) + 2H^+(aq) + 2e^- \longrightarrow H_3AsO_3(aq) + H_2O(l)$	+0.559
$Ba^{2+}(aq) + 2e^- \longrightarrow Ba(s)$	−2.90
$BiO^+(aq) + 2H^+(aq) + 3e^- \longrightarrow Bi(s) + H_2O(l)$	+0.32
$Br_2(l) + 2e^- \longrightarrow 2Br^-(aq)$	+1.065
$BrO_3^-(aq) + 6H^+(aq) + 5e^- \longrightarrow \frac{1}{2}Br_2(l) + 3H_2O(l)$	+1.52
$2CO_2(g) + 2H^+(aq) + 2e^- \longrightarrow H_2C_2O_4(aq)$	−0.49
$Ca^{2+}(aq) + 2e^- \longrightarrow Ca(s)$	−2.87
$Cd^{2+}(aq) + 2e^- \longrightarrow Cd(s)$	−0.403
$Ce^{4+}(aq) + e^- \longrightarrow Ce^{3+}(aq)$	+1.61
$Cl_2(g) + 2e^- \longrightarrow 2Cl^-(aq)$	+1.359
$HClO(aq) + H^+(aq) + e^- \longrightarrow \frac{1}{2}Cl_2(g) + H_2O(l)$	+1.63
$ClO^-(aq) + H_2O(l) + 2e^- \longrightarrow Cl^-(aq) + 2OH^-(aq)$	+0.89
$ClO_3^-(aq) + 6H^+(aq) + 5e^- \longrightarrow \frac{1}{2}Cl_2(g) + 3H_2O(l)$	+1.47
$Co^{2+}(aq) + 2e^- \longrightarrow Co(s)$	−0.277
$Co^{3+}(aq) + e^- \longrightarrow Co^{2+}(aq)$	+1.842
$Cr^{3+}(aq) + 3e^- \longrightarrow Cr(s)$	−0.74
$Cr^{3+}(aq) + e^- \longrightarrow Cr^{2+}(aq)$	−0.41
$Cr_2O_7^{2-}(aq) + 14H^+(aq) + 6e^- \longrightarrow 2Cr^{3+}(aq) + 7H_2O(l)$	+1.33
$CrO_4^{2-}(aq) + 4H_2O(l) + 3e^- \longrightarrow Cr(OH)_3(s) + 5OH^-(aq)$	−0.13
$Cu^{2+}(aq) + 2e^- \longrightarrow Cu(s)$	+0.337
$Cu^{2+}(aq) + e^- \longrightarrow Cu^+(aq)$	+0.153
$Cu^+(aq) + e^- \longrightarrow Cu(s)$	+0.521
$CuI(s) + e^- \longrightarrow Cu(s) + I^-(aq)$	−0.185
$F_2(g) + 2e^- \longrightarrow 2F^-(aq)$	+2.87
$Fe^{2+}(aq) + 2e^- \longrightarrow Fe(s)$	−0.440
$Fe^{3+}(aq) + e^- \longrightarrow Fe^{2+}(aq)$	+0.771
$Fe(CN)_6^{3-}(aq) + e^- \longrightarrow Fe(CN)_6^{4-}(aq)$	+0.36

Half-reaction	$E°$ (V)
$2H^+(aq) + 2e^- \longrightarrow H_2(g)$	0.000
$2H_2O(l) + 2e^- \longrightarrow H_2(g) + 2OH^-(aq)$	−0.83
$HO_2^-(aq) + H_2O(l) + 2e^- \longrightarrow 3OH^-(aq)$	+0.88
$H_2O_2(aq) + 2H^+(aq) + 2e^- \longrightarrow 2H_2O(l)$	+1.776
$Hg_2^{2+}(aq) + 2e^- \longrightarrow 2Hg(l)$	+0.789
$2Hg^{2+}(aq) + 2e^- \longrightarrow Hg_2^{2+}(aq)$	+0.920
$Hg^{2+}(aq) + 2e^- \longrightarrow Hg(l)$	+0.854
$I_2(s) + 2e^- \longrightarrow 2I^-(aq)$	+0.536
$IO_3^-(aq) + 6H^+(aq) + 5e^- \longrightarrow \frac{1}{2}I_2(s) + 3H_2O(l)$	+1.195
$K^+(aq) + e^- \longrightarrow K(s)$	−2.925
$Li^+(aq) + e^- \longrightarrow Li(s)$	−3.05
$Mg^{2+}(aq) + 2e^- \longrightarrow Mg(s)$	−2.37
$Mn^{2+}(aq) + 2e^- \longrightarrow Mn(s)$	−1.18
$MnO_2(s) + 4H^+(aq) + 2e^- \longrightarrow Mn^{2+}(aq) + 2H_2O(l)$	+1.23
$MnO_4^-(aq) + 8H^+(aq) + 5e^- \longrightarrow Mn^{2+} + 4H_2O(l)$	+1.51
$MnO_4^-(aq) + 2H_2O(l) + 3e^- \longrightarrow MnO_2 + 4OH^-(aq)$	+0.59
$HNO_2(aq) + H^+(aq) + e^- \longrightarrow NO(g) + H_2O(l)$	+1.00
$N_2(g) + 4H_2O(l) + 4e^- \longrightarrow 4OH^-(aq) + N_2H_4(aq)$	−1.16
$N_2(g) + 5H^+(aq) + 4e^- \longrightarrow N_2H_5^+(aq)$	−0.23
$NO_3^-(aq) + 4H^+(aq) + 3e^- \longrightarrow NO(g) + 2H_2O(l)$	+0.96
$Na^+(aq) + e^- \longrightarrow Na(s)$	−2.71
$Ni^{2+}(aq) + 2e^- \longrightarrow Ni(s)$	−0.28
$O_2(g) + 4H+(aq) + 4e^- \longrightarrow 2H_2O(l)$	+1.23
$O_2(g) + 2H_2O(l) + 4e^- \longrightarrow 4OH^-(aq)$	+0.40
$O_2(g) + 2H^+(aq) + 2e^- \longrightarrow H_2O_2(aq)$	+0.68
$O_3(g) + 2H^+(aq) + 2e^- \longrightarrow O_2(g) + H_2O(l)$	+2.07
$Pb^{2+}(aq) + 2e^- \longrightarrow Pb(s)$	−0.126
$PbO_2(s) + HSO_4^-(aq) + 3H^+(aq) + 2e^- \longrightarrow PbSO_4(s) + 2H_2O(l)$	+1.685
$PbSO_4(s) + H^+(aq) + 2e^- \longrightarrow Pb(s) + HSO_4^-(aq)$	−0.356
$PtCl_4^{2-}(aq) + 2e^- \longrightarrow Pt(s) + 4Cl^-(aq)$	+0.73
$S(s) + 2H^+(aq) + 2e^- \longrightarrow H_2S(g)$	+0.141
$H_2SO_3(aq) + 4H^+(aq) + 4e^- \longrightarrow S(s) + 3H_2O(l)$	+0.45
$HSO_4^-(aq) + 3H^+(aq) + 2e^- \longrightarrow H_2SO_3(aq) + H_2O(l)$	+0.17
$Sn^{2+}(aq) + 2e^- \longrightarrow Sn(s)$	−0.136
$Sn^{4+}(aq) + 2e^- \longrightarrow Sn^{2+}(aq)$	+0.154
$VO_2^+(aq) + 2H^+(aq) + e^- \longrightarrow VO^{2+}(aq) + H_2O(l)$	+1.00
$Zn^{2+}(aq) + 2e^- \longrightarrow Zn(s)$	−0.763

Appendix

Spreadsheets

A computer spreadsheet program is a powerful tool for manipulating quantitative information.* In addition to simplifying repetitive calculations, spreadsheets allow us to plot such results as calibration curves and acid-base or redox titrations. Such graphs are critical for our understanding and interpretation of quantitative relationships.

Though nearly any spreadsheet program could be used for these purposes, the instructions given herein apply to one of the more widely available spreadsheets—Microsoft Excel. You will need specific directions for other spreadsheet software, but they will not be very different from what is presented here.

GETTING STARTED: AN EXAMPLE

As an example we will prepare a spreadsheet to compute the density of water from the equation

$$\text{density (g/mL)} = a_0 + a_1 T + a_2 T^2 + a_3 T^3 \quad (1)$$

where $a_0 = 0.99989$, $a_1 = 5.3322 \times 10^{-5}$, $a_2 = -7.5899 \times 10^{-6}$, $a_3 = 3.6719 \times 10^{-8}$, and T = temperature in °C. The blank spreadsheet in Figure H.1a has columns labelled A, B, C, . . . and rows numbered 1, 2, 3 . . ., etc. The box in column B, row 4 is called *cell* B4.

It is helpful to begin each spreadsheet with a title to make it more readable and to remind you of what you are doing with it. In Figure H.1b, we highlighted cell A1 and typed **Calculating the Density of H_2O with Equation 1.** The computer automatically spreads the text into adjoining cells.

We placed the constants in column A by selecting cell A4 and typing **Constants:** as a column heading. We then selected cell A5 and typed **a_0=** to indicate that the constant a_0 will be written in the next cell. This we did by selecting cell A6 and typing the number **0.99989** (without extra spaces). In cells A7 to A12, we entered the remaining constants in the same way. You may need to increase the number of decimal places using the button on the toolbar or by FORMAT ⟶ cells . . . ⟶ Number, click on Number in the list and increase decimal places to five. Note, however, that powers of ten are written in an unusual format, for example, E-05 means 10^{-5}. Your spreadsheet should now look similar to Figure H.1b.

*Spreadsheets for analytical chemistry: R. de Levie, *Principles of Quantitative Analysis,* New York: McGraw-Hill, 1997; D. Diamond and V. Hanratty, *Spreadsheet Applications in Chemistry Using Microsoft Excel,* New York: John Wiley and Sons, 1997; H. Freiser, *Concepts and Calculations in Analytical Chemistry: A Spreadsheet Approach,* Boca Raton, FL: CRC Press, 1992; R. de Levie, *A Spreadsheet Workbook for Quantitative Chemical Analysis,* New York: McGraw-Hill, 1992; B.V. Liengme, *A Guide to Microsoft Excel for Scientists and Engineers,* New York: John Wiley and Sons, 1997.

Columns

Rows

	A	B	C
1			
2			
3			
4		cell B4	
5			
6			
7			
8			
9			
10			
11			
12			

(a)

	A	B	C
1	Calculating the Density of H_2O with Equation 1		
2			
3			
4	Constants:		
5	a_0 =		
6	0.99989		
7	a_1 =		
8	5.3322E-05		
9	a_2 =		
10	-7.5899E-06		
11	a_3 =		
12	3.6719E-08		

(b)

	A	B	C
1	Calculating the Density of H_2O with Equation 1		
2			
3			
4	Constants:	Temp (°C)	Density (g/mL)
5	a_0 =	5	0.99997
6	0.99989	10	
7	a_1 =	15	
8	5.3322E-05	20	
9	a_2 =	25	
10	-7.5899E-06	30	
11	a_3 =	35	
12	3.6719E-08	40	

(c)

	A	B	C
1	Calculating the Density of H_2O with Equation 1		
2			
3			
4	Constants:	Temp (°C)	Density (g/mL)
5	a_0 =	5	0.99997
6	0.99989	10	0.99970
7	a_1 =	15	0.99911
8	5.3322E-05	20	0.99821
9	a_2 =	25	0.99705
10	-7.5899E-06	30	0.99565
11	a_3 =	35	0.00103
12	3.6719E-08	40	0.99223
13			
14	Formula:		
15	C5 = $A$6+$A$8*B5+$A$10*B5^2+$A$12*B5^3		

(d)

FIGURE H.1 Evolution of a spreadsheet for computing the density of water.

In cell B4, we then typed the heading **Temp (°C)** and entered temperatures from 5 through 40 in cells B5 through B12. This completes our *data input* to the spreadsheet. The *data output* will be the computed values of the density and will be entered in column C. We now need to input the computation and where to put the result. First we will highlight cell C4 and enter the heading **Density (g/mL)**. Then in cell C5 we will type the formula to carry out the computation:

=$A$6 + $A$8*B5 + $A$10*B5^2 + $A$12*B5^3

(You may omit the spaces before and after the arithmetic operators.) When you hit the RETURN or ENTER key, the number **0.99997** appears in cell C5. The formula above is the spreadsheet translation of Equation 1. $A$6 refers to the constant value in cell A6. (We will explain the dollar signs shortly.) B5 refers to the temperature given in cell B5. The times sign is *, and the exponentiation sign is ^. Thus, for example, the term "$A$12*B5^3" means "(contents of cell A12) × (contents of cell B5)³."

Now comes the most magical and useful property of the spreadsheet. Highlight cell C5 and the empty cells below it from C6 to C12. Then select the FILL DOWN command from the Edit menu. This procedure copies the formula from C5 into the cells below it and computes the results from the above formula operating on the corresponding entries in columns A and B. Then the computed density of water at each temperature appears in column C as in Figure H.1d.

In this example, we made three types of entries. *Labels* such as **a$_0$=** were typed in as text. An entry that does not begin with a digit or an equals sign is treated as text. *Numbers,* such as 5 in cell B5, were typed in column B. The spreadsheet treats a number differently from text. In cell C5, we entered a *formula* to carry out the computation. Formulas necessarily begin with an equals sign. The mathematical operations of addition, subtraction, multiplication, division, and exponentiation have the symbols +, −, *, /, and ^, respectively. *Functions* such as Exp (.) can be typed by you or they can be selected from the Insert menu or function menu. Exp (.) raises *e* to the power in parentheses. Other functions such as ln (.), log (.), sin (.), and cos (.) are also available. In some spreadsheet programs, functions are written in all capital letters.

The order of arithmetic operations in formulas is ^ first, followed by * and / (evaluated in order from left to right as they appear), and finally followed by + and − (also evaluated from left to right). Make liberal use of parentheses to be sure that the computer does what you intend. The contents of parentheses are evaluated first, before carrying out operations outside the parentheses. Some examples follow:

$$9/5*100 + 32 = (9/5)*100 + 32 = (1.8)*100 + 32$$

$$= (1.8)*100 + 32 = (180) + 32 = 212$$

$$9/5*(100 + 32) = 9/5*(132) = (1.8)*(132) = 237.6$$

$$9 + 5*100/32 = 9 + (5*100)/32 = 9 + (500)/32$$

$$= 9 + (500)/32 = 9 + (15.625) = 24.625$$

$$9/5\hat{}2 + 32 = 9/(5\hat{}2) + 32 = (9/25) + 32 = (0.36) + 32 = 32.36$$

When in doubt about how an expression will be evaluated by the computer, use parentheses to get the program to do what you intend.

ABSOLUTE AND RELATIVE REFERENCES

The formula **=$A$8*B5** refers to cells A8 and B5 in different ways. The symbol $A$8 is an *absolute reference* to the contents of cell A8. Wherever cell $A$8 is called in the spreadsheet, the computer goes to cell A8 to look for a number. In contrast, B5 is a *relative reference* in the formula in cell C5. When called from cell C5, the computer goes to cell B5 to find a number, but when called from cell C6, the computer goes to cell B6 to look for a number. Likewise, if it were called from cell C19, the computer would look in cell B19. This is why the cell address written without dollar signs surrounding it is called a relative reference. If you want the computer to always look only in cell B5, then type **$B$5**.

A SECOND EXAMPLE: CALCULATING STANDARD DEVIATION

Suppose we wished to calculate the standard deviation of the following numbers: 17.4, 18.1, 18.2, 17.9 and 17.6. There are two ways in which we can use a spreadsheet to do this. The first way involves entering the numbers and the required equations and then directing the computer to do the requisite calculations. The second way involves entering the numbers and using "built-in" functions to do the calculations. The following illustrates both ways.

Writing the equations to do the calculations

Begin by reproducing the spreadsheet template shown in Figure H.2a. Cells B4 to B8 contain the data (x values) whose mean and standard deviation we will compute.

In cells B9 and B10 we type the formulas to compute the sum and mean value of the x values, respectively, as in Figure H.2b. In cell C4 type the formula to compute the deviations from the mean, where x is in cell B4 and the mean is in cell B10. This formula appears in cell B16 of Figure H.2b. Use the FILL DOWN command to compute values in cells C5 to C8. Type a formula in cell D4 to compute the square of the value in cell C4. This formula appears in cell B17 of Figure H.2b. Use the FILL DOWN command to compute val-

	A	B	C	D
1	Computing standard deviation			
2				
3		Data = x	x-mean	(x-mean)^2
4		17.4		
5		18.1		
6		18.2		
7		17.9		
8		17.6		
9	sum =			
10	mean =			
11	std dev =			
12				
13	Formulas:	B9 =		
14		B10 =		
15		B11 =		
16		C4 =		
17		D4 =		
18		D9 =		
19				
20	Calculations using built-in functions:			
21	sum =			
22	mean =			
23	std dev =			

(a)

	A	B	C	D
1	Computing standard deviation			
2				
3		Data = x	x-mean	(x-mean)^2
4		17.4	-0.44	0.1936
5		18.1	0.26	0.0676
6		18.2	0.36	0.1296
7		17.9	0.06	0.0036
8		17.6	-0.24	0.0576
9	sum =	89.2		0.452
10	mean =	17.84		
11	std dev =	0.3362		
12				
13	Formulas:	B9 = B4+B5+B6+B7+B8		
14		B10 = B9/5		
15		B11 = SQRT(D9/(5-1))		
16		C4 = B4-$B$10		
17		D4 = C4^2		
18		D9 = D4+D5+D6+D7+D8		
19				
20	Calculations using built-in functions:			
21	sum =	89.2		
22	mean =	17.84		
23	std dev =	0.3362		
24				
25	Formulas:	B21 = SUM(B4:B8)		
26		B22 = AVERAGE(B4:B8)		
27		B23 = STDEV(B4:B8)		

(b)

FIGURE H.2 A spreadsheet to calculate standard deviation.

ues in cells D5 to D8. Type a formula in cell D8 to compute the sum of the numbers in cells D4 to D8 (see cell B18 in Figure H.2b). Type a formula in cell B11 to compute the standard deviation (cell B15, Figure H.2b). Then use cells B13 through B18 to document your formulas. Remember that all the formulas start with an equals sign, not a cell number. For example, the formula used to calculate the sum is = B4 + B5 + B6 + B7 + B8, but your documentation is B9 = B4 + B5 + B6 + B7 + B8. The former would be typed in cell B9, and the latter just reminds us what was done to get the number that appears in cell B9.

Using built-in functions

The calculations described above can be greatly simplified by using formulas that are built into the spreadsheet program. In cell B21 type **=SUM(B4:B8)**, which means find the sum of the numbers in cells B4 to B8. Cell B21 should display the same number as cell B9. In general, you will not know what functions are available and how to write them. Find the function menu in your program and find SUM in this menu. Now select cell B22 and go to the function menu; find a function that might be called AVERAGE or MEAN. When you type **=AVERAGE(B4:B8)** in cell B22, the value that appears in cell B22 should be the same as the number in cell B10. For cell B23, to find the standard deviation, type **=STDEV(B4:B8)** in cell B23 and then press ENTER. The value that appears in cell B23 is the same as in cell B11. In future spreadsheets, you may make liberal use of built-in functions to save work.

STEPS FOR PRODUCING A GRAPH FROM A SPREADSHEET

1. Enter the Excel program.
2. Type the x-axis label in cell A1.
3. Type the y-axis label in cell B1.
4. Input x-axis data in Column A, beginning with cell A2.
5. Input y-axis data in Column B, beginning with cell B2.
6. Click and drag to select all data, including column headings (i.e., cells A1 through B12).
7. Click on the ChartWizard icon in the upper right of the screen. This will cause a crossbar with a small chart icon to appear as the mouse cursor.
8. Select the space where you want the graph to be drawn by clicking and dragging from the upper left to lower right; this will draw a box in which the graph will be placed. (The size of this box can be altered at a later time.)
9. ChartWizard will show a dialog box (Step 1 of 5), asking you to verify that the cells you selected in Step 6 above contain the data you wish to graph. Click Next>.
10. Select a chart type (generally XY Scatter). Click Next>.
11. Select a format for the XY Scatter plot (generally 1). Click Next>.
12. You will see a sample chart. Check to see that the graph looks correct and that the choices to the right of the graph are as follows: Data series are in columns; use first 1 column for x data, and use first 1 row for legend text. (At this point, the graph may have the y-axis label as the title; don't worry, this will be changed in the next step. Also, the legend can be deleted later if not desired). Click Next>.
13. In Step 5 of 5 you can delete the legend and give the chart a title and x- and y-axis labels. Click Finish.

The following are formatting preferences:

14. The graph size can be changed at any time by clicking once on the graph area (anywhere) to get small black squares at the corners and on the sides of the graph area; click on a square and drag in any direction until sized as desired.
15. The format of either axis can be changed by double-clicking on the axis. Things that can be changed are:
    Patterns: for tick marks
    Values: numerical values printed on the axis
    Font: for numbers
    Formatting: gives the numbers in general, scientific, etc. format
    Alignment of the numbers on the axis
16. Any text on the graph can be changed by clicking once to highlight the text and typing in changes. Text formatting may be changed by double-clicking on the text. Choices for changes are:
    Patterns: add borders and colors
    Font: type and size
    Alignment: placement on the graph
17. The format of the points displayed on the graph may be changed by double-clicking on any point. Choices for change are:
    Patterns: change marker style/color, add a connecting line, delete markers
    X Values: change your original data source for x values (from the spreadsheet)
    Name & Values: change your original data source for y values (from the spreadsheet)
    X/Y Error bars: add error bars
    Data Labels: show value will print y values next to each data point; show label will print x values next to each data point (If data labels are added, you can format them by double-clicking on one of them; choices are Patterns, Font, Number, and Alignment.)

To print data and graph:

18. Make sure the graph does not cover any data (move it according to Step 14 if necessary).
19. Go to File, Print Preview, arrange the page as desired, and print.

To print graph only:

20. Double-click in the graph space to get a highlighted border.
21. Go to File, Print Preview, arrange the page as desired, and print.

To add a linear least-squares line to an existing graph:

22. Calculate a new y data set using $y=mx+b$ and place this in Column C, with an appropriate heading in cell C1.
23. Click once in the graph area.
24. Click on the ChartWizard icon.
25. Change the range of data set to include the new data. Click Next>.
26. View the new graph. Click OK.

27. To remove markers and add in a line, double-click on any point of the new data set, set line to Automatic, and set Marker to None. Click OK. (To see the new line without the markers, click anywhere off of the line).

To create a graph with the linear least-squares data already calculated:

28. Make sure the x-axis data is in Column A.
29. Begin at Step 6 above and select all three columns of data (A, B, and C).
30. Follow Steps 7 to 16 as desired. At Step 17, select the linear least-squares data set, remove its markers, and add a line.
31. Use the embedded functions to obtain the SLOPE and INTERCEPT of the least-squares fit line of the form y = slope x + intercept.

Appendix I

Qualitative-Analysis Techniques

I. MIXING SOLUTIONS AND PRECIPITATION

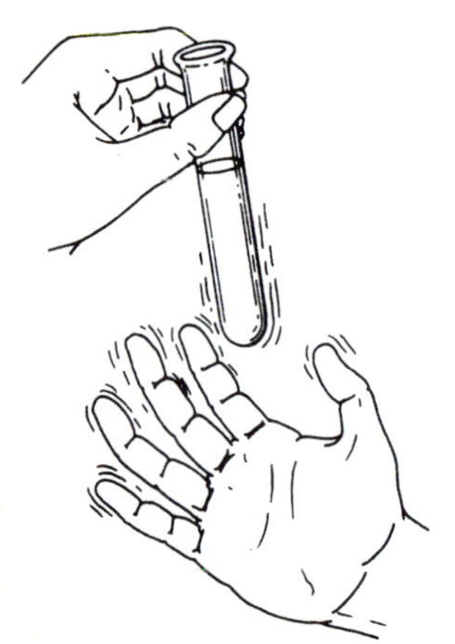

▲ **FIGURE I.1**

When one solution is added to another in a small test tube, it is important that the two be thoroughly mixed. Mixing can be accomplished by using a *clean* stirring rod, or it can be achieved by holding the test tube at the top in one hand and stroking or "tickling" it with the fingers of the other hand, as shown in Figure I.1. When precipitation reagents are added to solutions in the test tube and it is believed that precipitation is complete, centrifuge the sample. Always balance the centrifuge by placing a test tube filled with water to about the same level as your sample test tube directly *across* the centrifuge head from your sample. It usually requires only about 30 sec of centrifugation for the precipitate to settle to the bottom of the test tube.

II. DECANTATION AND WASHING OF PRECIPITATES

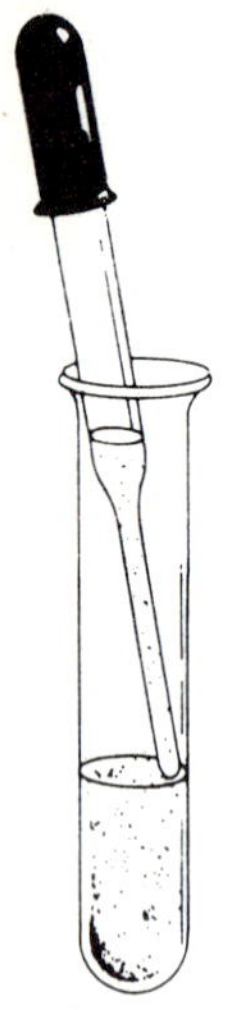

▲ **FIGURE I.2** Use of a dropper (also called a capillary pipet) to withdraw liquid from above a solid.

The liquid above the precipitate is the *supernatant liquid,* or the *decantate.* The best way to remove this liquid without disturbing the precipitate is to withdraw it by means of a capillary pipet, as shown in Figure I.2. We loosely refer to this operation as *decantation.* Because the preciptiate separated from the supernatant liquid by this technique will be wet with the decantate, it is necessary to *wash* the precipitate free of contaminating ions. Washing is usually accomplished by adding about 10 drops of distilled water to the precipitate, stirring with a stirring rod, and repeating the centrifuging and decanting.

III. TESTING ACIDITY

Instructions sometimes require making a solution acidic or basic to litmus by adding acid or base. Always be sure that the solution is thoroughly mixed after adding the acid or base; then, by means of a clean stirring rod, remove a drop of the solution and apply it to litmus paper. Do not dip the litmus paper directly into the solution. Remember, just because you have added acid (or base) to a solution does not ensure that it is acidic (or alkaline).

IV. HEATING SOLUTIONS IN SMALL TEST TUBES

The safest way to heat solutions in small test tubes is by means of a water bath, as described in Experiment 1 and shown in Figure I.3. A pipe cleaner wrapped around the test tube serves as a convenient handle for placing the tube into or removing it from the bath.

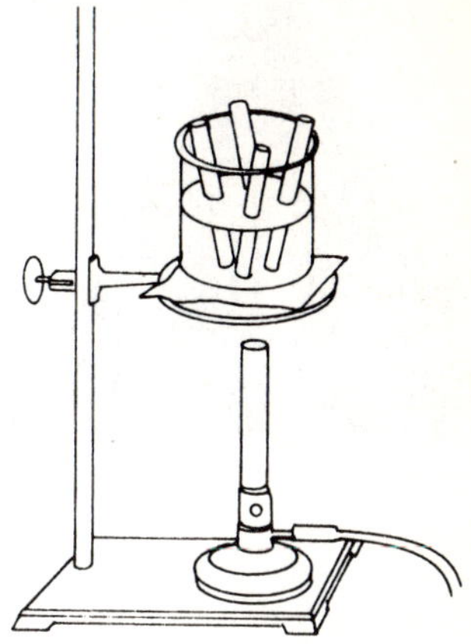

▲ **FIGURE I.3** Water bath for heating solutions in test tubes.

Appendix

J

Answers to Selected Review Questions

EXPERIMENT 1

1. The basic units of measurement in the SI system are mass, kilogram; temperature, K; volume, m^3; length, meter; energy, joule.
2. (1.35 L)(1000 mL/L) = 1350 mL = 1350 cm^3
4. Convection currents will tend to buoy the object and thus lead to an inaccurately low mass. Moreover, hot objects can damage the balance.
5. Thermometers and volumetric glassware are designed to be accurate, but their manufacture is subject to human error, and they should be calibrated.
6. Precision is a measure of the internal consistency of a replicate set of data.
7. Density is mass per unit volume. The determination of density requires a measurement of both mass and volume and so could not be done with a single unitary measurement.
9. $$\text{mean} = \frac{10.3 + 10.1 + 10.2}{3} = 10.2$$

   $$\text{average deviation from mean} = \frac{0.1 + 0.1 + 0.0}{3} = 0.07$$
11. $d = \frac{m}{V}$; $m = dV = (1.174\text{ g/mL})(850\text{ mL}) = 997.9\text{ g}$, or to three significant figures, 998 g = 0.998 kg.

EXPERIMENT 2

1. Melting points, boiling points, solubility properties, color, and densities are five physical properties.
3. Hexane, m.p. -94°C, is a liquid, and naphthalene, m.p. 80°C, is a solid at room temperature because room temperature is normally about 20°C.
4. The density of sodium acetate could not be determined in water because it is soluble in water. It could be determined in cyclohexane because it is not soluble in cyclohexane.
6. Thermometers, pipets, and other pieces of laboratory equipment are mass produced and subject to human error. Consequently, they should always be calibrated.
7. Chloroform is not miscible with water but is miscible with cyclohexane.
11. The liquid is hexane. (See Table 2.1.)

EXPERIMENT 3

1. A mixture differs from an impure substance only in the relative amounts of materials. If the substance contains primarily one component, it is called an impure substance. If, on the other hand, it contains several components in similar amounts, it is called a mixture. In reality, all impure substances are mixtures.
2. Sublimation is the process by which a substance changes physical states from the solid to the gaseous and back to the solid state without passing through the liquid state.
3. Filtration differs from decantation in that the liquid phase passes through a semipermeable substance such as a filter, whereas in decantation the liquid phase is separated from the solid phase by carefully pouring off the liquid. Decantation does not involve the use of another substance to achieve the separation, and is faster than filtration.
4. A hot object creates a buoyancy effect by radiating energy in the form of heat and will appear to have a reduced mass. In addition the hot object may cause damage to the delicate balance and so should be allowed to cool to room temperature before it is weighed.
5. All of the original sample should be recovered; thus the sum of the weights of the components, NH_4Cl, SiO_2, and NaCl, should precisely equal the total weight of the sample as no matter is being converted into energy in this experiment.
7. Zinc chloride is insoluble in cyclohexane and can be removed by filtering it from the cyclohexane.

EXPERIMENT 4

1. Before a chemical equation can be written, you must know the reactants and products of the reaction.
2. A color change or formation of a solid or a gas are indicative of a chemical reaction. Also, evolution or absorption of heat.
3. NO_2 is brown in color and has a pungent odor, while NO is colorless and has a slightly pleasant odor.
4. Metathesis reactions are atom (or group) transfer reactions. For example, $AgNO_3 + NaCl \longrightarrow AgCl + NaNO_3$ is a metathesis reaction in which chloride transfer occurs.
5. A precipitate is an insoluble substance that separates from a homogeneous chemical reaction as a solid.
6. $2KBrO_3(s) \longrightarrow 2KBr(s) + 3O_2(g)$; $MnBr_2(aq) + 2AgNO_3(aq) \longrightarrow Mn(NO_3)_2(aq) + 2AgBr(s)$

EXPERIMENT 5

1. A compound is composed of a definite number of whole atoms in a fixed proportion.
2. We use relative atomic weights because the actual weights of atoms are exceedingly small and are awkward numbers to manipulate.

3. Formula weights are used for ionic compounds that do not exist as discrete molecular entities. Molecular weights are used for covalent compounds that do exist as discrete molecular entities.
4. The formula weight of Na_2CO_3 is:

$$\begin{array}{c} \text{Na} \quad \text{C} \quad \text{O} \\ 2(23.0) + 12.0 + 3(16.0) = 106 \text{ amu} \end{array}$$

$$\%\text{Na} = \frac{46.0 \text{ amu}}{106 \text{ amu}} \times 100 = 43.4\%$$

$$\%\text{C} = \frac{12.0 \text{ amu}}{106 \text{ amu}} \times 100 = 11.3\%$$

$$\%\text{O} = \frac{48.0 \text{ amu}}{106 \text{ amu}} \times 100 = 45.3\%$$

5. Assuming 100 g of compound, this would contain 35.74 g of Ca and 63.36 g of Cl. Thus we have

$$\text{moles Ca} = (35.74 \text{ g})\left(\frac{1 \text{ mol Ca}}{140.0 \text{ g}}\right) = 0.894 \text{ mol}$$

$$\text{moles Cl} = (63.36 \text{ g})\left(\frac{1 \text{ mol Cl}}{35.45 \text{ g}}\right) = 1.79 \text{ mol}$$

or $Ca_{0.894}Cl_{1.79}$, which implies that the smallest whole-number combining ratio of the constituent elements (that is, the empirical formula) is $CaCl_2$.
6. The law of definite proportions states simply that compounds form from elements in definite proportions by weight such that all samples of the same compound will contain the same ratios of the constituent elements.
7. Molecular formulas are always identical to or multiples of the empirical formulas, because the empirical formulas represent the smallest whole-number combining ratios and the molecular formulas the actual numbers of the constituent elements in chemical compounds.

12. $$\text{Atoms potassium} = \left(6.02 \text{ x } 10^{23} \frac{\text{atoms}}{\text{mol}}\right)\left(\frac{0.01456 \text{ g}}{39.0983 \text{ g/mL}}\right)$$

$$= 3.73 \text{ x } 10^{19} \text{ atoms}$$

EXPERIMENT 6

1. Numerous examples of redox and metathesis reactions could be cited. Two are $CuO + H_2 \longrightarrow Cu + H_2O$ (redox) and $AgNO_3 + NaI \longrightarrow AgI + NaNO_3$ (metathesis).
2. Reactions will proceed to completion whenever one of the products is physically removed from the reaction medium. This often occurs when a gas or a solid is formed.
3. Percent yield = (actual yield/theoretical yield)(100).
4. Materials can be separated from one another by distillation, sublimation, filtration, decantation, sedimentation, chromatography, and extraction.
5. Exothermic reactions are usually accompanied by a temperature increase, while endothermic reactions are usually accompanied by a temperature decrease.

6. The balanced chemical equation is $Cu + 4HNO_3 \longrightarrow Cu(NO_3)_2 + 2H_2O + 2NO_2$. From this equation we have moles $Cu(NO_3)_2$ = moles Cu and

$$\text{moles Cu} = \frac{2.09\ \text{g}}{63.5\ \text{g/mol}} = 0.0329\ \text{mol}$$

The theoretical yield is $Cu(NO_3)_2$ = (0.0329 mol)(188 g/mol) = 6.19 g

$$\%\ \text{yield} = \frac{3.40\ \text{g} \times 100}{6.19\ \text{g}} = 54.9\%$$

7. The maximum percent yield in any reaction is 100%.

EXPERIMENT 7

1. Many chemicals are hazardous, and the haphazard mixing of chemicals could lead to the production of species with very dangerous properties.
2. Addition of a strong base to a solution containing NH_4^+ will release gaseous NH_3, which will cause a piece of moist, red litmus held above the reaction solution to turn blue.
3. Addition of an acid to a solution containing CO_3^{2-} will release gaseous CO_2, which will react with $Ba(OH)_2$ to precipitate $BaCO_3$.
4. Solid chloride salts will release gaseous HCl when heated with concentrated H_2SO_4, and solutions of chloride salts will precipitate AgCl if treated with $AgNO_3$ solution.
5. Solutions of sulfate salts will precipitate $BaSO_4$ when treated with $BaCl_2$.
6. Solutions of iodide salts will react with Cl_2 to liberate I_2, which will appear brown in H_2O and purple in mineral oil.
8. $2LiCl(s) + H_2SO_4(aq) \longrightarrow 2HCl(g) + Li_2SO_4(aq)$
   $NH_4^+(aq) + OH^-(aq) \longrightarrow NH_3(g) + H_2O(l)$
   $AgNO_3(aq) + I^-(aq) \longrightarrow AgI(s) + NO_3^-(aq)$
   $NaHCO_3(s) + H^+(aq) \longrightarrow CO_2(g) + H_2O(l) + Na^+(aq)$

EXPERIMENT 8

1. Gravimetric analyses involve a weighing as the determining measurement, whereas volumetric analyses involve a volume measurement as the determining measurement.
2. Stoichiometry is the mole ratio of atoms in a compound or compounds in a chemical reaction and refers to the amounts of substances involved in reactions.
3. Silver chloride is photosensitive and reacts with light to produce silver metal and chlorine gas, which will lead to a low result if the silver chloride is not protected from light.
4. Indeterminate errors are just that, indeterminant. They cannot be ascertained or eliminated, but rather occur in a random fashion and so are accounted for with statistics.
5. Standard deviation measures precision.
6. If the filter paper were opened in this way, the solution and precipitate would simply pass through, and no filtration would result.

7. Since photodecomposition liberates Cl_2 from AgCl according to the equation $2AgCl \xrightarrow{hv} 2Ag + Cl_2$, the Ag will weigh less than the AgCl and your results will be low.

EXPERIMENT 9

1. Only this level of accuracy is usually required for the analysis of a consumer chemical.
2. $MgSO_4 \cdot 7H_2O$ is Epsom salts, NH_3 is household ammonia, and $(CH_3)_2CHOH$ is isopropyl alcohol.
3. The label 23-19-17 means that the plant food is guaranteed to contain at least 23% nitrogen, 19% phosphorus (expressed as P_2O_5), and 17% of potassium (expressed as K_2O).
4. Assuming 100 g of plant food, we would have 19 g of P_2O_5.

$$(19\text{ g } P_2O_5)\left(\frac{1\text{ mol } P_2O_5}{141.9\text{ g } P_2O_5}\right)\left(\frac{2\text{ mol P}}{1\text{ mol } P_2O_5}\right)\left(\frac{30.97\text{ g P}}{1\text{ mol P}}\right) = 8.3\text{ g P} \therefore 8.3\%\text{ P}$$

5. Assuming 100 g of plant food, we would have 17 g of K_2O

$$(17\text{ g } K_2O)\left(\frac{1\text{ mol } K_2O)}{94.2\text{ g } K_2O}\right)\left(\frac{2\text{ mol K}}{\text{mol } K_2O}\right)\left(\frac{39.1\text{ g K}}{1\text{ mol K}}\right) = 14\text{ g K}$$

$$\%K = \frac{14\text{ g}}{100\text{ g}} \times 100 = 14\%$$

6. The formula weight of $CuSO_4 \cdot 5H_2O$ is $63.54 + 32.06 + 9 \times 16.0 + 10 \times 1.0 = 249.6$
7. Standard deviation measures precision.
8. $\text{mean} = \dfrac{12.1 + 12.4 + 12.6}{3} = 12.4$

   $\text{average deviation from the mean} = \dfrac{0.3 + 0.0 + 0.2}{3} = 0.17$

$$s = \sqrt{\frac{(0.3)^2 + (0.0)^2 + (0.2)^2}{2}} = 0.3$$

   relative average deviation from the mean = 0.17/12.4 = 0.014
9. Indeterminate errors are errors that cannot be detected. They occur randomly and are dealt with statistically.
10. Qualitative analysis determines what substances are present, whereas quantitative analysis tells us how much of the substance is present.
11. Accuracy is the closeness of a result to the true value.

EXPERIMENT 10

1. Chromatography is yet another technique that permits the separation of substances.
2. An R_f value is the distance that a substance moves on a chromatographic support relative to the distance that the solvent moves on the same support under the same conditions. It aids in the qualitative identification of substances.

3. The reactions are, for nickel, $Ni^{2+} + 2DMG^- \longrightarrow Ni(DMG)_2$, where DMG^- is the monoanion of dimethylglyoxime; and for copper, $Cu^{2+} + 4NH_3 \longrightarrow [Cu(NH_3)_4]^{2+}$.
4. The solvent moves along the chromatographic support because of capillary action and dipole-dipole attraction.
10. The petri dish (or beaker) should be covered so as to maintain equilibrium between the solvents in the liquid and gaseous phases. If the vessel were not covered, volatile liquid would escape.

EXPERIMENT 11

1. Ionic bonding involves the transfer of one or more electrons from an atom with a low ionization potential to an atom with a high electron affinity. Covalent bonding results from the sharing of electrons between two atoms. Metallic bonding is a type of covalent bonding, but it involves the sharing of electron density among several neighboring atoms in three dimensions.
2. HCl, HCN, and CO_2 possess polar covalent bonds.
3. HCl and HCN have molecular dipole moments.
4. AB_n molecules wherein the A atom is surrounded by 2, 3, 4, 5, and 6 electron pairs have linear, trigonal-planar, tetrahedral, trigonal-bipyramidal, and octahedral geometries, respectively.

EXPERIMENT 12

1. Red, orange, yellow, green, blue, indigo, violet.
3. $\lambda = c/v = 3 \times 10^8 \text{ ms}^{-1}/76 \text{ s}^{-1} = 3.9 \times 10^6 \text{ m}$;

$$3.9 \times 10^6 \text{ m} \left(\frac{1 \text{ mile}}{1.61 \text{ m}}\right) = 2.4 \times 10^6 \text{ miles}$$

4. $E = hv = 6.63 \times 10^{-34} \text{ J} \cdot \text{s} \times 76 \text{ s}^{-1} = 5.04 \times 10^{-32} \text{ J}$
5. Green light has the higher frequency and greater energy because it has the shorter wavelength.

EXPERIMENT 13

1. The pressure of an ideal gas increases as the temperature increases at constant volume.
2. The volume of an ideal gas decreases as the pressure increases at constant temperature.
3. The volume of an ideal gas increases as the number of molecules increases at constant temperature and pressure.
4. $PV = nRT$; when the units of R are L-atm/mol-K, P is in atm, V in L and T in K.
5. Combining the ideal-gas law and the definition of moles, we have

$$PV = nRT \text{ and } n = \frac{m}{\mathcal{M}}$$

$$n = \frac{PV}{RT} = \frac{m}{\mathcal{M}} \quad \text{or} \quad \mathcal{M} = \frac{mRT}{(PV)}$$

6.
$$\mathcal{M} = \frac{(0.80\text{ g})\left(\dfrac{0.082\text{ L-atm}}{\text{mol-K}}\right)(298\text{ K})}{\left(\dfrac{680\text{ mm}}{760\text{ mm/atm}}\right)(0.295\text{ L})}$$

$$= 74\text{ g/mol}$$

7. STP is 273 K and 1 atm, so that

$$V = (600\text{ mL})\left(\frac{273\text{ K}}{333\text{ K}}\right)\left(\frac{650\text{ mm}}{760\text{ mm}}\right)$$
$$= 420\text{ mL}$$

EXPERIMENT 14

1. Gases obey the ideal-gas law at high temperatures and low pressures.
2. From $PV = nRT$, $R = PV/nT$ and STP is 1 atm and 0°C, or 273 K, so that

$$R = \frac{(1\text{ atm})(22.4\text{ L})}{(1\text{ mol})(273\text{K})} = 0.082\,\frac{\text{L-atm}}{\text{mol-K}}$$

3. Equalizing the water levels equalizes the pressures and ensures that the total pressure in the bottle is atmospheric and does not contain a contribution from the pressure due to the height of the water column.
4. Since gaseous and liquid water are in dynamic equilibrium, there will always be some water vapor above a sample of liquid water. Since the vapor pressure of water is reasonably high at ambient temperature, it makes a significant contribution to the total pressure.
5. An error analysis allows you to judge the reliability of your data and gives an indication of the potential sources of error.
6. The ideal-gas law assumes that there are no forces of attraction between the individual gaseous molecules. Whenever this isn't so, real molecules will not obey the ideal-gas law. This would be expected to occur at very high pressures and at very low temperatures where molecules are so close to one another that they necessarily interact.

   The ideal-gas law also assumes that the gas particles have no volume. At high pressures their volume may become appreciable relative to the volume of the container.
7. $PV = nRT$

$$P = \frac{nRT}{V} = \frac{\left(\dfrac{0.05\text{ g}}{2\text{ g/mol}}\right)\left(\dfrac{0.082\text{ L-atm}}{\text{mol-K}}\right)(298\text{ K})}{0.100\text{ L}}$$
$$= 6\text{ atm}$$

   assuming no gas in the void volume. Clearly this does present a problem, as H_2 gas is extremely explosive. Because lead-storage batteries produce H_2, sealing them would be dangerous.

EXPERIMENT 15

1. Most metals have high thermal and electrical conductivities, high luster, malleability, and ductility, whereas nonmetals usually have low

thermal and electrical conductivities, low luster, low malleability, and low ductility. In addition, relative to nonmetals, metals have low ionization potentials and low electron affinities.

2. Ionization energy is the energy required to remove an electron from an atom in the gaseous state.
3. Electron affinity is the energy produced when an electron is added to a species in the gas phase and is the inverse of the ionization energy in a physical sense, but the values are not just of opposite sign because you are considering slightly different processes; in each case, they differ by one electron.
4. An oxidation must always be accompanied by a reduction because the species being oxidized must transfer an electron to some other species that is reduced. The electron cannot just be given up to free space.
5. By systematically observing the displacement reactions among metals and their cations, it is possible to determine the relative oxidation potentials of the metals. The metal with the lower reduction potential will reduce a cation of a metal with a higher reduction potential.
6. $2Mg + O_2 \longrightarrow 2MgO$; $Zn + 2HCl \longrightarrow ZnCl_2 + H_2$; $Zn + Cu^{2+} \longrightarrow Zn^{2+} + Cu$.

EXPERIMENT 16

1. Hydrogen is produced from H_2O at the cathode. Hydrogen thus undergoes a change in oxidation state from +1 in H_2O to 0 in H_2, and this is a reduction reaction.
2. A coulomb is the quantity of electrical charge passing a point in a circuit in 1 sec when the current is 1 ampere.
3. In general, reduction occurs at the cathode and oxidation at the anode in an electrochemical cell regardless of whether the cell is producing energy (a galvanic cell) or consuming energy (an electrolytic cell).
4. The water column exerts a pressure that is directly proportional to the height of the water column, just as atmospheric pressure and altitude are related. The higher the altitude above sea level, the less the atmosphere and the lower the atmospheric pressure.
5. The sulfuric acid is present to increase the rate of the reaction by increasing the ability of the solution to conduct an electrical current. According to Faraday's law, the more current that passes through this solution, the larger the number of species that will react. The more ions that are present in solution, the greater will be the conductivity of the solution and the greater the current flow. Sulfuric acid is a strong electrolyte and is almost completely dissociated in solution to produce the ions H_3O^+ and SO_4^{2-}. It does not directly enter into the electrochemical reaction.

EXPERIMENT 17

1. A faraday is the charge on 1 mol of electrons and equals 96,500 coulombs. A salt bridge is an ion-containing conducting medium used to physically and chemically separate two half-cells in an electrochemical cell. An anode is an electrode at which oxidation occurs, and a cathode is an electrode at which reduction occurs in an electrochemical cell. A voltaic cell is an electrochemical cell that produces electrical energy

by means of a spontaneous redox reaction. An electrolytic cell is one that requires energy in order to bring about a redox reaction.

2. The cell $Ag|Ag^+||Cu^{2+}|Cu$ represents the reaction $2Ag + Cu^{2+} \longrightarrow 2Ag^+ + Cu$.
3. $E° = 0.34 - 0.80\ V = -0.46\ V$
4. The cell $Zn|Zn^{2+}(0.10\ M)||Cu^{2+}(0.40\ M)|Cu$ represents the reaction $Zn + Cu^{2+} \longrightarrow Zn^{2+} + Cu$, for which

$$E = E° - (0.059/2)\log([Zn^{2+}]/[Cu^{2+}])$$

so that

$E = 1.100\ V - 0.0295\ \log([0.10]/[0.40]) = 1.100 + 0.0178 = 1.118\ V.$

5. $E° = -1.18 + 2.90 = 1.72\ V$. To calculate K_{eq}, we use the equation $E = E° - (0.0592/n)\log K_{eq}$. Because at equilibrium $E = 0$, we have

$$E° = \left(\frac{0.0592}{n}\right)\log K_{eq}$$

$$\log K_{eq} = \frac{nE°}{0.0592} = \frac{(2)(1.72)}{0.0592} = 58.1081$$

$$K_{eq} = \text{antilog}(58.1081) = 1.28 \times 10^{58}$$

The free energy for the given cell is

$$\Delta G° = -n\mathcal{F}E° = -(2\text{ mol } e-)\left(\frac{96{,}500\text{ J}}{\text{V-mol } e-}\right)(1.72\text{ V})$$

$$= -332{,}000\text{ J or } -332\text{ kJ}$$

6. Most spontaneous reactions (ΔG negative) are exothermic (ΔH negative). Because voltaic cells have spontaneous reactions, we would expect ΔH to be negative for most voltaic cells.

EXPERIMENT 18

1. Ionic bonding results from transfer of electrons from one species to another to form a chemical bond, whereas covalent bonding results from sharing electron density (equally or unequally) between the two partners forming the chemical bond. Ionic bonding will result from the union of atoms having low ionization energies with atoms of high electron affinity. Covalent bonding will result from the union of atoms with similar or identical ionization energies and electron affinities.
2. An anhydride is literally a compound without water. Thus, $H_2SO_4 - H_2O = SO_3$; SO_3 is the anhydride of H_2SO_4.
3. According to the Brønsted-Lowry definition, an acid is a proton donor and a base is a proton acceptor.
4. The autodissociation of water is $2H_2O \rightleftharpoons H_3O^+ + OH^-$, for which $K_{eq} = [H_3O^+][OH^-]/[H_2O]^2$ and $K_w = [H_3O^+][OH^-]$. Since the molar concentration of water changes so little in this reaction, it is essentially a constant and is incorporated in K_w. $K_w = 55.6\ K_{eq}$ for dilute aqueous solutions.
5. The concept of pH is introduced to avoid the manipulation of very small numbers, because the concentrations of H_3O^+ are usually very small in aqueous solution.

6. The following table illustrates when aqueous solutions are acidic, basic, or neutral in terms of $[H^+]$, $[OH^-]$, and pH:

	$[H^+]$	$[OH^-]$	pH
Acidic	$> 10^{-7}\,M$	$< 10^{-7}\,M$	< 7
Basic	$< 10^{-7}\,M$	$> 10^{-7}\,M$	> 7
Neutral	$= 10^{-7}\,M$	$= 10^{-7}\,M$	$= 7$

7. $pH = -\log[H^+] = -\log(10^{-4}) = 4$
   $[OH^-] = 10^{-14}/[H^+] = 10^{-14}/10^{-4} = 10^{-10}\,M$
8. $MgO + H_2O \longrightarrow Mg(OH)_2$; $SO_3 + H_2O \longrightarrow H_2SO_4$

EXPERIMENT 19

1. Solute is is the lesser component and solvent the greater component in a solution.
2. Three colligative properties are boiling point, freezing point, and vapor pressure. They are called colligative properties because they are related to the number and energy of collisions between particles and not to what the particles are.
3. A volatile substance has a high vapor pressure, and a nonvolatile substance has a low vapor pressure at room temperature. Obviously, volatility is a relative term and depends upon temperature and pressure. Volatility increases with increasing temperature and decreases with increasing pressure.
6. Supercooling involves the lowering of the temperature of a substance below its normal freezing point without the solidification of the substance. Supercooling can be minimized by cooling slowly with rapid stirring.
7. Mol of benzene is 2.50 g/(78 g/mol) = 0.032 mol. Consequently, a solution of 2.50 g of benzene in 120 g of $CHCl_3$ is 0.032 mol/0.120 kg or 0.267 molal. The freezing-point lowering is thus $\Delta T = K_{fp}m$, or $\Delta T = (4.68°C/m)(0.267\ m) = 1.25°C$, and the freezing point is $-63.5°C - 1.25°C = -64.8°C$.
8. $\Delta T = 80.6°C - 75.4°C = 5.2°C$; $m = \Delta T/K_{fp} = 5.2°C/(6.9°C/m) = 0.75\ m$; m = moles solute/1000 g solvent, so that 2.00 g/25.0 g = x g/1000 g, x = 80 g, and 80 g = 0.75 mol, or molar mass = 80 g/0.75 mol = 1.1×10^2 g/mol.

EXPERIMENT 20

1. Standardization is the process of determining the concentration of a solution. It is usually achieved by titrating the solution to be standardized against a known amount of a primary standard substance according to a known reaction.
2. Titration is the technique of accurately measuring the volume of a solution that is required to react with a known amount of another reagent.
3. Molarity is the number of moles of solute in a liter of solution.
4. You weigh by difference because this is in general more accurate than weighing directly.

5. An equivalence point is the point in a titration at which stoichiometrically equivalent amounts of the two reactants are brought together. An end point is the point in a titration where some indicator (such as a dye or electrode) undergoes a discernible change. Ideally, one hopes that these two points coincide, but in practice they differ slightly. It becomes of primary importance to minimize the difference if accuracy is desired.
6. Parallax is the apparent displacement or the difference in apparent direction of an object as seen from two different points not on a straight line with the object. It should be avoided because it introduces an error in the measurement.
7. Carbon dioxide should be removed from the water because it is an acid anhydride and reacts with water to produce carbonic acid according to the reaction $CO_2 + H_2O \rightleftharpoons H_2CO_3$. Its presence would lead to an erroneous determination of the amount of a particular acid or base in solution. It is particularly detrimental to sodium hydroxide solutions as these absorb carbon dioxide to produce sodium carbonate.
8. $$\text{Molarity} = \frac{\text{moles solute}}{\text{liter solution}}$$

$$= \frac{(3.78\text{ g})}{(126\text{ g/mol})(0.200\text{ L})} = 0.150\ M$$

EXPERIMENT 21

1. (a) Molecular: $2HNO_3(aq) + BaCO_3(s) \longrightarrow Ba(NO_3)_2(aq) + CO_2(g) + H_2O(l)$,
   ionic: $2H^+(aq) + 2NO_3^-(aq) + BaCO_3(s) \longrightarrow$
   $Ba^{2+}(aq) + 2NO_3^-(aq) + CO_2(g) + H_2O(l)$;
   net ionic: $BaCO_3(s) + 2H^+(aq) \longrightarrow Ba^{2+}(aq) + H_2O(l) + CO_2(g)$
   (b) As above; insoluble product $PbCl_2(s)$ is formed; net ionic:
   $Pb^{2+}(aq) + 2Cl^-(aq) \longrightarrow PbCl_2(s)$
   (c) Net ionic: $HC_2H_3O_2(aq) + OH^-(aq) \longrightarrow H_2O(l) + C_2H_3O_2^-(aq)$.
   (d) Net ionic: $Ca^{2+}(aq) + CO_3^{2-}(aq) \longrightarrow CaCO_3(s)$
   (e) Net ionic: $NH_4^+(aq) + OH^-(aq) \longrightarrow NH_3(g) + H_2O(l)$
2. The following are not water soluble: $CuCO_3$, ZnS.
5. $BaCl_2$, $AgNO_3$, HCl, and HNO_3 are strong electrolytes.
6. HF and NH_3 are weak electrolytes.

EXPERIMENT 22

1. According the the Brønsted-Lowry definition, an acid is a proton donor and a base is a proton acceptor.
2. A weak acid dissociates in aqueous solution according to the equilibrium $HA + H_2O \rightleftharpoons H_3O^+ + A^-$, for which the equilibrium constant is $K_{eq} = [H_3O^+][A^-]/[HA][H_2O]$, and the dissociation constant is $K_a = [H_3O^+][A^-]/[HA]$ or $K_a = [H_2O]K_{eq}$.
3. The pH at the equivalence point in an acid-base titration depends upon the nature of the species present. For the titration of a strong acid and a strong base, the pH will be 7 because a salt that does not hydrolyze will be formed (for example, NaOH + HCl). For the titration of a strong acid with a weak base (for example, HCl + NH_4OH), the pH will be less than 7; and for the titration of a weak acid with a strong base (e.g.,

HOAc + NaOH), the pH will be greater than 7. This is so because the salt formed in each case (NH_4Cl and NaOAc) will undergo hydrolysis reactions with water.

4. $pK_a = -\log K_a = -\log(7.5 \times 10^{-5}) = 5 - \log 7.5 = 5 - 0.88 = 4.12$.
5. Two electrodes are necessary for an electrical measurement because some current flow must occur and this requires both a donor and an acceptor for the electrons.
6. A buffer solution is a solution that is resistant to a pH change. It always contains a weak electrolyte and normally is composed of two species, such as a weak acid and one of its salts or a weak base and one of its salts. Two specific examples are acetic acid plus sodium acetate and ammonium hydroxide plus ammonium chloride. An example of a single-component buffer solution is disodium hydrogen phosphate, Na_2HPO_4.
7. At one-half equivalence point $[HA] = [A^-]$. Because $HA \rightleftharpoons H^+ + A^-$ and $K_a = [H^+][A^-]/[HA]$, at one-half equivalence point, $K_a = [H^+]$. Therefore, $pK_a = pH$, so $pK_a = 6.57$, $K_a = \text{antilog}(-6.57) = \text{antilog}(0.43) \times 10^{-7} = 2.7 \times 10^{-7}$.

EXPERIMENT 23

1. A polyprotic acid is one that possesses more than one ionizable proton. For example, H_2SO_4 is a diprotic acid and H_3PO_4 a triprotic acid because they possess two and three ionizable protons, respectively.
3. The number of mmol of NaOH required is twice the number of mmol of $H_2C_2O_4$:
   $2(50 \text{ mL} \times 3.6 \text{ mmol/mL}) = 360 \text{ mmol NaOH}$
4. $(0.250\text{L})(0.15 \text{ mol } H_2SO_4/\text{L})(2 \text{ mol } H^+/\text{mol } H_2SO_4) = 0.075 \text{ mol } H^+$
5. A pH meter must be standardized because it measures relative potentials and thus relative pH. It is necessary to know to what the measurement is relative; consequently, a standard must be measured and the meter set to the known value for this standard.
6. See the answer to question 7, Experiment 22, for the method: $pK_1 = 4.20$; $K_1 = 6.3 \times 10^{-5}$; $pK_2 = 7.34$; $K_2 = 4.6 \times 10^{-8}$.

EXPERIMENT 24

1. According to the Brønsted-Lowry definitions, an acid is a proton donor and a base is a proton acceptor.
2. Cu^{2+}, Zn^{2+}, and SO_3^{2-} will undergo hydrolysis.
3. For example, $Cu^{2+}(aq) + 2H_2O(l) \rightleftharpoons Cu(OH)_2(s) + 2H^+(aq)$.
4. $K_b = 1.0 \times 10^{-14}/2.1 \times 10^{-9} = 4.8 \times 10^{-6}$
5. The conjugate base is HPO_4^{2-} and the conjugate acid is H_3PO_4.
6. For example, $CaSO_4$ may be made from $Ca(OH)_2$ and H_2SO_4.
11. $pK_a = -\log (1.8 \times 10^{-5}) = 4.74$ and $pH = 4.74 + \log (0.40/0.20) = 4.74 + (0.30) = 5.04$

EXPERIMENT 25

1. $CaF_2 \rightleftharpoons Ca^{2+} + 2F^-$; $K_{sp} = [Ca^{2+}][F^-]^2$

2. $$\text{moles Ag}^+ = \left(\frac{5\text{ mL}}{1000\text{ mL/L}}\right)(0.004\text{ mol/L})$$
$$= 2 \times 10^{-5}\text{ mol}$$
$$\text{moles CrO}_4^{2-} = \left(\frac{5\text{ mL}}{1000\text{ mL/L}}\right)(0.0024\text{ mol/L})$$
$$= 1 \times 10^{-5}\text{ mol}$$

3. $2AgNO_3 + K_2CrO_4 \longrightarrow 2KNO_3 + Ag_2CrO_4$. The stoichiometry requires 2 mol of $AgNO_3$ for each mole of K_2CrO_4.
$$\text{millimoles AgNO}_3 = (10\text{ mL})(0.004\text{ mmol/mL})$$
$$= 0.04\text{ mmol}$$
$$\text{millimoles K}_2\text{CrO}_4 = (10\text{ mL})(0.0024\text{mmol/mL})$$
$$= 0.024\text{ mmol}$$
and 0.024 mmol K_2CrO_4 would require 0.048 mmol $AgNO_3$. Hence, K_2CrO_4 is in excess.

4. $[Ba^{2+}] = (20\text{ mL}/40\text{ mL})(1 \times 10^{-4}\text{ mol/L}) = 5 \times 10^{-5}\text{ mol/L}$; $[CrO_4^{2-}] = (20\text{ mL}/40\text{ mL})(1 \times 10^{-4}\text{ mol/L}) = 5 \times 10^{-5}\text{ mol/L}$. Since $K_{sp} = [Ba^{2+}][CrO_4^{2-}] = 1.2 \times 10^{-10}$ and ion product $= (5 \times 10^{-5})(5 \times 10^{-5}) = 2.5 \times 10^{-9}$. Because the ion product is more than K_{sp}, precipitation would occur.

9. A sparingly soluble salt will precipitate when the ion product exceeds K_{sp}.

EXPERIMENT 26

1. An exothermic reaction is a reaction that produces heat; its ΔH will be less than zero, or negative. An endothermic reaction is a reaction that absorbs heat; its ΔH will be greater than zero, or positive.
2. $(625\text{ mL})(1.0\text{ g/mL})(50.0°C - 10.0°C)(4.18\text{ J/g} - °C) = 104{,}500\text{ J} = 104\text{kJ}$.
3. The heat capacity of a substance is the amount of energy, usually in the form of heat, necessary to raise the temperature of a specified amount of the substance (usually 1 g) by 1°C.
4. $\Delta T = 38.8°C - 23.3°C = 15.5°C$;
heat required $= (60.0\text{ g})(4.18\text{ J/g-°C})(15.5°C) = 3.89 \times 10^3\text{ J}$

EXPERIMENT 27

1. The factors that influence the rate of a chemical reaction are temperature, pressure, concentration, particle size, catalysts, and the nature of the species undergoing the reaction.
2. The general form of the rate law for the reaction $A + B \longrightarrow$ is: Rate $= k[A]^x[B]^y$.
3. A reaction that obeys the rate law of the form rate $= k[A]^2[B]^3$ is second order in A and third order in B. This reaction is fifth order overall.
4. The chemical reactions involved in this experiment are $S_2O_8^{2-} + 2I^- \longrightarrow I_2 + 2SO_4^{2-}$, $I_2 + 2S_2O_3^{2-} \longrightarrow 2I^- + S_4O_6^{2-}$, and I_2 + starch $\longrightarrow$ starch $\cdot$ I_2(blue), for which the rate of disappearance of $S_2O_3^{2-}$ is $k[S_2O_8^{2-}]^x[I^-]^y$ and the rate of appearance of blue color equals the rate of formation of the starch-iodine complex, which is proportional to the

rate of appearance of iodine. Thus, the rate of appearance of iodine is $k[S_2O_8^{2-}]^x[I^-]^y$ and the rate of appearance of blue color is $k[I_2]^x[\text{starch}]^y \alpha k[S_2O_8^{2-}]^x[I^-]^y$.

5. Rate = $k[A]^2[B]^2$.

6. $$\text{Rate} = \left(\frac{2 \times 10^{-4}\ \text{mol}}{0.050\ \text{L}}\right)\left(\frac{1}{188\ \text{s}}\right)$$
$$= 2.1 \times 10^{-5}\ \frac{\text{mol}}{\text{L-s}}$$

7. From a knowledge of the rate of chemical reactions, chemists can determine how long to run a reaction in order to obtain the desired products. Clearly, this is of extreme practical value in the synthesis of new compounds.

EXPERIMENT 28

1. See the answer to question 1, Experiment 27.
2. (a) A reaction with a rate law of the form rate = $k[A]^2[B]$ is third order.
   (b) If the concentrations of both A and B are doubled, the rate will increase by a factor of 8.
   (c) k is known as the specific rate constant.
   (d) Changing the concentrations of the reactants may change the rate but does not change the specific rate constant. The rate will be quadrupled.
5. Rate = $k[A]^2[B]^2$.

EXPERIMENT 29

I Cations

CATIONS

1. Ammonium, NH_4^+; silver, Ag^+; ferric or iron(III), Fe^{3+}; aluminum, Al^{3+}; chromic or chromium(III), Cr^{3+}; calcium, Ca^{2+}; magnesium, Mg^{2+}; nickel or nickel(II), Ni^{2+}; zinc, Zn^{2+}; and sodium, Na^+.
2. Sometimes ions behave very similarly. For example, Ba^{2+} and Pb^{2+} both form yellow precipitates with K_2CrO_4. Hence, other additional confirmatory tests would be required to distinguish between these ions. For example, Pb^{2+} forms a white precipitate with HCl, whereas Ba^{2+} does not.
3. Only three of the ions are colored as follows: Fe^{3+}, rust to yellow; Cr^{3+}, blue-green; Ni^{2+}, green.
4. Because only AgCl precipitates on the addition of HCl to a solution of the ten cations, all chlorides of these ions except AgCl must be soluble.
5. Based on the group separation chart (Figure 29.1) Ag^+ can be separated from all of the other ions by the addition of HCl. Silver forms insoluble AgCl.
6. Examination of the group separation chart shows that in the presence of the buffer NH_3—NH_4Cl, Cr^{3+} precipitates as $Cr(OH)_3$ while Mg^{2+} remains in solution.
7. Addition of chloride ions to a solution containing Al^{+3} and Ag^+ would precipitate AgCl, while Al^{3+} would remain in solution. HCl would be a good source of chloride ions, and the H^+ would help retard hydrolysis of Al^{3+}.
8. $NH_4^+ + OH^- \rightleftharpoons NH_3 + H_2O$; $AgCl + 2NH_3 \rightleftharpoons [Ag(NH_3)_2]^+ + Cl^-$.

ANIONS

II Anions

1. Sulfate, SO_4^{2-}; nitrate, NO_3^-; carbonate, CO_3^{2-}; chloride, Cl^-; bromide, Br^-; and iodide, I^-.
2. See Table 29.1: behavior of anions with concentrated sulfuric acid.
3. No. Because HCl is pungent and forms on the addition of H_2SO_4 to chloride salts, it would be impossible to tell that a second gas that is odorless is also present. Hence you would not be aware of the presence of CO_3^{2-} in the mixture.

EXPERIMENT 30

Part I

1. Ag^+, Hg_2^{2+} and Pb^{2+}.
2. $PbCl_2$.
3. AgCl.
4. (a) $BaCl_2$ is water soluble and AgCl is not.
   (b) HCl forms a white precipitate with $AgNO_3$, whereas HNO_3 does not form a precipitate with $AgNO_3$.
8. With the use of a small pipet.

Part II

1. Pb^{2+}, Cu^{2+}, Bi^{3+}, Sn^{4+}
2. CuS does not dissolve in aqueous $(NH_4)_2S$, whereas SnS_2 does.
3. Aqueous NH_3 precipitates $Bi(OH)_3$ and $Cu(NH_3)_4^{2+}$ remains in solution

Part III

1. Fe^{3+}, Ni^{2+}, Mn^{2+}, Al^{3+}.
2. Cu^{2+} (blue), Fe^{3+} (yellow to reddish brown), Al^{3+} (colorless), Ni^{2+} (green).
3. $Fe(OH)_3$ (red-brown), MnS (salmon), $Al(OH)_3$ (colorless), $Ni(OH)_2$ (green).
4. Aqueous NaOH precipitates $Fe(OH)_3$, whereas $Al(OH)_4^-$ remains in solution.
7. (a) HCl or H_2SO_4.
   (b) Excess NaOH.

Part IV

1. Ba^{2+}, Ca^{2+}, NH_4^+, Na^+.
2. $BaCrO_4$.
3. $BaSO_4$ (white), $BaCrO_4$ (yellow), CaC_2O_4 (white).
4. (a) Ba^{2+} (green).
   (b) Ca^{2+} (orange-red).
   (c) Na^+ (yellow).

Part V

1. Sulfate, SO_4^{2-}; nitrate, NO_3^-; carbonate, CO_3^{2-}; chromate, CrO_4^{2-}; phosphate, PO_4^{3-}; chloride, Cl^-; bromide, Br^-; sulfide, S^{2-}; sulfite, SO_3^{2-}; and iodide, I^-.
2. See Table 30.1: behavior of anions with concentrated sulfuric acid.

3. No. Because HCl is pungent and forms on the addition of H_2SO_4 to chloride salts, it would be impossible to tell that a second gas that is odorless is also present. Hence, you would not be aware of the presence of CO_3^{2-} in the mixture.
4. (a) SO_4^{2-}.
   (b) CO_3^{2-}, PO_4^{3-}, S^{2-}, or OH^-.
   (c) Cl^-.
5. SO_4^{2-}, CO_3^{2-}, OH^-, CrO_4^{2-}, PO_4^{3-}, Cl^-, Br^-, and I^-.
6. CO_3^{2-}.

EXPERIMENT 31

1. The concentration, the cell path length, wavelength, and solvent.
2. Absorbance $A = \log I_0/I$ and percent transmittance $\%T = (I/I_0)100$, so $\%T/100 = I/I_0$ or $100/\%T = I_0/I$ and $\log(100/\%T) = \log(I_0/I) = A$. Or we may write $A = 2 - \log \%T$.
3. A spectrophotometer must possess a light source, a monochromator, a sample cell, a detector, and a meter.
4. The Beer-Lambert law is $A = abc$, where A is the absorbance, a is the absorptivity or extinction coefficient, b is the solution path length, and c is the molar concentration of absorbing species.
5. Not all substances precisely obey the Beer-Lambert law over all concentration ranges. A calibration curve will provide the relationship between concentration and absorbance under conditions that are similar to or identical with those used for the analysis.
6. Hydroxylamine is used to reduce Fe^{3+} to Fe^{2+}.
7. Because $\%T = (I/I_0)(100)$, $\log \%T = \log(I/I_0) + \log 100$, or $\log(I_0/I) = 2 - \log \%T$. $A = 2 - \log \%T = 2 - \log 80 = 2 - 1.9 = 0.1$.

EXPERIMENT 32

1. The volumes of 1×10^{-3} *M* solutions required to prepare 50 mL of solutions with the cited concentrations are

Concentration (*M*) $\times 10^{-5}$:	2	5	10	20	50	75
Volume of 10^{-3} *M* solution required (mL):	1	2.5	5	10	25	37.5

2. The species $(NH_4)_3PO_4 \cdot NH_4VO_3 \cdot Mo_{16}O_{48}$ is thought to be the light absorbing species in this experiment.
3. The reaction $(NH_4)_3PO_4 + NH_4VO_3 + 16MoO_3 \longrightarrow (NH_4)_4PVMo_{16}O_{55}$ is thought to be the reaction that forms the heteropoly acid that absorbs the light in this experiment.
5. A calibration curve is constructed by measuring the absorbances of solutions with varying known concentrations. It is constructed in order to ascertain the relationship between absorbance and concentration of the absorbing species.
6. If the measured absorbance of the unknown is greater than that of any of the calibration solutions, you should measure the absorbance of a more dilute solution. If the measured absorbance of the unknown is less than that of any of the calibration solutions, you should measure the absorbance of a more concentrated solution.

8. 8.64 mg $FeSCN^{2+}$/L = 8.64 × 10^{-3} g $FeSCN^{2+}$/(113.9 g/mol)L = 7.60 × 10^{-5} M; if $T = 0.30$, then $A = 2 - \log 30.0 = 2 - 1.48 = 0.52$ and from $A = abc$, $a = A/bc = 0.52/[1.00\text{ cm}][7.60 \times 10^{-5}\,M] = 6.84 \times 10^{+3}$ L/mol-cm.

EXPERIMENT 33

1. From the relation moles $Na_2S_2O_3$ = 6 moles KIO_3; moles $Na_2S_2O_3$ = (V)(M) and

$$\text{moles } KIO_3 = \text{g } KIO_3/(214.0 \text{ g } KIO_3/\text{mol})$$

one obtains

$$M_{Na_2S_2O_3} = \frac{6 \times \text{g } KIO_3}{(214.0 \text{ g } KIO_3/\text{mol})(V_{Na_2S_2O_3})}$$

Therefore,

$$M = \frac{6(0.1309 \text{ g})}{(214.0 \text{ g/mol})(0.02735 \text{ L})} = 0.1342\ M$$

2. Chloroform is an antibacterial agent and as such kills bacteria. Because the bacteria are killed, they do not consume oxygen by their growth, so the water sample is preserved by the addition of chloroform.
3. In the reaction $Mn(SO_4)_2 + 2KI \longrightarrow MnSO_4 + K_2SO_4 + I_2$, manganese is reduced (its oxidation state changes from +4 to +2) and iodide is oxidized (its oxidation state changes from −1 to 0).
4. The blue color returns because oxygen in the air oxidizes iodide to iodine according to the reaction $4I^- + O_2 + 4H^+ \longrightarrow 2I_2 + 2H_2O$.
5. (1 × 10^{-3} mol/L)(32 g/mol)(1000 mg/g) = 32 mg/L = 32 ppm.

EXPERIMENT 34

3. A coordination compound is formed by the reaction of a Lewis acid with a Lewis base; it contains one or more coordinate covalent bonds. An example is the compound $H_3N{:}BF_3$.
4. Geometric isomers have the same empirical and molecular formulas, but differ in their spatial arrangements of the constituent atoms.
5. The compounds $[Co(NH_3)_4Cl_2]$ and $[Co(NH_3)_3Cl_3]$ can each exist in two geometric isomeric forms, which are shown below.

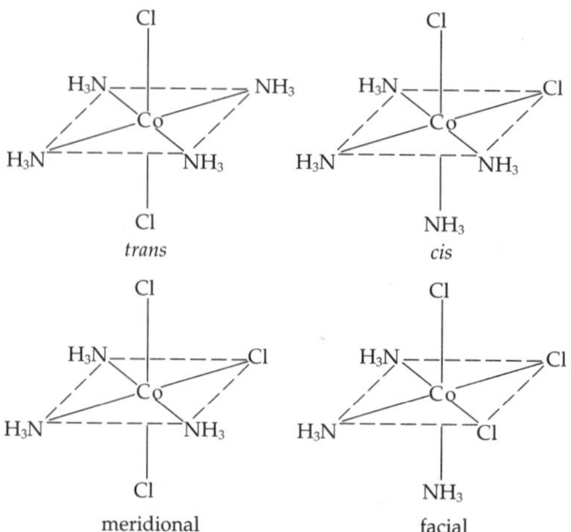

6. The chlorine atoms are in equivalent environments in *cis*- and *trans*-$[Co(NH_3)_4Cl_2]$ and facial $[Co(NH_3)_3Cl_3]$, but there are two different chlorine environments in meridional $[Co(NH_3)_3Cl_3]$.
7. Dichroism is the property according to which the colors of a crystal are different when the crystal is viewed in the direction of two different axes.
8. Triturate means to cause a semisolid to become a solid by crushing or grinding.
9. You must find the answer to question 9 in the library.
10. You must find the answer to question 10 in the library.
11. It works by forming a soluble iron(II) oxalate complex, $[Fe(C_2O_4)_3]^{4-}$, by reacting with iron oxide (rust), according to the reaction: $FeO + 3H_2C_2O_4 \longrightarrow H_4[Fe(C_2O_4)_3] + H_2O$. In the reactions of Fe_2O_3 or Fe_3O_4, iron is also reduced by oxalate and CO_2 is formed.

EXPERIMENT 35

1. One can calculate the molarity after recognizing that moles oxalate = 5/2 moles permanganate:

$$M = \frac{2}{5}\left(\frac{\text{g Na}_2\text{C}_2\text{O}_4}{\text{molar mass Na}_2\text{C}_2\text{O}_4}\right)\left(\frac{1}{V\,\text{KMnO}_4}\right)$$
$$= \frac{2(0.45868\text{ g})}{5(134.0\text{ g/mol})(0.03453\text{ L})} = 0.03965\ M$$

2. The average is 15.52 and the standard deviation is 0.26.
3. The solution slowly decolorizes after the equivalence point has been reached because permanganate undergoes photochemical reduction and because it is slowly reduced by water.
4. The permanganate is filtered in order to remove insoluble MnO_2. Because rubber reduces MnO_4^- to MnO_2, the solution should not contact rubber or other organic material.
5. $16H^+ + 2MnO_4^- + 5C_2O_4^{2-} \longrightarrow 10CO_2 + 2Mn^{2+} + 8H_2O$.

$$\text{moles MnO}_4^- = \left(\frac{2}{5}\right)\text{moles C}_2\text{O}_4^{2-}$$
$$= \left(\frac{2}{5}\right)\frac{(0.36\text{ g})\left(\dfrac{2\text{ mol C}_2\text{O}_4^{2-}}{\text{mole K}_2[\text{Cu(C}_2\text{O}_4)_2]\cdot 2\text{H}_2\text{O}}\right)}{(353.83\text{ g/mole K}_2[\text{Cu(C}_2\text{O}_4)_2]\cdot 2\text{H}_2\text{O}}$$
$$= 8.16 \times 10^{-4}\text{ mol} = (V_{\text{KMnO}_4})\,(M_{\text{KMnO}_4})$$
$$V_{\text{KMnO}_4} = \frac{8.16 \times 10^{-4}\text{ mol}}{0.1\text{ mol/L}}$$
$$= 8.16 \times 10^{-3}\text{ L} = 8.16\text{ mL}$$

EXPERIMENT 36

1. (a) $2NaHCO_3(s) \xrightarrow{\Delta} Na_2CO_3(s) + H_2O(g) + CO_2(g)$
(b) $CaCO_3(s) \xrightarrow{\Delta} CaO(s) + CO_2(g)$
(c) $CaO(s) + H_2O(l) \longrightarrow Ca^{2+}(aq) + 2OH^-(aq)$

(d) $NH_3(g) + H_2O(l) \rightleftharpoons NH_4^+(aq) + OH^-(aq)$
(e) $2NH_4^+(aq) + CaO(s) \longrightarrow Ca^{2+}(aq) + H_2O(l) + 2NH_3(g)$
(f) $H^+(aq) + HCO_3^-(aq) \longrightarrow H_2O(l) + CO_2(g)$

2. Henry's law states that the solubility of a gas in a liquid is directly proportional to the partial pressure of the gas above the solution.
5. The common ion affect of Cl^- would decrease the solubility of NaCl.

EXPERIMENT 37

1. NaOCl and $Ca(OCl)_2$
3. IO_3^- is an oxidizing agent
4. Iodide ions, I^- are oxidized by air to I_2, and the oxidation is accelerated in acid soluton. Hence there would be additional iodine to that formed from the KIO_3.
5. (a) Pale yellow
   (b) Blue-black
   (c) Colorless
7. moles KIO_3 = (0.02095 L) (0.0111 mol KIO_3/L) = 2.33×10^{-4} mol KIO_3

   moles $Na_2S_2O_3$ = 6 (mol KIO_3)
   = 6 (2.33×10^{-4} mol) = 1.40×10^{-3} mol $Na_2S_2O_3$

   M $Na_2S_2O_3$ = 1.40×10^{-3} mol $Na_2S_2O_3$/0.01692 L
   = 0.0827 M
8. A standard solution is one whose concentration is accurately known.

EXPERIMENT 38

1. Molecular formulas indicate the composition of molecules but do not indicate how the atoms are arranged; therefore, they do not distinguish between isomers. Structural formulas indicate how the atoms are arranged in molecules as well as the molecular composition. The molecular formula $C_2H_4Cl_2$ may refer to either isomer:

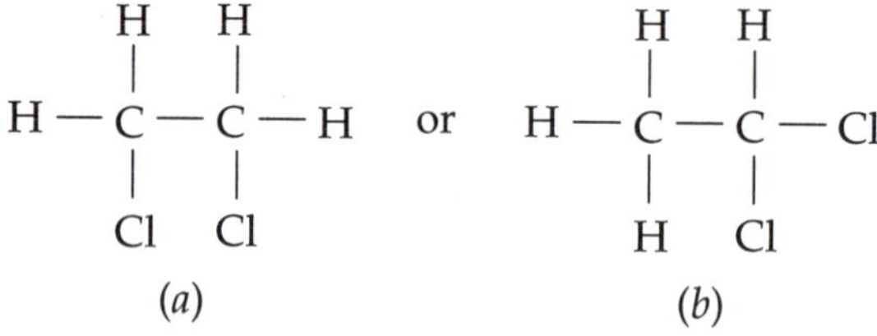

2. Condensed structural formulas are a convenient, simple way of writing formulas and are a "shorthand version" of structural formulas. For example, molecules (*a*) and (*b*) above may be written: (*a*) CH_2ClCH_2Cl and (*b*) CH_3CHCl_2.
3. Isomers are compounds with the same molecular formula but different structural formulas; for example, (*a*) and (*b*) above are isomers.
4. Because atoms are arranged differently in isomers, the molecules have different properties. For example, ethyl alcohol, CH_3CH_2OH, has an —OH group and in some ways resembles H—OH. Dimethyl ether,

CH_3OCH_3, is an isomer of ethyl alcohol and does not contain the —OH group. Hence, it differs from ethyl alcohol.

5.
```
      H   H
      |   |
  H — C — C — H   or  CH3CH3
      |   |
      H   H
```
Ethane

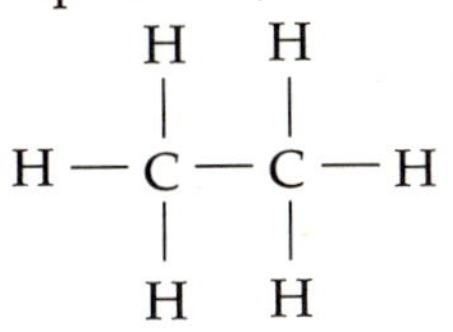

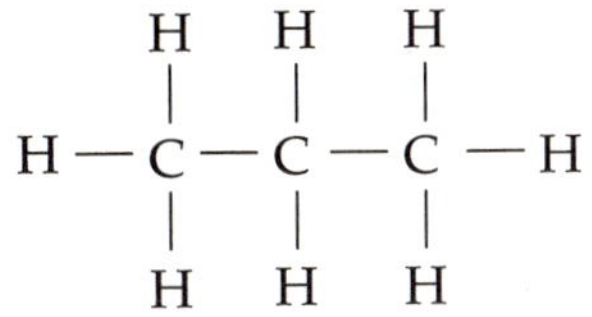

```
      H   H   H
      |   |   |
  H — C — C — C — H   or  CH3CH2CH3
      |   |   |
      H   H   H
```
Propane

EXPERIMENT 39

1. Esters are derivatives of carboxylic acids and have the general formula

```
       O
       ||
   R — C — O — R
```

2. The mineral acid functions as a catalyst. It lets you accomplish the esterification in a shorter time.
3. Esters can be prepared from carboxylic acids and alcohols. They can also be made from acid anhydrides and alcohols.
4. Esters generally have pleasant, fruitlike odors.
5. Phenols form colored complexes with ferric chloride. The colors may range from green through violet. Hence, $FeCl_3$ can be used to test for the presence of phenols.
6. The carboxylic acid group is

```
      O
      ||
   — C — OH
```

7. Recrystallization is often utilized to purify a solid compound by separating impurities that are soluble in the solvent that is used for the recrystallization.
8. Salicylic acid (benzene ring with OH and CO_2H) + excess acetic anhydride → acetylsalicylic acid (benzene ring with $OCCH_3$ (C=O) and CO_2H) + CH_3COOH

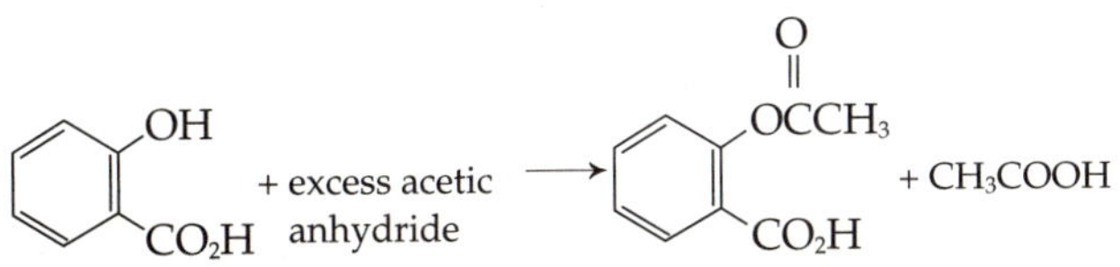

Moles acetylsalicylic acid = moles salicylic acid; moles salicylic acid = 1.00 g/(138 g/mol) = 7.25×10^{-3} mol. Theoretical yield of acetylsalicylic acid is (7.25×10^{-3} mol)(180 g/mol) = 1.30 g; % yield = (1.02 g/1.30 g)(100) = 78.5%.

EXPERIMENT 40

1. An *analyte* is the substance being determined in an analysis; a *standard solution* is one whose concentration is accurately known; *titration* is the operation of adding a standard solution (usually called the titrant) to another solution until the chemical reaction between the two solutes is complete; *standardization* is the process of accurately determining the concentration of a solution.

2. (a)

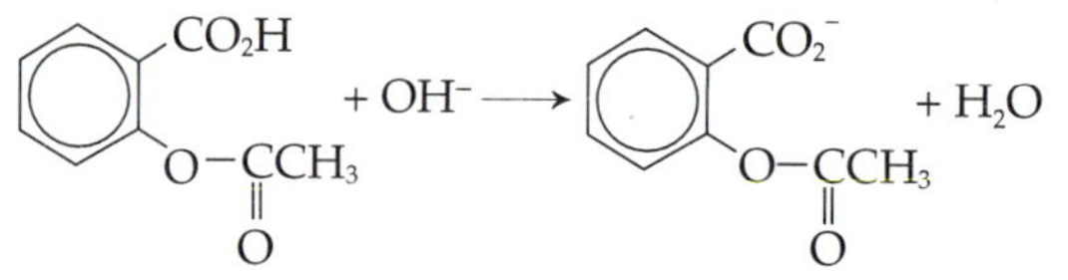

$$C_6H_4(CO_2H)(O{-}\overset{\displaystyle O}{\overset{\|}{C}}CH_3) + OH^- \longrightarrow C_6H_4(CO_2^-)(O{-}\overset{\displaystyle O}{\overset{\|}{C}}CH_3) + H_2O$$

(b)

$$C_6H_4(CO_2H)(O{-}\overset{\displaystyle O}{\overset{\|}{C}}CH_3) + 2OH^- \longrightarrow C_6H_4(CO_2^-)(OH) + CH_3CO_2^- + H_2O$$

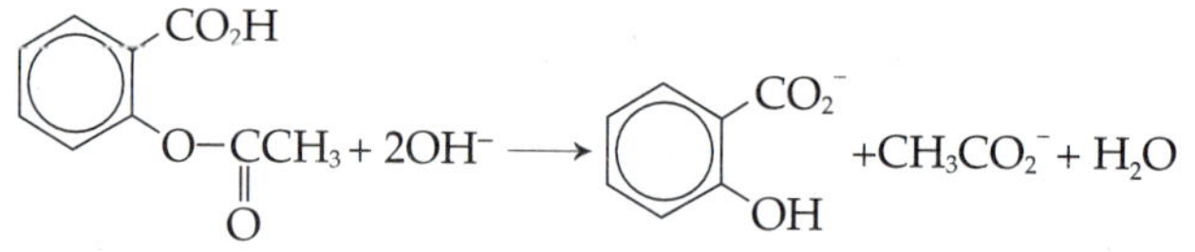

3.

$$R{-}\overset{\displaystyle O}{\overset{\|}{C}}{-}OCH_3 + OH^- \longrightarrow R{-}\overset{\displaystyle O}{\overset{\|}{C}}{-}O^- + CH_3OH$$

4. $33.5\text{ mL} \times 0.25\ M = 35.8\text{ mL} \times M$; and $M = \dfrac{33.5\text{ mL} \times 0.25\ M}{35.8\text{ mL}} = 0.23\ M$

5. $38.5\text{ mL} \times 0.125\text{ mmoles/mL} = 4.48\text{ mmol}$
$(40.0\text{ g/mol})\ (1\text{ mol}/1000\text{ mmol})(4.48\text{ mmol}) = 0.179\text{ g}$

EXPERIMENT 41

4. The strength of interaction between an ion and an ion-exchange resin increases with an increase in the charge-to-size ratio of the ion because these are electrostatic interactions.

6. Mg^{2+} should be eluted more slowly because it has the larger charge-to-size ratio.

Appendix

Vapor Pressure of Water at Various Temperatures

Temperature (°C)	Pressure (mm Hg)	Temperature (°C)	Pressure (mm Hg)
0	4.6	26	25.2
1	4.9	27	26.7
2	5.3	28	28.3
3	5.7	29	30.0
4	6.1	30	31.8
5	6.5	31	33.7
6	7.0	32	35.7
7	7.5	33	37.7
8	8.0	34	39.9
9	8.6	35	42.2
10	9.2	40	55.3
11	9.8	45	71.9
12	10.5	50	92.5
13	11.2	55	118.0
14	12.0	60	149.4
15	12.8	65	187.5
16	13.6	70	233.7
17	14.5	75	289.1
18	15.5	80	355.1
19	16.5	85	433.6
20	17.5	90	525.8
21	18.7	95	633.9
22	19.8	97	682.1
23	21.1	99	733.2
24	22.4	100	760.0
25	23.8	101	787.6

Appendix

Names, Formulas, and Charges of Common Ions

Positive ions (cations)	Negative ions (anions)
1+	**1−**
Ammonium (NH^{4+})	Acetate ($C_2H_3O_2^-$)
Cesium (Cs^+)	Bromide (Br^-)
Copper(I) or cuprous (Cu^+)	Chlorate (ClO_3^-)
Hydrogen (H^+)	Chloride (Cl^-)
Lithium (Li^+)	Cyanide (CN^-)
Potassium (K^+)	Dihydrogen phosphate ($H_2PO_4^-$)
Silver (Ag^+)	Fluoride (F^-)
Sodium (Na^+)	Hydride (H^-)
	Hydrogen carbonate or bicarbonate (HCO_3^-)
2+	Hydrogen sulfite or bisulfite (HSO_3^-)
Barium (Ba^{2+})	Hydroxide (OH^-)
Cadmium (Cd^{2+})	Iodide (I^-)
Calcium (Ca^{2+})	Nitrate (NO_3^-)
Chromium(II) or chromous (Cr^{2+})	Nitrite (NO_2^-)
Cobalt(II) or cobaltous (Co^{2+})	Perchlorate (ClO_4^-)
Copper(II) or cupric (Cu^{2+})	Permanganate (MnO_4^-)
Iron(II) or Ferrous (Fe^{2+})	Thiocyanate (SCN^-)
Lead(II) or plumbous (Pb^{2+})	
Magnesium (Mg^{2+})	**2−**
Manganese(II) or manganous (Mn^{2+})	Carbonate (CO_3^{2-})
Mercury(I) or mercurous (Hg_2^{2+})	Chromate (CrO_4^{2-})
Mercury(II) or mercuric (Hg^{2+})	Dichromate ($Cr_2O_7^{2-}$)
Strontium (Sr^{2+})	Hydrogen phosphate (HPO_4^{2-})
Nickel (Ni^{2+})	Oxide (O^{2-})
Tin(II) or stannous (Sn^{2+})	Peroxide (O_2^{2-})
Zinc (Zn^{2+})	Sulfate (SO_4^{2-})
	Sulfide (S^{2-})
3+	Sulfite (SO_3^{2-})
Aluminum (Al^{3+})	
Chromium(III) or chromic (Cr^{3+})	**3−**
Iron(III) or ferric (Fe^{3+})	Arsenate (AsO_4^{3-})
	Phosphate (PO_4^{3-})

Appendix

Some Molar Masses*

Formula	Mass	Formula	Mass	Formula	Mass
AgBr	187.78	HNO_3	63.01	$MgSO_4$	120.37
AgCl	143.32	H_2O	18.015	MnO_2	86.94
Ag_2CrO_4	331.73	H_2O_2	34.01	Mn_2O_3	157.88
AgI	234.77	H_3PO_4	98.00	NaBr	102.90
$AgNO_3$	169.87	H_2S	34.08	$NaC_2H_3O_2$	82.03
Al_2O_3	101.96	H_2SO_3	82.08	NaCl	58.44
$Al_2(SO_4)_3$	342.14	H_2SO_4	98.08	NaCN	49.01
B_2O_3	69.62	HgO	216.59	$Na_2C_2O_4$	134.0
$BaCO_3$	197.35	Hg_2Cl_2	472.09	Na_2CO_3	105.99
$BaCl_2$	208.25	$HgCl_2$	271.50	$NaNO_3$	84.99
$BaCrO_4$	253.33	KBr	119.01	Na_2O_2	77.98
$Ba(OH)_2$	171.36	$K_2C_2O_4 \cdot H_2O$	184.24	NaOH	40.00
$BaSO_4$	233.40	$K_3[Cr(C_2O_4)_3] \cdot 3H_2O$	487.42	NaSCN	81.07
CO_2	44.01	$K_2[Cu(C_2O_4)_2] \cdot 2H_2O$	353.83	Na_2SO_4	142.04
$CaCO_3$	100.09	$K_3[Fe(C_2O_4)_3] \cdot 3H_2O$	491.27	$Na_2S_2O_3 \cdot 5H_2O$	248.18
CaO	56.08	$K_3[Al(C_2O_4)_3] \cdot 3H_2O$	462.40	NH_3	17.03
$CaSO_4$	136.14	KCl	74.56	NH_4Cl	53.49
$[Co(C_5H_5N)_4(NCS)_2]$	491.05	$KClO_3$	122.55	NH_4NO_3	80.04
CuO	79.54	K_2CrO_4	194.20	$(NH_4)_2SO_4$	132.14
Cu_2O	143.08	$K_2Cr_2O_7$	294.19	$[Ni(C_5H_5N)_4(NCS)_2]$	490.83
$CuSO_4$	159.60	$KHC_8H_4O_4$ (phthalate)	204.23	$PbCrO_4$	323.18
$[Cu(C_5H_5N)_2(NCS)_2]$	337.54	K_2HPO_4	174.18	PbO	223.19
$Fe(NH_4)_2(SO_4)_2 \cdot 6H_2O$	392.14	KH_2PO_4	136.09	PbO_2	239.19
FeO	71.85	$KHSO_4$	136.17	$PbSO_4$	303.25
Fe_2O_3	159.69	KI	166.01	P_2O_5	141.94
Fe_3O_4	231.54	KIO_3	214.00	SiO_2	60.08
HBr	80.92	KIO_4	230.00	$SnCl_2$	189.60
$H_2C_2O_4$ (oxalic)	90.04	$KMnO_4$	158.04	SnO_2	150.69
$HC_2H_3O_2$ (acetic)	60.05	KNO_3	101.11	SO_2	64.06
$HC_7H_5O_2$ (benzoic)	122.12	KOH	56.11	SO_3	80.06
HCl	36.46	K_2SO_4	174.27	$ZnCl_2$	136.27
$HClO_4$	100.46	$MgNH_4PO_4$	137.35	$[Zn(C_5H_5N)_2(NCS)_2]$	339.37
$H_2C_2O_4 \cdot 2H_2O$	126.07	MgO	40.31		

*grams/mole

Appendix

Basic SI Units, Some Derived SI Units, and Conversion Factors

	SI unit	Conversion factors
Length	Meter (m)	1 m = 100 centimeters (cm) = 1.0936 yards (yd) 1 cm = 0.3937 inch (in.) 1 in. = 2.54 cm = 0.0254 m 1 angstrom (Å) = 10^{-10} m
Mass	Kilogram (kg)	1 kg = 1000 grams (g) = 2.205 pounds (lb) 1 lb = 453.6 grams (g) 1 atomic mass unit (amu) = 1.66053×10^{-24} g
Time	Second (s)	1 day (d) = 86,400 s 1 hour (hr) = 3600 s 1 minute (min)= 60 s
Electric current	Ampere (A)	
Temperature	Kelvin (K)	0 K = −273.15° Celsius (C) = −459.67° Fahrenheit (F) °F = (9/5)°C + 32° °C = (5/9)(°F − 32°) K = °C + 273.15°
Luminous intensity	Candela (cd)	
Amount of substance	Mole (mol)	
Volume (derived)	Cubic meter (m^3)	1 liter (L) = 10^{-3} m^3 = 1.057 quarts (qt) 1 $in.^3$ = 16.4 cm^3
Force (derived)	Newton (N = m-kg/s^2)	1 dyne (dyn) = 10^{-5} N
Pressure (derived)	Pascal (Pa = N/m^2)	1 atmosphere (atm) = 101,325 Pa = 760 mm Hg = 14.70 lb/$in.^2$ = 1.013×10^6 dyn/cm^2
Energy (derived)	Joule (J = N-m)	1 calorie (cal) = 4.184 J 1 electron volt (eV) = 1.6022×10^{-19} J 1 erg = 6.2420×10^{11} eV 1 J = 10^7 ergs

Appendix

Composition of Commercial Reagent Acids and Bases

	Molar mass (g)	Density (g/mL)	% Strength (w/w)	Molarity	Approx. mL conc. reagent/1 L 1 *M* solution
HCl	36.461	1.19	37.2	12.1	82.5
HNO_3	63.013	1.42	70.0	15.8	63.5
HF	20.006	1.19	48.8	29.0	34.5
$HClO_4$	100.458	1.67	70.5	11.7	85.5
$HC_2H_3O_2$	60.053	1.05	99.8	17.4	57.5
H_2SO_4	98.082	1.84	96.0	18.0	55.5
H_3PO_4	97.994	1.70	85.5	14.8	67.5
NH_4OH	35.046	0.90	57.6*	14.8	67.5
NaOH	39.997	1.53	50.5	19.3	52.0
KOH	56.106	1.46	45.0	11.7	85.5

* Equivalent to 28% NH_3.

Appendix

ANNOTATED INSTRUCTOR'S EDITION

Additional Answers to Review Questions for Instructors

EXPERIMENT 1

3. $(1.47\text{ g})\left(\frac{1000\text{ mg}}{1\text{ g}}\right) = 1.47 \times 10^3\text{ mg}$
8. Density = 3.66 g/0.4018 mL = 9.11 g/mL
10. °F = 9/5(37.0) + 32= 98.6° F
12. The accuracy of the balance is ±0.1 mg and because 1 mg = 0.001 g this weight should be reported as 5.0000 g or to five significant figures.

EXPERIMENT 2

2. $\text{Density} = \frac{9.02\text{ g}}{8.192\text{ mL}} = 1.101\text{ g/mL}$
5. Toluene boils at 111°C at 760 mm Hg. From Figure 2.5 the boiling point correction at 670 mm Hg is 3.5°C. Thus, toluene would boil at 111°C − 3.5°C = 107.5°C or 108°C at 670 mm Hg.
8. The bottom layer is water because water is more dense than toluene.
9. Cyclohexane may be used because cadmium nitrate is not soluble in it.
10. The solid is benzophenone. (see Table 2.1).
12. The volume of 10 g of hexane (10 g/0.66 g-mL^{-1} = 15 mL) is greater than the volume of 10 g of chloroform (10 g/1.49 g-mL^{-1} = 6.7 mL).

EXPERIMENT 3

6. The total weight of the mixture is 2.10 g + 1.38 g + 6.52 g = 10.00 g; %SiO_2 = (2.10 g/10.0 g)(100) = 21.0%
8. Zinc chloride could be extracted from SiO_2 with water.
9. This student's analyses are greater than 100% (15% + 20% + 75% = 110%). She probably did not dry the NaCl and SiO_2 sufficiently.
10. One extraction may not completely remove the NaCl. Multiple extractions with small volumes are more efficient than a single extraction with a large volume of solvent. Consider, for example, the dilution that results from three successive 10-mL extractions and compare it with one 30-mL extraction.

EXPERIMENT 4

7. H_2 is odorless and H_2S smells like rotten eggs. Moreover, H_2S reacts with solutions of many metal salts to form colored precipitates.
8. KCl is soluble in cold water while $PbCl_2$ is insoluble.
9. $H_2CO_3(aq) \longrightarrow H_2O(l) + CO_2(g)$

   $H_2SO_3(aq) \longrightarrow H_2O(l) + SO_2(g)$

EXPERIMENT 5

8. 107.868 g/(6.02×10^{23} atoms) = 1.79×10^{-22} g/atom
9. mol $Na_2CaSi_6O_{14}$ = (500)(400 g)/(478 g/mol) = 4180 mol glass
   gram SiO_2 = (4180 mol glass)(6 mol SiO_2/mol glass)(60.09 g/mol SiO_2)
   = 1,510,000 g or 1510 kg
10. Assume 100 g caffeine, then

    mol C = 49.5 g/(12.0 g/mol) = 4.13 mol

    mol H = 5.15 g/(1.01 g/mol) = 5.15 mol

    mol N = 28.9 g/(14.0 g/mol) = 2.06 mol

    mol O = 16.5 g/(16.0 g/mol) = 1.03 mol

    Hence, $C_{4.13}H_{5.15}N_{2.06}O_{1.03}$, and dividing through by 1.03 gives $C_4H_5N_2O$. The formula weight of $C_4H_5N_2O$ is 97 amu, and the factor to multiply the empirical formula subscripts by is:

    $$\frac{MW}{FW} = \frac{195}{97} = 2.01$$

11. One mole of substance contains (92.02 g)(.6957)/16 g/mol = 4.00 moles O and (92.02 g)(.3043)/14 gmol = 2.00 moles N. Hence, the molecular formula is N_2O_4 and the empirical formula is NO_2. $N_2 + 2O_2 \longrightarrow N_2O_4$.

EXPERIMENT 6

8. Decantation is the process of separation of a liquid from a solid by gently pouring the liquid from the solid so as to not disturb the solid. Filtration is the process of separating a solid from a liquid by means of a porous substance, a filter, which allows the liquid to pass through but not the solid. Consult Experiment 3.
9. $Cu(OH)_2(s) \longrightarrow CuO(s) + H_2O(g)$
10. $CuO(s) + H_2SO_4(aq) \longrightarrow CuSO_4(aq) + H_2O(l)$

EXPERIMENT 7

7. Add NaCl (or other source of Cl^-) to form white precipitate of AgCl.
9. Tap water contains impurities such as Cl^-, Fe^{3+}, and Ca^{2+} that might interfere with the chemical tests.
10. No. Both Na_2CO_3 and NaCl react with H_2SO_4 to form gases. The former forms CO_2 (odorless) and the latter HCl (pungent odor). The presence of NaCl would thus mask the presence of Na_2CO_3. (Another test is necessary to confirm the presence of Na_2CO_3, such as allowing the evolved gases to react with a $Ba(OH)_2$ solution, causing insoluble $BaCO_3$ to form.)

11. Treatment of the solid with dilute HNO_3 would cause effervescence and the formation of CO_2. Holding a drop of $Ba(OH)_2$ solution above the reaction would confirm CO_3^{2-} by formation of $BaCO_3$ as a precipitate. If $AgNO_3$ were added to the HNO_3 solution, AgCl would precipitate.
12. Divide the mixture into two portions. Treat one portion with concentrated H_2SO_4. Violet vapors of I_2 would indicate I^-. Dissolve the other portion in water and add $BaCl_2$ to precipitate $BaSO_4$.

EXPERIMENT 8

8.
$$\overline{x} = \frac{10.1 + 10.4 + 10.6}{3} = 10.4$$

average deviation = (0.3 + 0.0 + 0.2)/3 = 0.2

$$SD = \sqrt{\frac{(0.3)^2 + (0.0)^2 + (0.2)^2}{2}} = 0.3$$

relative average deviation = 0.2/10.4 = 0.02

9. A gravimetric factor is a multiplier which converts the mass of substance weighed to the mass of substance sought. It is a ratio that involves two molar masses and the stoichiometric relationship between them.

EXPERIMENT 9

All answers are given in Appendix J.

EXPERIMENT 10

5. Since the solvent will evaporate and leave no residue, the solvent front cannot be located unless it is marked immediately.
6. R_f = 48 mm/76 mm = 0.63
7. Changing the solvent will change the R_f value because it depends on the solubility of the compound in the solvent.
8. The R_f values depend on the nature and strength of interaction between the support and the substance being eluted. Fe^{3+} has a greater charge than Ni^{2+} and would be expected to interact more strongly with the support and have a smaller R_f value.
9. No, but being colored facilitates detection.

EXPERIMENT 11

5. For sp, sp^2, sp^3, and sp^3d^2 hybridization there are 2, 3, 4, and 6 equivalent orbitals, respectively; sp^3d hybridization is different from the others. The three equatorial orbitals of a trigonal-bipyramid are all equivalent to each other, but they are different from the two equivalent axial orbitals.
6. Formal charge is the difference between the number of electrons around an atom in the free state and an atom in a chemical compound.

7. To calculate formal charges, we first draw Lewis electron-dot formulas. The formal charges of the atoms in CO, CO_2, and CO_3^{2-} are illustrated below:

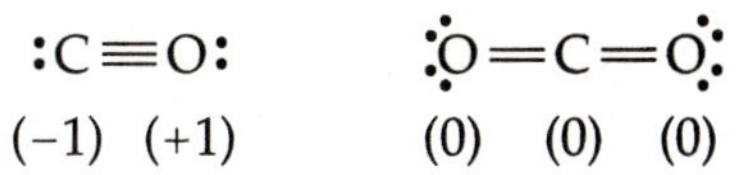

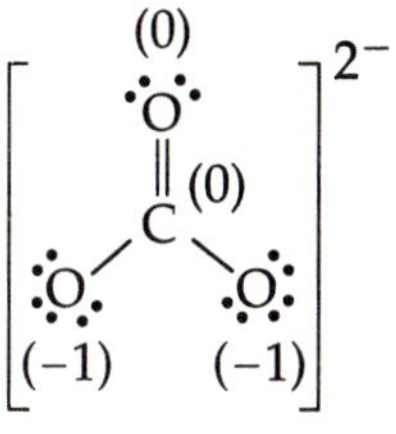

:C≡O: (−1) (+1)

:Ö=C=Ö: (0) (0) (0)

$[CO_3]^{2-}$: O (0) double-bonded to C (0); two singly bonded O atoms (−1) (−1)

EXPERIMENT 12

2. For atomic species, absorption of energy involves excitation of electrons from a lower to a higher potential energy. Emission of energy involves the reverse process, relaxation of electrons from a higher to a lower potential energy.
6. The visible range is from 500 to 700 nm, and since 285 nm is outside this range a visible spectroscope could not detect 285 nm radiation.
7. A 518 nm emission is blue-green.
8.

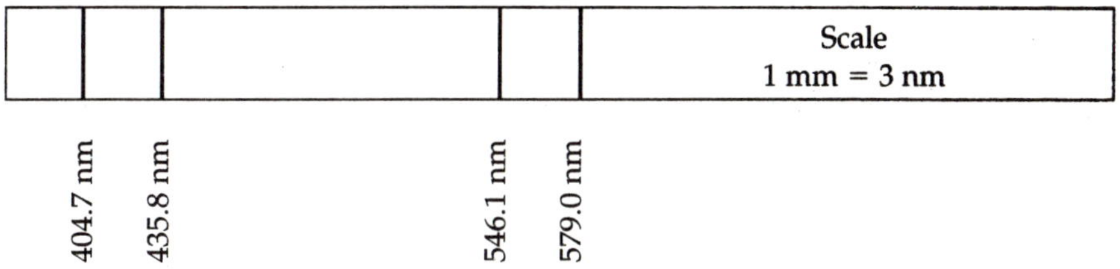

Scale
1 mm = 3 nm

404.7 nm 435.8 nm 546.1 nm 579.0 nm

EXPERIMENT 13

8. Pressure = 650 mm Hg + 40 mm Hg = 690 mm Hg
9. As the level of the mercury column is lowered, the pressure decreases, the gas expands and the volume increases.
10. From the ideal-gas law, $PV = nRT$ [1] Boyle's law: For a fixed amount of gas at constant temperature, the product nRT is a constant, or $PV = k$ or $V = k/P$. Charles's law: For a fixed amount of gas at a constant pressure, equation [1] becomes $V = (nR/P)T = kT$, where $(nR/P) = k$ = constant. Avogadro's law: Consider equal volumes of gases under the same conditions: $P_1V_1/RT = n_1$ and $P_2V_2/RT_2 = n_2$. Then it follows that $P_1V_1/RT_1 = P_2V_2/RT_2$ or $n_1 = n_2$.
11. From the stoichiometry of the reaction, two volumes of O_2 are required for each volume of CH_4. Hence, 2.5 L of CH_4 requires 5.0 L of O_2, and 2.5 L of CO_2 are produced.

12. (a) from $PV = nRT$ and using one mole = 32.0 g of O_2 we have

$$D = \frac{M}{V} = \frac{MW}{V} = \frac{(32.0\ \text{g/mol})(1\ \text{atm})}{(0.0821\ \text{L-atm/mol-K})(273\ \text{K})} = 1.43\ \text{g-L}^{-1}$$

(b) The molar mass is 32.0 g and the molar volume at STP is 22.4 L. Hence, the density = 32.0 g/22.4 L = 1.43 g-L^{-1}. Yes, they do give the same result.

EXPERIMENT 14

8. Increased intermolecular forces cause the molecules to approach one another more closely and decrease the volume. To account for this, the corrective term is subtracted.

9. From $PV = nRT$, $n = \dfrac{PV}{RT} =$

$$\frac{\left(\dfrac{670\ \text{mm Hg}}{760\ \text{mm Hg/atm}}\right)\left(\dfrac{562\ \text{cm}^3}{1000\ \text{cm}^3/\text{L}}\right)}{(0.0821\ \text{L-atm/mol-K})(293\ \text{K})} = 0.0206\ \text{mol}$$

10. Solving the ideal-gas law for molecular weight (MW): $PV = (g/MW)RT$ gives MW = gRT/PV. Consider 1 L gas at 1 atm and 100°C:

$$\text{MW} = \frac{(2.550\ \text{g})(0.0821\ \text{L-atm/mol-K})(373\ \text{K})}{(1.00\ \text{atm})(1.00\ \text{L})} = 78.1\ \text{g/mol}$$

Formula weight of CH is 13.0 and because 78.1/13.0 − 6.00, the molecular formula is C_6H_6.

11. Neon (Ne), because it is nonpolar. The intermolecular forces between HBr molecules are relatively large.

EXPERIMENT 15

7.

Reaction	Species oxidized	Species reduced
$Cl_2 + 2I^- \longrightarrow I_2 + 2Cl^-$	I^-	Cl_2
$WO_2 + 2H_2 \longrightarrow W + 2H_2O$	H_2	WO_2
$Ca + 2H_2O \longrightarrow H_2 + Ca(OH)_2$	Ca	H_2O
$4Al + 3O_2 \longrightarrow 2Al_2O_3$	Al	O_2

8.

Reaction	More-active metal
$2Li + Cu^{2+} \longrightarrow 2Li^+ + Cu$	Li
$Cr + 3V^{3+} \longrightarrow 3V^{2+} + Cr^{3+}$	Cr
$Cd + 2Ti^{3-} \longrightarrow 2Ti^{2+} + Cd^{2+}$	Ti^{3+}

EXPERIMENT 16

6. Coulombs = $(40 \times 10^{-6}$ A)(6 months)(30 days/month)(24 h/day). (60 min/h)(60 s/min) = 622 C; Faradays = 622 C/(96,500 C/$\mathcal{F}$) = $6.45 \times 10^{-3}\ \mathcal{F}$.

7. (0.100 mol Cl_2)(2 mol e^-/mol Cl_2)(1 $\mathcal{F}$/mol e^-) = 0.200 $\mathcal{F}$.

8. From $PV = nRT$,

$$n = PV/RT = \frac{(1\ \text{atm})(0.200\ \text{L})}{(0.0821\ \text{L-atm/mol-K})(273\ \text{K})}$$

$$= 8.92 \times 10^{-3}\ \text{mol}$$

$(8.92 \times 10^{-3}\ \text{mol}\ H_2)(2\ \text{mol}\ e^-/\text{mol}\ H_2) = 17.84 \times 10^{-3}\ \text{mol}\ e^-$

$$(96{,}500\ \text{C/mol}\ e^-)(17.84 \times 10^{-3}\ \text{mol}\ e^-)\left(\frac{1\text{A}\cdot\text{s}}{\text{C}}\right)\left(\frac{1}{0.50\ \text{A}}\right) = 3.44 \times 10^3\ \text{s}$$

9. $$\text{Faraday} = \frac{(0.50\ \text{h})(60\ \text{min/h})(60\ \text{s/min})(1.00 \times 10^{-3}\ \text{A})}{(96{,}500\ \text{C}/\mathcal{F})} = 1.9 \times 10^{-5}\ \mathcal{F}$$

For $Ag^+ + e^- \longrightarrow Ag$, one mole Ag will be precipitated for each Faraday. Hence, 1.9×10^{-5} moles or $(1.9 \times 10^{-5}\ \text{moles})(108\ \text{g/mol}) = 2.0 \times 10^{-3}$ g Ag was in solution.

10. $$\text{Faraday} = \frac{(8.00\ \text{hr})(60\ \text{min/h})(60\ \text{s/min})(4.0 \times 10^4\ \text{A})}{(96{,}500\ \text{C}/\mathcal{F})} = 11{,}940\ \mathcal{F}$$

For $Na^+ + e^- \longrightarrow Na$, one mole of Na may be produced for each Faraday.
Hence, g Na = $(11940\ \mathcal{F})(22.99\ \text{g}/\mathcal{F}) = 2.75 \times 10^5$ g. For $2Cl^- \longrightarrow Cl_2 + 2e^-$, one-half mole of Cl_2 may be produced for each Faraday.
Hence, g Cl_2 = $(11940\ \mathcal{F})(35.45\ \text{g}/\mathcal{F}) = 4.24 \times 10^5$ g

EXPERIMENT 17

7. $Pd^{2+} + H_2 = Pd + 2H^+$ $E° = 0.987V > 0$ spontaneous
$Sn^{4+} + H_2 = Sn^{2+} + 2H^+$ $E° = 0.154V > 0$ spontaneous
$Ni^{2+} + H_2 = Ni + 2H^+$ $E° = -0.250V < 0$ not spontaneous
$Cd^{2+} + H_2 = Cd + 2H^+$ $E° = -0.403V < 0$ not spontaneous
Pd^{2+} and Sn^{4+} can be reduced by hydrogen

8.

Spontaneous reaction	Oxidizing agent	Reducing agent
$Pd^{2+} + H_2 = Pd + 2H^+$	Pd^{2+}	H_2
$Sn^{4+} + H_2 = Sn^{2+} + 2H^+$	Sn^{4+}	H_2
$Ni + 2H^+ = Ni^{2+} + H_2$	H^+	Ni
$Cd + 2H^+ = Cd^{2+} + H^2$	H^+	Cd

Pd^{2+} and Sn^{4+} could be reduced by hydrogen.

EXPERIMENT 18

9. pH = 7.4 = $-\log[H^+]$; $\log[H^+] = -7.4$; $[H+] = 3 \times 10^{-8}$ M
$[OH^-] = 10^{-14}/[H^+] = 10^{-14}/(3 \times 10^{-9}) = 3.3 \times 10^{-6}$

10. $Ca(OH)_2 \longrightarrow Ca^{2+} + 2OH^-$, and since $Ca(OH)_2$ is a strong electrolyte $[OH^-] = 2(0.024\ M) = 0.048\ M$; $[H^+] = 10^{-14}/0.048 = 2.1 \times 10^{-13}\ M$.
11. pH is approximately 6.0
12. For 0.01 M HCl, $[H^+] = 1 \times 10^{-2}$, pH = 2.0 and for 0.02 M NaOH, $[OH^-] = 2 \times 10^{-2}\ M$; $[H^+] = 10^{-14}/2 \times 10^{-2} = 5 \times 10^{-13}\ M$, pH = 12.3. Hence, 0.02 M NaOH has a higher pH. The HCl solution is acidic and the NaOH solution is basic.

EXPERIMENT 19

4. A nonvolatile solute lowers the vapor pressure and freezing point and raises the boiling point of a solvent.
5. 3.0 g in 100 g = 30g in 1000g benzene. Thus 30 g/(60.0 g/mol) = 0.50 molal.
9. This solution requires:
   $(0.500\ \text{mol}\ NaNO_3/10^3\ \text{g}\ H_2O)(250\ \text{g}\ H_2O)(85\ \text{g/mol}\ NaNO_3) = 10.8\ \text{g}\ NaNO_3$
10. molality = moles solute/kg solvent
    molarity = moles solute/liter solution

EXPERIMENT 20

9. mol NaOH = mol KHP. M(0.0500 L) = (0.620 g/204.2 g/mol)
   M = 0.060 M
10. mol KHP = mol NaOH = (0.0306 L)(0.100 M) = 3.06×10^{-3} mol
    mass KHP = $(3.06 \times 10^{-3}$ mol)(204.2 g/mol) = 0.625 g
    % KHP = (0.625 g/0.745 g) × 100 = 83.9%

EXPERIMENT 21

3. $H_2CO_3(aq) \longrightarrow H_2O(l) + CO_2(g)$
   $H_2SO_3(aq) \longrightarrow H_2O(l) + SO_2(g)$
4. 44°C (graph should be plotted before beginning experiment).
7. NaI: Na^+, I^-; K_2SO_4: K^+, SO_4^{2-}; NaCN: Na^+, CN^-; $Ba(OH)_2$: Ba^{2+}, OH^-; $(NH_4)_2SO_4$; NH_4^+, SO_4^{2-}.
8. $NaOH + HNO_3 \longrightarrow NaNO_3 + H_2O$

   $KOH + HCl \longrightarrow KCl + H_2O$

   $Ba(OH)_2 + H_2SO_4 \longrightarrow BaSO_4 + 2H_2O$

EXPERIMENT 22

8. mol HA = mol NaOH = (0.03015 L)(0.0995 *M*) = 3.00×10^{-3} mol
   0.216 g/(3.00×10^{-3}) mol = 72.0 g/mol
9. At the one-half equivalence point, [HOAc] = $[OAc^-]$. From the dissociation constant, $1.85 \times 10^{-5} = [H^+][OAc^-]/[HOAc]$, it follows that $1.85 \times 10^{-5}\ M = [H^+]$ and pH = 4.73. At the equivalence point, the solution contains 0.050 *M* NaOAc and the pH depends upon hydrolysis of the salt:

   $OAc^- + H_2O = HOAc + OH^-$ for which

$K_b = [HOAc][OH^-]/[OAc^-] = K_w/K_a = (1 \times 10^{-14})/(1.85 \times 10^{-5})$

$= 5.41 \times 10^{-10}$

Let $x = [OH^-] = [HOAc]$, then

$$[OAc^-] = 0.050 - x \simeq 0.50\ M$$

Substituting into the expression for K_b and solving for x yields:

$$5.41 \times 10^{-10} = x \cdot x/0.050 \text{ and } x = 5.2 \times 10^{-6}\ M$$

The hydrogen ion concentration is $[H^+] = 10^{-14}/(5.2 \times 10^{-6}) = 1.92 \times 10^{-9}$ and pH = 8.72.

EXPERIMENT 23

2. Because the acid contains three acidic protons and only one third of the protons have been neutralized by 16.5 mL of base, 3 × 16.5 mL = 49.5 mL are required to completely neutralize the acid.
7. For a weak acid HA, $K_a = [H_3O^+][A^-]/[HA]$ and when the acid is half neutralized, $[A^-] = [HA]$. Thus $K_a = [H_3O^+]$ and $-\log K_a = -\log [H_3O^+]$ or pK_a = pH at the half-neutralization point.
8. Yes; K_b could be determined in a similar manner but a strong acid should be used in the titration.
9. The equivalence-point volumes of a polyprotic acid are multiples of the first equivalence-point volume: the second equivalence-point volume is twice the first, and the third is three times the first.
10. For a diprotic acid, one-quarter neutralized means that we are halfway to the first equivalence point. Thus, pH = pK_a = 2.90.

EXPERIMENT 24

7. Salts are the products formed by the reactions of acids and bases.
8. $BaCl_2$ would be neutral; it is a salt of a strong acid and base. $CuSO_4$, $ZnCl_2$, and $(NH_4)_2SO_4$ would be acidic; these are salts of weak bases and strong acids. NaCN would be basic; it is a salt of a strong base and weak acid.
9. If pH = 6, then $[H^+] = 10^{-6}\ M$ and $[OH^-] = 10^{-14}/10^{-6} = 10^{-8}\ M$.
10. $$M^+(aq) + H_2O(l) = MOH(aq) + H^+(aq)$$

$$K_b = ([MOH][H^+]/[M^+]. \text{ pH} = 5.3 = -\log[H^+]$$

$$[H^+] = \text{antilog}(-5.3) = 5.0 \times 10^{-6}\ M. \ [M^+] = 0.1\ M$$

$$K_b = (5 \times 10^{-6})(5 \times 10^{-6})/0.1 = 2.5 \times 10^{-10}$$

EXPERIMENT 25

5. $BaCO_3 \longrightarrow Ba^{2+} + CO_3^{2-}$ and $5.1 \times 10^{-9} = [Ba^{2+}][CO_3^{2-}]$
If s = the molar solubility
$s = [Ba^{2+}] = [CO_3^-]$ and $5.1 \times 10^{-9} = s^2$ or $s = 7.1 \times 10^{-5}$ mol/L
The grams that dissolve in 1000 mL
grams $BaCO_3$ =
$(1L) \times (7.1 \times 10^{-5} \text{ mol L}^{-1})(197 \text{ g BaCO}_3/\text{mol}) = 1.4 \times 10^{-2}$ g

6. The solubility-product constant does not explicitly contain the concentration of the insoluble salt, whereas the expression for the equilibrium constant has this concentration in the denominator.
7. (a) The K_{sp} of a substance may be determined by evaporating a given volume of a saturated solution of a salt and finding the weight of the salt that dissolved. (b) Concentration of ions in a saturated solution may be determined by colorimetric, emf or electrical conductivity measurements and then these values can be used to calculate the K_{sp}.
8. If there is no solid present, one does not know that the solution is truly saturated. Solid must be present for the equilibrium to be established.
10. Insoluble bases react with acids and thus go into soluton as do insoluble carbonates as illustrated by the following equations:

$$Fe(OH)_2(s) + 2HCl(aq) \longrightarrow 2H_2O(l) + FeCl_2(aq)$$

$$FeCO_3(s) + 2HCl(aq) \longrightarrow H_2O(l) + CO_2(g) + FeCl_2(aq)$$

EXPERIMENT 26

5. Specific heat is the amount of heat required to produce a temperature change of 1°C for 1 gram of a substance.
6. Assume density of water is 1.00 g/mL and no heat is lost to the calorimeter. Then the heat lost by the hot water equals the heat gained by the cold.

$$50\text{ g}(60°\text{C} - T_f) \times 4.18\text{ J/K-g} = 25\text{ g}(T_f - 20°\text{C}) \times 4.18\text{ J/K-g}$$

$$3000°\text{C} - 50\ T_f = 25\ T_f - 500°\text{C}$$

$$T_f = 47°\text{C}$$

7. A piece of metal of known mass at some known elevated temperature could be dropped into water with known mass and temperature in a calorimeter and noting the rise in the temperature of the water. The heat lost by the metal equals the heat gained by the water and the calorimeter.
8. heat lost by metal = heat gained by water

$$5.10\text{ g} \times (48.6°\text{C} - 28.2°\text{C}) \times \text{S.H.} = 20.0\text{ g}(28.2°\text{C} - 22.1°\text{C}) \times 4.18\text{ J/K-g}$$

$$\text{S.H.} = 4.90\text{ J/K-g}$$

9. Heat required = 2.51 J/K-g × 250 g × (33°C − 18°C) = 9.40 kJ
10. $\Delta H = (103.25\text{ g})(32.0°\text{C} - 23.9\text{C°})\left(\frac{\text{K}}{°\text{C}}\right)(4.18\text{ J/K-g})(40.0\text{ g/mol})(1/3.25\text{ g}) = 43.1\text{ kJ/mol}$

EXPERIMENT 27

8. This is a heterogeneous reaction, therefore the smaller the zinc particles, the more surface area, and the faster the reaction with hydrochloric acid.
9. If the rate doubles for each 10° temperature increase, an increase of 40° would cause the rate to increase by a factor of 16 (i.e., 2 × 2 × 2 × 2).

10. Tripling the concentration of B causes a rate increase of 27, hence the reaction is third order in B. Doubling the concentration of A causes a rate increase of 4 and the order of the reaction in A is two. Thus, rate = $k[A]^2[B]^3$.

$$k = \frac{\text{rate}}{[A]^2[B]^3} = \frac{0.3 \times 10^{-6}\ \text{M/h}}{(1.0 \times 10^{-2}\ \text{M})^2(1\ \text{M})^3} = 3 \times 10^{-3}\ \text{M}^{-4}\,\text{h}^{-1}$$

EXPERIMENT 28

1. (a) Nature of reactants.
   (b) Concentration of reactants.
   (c) Temperature.
   (d) Catalysts.
   (e) Pressure.
   (f) Particle size.
3. A catalyst is a substance that increases the rate of a reaction without itself undergoing a net chemical change.
4. $2H_2O_2(aq) \xrightarrow{I^-} O_2(g) + 2H_2O(l)$
6. (a) No effect.
   (b) Double.
   (c) Quadruple.
   (d) Eight-fold increase.
   (e) Increase by $\sqrt{2}$.
7. Rate = (2.0 mL − 0.0 mL)/(45 s − 0.0 s) = 4.4×10^{-2} mL/s

   Rate = (3.9 mL − 2.0 mL)/(88 s − 45 s) = 4.4×10^{-2} mL/s

   Rate = (5.8 mL − 3.9 mL)/(131 s − 88 s) = 4.4×10^{-2} mL/s

   Average rate = 4.4×10^{-2} mL/s

EXPERIMENT 29

All answers are in Appendix J.

CATIONS

EXPERIMENT 30

Part I

5. (a) $AgCl(s) + 2NH_3(aq) \longrightarrow Ag(NH_3)_2^+(aq) + Cl^-(aq)$

   (b) $Pb^{2+}(aq) + CrO_4^{2-}(aq) \longrightarrow PbCrO_4(s)$

   (c) $Hg_2Cl_2(s) + 2NH_3(aq) \longrightarrow HgNH_2Cl(s) + Hg(l) + NH_4^+(aq) + Cl^-(aq)$

   (d) $Ag(NH_3)_2^+(aq) + 2H^+(aq) + Cl^-(aq) \longrightarrow AgCl(s) + 2NH_4^+(aq)$
6. Ag^+, Pb^{2+} and Hg_2^{2+} are absent.
7. Precipitates are washed to free them of impurities that may be adhering to them.

Part II

4. Add HCl; AgCl precipitates while Cu^{2+} remains in solution.
5. (a) $2Bi^{3+}(aq) + 2S^{2-}(aq) \longrightarrow Bi_2S_3(s)$

   (b) $SnS_2(s) + S^{2-}(aq) \longrightarrow SnS_3^{2-}(aq)$

(c) $3PbS(s) + 8H^+(aq) + 2NO_3^-(aq) \longrightarrow$
$3Pb^{2+}(aq) + 3S(s) + 2NO(g) + 4H_2O(l)$

(d) $Bi^{3+}(aq) + 3NH_3(aq) + 3H_2O(l) \longrightarrow 3NH_4^+(aq) + Bi(OH)_3(s)$

(e) H_2O_2 is added to ensure that Sn is in the +4 oxidation state so it can form the soluble SnS_3^{2-} ion.

6. H_2O_2 is added to oxidize Sn^{2+} to Sn^{4+}.
7. CuS—black; SnS_2—yellow; $PbSO_4$—white.

Part III

5. $Al(OH)_3$ can be separated from $Ni(OH)_2$ by adding NaOH. $Al(OH)_3$ will form the soluble $Al(OH)_4^-$ ion while $Ni(OH)_2$ remains insoluble.
6. (a) $Fe^{3+}(aq) + 3OH^-(aq) \longrightarrow Fe(OH)_3(s)$

   (b) $Al(OH)_3(s) + 3H^+(aq) \longrightarrow Al^{3+}(aq) + 3H_2O(l)$

   (c) $FeS(s) + 2H^+ \longrightarrow Fe^{2+} + H_2S(g)$

   (d) $3NiS(s) + 2NO_3^-(aq) + 8H^+(aq) \longrightarrow$
   $3Ni^{2+}(aq) + 2S(s) + 2NO(g) + 4H_2O(l)$
8. Fe^{3+} forms the blood-red $Fe(SCN)_6^{3-}$ ion; Ni^{2+} forms a red precipitate with H_2DMG.
9. NH_4Cl is an acidic salt and it will decrease the pH.

ANIONS | Part IV

5. (a) H_2S will precipitate CuS but not Ba^{2+}.
   (b) HCl will precipitate AgCl but not Ca^{2+}
   (c) K_2CrO_4 will precipitate $BaCrO_4$ but not NH_4^+.
6. (a) $Ba^{2+}(aq) + CrO_4^{2-}(aq) \longrightarrow BaCrO_4(s)$

   (b) $NH_4^+(aq) + OH^-(aq) \longrightarrow NH_3(aq) + H_2O(l)$

Part V

7. (a) Sulfuric acid would yield a brown gas when heated with $Hg(NO_3)_2$ but would have no effect on $BaSO_4$.
   (b) Calcium bromide would yield a red-brown gas (Br_2). There would be no evidence of reaction with Na_3PO_4.
   (c) ZnS would yield a gas with a rotten-egg odor and no observable change occurs with $BaSO_4$.
   (d) K_2CrO_4 would change from yellow to orange, whereas HgI_2 would yield violet vapors.
8. (a) Na_2CO_3.
   (b) NaBr.
   (c) Na_2CrO_4.
   (d) NaCl.
   (e) Na_2SO_3.

9. Add $Hg(C_2H_3O_2)_2$ to the unknown to remove iodide. Then, to the supernatant liquid, add $FeSO_4$, mix and then add H_2SO_4 down the side of the test tube. A brown ring confirms NO_3^-.
10. Remove SO_3^{2-} by oxidizing with Na_2O_2. Acidify the solution and test the evolved gas by holding a drop of $Ba(OH)_2$ near it to determine if a white precipitate of $BaCO_3$ forms.

EXPERIMENT 31

8. If the absorbance of a solution is greater than 1, dilute the solution to reduce the absorbance.
9. $(4.0\text{ mL} \times 0.050\text{ mg Fe/mL})/50\text{ mL} = 4.0 \times 10^{-3}\text{ mg/mL}$
10. $A = abc$, $A = 5.1\text{ L/mol–cm} \times 1\text{ cm} \times 0.0400\text{ mol/L} = 0.20$

EXPERIMENT 32

4. The Beer-Lambert law states that absorbance *(A)* is related to concentration *(c)* by the relation $A = abc$, where a is molar absorbtivity and b is the solution path length.
7. $a = A/bc = 0.76/(0.1500\text{ mol/L} \times 1.0\text{ cm}) = 5.1\text{ L/mol-cm}$.
   Using this value for the extinction coefficient, $c = A/ab$ $0.52/(5.1\text{ L/mol-c} \times 1.0\text{ cm}) = 0.10\ M$.
9. Eutrophication is the enrichment of a body of water with nutrients such that growth of organisms in water makes it unfit for human purposes.
10. No; $30.0\text{ mg/L} - 0.25(30.0\text{ mg/L}) = 22.5\text{ mg/L}$, which exceeds the allowable 0.3 mg/L.
11. $P_3O_{10}^{5-} + 2H_2O = 3PO_4^{3-} + 4H^+$

EXPERIMENT 33

6. Consider the dilution of 1 L of the water: 1 L × 400 ppm/5 ppm = 80 L, thus an 80-fold dilution is required if the BOD is 400 ppm. Similarly a 2000- and 20-fold dilution are required for BOD's of 10,000 and 100 respectively.
7. $$\text{moles } KIO_3 = \frac{0.150\text{ g}}{214.0\text{ g/mol}} = 7.01 \times 10^{-4}\text{ mol}$$

   $$\text{moles } Na_2S_2O_3 = 6 \times \text{moles } KIO_3$$
   $$= 6 \times 7.01 \times 10^{-4}\text{ mol} = 4.21 \times 10^{-3}\text{ mol}$$
   $$\text{Therefore, } M = 4.21 \times 10^{-3}\text{ mol}/1.06 \times 10^{-2}\text{ L}$$
   $$= 0.397\ M$$
8. BOD = (8.4 − 5.6) ppm = 2.8 ppm
9. Yes, though this is close to borderline.

EXPERIMENT 34

1. A Lewis base is an electron pair donor, and a Lewis acid is an electron pair acceptor.
2. Molecules or ions that surround a metal ion in a complex are known as ligands and are Lewis bases. The central metal and the ligands bound to it consitute the coordination sphere.

EXPERIMENT 35

6. $\%C_2O_4^{2-} = [g\ C_2O_4^{2-}/(g\text{-compound}/mol)]100$:

   (a) $H_2C_2O_4$; $\%C_2O_4^{2-} = (88.02\ g)(100)/(90.04\ g) = 97.76\%$

   (b) $Na_2C_2O_4$: $\%C_2O_4^{2-} = (88.02\ g)(100)/(134.0\ g) = 65.69\%$

   (c) $K_2C_2O_4$: $\%C_2O_4^{2-} = (88.02\ g)(100)/(166.2\ g) = 52.96\%$

   (d) $K_3[Al(C_2O_4)_3] \cdot 3H_2O$: $\%C_2O_4 = (264.06\ g)(100)/(462.34\ g) = 57.11\%$

7. $$g\,C_2O_4^{2-} = \left(\frac{5\ mol\ C_2O_4^{2-}}{2\ mol\ MnO_4^-}\right)(0.02876\ L)(0.0200\ mol/L)(88.02\ g\,C_2O_4^{2-}/mol)$$

   $$= 0.127\ g$$

   $\%C_2O_4^{2-} = (0.127\ g)(100)/(0.250\ g) = 50.8\%$

8. Weight of $K_3[Fe(C_2O_4)_3] \cdot 3H_2O$ that contains $0.127\ g\ C_2O_4^{2-}$:

   $$= (0.127\ g\ C_2O_4^{2-})\frac{(1\ mol\ C_2O_4^{2-})}{(88.02\ g/mol)}$$

   $$\times \frac{(1\ mol\ K_3[Fe_3(C_2O_4)_3] \cdot 3H_2O)}{(3\ mol\ C_2O_4^{2-})(491.2\ g\ K_3[Fe(C_2O_4)_3] \cdot 3H_2O/mol)} = 0.236\ g$$

   $\%$ purity $= (0.236\ g)(100)/(0.250\ g) = 94.4\%$

EXPERIMENT 36

3. Add acid and identify the gas, CO_2, by the clouding of a suspended drop of $Ba(OH)_2$.
4. Add NaOH and place a piece of moist litmus paper above the solution. It will turn blue if NH_3 gas is present.
6. mol NaCl $= 15\ g/58\ g/mol = 0.26$ mol; mol $(NH_4)_2CO_3 = 10\ g/96\ g/mol = 0.10$ mol; $(NH_4)_2CO_3$ is the limiting reagent, therefore, 0.10 mol $(NH_4)_2CO_3 \times 2$ mol $NaHCO_3$/mol $(NH_4)_2CO_3 \times 84$ g $NaHCO_3$/mol $= 17$ g $NaHCO_3$.
7. The thistle tube must extend beneath the hydrochloric acid surface to prevent the escape of CO_2.

EXPERIMENT 37

2. a. $HOCl + 2I^- + H^+ \longrightarrow Cl^- + I_2 + H_2O$
   b. $2\,S_2O_3^{2-} + I_2 \longrightarrow S_4O_6^{2-} + 2I^-$
6. For each mole of KIO_3, one sixth of a mol of $Na_2S_2O_3$ react.
9. Oxidation state of (a) Cl in OCl^- is +1, and (b) S in $S_2O_3^{2-}$ is +2.

EXPERIMENT 38

6. Empirical formulas are the simplest formulas showing the simplest ratios among atoms. Molecular formulas are true formulas showing the correct ratios among atoms as well as the number of atoms per molecule.
7. CH
8. Substances with the same molecular formula but different structural formulas are called structural isomers. Geometric isomers have the same connectivity but a different orientation of atoms in space.
9.

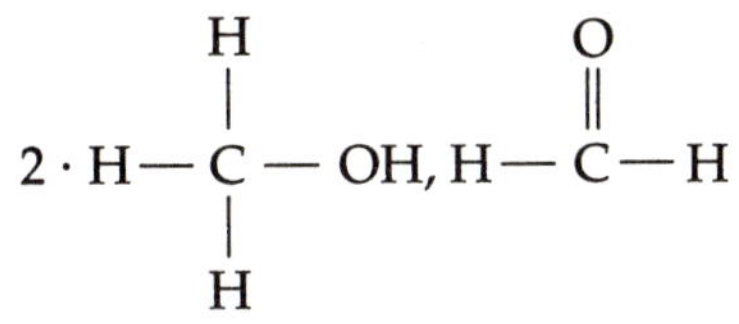

$$2 \cdot \mathrm{H{-}\overset{\displaystyle H}{\underset{\displaystyle H}{\overset{|}{\underset{|}{C}}}}{-}OH,\ H{-}\overset{\displaystyle O}{\overset{\|}{C}}{-}H}$$

10. Methyl alcohol or methanol,

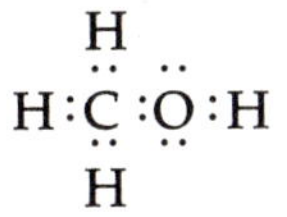

$$\mathrm{H:\overset{\displaystyle H}{\underset{\displaystyle H}{\ddot{\underset{..}{C}}}}:\ddot{\underset{..}{O}}:H}$$

formaldehyde or methanal,

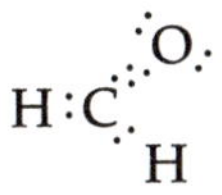

$$\mathrm{H:C(:H)::\ddot{O}:}$$

EXPERIMENT 39

9. $\mathrm{CH_3C({=}O){-}OCH_3}$, methyl acetate; $\mathrm{H{-}C({=}O){-}OCH_2CH_3}$, ethyl formate.
10. (a) methyl alcohol + acetate acid ⟶ methyl acetate
    (b) ethyl alcohol + formic acid ⟶ ethyl formate
11. CH_3CO_2Na, sodium acetate, and methyl alcohol.
12. 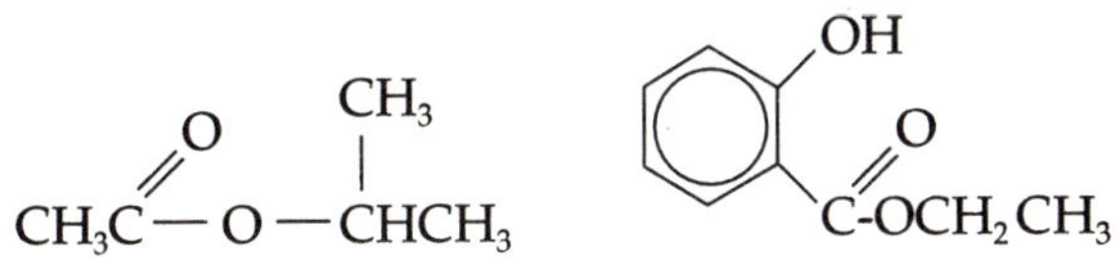 $\mathrm{CH_3C({=}O){-}O{-}CH(CH_3)CH_3}$ $\mathrm{C_6H_4(OH)C({=}O){-}OCH_2CH_3}$ (benzene ring with OH and ortho $\mathrm{C({=}O){-}OCH_2CH_3}$)

EXPERIMENT 40

6. (0.65 g aspirin)/(180 g/mol) × (1 mol NaOH/mol aspirin) × (1 L/0.15 mol NaOH) × (1000 mL/1 L) = 24 mL
7. Some of the aspirin may decompose to form acetic acid.
8. Buffers resist changes in pH and usually consist of a weak acid (or base) and a salt of the corresponding weak acid (or base).
9. An alkaline buffer might be added to reduce the acidity of the aspirin.
10. Buffered aspirin contains, in addition to the aspirin, other titratable substances and the "impurities" are not neutral substances.

EXPERIMENT 41

1. See discussion.
2. Mobile phase is a liquid or gas that passes over a solid or stationary phase.
3. Eluate is the solution that comes off a column and eluent is the solution added to the column.
5. Hard water contains mineral impurities, e.g., Ca^{2+}, Fe^{3+}, or Mg^{2+}. Deionized water is water that has both cations and anions removed from it by means of ion exchangers.
7. Grams Zn^{2+} =

$$(0.03395 \text{ L})\ (0.1024 \text{ mol NaOH/L})(1 \text{ mol Zn}^{2+}/2 \text{ mol NaOH}) \times (65.38 \text{ g Zn}^{2+}/\text{mol Zn}^{2+}) = 0.1136 \text{ g}$$

$$\%\text{Zn} = (0.1136 \text{ g})(100)/(0.300 \text{ g}) = 37.9\%.$$

8. $\text{mol Zn} = \text{mol ZnCl}_2 \cdot \text{n H}_2\text{O}$
$= (0.436 \text{ g})/(65.38 \text{ g/mol}) = 0.001738 \text{ mol}$

$1 \text{ mol ZnCl}_2 \cdot n\ \text{H}_2\text{O} = 0.300 \text{ g}/0.001738 \text{ mol} = 172.6 \text{ g}$

$1 \text{ mol ZnCl}_2 = 136.3 \text{ g}$

$\text{gram H}_2\text{O} = 172.6 \text{ g} - 136.3 \text{ g} = 36.3 \text{ g}$

$\text{mol H}_2\text{O} = (36.3 \text{ g})/(18.0 \text{ g/mol}) = 2 = n$

9. $\text{mol NaCl} = \text{mol NaOH required} = (0.135 \text{ g})/(58.45 \text{ g/mol})$
$= 2.31 \times 10^{-3} \text{ mol}$
$(2.31 \times 10^{-3} \text{ mol}/(0.1231 \text{ mol/L}) = 0.0187 \text{ L} = 18.7 \text{ mL}$